Methods in Enzymology

Volume 209
PHOSPHOLIPID BIOSYNTHESIS

METHODS IN ENZYMOLOGY

EDITORS-IN-CHIEF

John N. Abelson   Melvin I. Simon

DIVISION OF BIOLOGY
CALIFORNIA INSTITUTE OF TECHNOLOGY
PASADENA, CALIFORNIA

FOUNDING EDITORS

Sidney P. Colowick and Nathan O. Kaplan

Methods in Enzymology

Volume 209

Phospholipid Biosynthesis

EDITED BY

Edward A. Dennis

DEPARTMENT OF CHEMISTRY
UNIVERSITY OF CALIFORNIA, SAN DIEGO
LA JOLLA, CALIFORNIA

Dennis E. Vance

LIPID AND LIPOPROTEIN RESEARCH GROUP
DEPARTMENT OF BIOCHEMISTRY
UNIVERSITY OF ALBERTA
EDMONTON, ALBERTA, CANADA

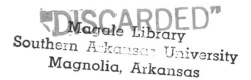

ACADEMIC PRESS, INC.
Harcourt Brace Jovanovich, Publishers
San Diego   New York   Boston
London   Sydney   Tokyo   Toronto

Academic Press, Inc.
San Diego, California 92101

United Kingdom Edition published by
Academic Press Limited
24–28 Oval Road, London NW1 7DX

Library of Congress Catalog Number: 54-9110

International Standard Book Number: 0-12-182110-2

PRINTED IN THE UNITED STATES OF AMERICA
92 93 94 95   9 8 7 6 5 4 3 2 1

Table of Contents

Section I. Strategies for Generating Phospholipid Synthesis Mutants

Section II. Acyltransferases

Section III. Kinases

Section IV. Phosphatases

Section V. Cytidylyltransferases

Section VI. Phosphocholine/Phosphoethanolamine Phosphotransferases

Section VII. Synthases

Section VIII. Phospholipid Transformations

Section IX. Alkyl Ether/Plasmalogen Biosynthetic Enzymes

Section X. Sphingolipid Biosynthesis Enzymes

Section XI. Enzymes of Lipid A Synthesis

Section XII. Phospholipid Transfer Processes

Contributors to Volume 209

Article numbers are in parentheses following the names of contributors.
Affiliations listed are current.

MATT S. ANDERSON (53), *Department of Biochemistry, Merck Sharp & Dohme Research Laboratories, Rahway, New Jersey 07065*

RICHARD A. ANDERSON (21), *Department of Pharmacology, University of Wisconsin Medical School, Madison, Wisconsin 53706*

VINCENT ARONDEL (61), *Laboratory of Plant Cell and Molecular Physiology, Unité de Recherche Associée au Centre National de la Recherche Scientifique, Université Pierre et Marie Curie, 75252 Paris Cedex 05, France*

MYONGSUK BAE-LEE (35), *Yukong Limited, Fairfield, New Jersey 07006*

CHANTAL E. BAZENET (21), *Department of Pharmacology, University of Wisconsin Medical School, Madison, Wisconsin 53706*

ROBERT M. BELL (5, 17, 32), *Department of Biochemistry, Duke University Medical Center, Durham, North Carolina 27710*

CHARLES J. BELUNIS (20), *Department of Biochemistry, Merck Sharp & Dohme Research Laboratories, Rahway, New Jersey 07065*

MERLE L. BLANK (46), *Department of Biological Chemistry, Medical Sciences Division, Oak Ridge Associated Universities, Oak Ridge, Tennessee 37831*

ALEX BROWN (44), *Renal Division, Washington University School of Medicine, St. Louis, Missouri 63110*

KATHRYN A. BROZEK (56), *Department of Biochemistry, University of Wisconsin-Madison, Madison, Wisconsin 53706*

GEORGE M. CARMAN (20, 24, 28, 35, 36), *Department of Food Science, Rutgers University, New Brunswick, New Jersey 08903*

PATRICK C. CHOY (8, 9), *Department of Biochemistry and Molecular Biology, University of Manitoba, Winnipeg, Manitoba R3E 0W3, Canada*

ROSALIND A. COLEMAN (11), *Departments of Nutrition and Pediatrics, University of North Carolina, Chapel Hill, North Carolina 27599*

ROSEMARY B. CORNELL (31), *Department of Chemistry and Biochemistry, Simon Fraser University, Burnaby, British Columbia V5A 1S6, Canada*

T. P. DALTON (59), *Department of Biochemistry and Molecular Biology, University of Kansas Medical Center, Kansas City, Kansas 66103*

ARUN K. DAS (45), *Neuroscience Laboratory, University of Michigan, Ann Arbor, Michigan 48109*

SALIL C. DATTA (48), *Indian Institute of Chemical Biology, Calcutta 700 032, India*

GÜNTHER DAUM (60), *Department of Biochemistry, Technische Universität Graz, A-8010 Graz, Austria*

PIERRE N. E. DE GRAAN (22), *Institute of Molecular Biology and Medical Biotechnology, University of Utrecht, 3584 CH Utrecht, The Netherlands*

EDWARD A. DENNIS (1), *Department of Chemistry, University of California, San Diego, La Jolla, California 92093*

WILLIAM DOWHAN (2, 25, 34, 37, 41), *Department of Biochemistry and Molecular Biology, University of Texas Medical School at Houston, Houston, Texas 77030*

A. L. ERWIN (57), *Laboratory of Bacteriology and Immunology, The Rockefeller University, New York, New York 10021*

DOUGLAS A. FELDMAN (29), *Department of Biological Chemistry, University of Mich-*

igan Medical School, and the Veterans Administration Medical Center, Ann Arbor, Michigan 48105

ANTHONY S. FISCHL (36), *Department of Food Science and Nutrition, University of Rhode Island, West Kingston, Rhode Island 02892*

DAVID A. FORD (50), *Department of Molecular and Cellular Cardiovascular Biochemistry, Washington University School of Medicine, St. Louis, Missouri 63110*

CINDEE R. FUNK (25), *Department of Cell Biology, Baylor College of Medicine, Houston, Texas 77030*

ANTHONY H. FUTERMAN (52), *Department of Membrane Research and Biophysics, Weizmann Institute of Science, Rehovot 76100, Israel*

MRIDUL K. GHOSH (48), *Georgetown University Medical Center, Rockville, Maryland 20850*

JOHN A. GLOMSET (19), *Glomset Laboratories, Howard Hughes Medical Institute, University of Washington, Seattle, Washington 98195*

ELISABETH BAKER GOLDEN (42), *Department of Medicine, National Jewish Center for Immunology and Respiratory Medicine, Denver, Colorado 80206*

RICHARD W. GROSS (50), *Department of Molecular and Cellular Cardiovascular Biochemistry, Washington University School of Medicine, St. Louis, Missouri 63110*

AMIYA K. HAJRA (10, 45, 48), *Neuroscience Laboratory, University of Michigan, Ann Arbor, Michigan 48109*

DIPAK HALDAR (6), *Department of Biological Sciences, St. John's University, Jamaica, New York 11439*

RANDOLPH Y. HAMPTON (55), *Department of Biochemistry, University of Wisconsin-Madison, Madison, Wisconsin 53706*

G. M. HELMKAMP, JR. (59), *Department of Biochemistry and Molecular Biology, University of Kansas Medical Center, Kansas City, Kansas 66103*

SUSAN A. HENRY (3), *Department of Biological Sciences, Carnegie Mellon University, Pittsburgh, Pennsylvania 15213*

SHUICHI HIRAOKA (38), *Department of Biochemistry, Saitama University, Urawa, Saitama 338, Japan*

RUSSELL H. HJELMSTAD (32), *Department of Pathology, Duke University Medical Center, Durham, North Carolina 27710*

SHUICHI HORIE (45), *Faculty of Pharmaceutical Sciences, Teikyo University, Sagamiko, Kanagawa 199-01, Japan*

KARL Y. HOSTETLER (12, 39), *Department of Medicine, Division of Endocrinology and Metabolism, University of California, San Diego, and the Veterans Administration Medical Center, La Jolla, California 92161*

LI HSU (13), *Department of Biochemistry, St. Jude Children's Research Hospital, Memphis, Tennessee 38101*

KIMBERLY A. HUDAK (3), *Department of Biological Sciences, Carnegie Mellon University, Pittsburgh, Pennsylvania 15213*

S. JULIA HUTERER (12), *Department of Neurology, University of Toronto, Toronto, Ontario M5G 1A5, Canada*

KOZO ISHIDATE (14), *Department of Chemical Toxicology, Medical Research Institute, Tokyo Medical and Dental University, Tokyo 101, Japan*

SUZANNE JACKOWSKI (13), *Department of Biochemistry, St. Jude Children's Research Hospital, Memphis, Tennessee 38101*

ALAIN JOLLIOT (61), *Laboratory of Plant Cell and Molecular Physiology, Unité de Recherche Associée au Centre National de la Recherche Scientifique, Université Pierre et Marie Curie, 75252 Paris Cedex 05, France*

JEAN-CLAUDE KADER (61), *Laboratory of Plant Cell and Molecular Physiology, Unité de Recherche Associée au Centre National de la Recherche Scientifique, Université Pierre et Marie Curie, 75252 Paris Cedex 05, France*

JULIAN N. KANFER (40), *Department of Biochemistry and Molecular Biology, University of Manitoba, Winnipeg, Manitoba R3E 0W3, Canada*

HIDEO KANOH (18), *Department of Biochemistry, Sapporo Medical College, Sapporo, Hokkaido 060, Japan*

JUHA KASURINEN (58), *Department of Medical Chemistry, University of Helsinki, SF-00170 Helsinki, Finland*

MICHAEL J. KELLEY (28), *Department of Chemistry, University of California, San Diego, La Jolla, California 92093*

CLAUDIA KENT (15), *Department of Biological Chemistry, University of Michigan Medical School, Ann Arbor, Michigan 48109*

TEN-CHING LEE (26, 33, 47, 49), *Department of Biological Chemistry, Medical Sciences Division, Oak Ridge Associated Universities, Oak Ridge, Tennessee 37831*

ROZENN N. LEMAITRE (19), *Department of Biochemistry, Howard Hughes Medical Institute, University of Washington, Seattle, Washington 98195*

QIAO-XIN LI (41), *Department of Pathology, University of Melbourne, Parkville, Victoria 3052, Australia*

JOHN M. LOPES (3), *Department of Molecular and Cellular Biochemistry, Stritch School of Medicine, Loyola University of Chicago, Maywood, Illinois 60153*

BOYD MALONE (49), *Department of Biological Chemistry, Medical Sciences Division, Oak Ridge Associated Universities, Oak Ridge, Tennessee 37831*

CHRISTOPHER R. MCMASTER (9), *Department of Biochemistry and Molecular Biology, University of Manitoba, Winnipeg, Manitoba R3E 0W3, Canada*

ALFRED H. MERRILL, JR. (51), *Department of Biochemistry, Emory University School of Medicine, Atlanta, Georgia 30322*

ALBRECHT MORITZ (22), *Center for Biomembranes and Lipid Enzymology, University of Utrecht, 3584 CH Utrecht, The Netherlands*

J. J. MUKHERJEE (8), *Department of Biochemistry and Molecular Biology, University of Manitoba, Winnipeg, Manitoba R3E 0W3, Canada*

R. S. MUNFORD (57), *Departments of Internal Medicine and Microbiology, University of Texas Southwestern Medical Center, Dallas, Texas 75235*

YASUO NAKAZAWA (14), *Department of Chemical Toxicology, Medical Research Institute, Tokyo Medical and Dental University, Tokyo 101, Japan*

JOSEPH T. NICKELS, JR. (20), *Department of Food Science, Rutgers University, New Brunswick, New Jersey 08903*

BERNADETTE C. OSSENDORP (63), *Center for Biomembranes and Lipid Enzymology, University of Utrecht, 3584 CH Utrecht, The Netherlands*

RICHARD E. PAGANO (52), *Department of Embryology, Carnegie Institution of Washington, Baltimore, Maryland 21210*

FRITZ PALTAUF (60), *Department of Biochemistry, Technische Universität Graz, A-8010 Graz, Austria*

THOMAS J. PORTER (15), *Division of Hematology, New England Medical Center, Boston, Massachusetts 02111*

JENNIFER J. QUINLAN (24), *Department of Food Science, Rutgers University, New Brunswick, New Jersey 08903*

CHRISTIAN R. H. RAETZ (4, 53, 54, 55, 56), *Department of Biochemistry, Merck Sharp & Dohme Research Laboratories, Rahway, New Jersey 07065*

NEALE D. RIDGWAY (43), *Department of Pediatrics and Biochemistry, Dalhousie University, Halifax, Nova Scotia B3H 4H7, Canada*

CHARLES O. ROCK (13), *Department of Biochemistry, St. Jude Children's Research Hospital, Memphis, Tennessee 38101*

FUMIO SAKANE (18), *Department of Biochemistry, Sapporo Medical College, Sapporo, Hokkaido 060, Japan*

MARK A. SCHEIDELER (5), *CNS Division, NovoNordisk A/S, 2670 Malov, Denmark*

MICHAEL SCHLAME (39), *Department of Medicine, Division of Endocrinology and Metabolism, University of California, San Diego, and the Veterans Administration Medical Center, La Jolla, California 92161*

ISAO SHIBUYA (38), *Department of Biochemistry, Saitama University, Urawa, Saitama 338, Japan*

FRED SNYDER (23, 26, 33, 44, 46, 47, 49), *Department of Biological Chemistry, Medical Sciences Division, Oak Ridge Associated Universities, Oak Ridge, Tennessee 37831*

PENTTI J. SOMERHARJU (58), *Department of Medical Chemistry, University of Helsinki, SF-00170 Helsinki, Finland*

CARL P. SPARROW (27), *Department of Atherosclerosis Research, Merck Sharp & Dohme Research Laboratories, Rahway, New Jersey 07065*

TAKAYUKI SUGIURA (7), *Faculty of Pharmaceutical Sciences, Teikyo University, Sagamiko, Kanagawa 199-01, Japan*

MARCI J. SWEDE (3), *Department of Biological Sciences, Carnegie Mellon University, Pittsburgh, Pennsylvania 15213*

PAUL G. TARDI (8), *Department of Biochemistry and Molecular Biology, University of Manitoba, Winnipeg, Manitoba R3E 0W3, Canada*

LILIAN B. M. TIJBURG (30), *Laboratory of Veterinary Biochemistry, University of Utrecht, 3508 TD Utrecht, The Netherlands*

TSUTOMU UCHIDA (16), *Department of Biochemistry, Gunma University School of Medicine, Maebashi 371, Japan*

DAVID S. VALLARI (47), *Hepatitis Diagnostic Products Research and Development, Abbott Laboratories, Abbott Park, Illinois 60064*

LAMBERT M. B. VAN GOLDE (30), *Laboratory of Veterinary Biochemistry, University of Utrecht, 3508 TD Utrecht, The Netherlands*

G. PAUL H. VAN HEUSDEN (63), *Department of Cell Biology and Genetics, University of Leiden, 2333 AL Leiden, The Netherlands*

DENNIS E. VANCE (43), *Lipid and Lipoprotein Research Group, Department of Biochemistry, University of Alberta, Edmonton, Alberta T6G 2S2, Canada*

ALES VANCURA (6), *Department of Biological Sciences, St. John's University, Jamaica, New York 11439*

S. E. VENUTI (59), *Department of Biochemistry and Molecular Biology, University of Kansas Medical Center, Kansas City, Kansas 66103*

CHANTAL VERGNOLLE (61), *Laboratory of Plant Cell and Molecular Physiology, Unité de Recherche Associée au Centre National de la Recherche Scientifique, Université Pierre et Marie Curie, 75252 Paris Cedex 05, France*

PIETER S. VERMEULEN (30), *Laboratory of Veterinary Biochemistry, University of Utrecht, 3508 TD Utrecht, The Netherlands*

DENNIS R. VOELKER (42, 62), *Department of Medicine, National Jewish Center for Immunology and Respiratory Medicine, Denver, Colorado 80206*

KEIZO WAKU (7), *Faculty of Pharmaceutical Sciences, Teikyo University, Sagamiko, Kanagawa 199-01, Japan*

JAMES P. WALSH (17), *Department of Biochemistry, University of Washington, Seattle, Washington 98195*

ELAINE WANG (51), *Department of Biochemistry, Emory University School of Medicine, Atlanta, Georgia 30322*

KEITH O. WEBBER (10), *Laboratory of Biochemical Genetics, National Heart, Lung, and Blood Institute, National Institutes of Health, Bethesda, Maryland 20892*

PAUL A. WEINHOLD (29), *Department of Biological Chemistry, University of Michigan Medical School, and the Veterans Administration Medical Center, Ann Arbor, Michigan 48105*

JAN WESTERMAN (22), *Center for Biomembranes and Lipid Enzymology, University* of Utrecht, 3584 CH Utrecht, The Netherlands

JOHN R. WHERRETT (12), *Department of Neurology, University of Toronto, Toronto, Ontario M5G 1A5, Canada*

KAREL W. A. WIRTZ (22, 58, 63), *Center for Biomembranes and Lipid Enzymology, University of Utrecht, 3584 CH Utrecht, The Netherlands*

KEIKO YAMADA (18), *Department of Biochemistry, Sapporo Medical College, Sapporo, Hokkaido 060, Japan*

SATOSHI YAMASHITA (16), *Department of Biochemistry, Gunma University School of Medicine, Maebashi 371, Japan*

RAPHAEL A. ZOELLER (4), *Department of Biophysics, Boston University School of Medicine, Boston, Massachusetts 02118*

Preface

The enzymology of phospholipid biosynthesis was covered as a small section in Volume 71 of *Methods in Enzymology* published in 1981. In the past decade, there have been many rapid advances in the purification and cloning of the enzymes involved in phospholipid biosynthesis. In 1981, only a few of the enzymes described had been purified and none cloned or sequenced.

This volume reports on the purification and cloning of many of the phospholipid biosynthetic enzymes from bacteria, yeast, and animal tissues. In addition, new strategies have been developed for the selection of mutants defective in phospholipid biosynthesis. Moreover, there has been significant progress in closely related areas of research that impinge on phospholipid metabolism: platelet activating factor, sphingolipid biosynthesis, phospholipid transfer proteins, and lipid A biosynthesis in *Escherichia coli*. Thus, it is timely and important to collate in one volume the new methods and procedures for studies of the enzymes of phospholipid biosynthesis.

This book will be a useful reference source for scientists in the field and will be a catalyst for further advancements on the structure and function of the enzymes of phospholipid biosynthesis. It should also be highly useful for studies on the regulation of phospholipid metabolism and the elucidation of the function of phospholipids.

This book is dedicated to Professor Eugene P. Kennedy, Hamilton Kuhn Professor Emeritus at Harvard Medical School, who discovered the central *de novo* pathways for the biosynthesis of phospholipids involving cytidine nucleotides and who has continued to make seminal contributions to advancing the understanding of phospholipid and membrane biochemistry and enzymology.

EDWARD A. DENNIS
DENNIS E. VANCE

METHODS IN ENZYMOLOGY

VOLUME 67. Vitamins and Coenzymes (Part F)
Edited by DONALD B. MCCORMICK AND LEMUEL D. WRIGHT

VOLUME 68. Recombinant DNA
Edited by RAY WU

VOLUME 69. Photosynthesis and Nitrogen Fixation (Part C)
Edited by ANTHONY SAN PIETRO

VOLUME 70. Immunochemical Techniques (Part A)
Edited by HELEN VAN VUNAKIS AND JOHN J. LANGONE

VOLUME 71. Lipids (Part C)
Edited by JOHN M. LOWENSTEIN

VOLUME 72. Lipids (Part D)
Edited by JOHN M. LOWENSTEIN

VOLUME 73. Immunochemical Techniques (Part B)
Edited by JOHN J. LANGONE AND HELEN VAN VUNAKIS

VOLUME 74. Immunochemical Techniques (Part C)
Edited by JOHN J. LANGONE AND HELEN VAN VUNAKIS

VOLUME 75. Cumulative Subject Index Volumes XXXI, XXXII, XXXIV–LX
Edited by EDWARD A. DENNIS AND MARTHA G. DENNIS

VOLUME 76. Hemoglobins
Edited by ERALDO ANTONINI, LUIGI ROSSI-BERNARDI, AND EMILIA CHIANCONE

VOLUME 77. Detoxication and Drug Metabolism
Edited by WILLIAM B. JAKOBY

VOLUME 78. Interferons (Part A)
Edited by SIDNEY PESTKA

VOLUME 79. Interferons (Part B)
Edited by SIDNEY PESTKA

[1] The Biosynthesis of Phospholipids

By Edward A. Dennis

Introduction

The study of the biosynthesis of phospholipids has become more complex as more and more phospholipids have been identified. Although the central *de novo* step was discovered by Eugene P. Kennedy and Samuel B. Weiss and reported in 1955[1] and no exceptions have been reported to date, new important lipids are constantly being identified and their biosynthetic transformations described. It is convenient to think of phospholipid biosynthesis as consisting of (1) the basic pathways leading to *de novo* formation of phospholipid and (2) special transformations that interconvert one phospholipid to another.

Phospholipid Pathways

There is clearly more than one way that nature has developed to make specific phospholipids, and often different species or subcellular organelles have developed different approaches. For example, phosphatidylserine is made in prokaryotes such as *Escherichia coli* by the reaction of L-serine with CDPdiglyceride, a *de novo* step, whereas mammalian cells such as rat liver utilize an L-serine–phosphatidylethanolamine exchange enzyme in a special transformation or transesterification of a preformed phospholipid not requiring the expenditure of a high-energy phosphate.

With regard to the source of the polar group on phospholipids, the literature is confusing because the term *de novo* has been used when a special transformation of one phospholipid into another occurs, such as the methylation of phosphatidylethanolamine to produce phosphatidylcholine, as occurs in mammalian liver. This also has the resultant effect of synthesizing choline from ethanolamine. A parallel example would be the decarboxylation of phosphatidylserine to produce phosphatidylethanolamine, which has the resultant effect of synthesizing ethanolamine from serine. When a preformed polar group is incorporated directly, it has sometimes been referred to as a salvage pathway in the literature. An example would be the incorporation of preformed choline in most mamma-

---

[1] E. P. Kennedy and S. B. Weiss, *J. Am. Chem. Soc.* **77**, 250 (1955).

lian tissues (see Ref. 2) by the reaction of cytidine diphosphate choline with 1,2-diglyceride to produce phosphatidylcholine, which is really a *de novo* step as defined above. Perhaps, *de novo* rather than "salvage" should always be used to describe the cytidine step, which is the key reaction in making new phosphodiester bonds.

Clearly established pathways for the biosynthesis of the glycerophosphatides are shown in Fig. 1. Although this is a complex diagram, it attempts to show, in the least redundant manner possible, the many interconversions which occur. Of course, all interconversions do not occur in every species. For example, prokaryotes do not generally make phosphatidylcholine, and both the prokaryotic and eukaryotic pathways for phosphatidylserine biosynthesis are included. The scheme also attempts to use the stereospecific (*sn*) numbering convention, especially to include the correct stereochemistry of the more complex lipids such as diphosphatidylglycerol (cardiolipin). Of course, all phospholipids cannot easily be shown on a single diagram without resulting in the complex metabolic charts we are familiar with, which contain so many arrows that they are indecipherable. No attempt is made to include in Fig. 1 the serine-derived sphingolipids, lipopolysaccharrides, or the alkyl ether- and plasmologen-containing phospholipids. Yet the latter compounds are related biosynthetically and/or functionally to the glycerol-containing phospholipids and should be fully considered as phospholipids from both viewpoints.

Importance and Discovery of Cytidine Nucleotides

It has been more than 35 years since the seminal discovery by Kennedy and Weiss that cytidine nucleotides rather than adenine nucleotides are responsible for the *de novo* step in phospholipid biosynthesis.[1] When considering methods employed to study phospholipid biosynthesis, it is worthwhile to examine what is perhaps the simplest lesson biochemistry professors emphasize: use pure reagents and know what they contain. Amusingly, being careful and intelligent biochemists almost led Kennedy and Weiss to miss the discovery of CTP and the cytidine pathway. It is illustrative to review the steps which led to that seminal discovery.

In 1955, Kennedy and Weiss[1] reported their observation that phosphorylcholine was incorporated into phosphatidylcholine (lecithin) by washed rat liver mitochondrial preparations in the presence of ATP, as had been previously reported by Kornberg and Pricer.[3] However, Ken-

[2] E. P. Kennedy, *in* "Phosphatidylcholine Metabolism" (D. Vance, ed.), p. 1. CRC Press, Boca Raton, Florida, 1989.
[3] A. Kornberg and W. E. Pricer, Jr., *Fed. Proc.* **11,** 242 (1952).

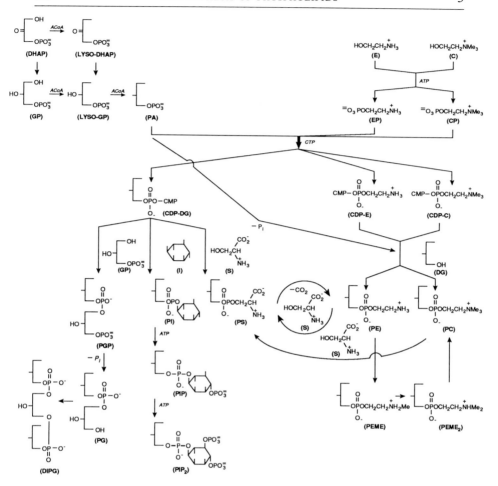

FIG. 1. Enzymes of glycerol phospholipid biosynthesis.

nedy and Weiss[1] could only obtain their result when they used a commercial preparation of ATP which was "amorphous" and impure. When they repeated the experiment with crystallized ATP, no incorporation was observed. This was also the case when ATP was generated *in situ* via oxidative phosphorylation.[2] This led Kennedy and Weiss to reason that a contaminant in commercial ATP was responsible for the observed reaction. They then examined a number of nucleotides and discovered that CTP, not ATP, supported phosphatidylcholine biosynthesis.

Kennedy and Weiss[1] (see also Ref. 4) went on to reason that the CTP functioned analogously to UTP in carbohydrate synthesis and that CTP could react with phosphorylcholine to produce cytidine diphosphate choline, which could react with an acceptor (α,β-diglyceride) to produce phosphatidylcholine. To prove this, they synthesized this compound from cytidine 5'-phosphate and radiolabeled phosphorylcholine using N,N-dicyclohexylcarbodiimide as a condensing agent.[5] They then demonstrated that the dinucleotide derivative supported phosphatidylcholine formation by rat liver mitochondrial preparations at rates far exceeding those for CTP. At the time, cytidine diphosphate choline was not known to exist in mammalian tissues, so Kennedy and Weiss had actually postulated the existence of this compound in cells before it was isolated and identified. In retrospect, the discovery of the role of cytidine nucleotides constituted the seminal event that led to the elucidation of the *de novo* pathway for the biosynthesis of phosphatidylcholine as well as other phospholipids. The discovery depended on the impurity of commercial reagents used in studying enzymatic reactions and constitutes an important lesson in methods in enzymology!

Acknowledgments

I thank Sam Weiss for sharing with me his personal recollections as a postdoctoral student with Eugene Kennedy when they discovered the requirement for CTP in phospholipid biosynthesis.

[4] E. P. Kennedy and S. B. Weiss, *J. Biol. Chem.* **222,** 193 (1956).
[5] H. G. Khorana, *J. Am. Chem. Soc.* **76,** 3517 (1954).

Section I

Strategies for Generating Phospholipid Synthesis Mutants

[2] Strategies for Generating and Utilizing Phospholipid Synthesis Mutants in *Escherichia coli*

By WILLIAM DOWHAN

Introduction

Phospholipids are generally recognized as essential for cell and organelle barrier function and as a matrix for membrane-associated enzymes and complexes. The molecular diversity of phospholipids is too extensive to be explained simply by a role in maintaining the proper physical state of the membrane bilayer.[1,2] Therefore, individual phospholipids must play a more dynamic role in the cell as important metabolic intermediates and precursors, as regulatory molecules, and as components of multimolecular complexes. Since there is generally no direct assay for phospholipid function, as there is for an enzyme, the functions of individual phospholipids have been determined incidentally to the study of a particular cellular process *in vitro* rather than by direct study of a particular phospholipid. The physiological significance of such functions cannot be substantiated by biochemical approaches alone.

The use of genetic approaches has great potential for direct study of the role of membrane phospholipid composition and individual phospholipid species in normal cellular processes. The genetic aspects of phospholipid biosynthesis and the use of molecular genetic techniques to study the function of phospholipids in *Escherichia coli* have been extensively reviewed.[1,3–5] Mutations in the structural genes encoding the various phospholipid biosynthetic and degradative enzymes have been sought in order to (1) establish metabolic steps and the enzymes involved, (2) clone genes related to phospholipid metabolism, (3) understand the organization of these genes and the regulation of their expression, (4) understand the regulation of phospholipid metabolism, and (5) construct strains with altered phospholipid composition to elucidate phospholipid function.

Because a unique phenotype for cells lacking a particular phospholipid is not easy to predict, the general strategy employed has been to isolate

[1] C. R. H. Raetz and W. Dowhan, *J. Biol. Chem.* **265**, 1235 (1990).
[2] L. Rilfors, G. Lindblom, A. Wieslander, and A. Christiansson, *in* "Membrane Fluidity" (M. Kates and L. A. Manson, eds.), p. 205. Plenum, New York, 1984.
[3] T. V. Boom and J. E. Cronan, Jr., *Annu. Rev. Microbiol.* **43**, 317 (1989).
[4] C. R. H. Raetz, *Annu. Rev. Genet.* **20**, 253 (1986).
[5] W. R. Bishop and R. M. Bell, *Annu. Rev. Cell Biol.* **4**, 579 (1988).

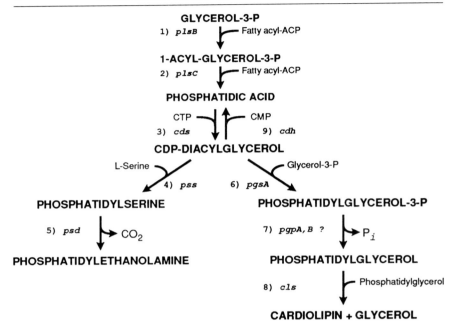

FIG. 1. Pathway of phospholipid metabolism in *E. coli* and the associated genes. The name of each gene is listed with the respective step catalyzed by the following enzymatic activities: (1) glycerophosphate acyltransferase; (2) acylglycero-3-phosphate acyltransferase; (3) CDPdiacylglycerol synthase; (4) phosphatidylserine synthase; (5) phosphatidylserine decarboxylase; (6) phosphatidylglycerophosphate synthase; (7) phosphatidylglycerophosphate phosphatase; (8) cardiolipin synthase; (9) CDPdiacylglycerol hydrolase. Except for the *plsC* gene [J. Coleman, *J. Biol. Chem.* **265**, 17215 (1990)] and the *cls* gene [G. Pluschke and P. Overath, *J. Biol. Chem.* **256**, 3207 (1981); S. Nishijima, Y. Asami, N. Uetake, S. Yamagoe, A. Ohta, and I. Shibuya, *J. Bacteriol.* **170**, 775 (1988)], all the genes are discussed in the text.

mutations in the structural genes encoding the known enzymes of phospholipid metabolism (Fig. 1). A standard approach for isolating mutants in a biosynthetic pathway is to identify a particular mutant by its requirement for a metabolite lying after the point of the mutation in the pathway. Unfortunately, phospholipid auxotrophs (strains requiring a particular phospholipid for growth) cannot be rescued by addition of the necessary phospholipid to the growth medium. There has been limited success in isolating auxotrophs dependent on the water-soluble precursors to phospholipids. A mutant in the committed step to phospholipid biosynthesis (Fig. 1, Step 1) with an increased K_m for glycerophosphate has been isolated.[6] Serine and cytidine auxotrophs specifically defective in phospho-

[6] R. M. Bell, *J. Bacteriol.* **117**, 1065 (1974).

lipid metabolism have never been isolated. Cytidine auxotrophs do accumulate phosphatidic acid and are limited in the synthesis of the major phospholipid classes (Fig. 1, Step 3), but these effects are probably secondary to reduced nucleic acid biosynthesis.[7]

Lacking the ability to use growth phenotypes or metabolite auxotrophy to isolate mutants, the challenge has been to design strategies to identify mutants defective in specific genes encoding enzymes responsible for the metabolic steps outlined in Fig. 1. The majority of mutants in phospholipid metabolism (Fig. 1, Steps 3–7, and 9) were isolated by using procedures efficient in the screening of large numbers of bacterial colonies (colony autoradiography[8,9]) for those defective in the *in vitro* activity of an enzyme responsible for a specific metabolic step in phospholipid metabolism; some of these mutants proved also to be defective in *in vivo* function as well. Once the initial sets of mutants were isolated, they were used to clone the structural genes encoding the respective enzymatic activities.[1] The cloned genes have been used more recently to tailor make additional mutants by a combination of *in vitro* molecular biological techniques and *in vivo* molecular genetic approaches to construct more defined mutants in phospholipid metabolism.[10,11]

Colony Autoradiography

Raetz[8,9] developed a procedure for screening bacterial colonies immobilized on filter papers for defects in phospholipid synthesis. This approach results in the isolation of mutants defective in an *in vitro* measured function which in many cases is not reflected in an *in vivo* recognizable phenotype, that is, the isolation of silent mutations. Many of the mutants isolated by this approach have reduced enzymatic activity or enzymatic activities which are sensitive to the *in vitro* assay conditions but are sufficiently active *in vivo* to support normal growth. In some cases little or no activity can be detected *in vitro* at any temperature, but the cells appear to grow normally. On the other hand, the growth pattern of mutants may show unusual temperature dependence, hypersensitivity to detergents, increased resistance or sensitivity to antibiotics, or uncharacteristic responses to the ionic strength, osmolarity, or pH of the growth medium. These growth phenotypes can be screened for once mutants, as defined

---

[7] B. R. Ganong and C. R. H. Raetz, *J. Biol. Chem.* **257**, 389 (1982).

[8] C. R. H. Raetz, *Proc. Natl. Acad. Sci. U.S.A.* **72**, 2274 (1975).

[9] C. E. Bulawa, B. R. Ganong, C. P. Sparrow, and C. R. H. Raetz, *J. Bacteriol.* **148**, 391 (1981).

[10] P. N. Heacock and W. Dowhan, *J. Biol. Chem.* **262**, 13044 (1987).

[11] P. N. Heacock and W. Dowhan, *J. Biol. Chem.* **264**, 14972 (1989).

by colony autoradiography, are isolated. Although the lack of correlation of the *in vitro* properties of a particular activity with a growth phenotype can complicate interpretation of the genetic and biochemical properties of mutants, this technique has been a powerful tool in the initial isolation of many of the mutations affecting phospholipid metabolism. Using the *in vitro* phenotype uncovered by colony autoradiography has made possible the genetic mapping of the defects, cloning of genes by complementation, and designing experiments which have lead to the isolation of better defined mutants. Below are several examples of how this technique has been used to isolate mutations in specific steps of phospholipid metabolism.

Preparation of Bacterial Colonies for Screening

Details of mutagenesis and plating of cells on agar plates can be found elsewhere.[12] Following mutagenesis, master LB (10 g/liter Bacto-yeast extract, 5 g/liter NaCl, and 5 g/liter Bacto-tryptone, Difco, Detroit, MI) agar plates are made by spreading about 6 surviving cells per 1 cm² of agar surface area followed by growth at 30° for 36 hr (optimum colony size of 1 mm); usually 9-cm circular petri dishes are employed, but larger circular or square petri dishes (limited only by the size of X-ray film available, see below) can be employed.[8] An immobilized replica of the colonies on the agar plate is made by gently pressing down a dry Whatman (Clifton, NJ) No. 42 filter paper (sterile) of the appropriate size on the plate. The filter paper is then slowly lifted off; all subsequent manipulations of the replica prints are performed with the colony side up. Several additional replicas can be made by repeating the above procedure; the size of the colonies on the replica print can be amplified by placing the filters on fresh LB plates and incubating at 30° until the desired colony size is reached. The colony pattern will reappear on the original agar plate after incubation overnight at 30°; the agar plate can be sealed with Parafilm and stored inverted for several weeks at 4° for later rescue of colonies of interest. One replica is stained for 10 min at room temperature with 0.05% Coomassie Brilliant blue in 10% acetic acid. After destaining for 48 hr in methanol–water–acetic acid (45 : 45 : 10) the blue-stained protein pattern corresponding to the bacterial colony pattern will aid colony identification (see below).

Those filters which will be used to assess *in vitro* enzymatic activities are first incubated with 1 ml of a solution of lysozyme (10 mg/ml) and EDTA (10 m*M*, pH 7.0) for 30 min at room temperature. After blotting the filters on adsorbent paper towels, they are placed in petri dishes with 1 ml

[12] J. H. Miller, "Experiments in Molecular Genetics." Cold Spring Harbor Laboratory, Cold Spring Harbor, New York, 1972.

of a solution of metabolic poisons (2 mM 2,4-dinitrophenol, 20 mM sodium azide, 10 mM KF, 1 mM Na$_2$HAsO$_4$, and 4 mM Tris-HCl at pH 8.0) and frozen at $-20°$ for at least 3 hr. The filters are then thawed, blotted, and desiccated at room temperature either by using a fan for 15 min or a vacuum cabinet for 20 min. This procedure effects lysis of the colonies on the filter papers and inactivates energy-dependent processes which might incorporate metabolic precursors into nonlipid material while minimizing the inactivation of temperature-sensitive enzymes of interest. The frozen filters can be stored for several weeks before they are used.

Incorporation of Radiolabel into Immobilized Colonies

Colonies with mutations in the *pss*,[8] *pgsA*,[13] or *cds*[14] loci can be detected by their inability to incorporate radioactive serine, glycero-3-phosphate, or CTP, respectively. The immobilized colonies can be assayed under different conditions to screen for different types of mutants. Temperature-sensitive or cold-sensitive mutants (the latter have never been reported) could be detected by assaying at temperatures either above (42°) or below (15° to 20°) the temperature at which the colonies were grown (30°). Silent mutants, namely, those which express enzymes sensitive to the *in vitro* assay conditions but are sufficiently functional *in vivo* to sustain growth, can be detected at the growth temperature of the colonies. Other special conditions can be designed such as screening for activities that are hypersensitive to changes in the normal conditions of the assay mixture or sensitive to detergents or other additions to the assay mixture.

The desiccated filters are incubated with 1.5 ml (based on 9-cm circular filters) of one of the following assay mixtures (containing the above metabolic inhibitors) for 10 to 60 min:

Phosphatidylserine synthase[8] (CDPdiacylglycerol–serine *O*-phosphatidyltransferase) (*pss* gene): 0.3 mM CDPdiacylglycerol (added as a sonically dispersed concentrated stock solution of the ammonium salt synthesized[15] from egg lecithin-derived phosphatidic acid), 2 mM DL-[3-^{14}C]serine (10 μCi/μmol), 0.1 M potassium phosphate (pH 7.4), and 0.1% Triton X-100

Phosphatidylglycerophosphate synthase[13] (CDPdiacylglycerol–glycerol-3-phosphate 3-phosphatidyltransferase) (*pgsA* gene): 0.13 mM CDPdiacylglycerol (as above), 0.8 mM *sn*-[U-^{14}C]glycero-3-

[13] M. Nishijima and C. R. H. Raetz, *J. Biol. Chem.* **254**, 7837 (1979).
[14] B. R. Ganong, J. M. Leonard, and C. R. H. Raetz, *J. Biol. Chem.* **255**, 1623 (1980).
[15] G. M. Carman and A. S. Fischl, this volume [36].

phosphate (10 μCi/μmol), 30 mM MgCl$_2$, 0.25 M Tris-HCl (pH 8),
1 mM dithiothreitol (DTT), and 0.1 M glycerol
CDPdiacylglycerol synthase[14] (CTP: phosphatidate cytidylytransfer-
ase) (*cds* gene): 0.8 mM phosphatidic acid (added as a sonically
dispersed concentrated stock solution of fully protonated egg lecithin-derived phosphatidic acid), 2.0 mM [α-^{32}P]CTP (1 μCi/μmol),
0.2% Triton X-100, 10 mM MgCl$_2$, 17 μg/ml of heat-treated (5 min
of boiling) bovine pancreatic RNase (eliminates nonlipid-related
incorporation of label), and 0.1 M potassium phosphate, pH 7.5

Control assays, which should show no incorporation of label, can be done
for all of the above screenings by eliminating the phospholipid substrate
from the mixtures. The reactions are terminated by immersing the filter
papers at 4° in 10% trichloroacetic acid which is 10 mM (unlabeled) in the
respective radiolabeled substrate employed. After 10 min the filters are
washed 5 times with 50 ml of 2% trichloroacetic acid on a Büchner funnel
under gentle suction. The filters are dried at 120° for 60 min and subjected
to autoradiography at −80° for 3 to 7 days.

Although initial screenings for the above examples were carried out at
30° and 42° in hopes of identifying temperature-sensitive mutations (for *in
vitro* activity and/or growth), such mutations were only isolated for the
pss gene.[8] In all cases mutations were isolated in which *in vitro* activities
were significantly reduced as indicated by colony autoradiography and
direct assay of cell-free extracts. In the case of the *cds* gene several
mutants were unable to grow above pH 8.0, which may be related to a
change in the pH optimum for the gene product.[7] Both *pss*[16] and *cds*[7]
mutants showed altered sensitivity to antibiotics. Temperature-sensitive
mutants in the *pgsA* gene have never been isolated, but *pgsA* mutants
result in the expression of temperature sensitivity for growth in cells
carrying certain alleles of mutants (*lpxB*) in lipopolysaccharide biosynthesis.[1,17]

Release of Radiolabel from Immobilized Colonies

Colonies with mutations in the *cdh* locus (CDPdiacylglycerol phospha-
tidylhydrolase) can be detected by their failure to release radiolabel from
CDPdiacylglycerol made *in situ* using a two-stage assay system.[9] The
desiccated filters are prewashed on a Büchner funnel with 50 ml of 10 mM
potassium phosphate (pH 7.5) first with and then without 2% streptomycin
sulfate. The cell contents are retained during the washings while endoge-

[16] C. R. H. Raetz and J. Foulds, *J. Biol. Chem.* **252,** 5911 (1977).
[17] M. Nishijima, C. E. Bulawa, and C. R. H. Raetz, *J. Bacteriol.* **145,** 113 (1981).

nous water-soluble metabolites are removed, which reduces the loss of the radiolabeled CDPdiacylglycerol made *in situ* to unrelated metabolic steps (phosphatidylserine synthesis, for example). Such a prewash step may be advisable to increase the incorporation of label in the screening procedures described above. The washed filters are first incubated for 40 min at 42° and pH 7.5 as for the CDPdiacylglycerol synthase screening above. The filters are then transferred to 1 ml of 30 mM EDTA (an inhibitor of the synthase) in 100 mM potassium phosphate, pH 7.5, for 40 min at 42°. The filters are then washed with trichloroacetic acid and processed as above. In the first reaction radiolabeled CDPdiacylglycerol is made *in situ*, whereas during the subsequent incubation radiolabeled phospholipid is converted to diacylglycerol and [^{32}P]CMP, resulting in unlabeled colonies. In this assay those colonies retaining radiolabel are potential mutants in the *cdh* locus.

A similar approach was used by Icho and Raetz[18] to detect mutations in phospholipid phosphatase activities. They used assay conditions similar to those for screening potential *pgsA* mutants but employed glycero-3-[^{32}P]phosphate as the substrate to effect *in situ* synthesis of phosphatidyl-glycero[^{32}P]phosphate. Under these conditions the endogenous phosphati-dylglycerophosphate phosphatases are sufficiently active to release nearly all of the incorporated label. Again those colonies which retained radiola-bel were candidates for phosphatase mutants. From such screenings Icho and Raetz isolated mutations in two phospholipid phosphatases, one spe-cific for phosphatidylglycerophosphate and one acting on phosphatidyl-glycerophosphate, phosphatidic acid, and lysophosphatidic acid, inclu-sively. A similar screen could be employed for other phosphatidic acid phosphatases by first labeling colonies with diacylglycerol and [γ-^{32}P]ATP using the endogenous diacylglycerol kinase.[19]

Analysis of Results

Colonies which have specifically incorporated radiolabel into acid-insoluble material form dark halos with light centers on the X-ray film (see Fig. 2). Labeled colonies among a large number of unlabeled colonies are easily detected, and detection of unlabeled colonies among a large number of labeled colonies is simplified by overlaying the exposed autoradiogram on the replica colony pattern as revealed by Coomassie blue staining of either an unscreened replica filter (see above) or the test filter. Once candidate colonies are identified and marked on the stained filter, viable colonies are identified and retrieved from the master agar plate by overlay-

[18] T. Icho and C. R. H. Raetz, *J. Bacteriol.* **153**, 722 (1983).
[19] C. R. H. Raetz and K. F. Newman, *J. Biol. Chem.* **253**, 3882 (1978).

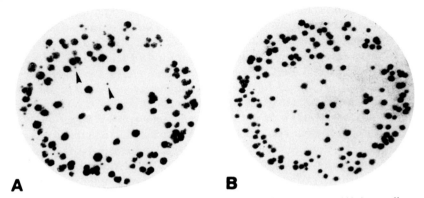

FIG. 2. Colony autoradiographic screening for potential *pss* mutants. (A) Autoradiogram generated from a filter paper replica of colonies of *E. coli* subjected to the assay mixture for phosphatidylserine synthase as described in the text. The arrows indicate two of several colonies [detected by Coomassie blue staining (B) of the same filter paper] which did not incorporate radiolabel. [From C. R. H. Raetz, *J. Biol. Chem.* **251**, 3242 (1976).]

ing the plate on the stained filter. The colonies can be grown up, replated, and the assay repeated, under both permissive and restrictive conditions if desired, to both purify the colonies and confirm their phenotype. If the candidate still shows promise, then growth phenotypes, analysis of extracts, etc., can be carried out. If the mutant displays an easily determined phenotype (such as temperature sensitivity for growth), then moving the mutation to fresh genetic backgrounds by P1 transduction,[12] mapping of the gene, and cloning the gene by complementation using a plasmid-borne *E. coli* DNA library are easily done.[20] Alternatively, the colony autoradiography technique can be used to follow transfer of the mutant gene between strains or correction of the mutation[21] by complementation. It is imperative that a direct correlation between a phenotype and a genetic locus be maintained before extensive time is invested in studying the mutations. There are many examples of phenotypes, such as temperature sensitivity for growth, which result from multiple mutations and are not directly related to the apparent temperature sensitivity of a particular enzyme under *in vitro* assay techniques. Alternatively, interesting phenotypes may not be expressed *in vivo* because of other mutations in a particular genetic background which suppress the phenotype by compensating

[20] C. R. H. Raetz, T. J. Larson, and W. Dowhan, *Proc. Natl. Acad. Sci. U.S.A.* **74**, 1412 (1977).
[21] A. Ohta, K. Waggoner, A. Radominska-Pyrek, and W. Dowhan, *J. Bacteriol.* **147**, 552 (1981).

for the mutation. These problems are usually resolved by transferring the DNA from the isolated mutant colonies to several different recipient strains by phage P1 transduction and scoring for simultaneous transfer of the enzyme defect and the associated phenotype.[12]

Site-Specific Mutation Techniques

The strength of generalized mutagenesis followed by screening lies in the possibility of uncovering previously unrecognized cellular processes or mechanisms,[1] but, owing to the spectrum of mutants possible, the technique does not always produce the specific type of mutant desired. Although colony autoradiography has allowed the identification of many of the genes associated with phospholipid metabolism, the isolation of null mutations (complete inactivation of the gene by deletion or disruption) is impractical since this may result in a lethal phenotype under all growth conditions with no colonies appearing for assay. The mutations isolated by colony autoradiography result in leaky (partial) expression of enzymatic activity which has complicated determination of whether certain phospholipids, namely, the major acidic phospholipids,[1] are essential and whether there are multiple genes responsible for encoding some of the activities important to phospholipid metabolism.[22] Finally, the ability to alter phospholipid composition in a systematic manner in order to correlate composition or the level of a particular phospholipid with cell function has not been realized through the isolation of mutants by the above technique. The techniques outlined below overcome the shortcomings in the colony autoradiography technique but are also dependent on this technique for initial identification of gene loci and cloning of genes by complementation of existing mutations.

Construction of Disrupted Genes

Because the approach to construction of disrupted genes has to be somewhat tailored to each situation and uses standard techniques of molecular biology[23] and bacterial genetics[12] described elsewhere, only the overall strategy is outlined here using the *pgsA*[10] gene disruption as an example. Variations of the following protocol have been applied to the *pss*,[22] *pgpA*,[24] and *pgpB*[24] gene loci. The approach makes no assumption of whether a null allele of the gene is lethal or not and allows for little selective advantage

[22] A. DeChevigny, P. N. Heacock, and W. Dowhan, *J. Biol. Chem.* **266**, 5323 (1991).
[23] L. G. Davis, M. D. Dibner, and J. F. Battey, "Basic Methods in Molecular Biology." Elsevier, New York, 1986.
[24] W. Dowhan and C. R. Funk, this volume [25].

or disadvantage of a strain carrying the null allele, thus minimizing the generation of additional unrecognized mutations which might compensate for the null mutation and complicate the analysis of the mutants.

Two plasmids carrying the cloned structural gene (preferably with no additional flanking genes) are constructed with the following properties. The *pgsA* gene is placed on a plasmid such as pBR322 with a ColE1 origin of replication and at least one selectable drug marker (ampicillin in this case). By *in vitro* techniques another selectable drug marker (kanamycin) is inserted into the coding region of the *pgsA* gene, which results in a null allele of the *pgsA* gene (Fig. 3, plasmid pPG4015). A second plasmid (pHD101) is constructed carrying a functional copy of the *pgsA* gene and a third selectable drug marker (chloramphenicol). This plasmid has the added features of an origin of replication which is both stably maintained in the presence of plasmids with the ColE1 origin of replication and unable to direct its own plasmid replication at 42°; therefore, plasmid pHD101 is stably maintained in the presence of plasmid pPG4015 and will propagate in cells grown at 30°. However, shifting the growth temperature to 42° results in cessation of plasmid replication and failure in the passage of the plasmid-borne copy of the functional *pgsA* gene to daughter cells.

Disruption of the chromosomal copy of the *pgsA* gene is carried out as outlined in Fig. 3 in strain G122, which is temperature sensitive in DNA polymerase I (*polA* gene). This enzyme is required for the autonomous replication and stable maintenance of plasmids like pPG4015 with a ColE1 origin of replication. Plasmid pPG4015 is introduced into strain G122 by transformation followed by selection on LB agar plates containing kanamycin at 30°. Since the plasmid cannot autonomously replicate in strain G122 at elevated temperatures, replating these transformants at 42° on kanamycin results in the selection of colonies in which the plasmid has integrated into the chromosome by homologous recombination at the *pgsA* locus resulting in the tandem arrangement of a functional and disrupted (*pgsA*::*kan*) copy of the genes separated by the plasmid sequence (Fig. 3). The tandem arrangement of the genes is then moved to a *polA*[+] strain using phage P1 transduction and selection for growth on kanamycin. To guarantee that a segment of the chromosome is transduced and not simply plasmid pPS4015, a marker in close proximity to the *pgsA* locus is also scored for. In this case the loss of tetracycline resistance (transposon Tn*10* carried in the *uvrC* gene which borders on *pgsA*) in the recipient strain is used. In the *polA*[+] background the plasmid will recombine out of the chromosome in either of the two manners shown in Fig. 3 resulting in a mixture of cells in the culture, some of which have the *pgsA*::*kan* allele in the chromosome and the functional gene on the plasmid in the cytoplasm. These two alleles of the *pgsA* gene are separated by a second

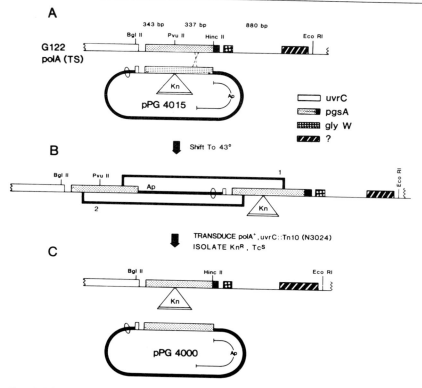

FIG. 3. Disruption of the *pgsA* gene. (A) The dashed lines indicate the point of a potential homologous recombination between the chromosome of strain G122 and plasmid pPG4015. (B) The predicted chromosomal structure after integration of plasmid pPG4015. The numbers 1 and 2 indicate the two possible homologous recombination events which would result in excision of the plasmid from the chromosome. Path 1 would result in the starting situation, and path 2 would lead to the configuration shown in (C). (C) Structures of the chromosome and the plasmid resulting from the excision of the plasmid leaving the disrupted gene in the chromosome. [From P. N. Heacock and W. Dowhan, *J. Biol. Chem.* **262**, 13044 (1987).]

phage P1 transduction into a *polyA*[+] *uvrC*::Tn*10* strain carrying plasmid pHD101. Tranductants are selected on kanamycin/LB agar plates at 30° (transfer of interrupted allele) and screened for ampicillin sensitivity (absence of plasmids pPG4000 or pPG4015), chloramphenicol resistance at 30° (retention of plasmid pHD101), and the loss of tetracycline resistance (chromosome exchange).

Strains with the above characteristics (HD30/pHD101,[10] *pgsA30*::*kan*/ *pgsA*[+] *cam*[R] replication$_{ts}$) carry the *pgsA*::*kan* allele in the chromosome and are dependent on plasmid pHD101 for the *pgsA* gene product. This strain grows normally at 30°, but at 42° the loss of plasmid pHD101 from the culture is correlated with the loss of phosphatidylglycerophosphate synthase activity, reduction of the acidic phospholipid content from 20 to 2% of the total, and cessation of growth. Therefore, the construction of the null allele demonstrated that a single gene product is responsible for this enzymatic activity and that acidic phospholipids are essential for cell growth. Similar conclusions were drawn from experiments which generated a *pss::kan* allele, although the *pss* gene was found not to be required if the growth medium was supplemented with millimolar levels of calcium or magnesium ion.[22] Disruption[24] of the *pgpA* and *pgpB* loci has shown that these genes do not encode essential gene products and are not responsible for the synthesis of the biosynthetic phosphatidylglycerophosphate phosphatase activity.

Regulation of Membrane Phospholipid Composition

To study the effect of membrane phospholipid composition on cell function, a set of mutants is needed in the *pss* and *pgsA* loci with progressively reduced expression of each gene product. Since these gene products catalyze the committed steps (Fig. 1, Steps 4 and 6) to the synthesis of major phospholipid classes (zwitterionic and acidic, respectively), such a set of mutants could be used to adjust systematically the phospholipid composition over a broad range. Genetic manipulation of the *pgsA* gene, as described below,[11] has made available a strain of *E. coli* in which the level of acidic phospholipids can be regulated. Similar genetic manipulation of the *pss* and *plsB* genes would make possible the systematic regulation of the zwitterionic phospholipid content and the membrane protein to phospholipid ratio, respectively.

Strain HDL1001[11] [*pgsA30::kan* Φ(*lacOP-pgsA*+)1 *lacY::Tn9 lacZ' recA srl::Tn10*] carries the *pgsA* null allele but in addition carries a functional copy of the *pgsA* gene in single copy integrated into the chromosome under the regulation of the inducible promoter for the *lac* operon. Therefore, the level of phosphatidylglycerophosphate synthase expressed is dependent on the level of isopropylthiogalactoside (IPTG, the gratuitous inducer of the *lac* operon) in the growth medium. The strain is absolutely dependent on IPTG for growth; the growth rate, specific activity of the gene product, and the acidic phospholipid content are proportional to the level (between 0 and 50 μM) of IPTG in the growth medium. This strain has been used to correlate the dependence of translocation of outer membrane precursor proteins across the inner membrane of *E. coli* with the level of acidic phospholipids in the inner membrane.[25–27]

---

[25] T. de Vrije, R. L. de Swart, W. Dowhan, J. Tommassen, and B. de Kruijff, *Nature (London)* **334**, 173 (1988).

There are several precautions which must be taken in working with strain HDL1001. Cells grown in the absence of IPTG generate two classes of mutants which are independent of IPTG for growth. One class is made up of constitutive mutations in the *lac* promoter, making the strain insensitive to the *lac* repressor so that the *pgsA* gene product is expressed and phospholipid composition is normal. This class of mutations spontaneously arises even in the presence of IPTG. The number of such mutants in a culture can be determined by plating serial dilutions of the culture on LB plates with and without IPTG. These mutations complicate analysis only when they become a significant proportion of a culture and mask the phospholipid composition or growth phenotype of strain HDL1001; this usually occurs if the starting culture has a large proportion of cells with this mutation or after a culture which has arrested for several hours due to lack of IPTG becomes overgrown by the mutation. The second class of mutants allows cells to grow at what are normally limiting levels of acidic phospholipids and is the result of mutations in genes not directly related to *pgsA* or its expression. Although the dependence on IPTG for growth is absolute, strain HDL1001 still expresses a low level of phosphatidylglycerophosphate synthase because the *lac* operon is never completely repressed. Thus far the only such "suppressor" mutations have been in the *lpp* locus,[28,29] which codes for the major outer membrane lipoprotein of *E. coli*: about 10% of the steady-state phosphatidylglycerol pool is used per generation as the source of a glycerol moiety covalently attached to the lipoprotein.[30] The relief brought about by an *lpp* mutation of this apparent drain on the limited phosphatidylglycerol pool in a leaky but not null *pgsA* mutant allows near normal growth. Although such double mutants still have low levels of acidic phospholipids, they can complicate experiments which depend on the growth dependence of the strain on IPTG. However, introduction of a null mutation at the *lpp* locus into strain HDL1001 (strain HDL11[27]) can be advantageous in studying the effect of phospholipid composition on nonessential cellular processes since the growth rate of this strain is not dependent on IPTG.

Cultures of either strain HDL1001 or strain HDL11 with different steady-state levels of phosphatidylglycerol are prepared as follows. A single colony from an LB agar plate containing 1 m*M* IPTG is used to prepare a fresh 5 ml overnight culture grown at 37° in LB medium con-

[26] R. Lill, W. Dowhan, and W. Wickner, *Cell* (*Cambridge, Mass.*) **60,** 271 (1990).

[27] R. Kusters, W. Dowhan, and B. de Kruijff, *J. Biol. Chem.* **266,** 8659 (1991).

[28] Y. Asai, Y. Katayose, C. Hikita, A. Ohta, and I. Shibuya, *J. Bacteriol.* **171,** 6867 (1989).

[29] P. N. Heacock and W. Dowhan, unpublished observation (1989).

[30] H. C. Wu, *in* "Bacterial Outer Membrane as Model Systems" (M. Inouye, ed.), p. 37. Wiley, New York, 1986.

taining 1 mM IPTG and 50 μg/ml kanamycin. The cells are harvested, washed 3 times with 5 ml of fresh LB medium, and diluted 1 : 10^4 into fresh LB medium containing no IPTG (antibiotic, which slows the growth of cells, is optional). The cells are incubated with shaking at 37° for 3 hr. This is sufficient time for reversible growth arrest to occur in the case of strain HDL1001 and limiting levels of acidic phospholipids to be reached by both strains. The cultures are divided into several individual subcultures and supplemented with IPTG between 0 and 100 μM. After an additional 4 hr of shaking the cultures should attain their respective steady-state phospholipid composition (both strains) and growth rate (for strain HDL1001). The cells can then be analyzed for the effect of phospholipid composition on a particular cellular process. Alternatively, such an analysis can be carried out as a function of the changing phospholipid composition either as IPTG is deprived from a fully induced culture or as IPTG is added back to a fully repressed culture. At the end of each experiment cultures should be checked for their dependence on IPTG either for growth (strain HDL1001) or phosphatidylglycerophosphate synthase specific activity (either strain).

Utilization of Existing Mutations

With the possible exception of a mutation in the biosynthetic phosphatidylglycerophosphate phosphatase,[24] a structural gene mutation and the complementing cloned DNA have been identified for every step in the synthesis of the major phospholipids of *E. coli* (Fig. 1). Using the above approach of constructing null alleles in these loci and placing the structural genes under regulation, each gene locus can be engineered to study the effect of the level of the respective gene product on cell processes. In addition, site-specific mutations,[31] first directed at the cloned structural gene followed by integration into the chromosome as outlined above for constructing null alleles, can be used to construct additional mutations in a specific gene locus known to be involved in phospholipid metabolism. Changes in specific amino acids or in domains of these gene products are now possible.[31] Such approaches, which link the techniques of molecular biology to the study of phospholipid metabolism, will be powerful tools in elucidating the role of phospholipids in cell function.

[31] R. Wu and L. Grossman (eds.), this series, Vol. 154.

[3] Strategies for Generating Phospholipid Synthesis Mutants in Yeast

By MARCI J. SWEDE, KIMBERLY A. HUDAK, JOHN M. LOPES, and SUSAN A. HENRY

Introduction

Many of the methods used for generating phospholipid biosynthesis mutants in yeast are based on methods that were developed for the isolation of similar mutants in *Escherichia coli*.[1] For example, rapid colony autoradiographic screening techniques originally developed for bacteria have been adapted for the identification of yeast mutants. As a simple microorganism, yeast provides many of the technical advantages of bacterial systems. In addition, it provides the opportunity to explore the regulation of phospholipids that are unique to eukaryotes such as phosphatidylinositol (PI) and phosphatidylcholine (PC). Recent success in isolating yeast mutants defective in the synthesis of these lipids has led to important insights concerning the mechanisms by which the cell regulates the proportional synthesis of its major membrane phospholipids.

Mutants Auxotrophic for Phospholipid Precursors

Isolation of mutants auxotrophic for phospholipid precursors, such as inositol and choline, has proved to be very successful for identifying both structural and regulatory gene mutants in yeast (Table I). For example, mutations at the *INO1* locus,[2] the structural gene for inositol-1-phosphate synthase (Fig. 1; I1PS), lead to inositol auxotrophy. Cells carrying an *ino1* mutation, when starved for inositol, rapidly cease synthesis of phosphatidylinositol (PI) and, shortly thereafter, stop dividing and die.[3,4] "Inositolless" death has been used in a number of fungal species[5–7] to devise enrichment procedures for the isolation of mutants with defects in a variety of other cellular functions including macromolecular synthesis.[8]

---

[1] W. Dowhan, this volume [2].

[2] M. R. Culbertson and S. A. Henry, *Genetics* **80**, 23 (1975).

[3] G. Becker and R. L. Lester, *J. Biol. Chem.* **252**, 8684 (1977).

[4] S. A. Henry, K. D. Atkinson, A. I. Kolat, and M. R. Culbertson, *J. Bacteriol.* **130**, 472 (1977).

[5] H. E. Lester and S. R. Gross, *Science* **139**, 572 (1959).

[6] R. Holliday, *Microb. Genet. Bull.* **13**, 28 (1962).

[7] P. L. Thomas, *Can. J. Genet. Cytol.* **14**, 785 (1972).

[8] S. A. Henry, T. F. Donahue, and M. R. Culbertson, *Mol. Gen. Genet.* **143**, 5 (1975).

TABLE I
YEAST PHOSPHOLIPID MUTANTS

Mutant	Description	Refs.[a]
cct	Defect in choline-phosphate cytidylyltransferase	1
	also isolated as suppressor of sec14	2
cdg1	Defect in CDPdiacylglycerol synthase	3
cho1 (pss, eth)	Mutations in structural gene for phosphatidylserine synthase (CDP-1,2-diacyl-sn-glycerol : L-serine O-phosphatidyltransferase, EC 2.7.8.8)	4–9
cki	Mutation in choline kinase also isolated as suppressor of sec14	10, 11
cpe1	Mutation resulting in constitutive expression of phospholipid biosynthetic genes	12
cpt1	Mutation in structural gene for sn-1,2-diacylglycerol–cholinephosphotransferase	13, 14
eam1 and eam2	Isolated as suppressors of ethanolamine auxotrophy of cho1 mutants	15, 16
ept1	Mutation in structural gene for sn-1,2-diacylglycerol–ethanolaminephosphotransferase	17
ino1	Mutation in structural gene for L-myo-inositol-1-phosphate synthase (EC 5.5.1.4)	18–20
ino2 and ino4	Mutations in regulatory genes encoding transcriptional activators of phospholipid biosynthetic genes	18, 20–22
lcb1	Defect in serine palmitoyltransferase	23
opi1	Mutation in gene encoding negative regulator of phospholipid biosynthetic genes	24
opi2 and opi4	Mutations resulting in constitutive inositol-1-phosphate synthase expression	24
opi5	Suppressor of ino2, ino4 double mutant	25
pem1 (cho2)	Mutations in structural gene for phosphatidyl-ethanolamine methyltransferase	26, 27
pem2 (opi3)	Mutations in structural gene for phospholipid methyltransferase	28–30
pim1 and pim2	Defective in phosphatidylinositolphosphate kinase and phosphatidylinositol kinase activities, respectively	31
pis	Defect in phosphatidylinositol synthase (CDPdiacylglycerol–inositol 3-phosphatidyltransferase [EC 2.7.8.11])	32

[a] Key to references: (1) J. Nikawa, K. Yonemura, and S. Yamashita, *Eur. J. Biochem.* **131,** 223 (1983); (2) L. S. Klig, M. J. Homann, S. D. Kohlwein, M. J. Kelley, and S. A. Henry, *J. Bacteriol.* **170,** 1878 (1988); (3) G. Lindegren, Y. L. Hwang, Y. Oshima, and C. C. Lindegren, *Can. J. Genet. Cytol.* **7,** 491 (1965); (4) K. D. Atkinson, B. Jensen, A. I. Kolat, E. M. Storm, S. A. Henry, and S. Fogel, *J. Bacteriol.* **141,** 558 (1980); (5) L. Kovac, I. Gbelska, V. Poliachova, J. Subik, and V. Kovacova, *Eur. J. Biochem.* **111,** 491 (1980); (6) J.-I. Nikawa and S. Yamashita, *Biochim. Biophys. Acta* **665,** 420 (1981); (7) V. A. Letts and I. A. Dawes, *Biochem. Soc. Trans.* **7,** 976 (1983); (8) K. Kiyono, K. Miura, Y. Kushima, T. Hikiji, M. Fukushima, I. Shibuya, and A. Ohta,

Isolation of inositol auxotrophs (Ino⁻) has also resulted in the identification of several classes of mutants defective in regulation of phospholipid biosynthesis. In wild-type yeast cells, the *INO1* gene is repressed[9] in response to the presence of the phospholipid precursors inositol and choline. A similar pattern of regulation has been observed for a number of enzymes of phospholipid biosynthesis in yeast.[10–14] Strains harboring a mutation in the *INO2* or *INO4* gene fail to derepress the *INO1* gene even in the absence of inositol. Since basal (repressed) levels of expression of the *INO1* gene are insufficient to allow growth in the absence of inositol, *ino2* and *ino4* mutants are inositol auxotrophs. The *ino2* and *ino4* mutants were subsequently found to have pleiotropic defects in phospholipid biosynthesis including decreased synthesis of PC.[15] The *INO2* and *INO4* genes are now known to encode positive regulators of *INO1* and other coregulated structural genes (Refs. 16–18; Table I).

[9] J. P. Hirsch and S. A. Henry, *Mol. Cell. Biol.* **6**, 3320 (1986).

[10] M. A. Carson, M. Emala, P. Hogsten, and C. J. Waechter, *J. Biol. Chem.* **259**, 6267 (1984).

[11] M. J. Homann, S. A. Henry, and G. M. Carman, *J. Bacteriol.* **163**, 1265 (1985).

[12] L. S. Klig, M. J. Homann, G. M. Carman, and S. A. Henry, *J. Bacteriol.* **162**, 1135 (1985).

[13] A. M. Bailis, M. A. Poole, G. M. Carman, and S. A. Henry, *Mol. Cell. Biol.* **7**, 167 (1987).

[14] M. Greenberg, S. Hubbell, and C. Lam, *Mol. Cell. Biol.* **8**, 4773 (1988).

[15] B. S. Loewy and S. A. Henry, *Mol. Cell. Biol.* **4**, 2479 (1984).

[16] G. M. Carman and S. A. Henry, *Annu. Rev. Biochem.* **58**, 635 (1989).

[17] D. K. Hoshizaki, J. E. Hill, and S. A. Henry, *J. Biol. Chem.* **265**, 4736 (1990).

[18] M. White, J. Lopes, and S. A. Henry, *Adv. Microb. Physiol.* **32**, 1 (1991).

J. Biochem. **102**, 1089 (1987); (9) K. Hosaka and S. Yamashita, *J. Bacteriol.* **143**, 176 (1980); (10) A. E. Cleves, T. P. McGee, K. Champion, M. Goebl, W. Dowhan, and V. A. Bankaitis, *Cell (Cambridge, Mass.)* **64**, 789 (1991); (11) J. Lopes, unpublished data (1991); (12) R. H. Hjelmstad and R. M. Bell, *J. Biol. Chem.* **262**, 3909 (1987); (13) K. Hosaka and S. Yamashita, *Eur. J. Biochem.* **162**, 7 (1987); (14) V. Bankaitis, personal communication (19): (15) K. D. Atkinson, *Genetics* **108**, 533 (1984); (16) K. D. Atkinson, *Genetics* **111**, 1 (1985); (17) R. H. Hjelmstad and R. M. Bell, *J. Biol. Chem.* **263**, 19748 (1988); (18) M. R. Culbertson and S. A. Henry, *Genetics* **80**, 23 (1975); (19) T. S. Donahue and S. A. Henry, *J. Biol. Chem.* **256**, 7077 (1981); (20) T. S. Donahue and S. A. Henry, *Genetics* **98**, 491 (1981); (21) B. S. Loewy and S. A. Henry, *Mol. Cell. Biol.* **4**, 2479 (1984); (22) J. P. Hirsch and S. A. Henry, *Mol. Cell. Biol.* **6**, 3320 (1986); (23) G. B. Wells and R. L. Lester, *J. Biol. Chem.* **258**, 10200 (1983); (24) M. Greenberg, B. Reiner, and S. Henry, *Genetics* **100**, 19 (1982); (25) B. S. Loewy, Ph.D. thesis, Albert Einstein College of Medicine, Bronx, NY (1985); (26) S. Yamashita and A. Oshima, *Eur. J. Biochem.* **104**, 611 (1980); (27) E. F. Summers, V. A. Letts, P. McGraw, and S. A. Henry, *Genetics* **120**, 909 (1988); (28) M. L. Greenberg, L. S. Klig, V. A. Letts, B. S. Loewy, and S. A. Henry, *J. Bacteriol.* **153**, 791 (1983); (29) S. Yamashita, A. Oshima, J.-I. Nikawa, and K. Hosaka, *Eur. J. Biochem.* **128**, 589 (1982); (30) P. McGraw and S. A. Henry, *Genetics* **122**, 317 (1989); (31) I. Uno, K. Fukamaki, H. Kato, T. Takenawa, and T. Ishikawa, *Nature (London)* **333**, 188 (1988); (32) J.-I. Nikawa and S. Yamashita, *Eur. J. Biochem.* **125**, 445 (1982).

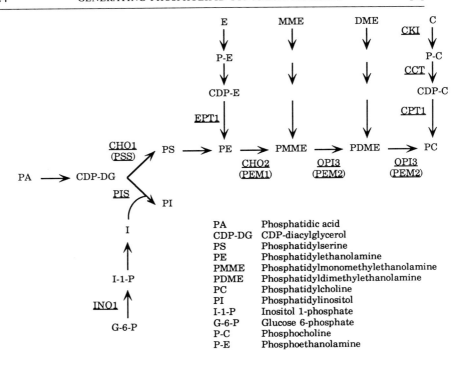

PA	Phosphatidic acid
CDP-DG	CDP-diacylglycerol
PS	Phosphatidylserine
PE	Phosphatidylethanolamine
PMME	Phosphatidylmonomethylethanolamine
PDME	Phosphatidyldimethylethanolamine
PC	Phosphatidylcholine
PI	Phosphatidylinositol
I-1-P	Inositol 1-phosphate
G-6-P	Glucose 6-phosphate
P-C	Phosphocholine
P-E	Phosphoethanolamine

Structural Genes and Products Encoded

INO1	Inositol-1-Phosphate Synthase
CHO1(PSS)	Phosphatidylserine Synthase
CHO2(PEM1)	Phosphatidylethanolamine Methyltransferase
OPI3(PEM2)	Phospholipid Methyltransferase
CKI	Choline Kinase
CCT	Cholinephosphate Cytidylyltransferase
CPT1	sn-1,2-Diacylglycerol Cholinephosphotransferase
EPT1	sn-1,2-Diacylglycerol Ethanolaminephosphotransferase
PIS	Phosphatidylinositol Synthase

FIG. 1. Pathway of phospholipid biosynthesis in *Saccharomyces cerevisiae*. The designations for structural genes are given alongside the relevant reactions. When more than one gene designation has been used in the literature, the designation used in this chapter is given first with other designations listed in parentheses.

The *pis* mutant requires inositol at concentrations of 100 μM or greater,[19] whereas the *ino1* mutants can grow in the presence of 10 μM inositol.[20] The *PIS1* gene reportedly encodes PI synthase (Fig. 1), and the requirement of the *pis* mutant for high levels of inositol is believed to be due to an alteration in the apparent K_m of PI synthase for inositol.[19] In addition, Uno *et al.*[21] isolated mutants (*pim1* and *pim2*; Table I) that are supersensitive to PI 1-phosphate antibody introduced into the cell by electroporation. The mutants appear to be defective in PI and PI-phosphate kinase activity.

Inositol auxotrophy is also a common phenotype of mutants with global defects in transcription. For example, mutants with defects in the large subunit of RNA polymerase II are inositol auxotrophs owing to a failure to derepress the *INO1* gene.[22] Similar Ino⁻ phenotypes have been reported for mutants with defects in a variety of other genes involved in RNA transcription.[23] However, in each of these cases, the defects in transcription involve many functions in addition to the expression of *INO1* that are unrelated to phospholipid biosynthesis. In contrast, the defects in the *ino2* and *ino4* mutants appear to be confined to phospholipid biosynthesis[17] (M. Nikoloff, personal communication, 1991).

Cells unable to grow in the absence of choline and/or ethanolamine define a single genetic locus designated *CHO1* (*PSS*) (Table I).[24–29] Letts and Dawes used a density centrifugation procedure as an enrichment for mutants defective in membrane biogenesis in the isolation of the *cho1* mutants.[28] The *CHO1* gene encodes phosphatidylserine (PS) synthase (CDPdiacylglycerol–serine *O*-phosphatidyltransferase),[30] the enzyme that catalyzes the synthesis of PS (Fig. 1). The *cho1* mutant strains grow when supplemented with ethanolamine or choline, synthesizing PE (or PC) from

[19] J.-I. Nikawa and S. Yamashita, *Eur. J. Biochem.* **125**, 445 (1982).

[20] M. R. Culbertson, T. F. Donahue, and S. A. Henry, *J. Bacteriol.* **126**, 243 (1976).

[21] I. Uno, K. Fukamaki, H. Kato, T. Takenawa, and T. Ishikawa, *Nature (London)* **333**, 188 (1988).

[22] M. L. Nonet and R. A. Young, *Genetics* **123**, 715 (1989).

[23] K. T. Arndt, C. A. Styles, and G. R. Fink, *Cell (Cambridge, Mass.)* **56**, 527 (1989).

[24] G. Lindegren, Y. L. Hwang, Y. Oshima, and C. C. Lindegren, *Can. J. Genet. Cytol.* **7**, 491 (1965).

[25] K. D. Atkinson, B. Jensen, A. I. Kolat, E. M. Storm, S. A. Henry, and S. Fogel, *J. Bacteriol.* **141**, 558 (1980).

[26] L. Kovac, I. Gbelska, V. Poliachova, J. Subik, and V. Kovacova, *Eur. J. Biochem.* **111**, 491 (1980).

[27] J.-I. Nikawa and S. Yamashita, *Biochim. Biophys. Acta* **665**, 420 (1981).

[28] V. A. Letts and I. A. Dawes, *Biochem. Soc. Trans.* **7**, 976 (1983).

[29] V. A. Letts and S. A. Henry, *J. Bacteriol.* **163**, 560 (1985).

[30] K. Kiyono, K. Miura, Y. Kushima, T. Hikiji, M. Fukushima, I. Shibuya, and A. Ohta, *J. Biochem. (Tokyo)* **102**, 1089 (1987).

free ethanolamine (or choline[31]), thus bypassing PS as a precursor (Fig. 1). The tightest *cho1* mutants synthesize no PS, suggesting that yeast cells can grow and function without PS as a structural component of membranes.[25] In contrast, it was necessary to isolate mutants of *E. coli* defective in PS synthase as conditional lethals.[32]

In other fungi such as *Neurospora*,[33] mutants defective in the phospholipid *N*-methyltransferases that catalyze the reaction series phosphatidylethanolamine (PE) → phosphatidylmonomethylethanolamine (PMME) → phosphatidyldimethylethanolamine (PDME) → phosphatidylcholine (PC) were isolated as choline auxotrophs. In addition, *pem1*[34] and *pem2*[35] mutants (Table I) were isolated as choline auxotrophs and found to be defective in PE methyltransferase and phospholipid (PL) methyltransferase activity, respectively (Fig. 1). However, the *opi3* (Table I) mutants have biochemical defects similar to *pem2* mutants but are not choline auxotrophs.[36] Likewise, *cho2* mutants (Table I) which have biochemical defects identical to *pem1* mutants are not auxotrophic for choline.[37] The *cho2* and *opi3* mutants also exhibit an inositol excretion phenotype (Opi⁻) which will be discussed in detail subsequently. Genes complementing the *pem1*, *pem2*,[38] *cho2*[37], and *opi3*[36] mutants were isolated independently. The restriction map of *CHO2* is identical to that of *PEM1*, and that of *OPI3* is identical to *PEM2*.[36,37] The *cho2* and *opi3* gene disruption mutants are not auxotrophic for choline and have phenotypes identical to the original *cho2* and *opi3* point mutants.[36,37] The tightest *opi3* mutants including the gene disruptant make virtually no PC but completely substitute PMME and some PDME for PC in their membranes.[36] The *cho2* mutants including the gene disruptants synthesize reduced but detectable PC.[37] It is believed that the residual PL methyltransferase activity in *cho2* mutants (corresponding to the gene product of the *OPI3/PEM2* gene) is capable of catalyzing the conversion of PE to PMME, although at a low efficiency. The presence of PMME and PDME in *opi3* strains and the remaining PC in *cho2* strains is believed to account for the lack of choline auxotrophy.

It has also been possible to isolate mutant strains defective in the synthesis of sphingosine-containing lipids by isolating mutants auxotrophic for sphingosine. The *lcb1* mutant (Table I) strains were identified on

[31] E. P. Kennedy and S. B. Weiss, *J. Biol. Chem.* **222,** 193 (1956).
[32] C. R. H. Raetz, *Microbiol. Rev.* **42,** 614 (1978).
[33] G. Scarborough and J. Nyc, *J. Biol. Chem.* **242,** 238 (1967).
[34] S. Yamashita and A. Oshima, *Eur. J. Biochem.* **104,** 611 (1980).
[35] S. Yamashita, A. Oshima, J.-I. Nikawa, and K. Hosaka, *Eur. J. Biochem.* **128,** 589 (1982).
[36] P. McGraw and S. A. Henry, *Genetics* **122,** 317 (1989).
[37] E. F. Summers, V. A. Letts, P. McGraw, and S. A. Henry, *Genetics* **120,** 909 (1988).
[38] T. Kodaki and S. Yamashita, *J. Biol. Chem.* **262,** 15428 (1987).

the basis of a growth requirement for 200 mM DL-erythrodihydrosphingo-sine[39] and lack serine palmitoyltransferase activity.

Mutants That Overproduce Inositol

The technique of isolating analog-resistant or analog-sensitive mutants has been used quite successfully in identifying regulatory mutants defective in various metabolic processes in yeast, including amino acid biosynthesis.[40,41] Initially, mutants constitutive for I1PS were sought by screening for mutants resistant to chemical analogs of inositol. However, since none of the 18 inositol analogs tested had any effect on the growth of wild-type yeast strains,[41a] another strategy had to be developed for identifying mutants constitutive for I1PS expression. Mutants that express I1PS constitutively were ultimately identified using a novel bioassay for inositol excretion.[42] The mutants possessing the inositol excretion phenotype are called Opi⁻ (for overproduction of inositol).

The assay involves growing inositol prototrophic strains that are to be tested on agar plates in small, equally spaced patches on inositol-deficient medium. The plates are incubated for several days (preincubation period) and are then sprayed with an even lawn of the inositol-requiring tester strain (Fig. 2). The tester is a diploid strain homozygous for *ino1* and *ade1* mutations and is therefore an inositol auxotroph and phenotypically red. Strains that excrete inositol are identified by spraying the plate with a light coat of an aerosol of the tester strain. The plate is allowed to dry for 5 to 10 sec. The spraying is repeated 2 or 3 times. An even coat of the tester strain is essential as the colonies may run if the plates are oversprayed. A red halo, indicating growth of the tester strain, surrounds putative Opi⁻ colonies and indicates an inositol excretion phenotype. Generally, the radius of the halo is 1 mm to 1 cm. However, the size of the halo is variable from trial to trial even for the same strain and can be affected by factors such as the length of preincubation, the temperature at which the plates are incubated, and the condition (age, dryness, etc.) of the plates.

The original mutagenesis[42] resulted in the isolation of five inositol-overproducing mutants representing four unlinked loci: *opi1–opi4* (Table I). Further characterization of strains carrying *opi1, opi2,* and *opi4* mutations revealed the constitutive expression of I1PS.[43] The *opi2* and *opi4*

[39] G. B. Wells and R. L. Lester, *J. Biol. Chem.* **258,** 10200 (1983).

[40] A. Schurch, J. Miozzari, and R. Vutter, *J. Bacteriol.* **117,** 1131 (1974).

[41] A. Wolfner, D. Yep, F. Mezzenguy, and G. R. Fink, *J. Mol. Biol.* **96,** 273 (1975).

[41a] M. Greenberg, Ph.D. Thesis, Albert Einstein College of Medicine, Bronx, NY (1980).

[42] M. Greenberg, B. Reiner, and S. Henry, *Genetics* **100,** 19 (1982).

[43] M. Greenberg, P. Goldwasser, and S. Henry, *Mol. Gen. Genet.* **186,** 157 (1982).

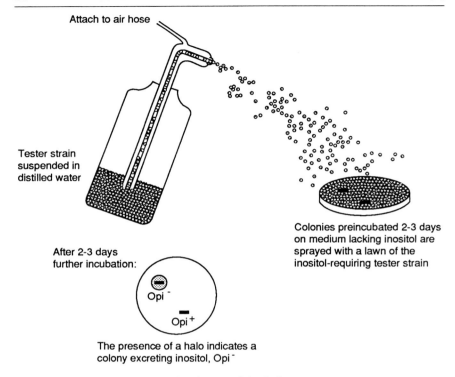

FIG. 2. Diagram of the Opi⁻ test.

mutants have weak inositol excretion phenotypes, and since only one allele of each locus has been isolated, these mutants have not been analyzed in detail. A mutant isolated in a subsequent screening for Opi⁻ mutants[44] was found to be pleiotropic and to have reduced activity of CDPdiacylglycerol synthase and was consequently named *cdg1* (Table I). The *opi1* mutants express I1PS, PS synthase, and the phospholipid *N*-methyltransferases constitutively.[12] The *OPI1* gene has been isolated, sequenced, and shown to encode a specific negative regulator of the *INO1* gene and other structural genes encoding enzymes of phospholipid biosynthesis[45] (Table I).

The *opi3* mutants[42] were later found to be defective in the final two methylations in the synthesis of PC.[46] Likewise, *cho2* mutants which are

[44] L. S. Klig, M. J. Homann, S. D. Kohlwein, M. J. Kelley, and S. A. Henry, *J. Bacteriol.* **170,** 1878 (1988).

[45] M. J. White, J. P. Hirsch, and S. A. Henry, *J. Biol. Chem.* **266,** 863 (1991).

[46] M. L. Greenberg, L. S. Klig, V. A. Letts, B. S. Loewy, and S. A. Henry, *J. Bacteriol.* **153,** 791 (1983).

defective in PE methylation[37] and *chol* mutants which have lesions in PS synthase[29] also possess an Opi⁻ phenotype. The inositol excretion phenotype of *opi3, cho2,* and *chol* strains is conditional; when *opi3* mutants are grown in medium containing dimethylethanolamine (DME) or choline, the Opi⁻ phenotype is eliminated.[36] Likewise, in *cho2* mutants the Opi⁻ phenotype is eliminated and PC biosynthesis is restored when monomethylethanolamine (MME), DME, or choline is added to the growth medium.[37] The *chol* mutants exhibit an Opi⁻ phenotype that is eliminated when ethanolamine, MME, DME, or choline is added to the growth medium.[29] In each case, only those precursors that enter the PC biosynthetic pathway downstream of the metabolic lesion in the particular mutant result in elimination of the Opi⁻ phenotype (Fig. 1). The elimination of the Opi⁻ phenotype correlates in each case with the restoration of *INO1* regulation in response to inositol.[9,29,36,37] Thus, the *INO1* gene is regulated in response to inositol only when PC or PDME biosynthesis is ongoing.

The nature of the cellular signal coordinating ongoing PC (or PDME) biosynthesis with regulation in response to inositol is not yet understood.[16,18] However, it is clear that expression of the *INO1* gene is a sensitive indicator of many perturbations in phospholipid metabolism in yeast. Many new Opi⁻ mutants with a variety of conditional Opi⁻ phenotypes have been isolated (P. McGraw, unpublished data, 1991). These include mutants that have an Opi⁻ phenotype unless supplemented with MME or choline, but not DME (as well as mutants that lose the Opi⁻ phenotype when supplemented with serine). In addition, mutants with a temperature-sensitive Opi⁻ phenotype have been isolated. These mutants presumably represent additional genes involved in phospholipid biosynthesis and/or its regulation, and they are currently undergoing further genetic and biochemical analysis (M. Swede, unpublished data, 1991).

Colony Autoradiography

The colony autoradiography technique has been successfully used in isolating mutants of phospholipid biosynthesis in *E. coli,*[47] and it has been adapted for use in yeast.[48] This technique provides a direct means of screening for mutants with specific biochemical defects without any prior knowledge about potential growth phenotypes. The technique involves recreating an enzymatic activity *in situ* in permeabilized colonies immobilized on a filter paper. The assay conditions will, of course, depend on the enzyme to be tested. Colonies are replica plated onto Whatman (Clifton,

---

[47] C. R. H. Raetz, *Proc. Natl. Acad. Sci. U.S.A.* **72,** 2274 (1975).
[48] M. J. Homann and G. M. Carman, *Anal. Biochem.* **135,** 447 (1983).

NJ) No. 42 paper and frozen at $-70°$ for 1 hr. They are then air dried to promote permeabilization. Filters are incubated in petri dishes containing the radiolabeled substrate. The reaction is quenched and the products precipitated by transferring the filters to a solution containing trichloroacetic acid (TCA). The filters are washed with cold TCA and air dried. The presence of the radiolabeled product is visualized by autoradiography. The filters are then stained with Coomassie blue for comparison with the autoradiograms.

Hjelmstad and Bell[49,50] used a modification of the colony autoradiograph technique to isolate yeast mutants defective in sn-1-diacylglycerol–cholinephosphotransferase (CPT) activity and sn-1-diacylglycerol–ethanolaminephosphotransferase activity (EPT). The mutants defective in CPT activity fell into three complementation groups. The *CPT1* locus is believed to be the structural gene for sn-1-diacylglycerol–cholinephosphotransferase.[9] Mutants defective in EPT activity fell into five complementation groups. The *EPT1* locus is believed to be the structural gene for sn-1-diacylglycerol–ethanolaminetransferase.[50] The *cpt1* and *ept1* mutants have no detectable auxotrophy or other growth phenotype, indicating the value of using colony autoradiography when screening for mutants defective in specific biochemical reactions.

Suppressors

Suppressors of a mutant phenotype mapping to a second locus may represent several types of mutational events. Some of the most interesting suppressors result from mutations in proteins that interact with the protein encoded by the locus represented by the original mutation. It is presumed that suppression by this mechanism involves interaction of the two mutant proteins in such a fashion that the original defect is compensated by mutation in the second protein. Another potential suppression mechanism involves creation of a biochemical or physiological bypass of the mutated function.

A suppressor analysis may provide unexpected opportunities to define relationships among cellular processes. For example, the yeast *SEC14* gene was recently found to encode the PI transfer protein.[51] Temperature-sensitive *sec14* mutants are blocked in transport through the Golgi complex. Thus, phospholipid transfer appears necessary for secretion in yeast. Furthermore, the *sec14* null mutants are inviable, suggesting that the

[49] R. H. Hjelmstad and R. M. Bell, *J. Biol. Chem.* **262**, 3909 (1987).
[50] R. H. Hjelmstad and R. M. Bell, *J. Biol. Chem.* **263**, 19748 (1988).
[51] V. A. Bankaitis, J. R. Aitken, A. E. Cleves, and W. Dowhan, *Nature (London)* **347**, 561 (1990).

function of the phospholipid carrier protein is indispensable to the yeast cell.[52] There is evidence that certain mutants that suppress the temperature-sensitive phenotype and the secretory defect of the *sec14* mutant have defects that lie within genes defining steps in the CDP-choline pathway for PC biosynthesis.[53] These suppressor strains have defects in the incorporation of ^{14}C-choline into PC via this pathway.[53] It has been determined that in one such suppressor mutant, the mutation lies within the *CKI* locus that encodes choline kinase that catalyzes the initial step of the CDP-choline pathway (Fig. 1).[53] Furthermore, a second suppressor mutation appears to be linked to the *CCT* locus that encodes cytidylyltransferase, the enzyme that catalyzes the second step in the CDP-choline pathway (Fig. 1) (V. Bankaitis, personal communication, 1991). The *sec14* null mutation is suppressed by the *cpt1* and *cki* mutations, and null mutations of the *cpt1* and *cki* loci suppress the *sec14* gene disruption mutation, indicating that the mechanism of suppression involves bypass of *sec14* function. Thus, it appears that the function of the PI transfer protein is essential only if the Kennedy pathway for PC biosynthesis is functional. However, *ept1*, *cho2*, and *opi3* mutations do not suppress the *sec14* phenotype. Thus, the *sec14* phenotype is suppressed by some defects in PC biosynthesis, but not by others. It is thought that this phenomenon may be related to a requirement to maintain the relative concentration of certain phospholipids within localized regions of the Golgi apparatus or endoplasmic reticulum (V. Bankaitis and T. McGee, personal communication, 1991).

Another example of bypass suppression in phospholipid biosynthesis involves suppression of *cho1* ethanolamine/choline auxotrophy. Atkinson[54,55] observed that mutations at two loci (*eam1* and *eam2*) unlinked to *CHO1* could alleviate the auxotrophy of *cho1* mutants without restoring PS biosynthesis. It is believed that the *eam1* and *eam2* mutations result in excess turnover or degradation of sphingolipids, thereby liberating sufficient ethanolamine to support growth in the absence of PS biosynthesis.

Currently, an analysis of suppressors of *ino2* and *ino4* inositol auxotrophic phenotypes is underway. A number of second-site suppressors that alleviate the inositol auxotrophy of specific alleles of each locus have been isolated (M. Nikoloff and J. Ambroziak, personal communication, 1991). In addition, dominant mutants mapping to a single locus, *OPI5*, have been found to bypass the requirement for *INO2* and *INO4* in the derepression

[52] V. A. Bankaitis, D. E. Malehorn, S. D. Emr, and R. J. Greene, *J. Cell Biol.* **108**, 1271 (1989).

[53] A. E. Cleves, T. P. McGee, K. Champion, M. Goebl, W. Dowhan, and V. A. Bankaitis, *Cell (Cambridge, Mass.)* **64**, 789 (1991).

[54] K. D. Atkinson, *Genetics* **108**, 533 (1984).

[55] K. D. Atkinson, *Genetics* **111**, 1 (1985).

of the *INO1* structural gene.[56] All of these suppressor mutants are currently undergoing further genetic and biochemical analysis.

Strategies Involving Molecular Biology

Gene disruptions, gene fusions, and *in vitro* mutagenesis have all been used to generate yeast phospholipid biosynthetic mutants. The construction of gene disruptions has been useful in the production of null mutations in cloned genes. This technique has been used to examine the phenotype of null mutations in many of the phospholipid biosynthetic genes. The one-step gene disruption method of Rothstein[57] was used to create insertion alleles of the *CHO1*,[13] *CHO2*,[37] *OPI3*,[36] and *OPI1*[45] genes. In addition, null alleles of *INO1*,[58] *INO4*,[17] *INO2* (M. Nikoloff, unpublished data, 1991), and *OPI1*[45] were constructed by replacement of coding sequences with a selectable marker.[57] In each of these cases, the gene disruption or null mutants had phenotypes identical to the original point mutants.

A fusion between the promoter of the *INO1* gene of *Saccharomyces cerevisiae* and the *lacZ* reporter gene of *E. coli* has been used successfully in the identification of phospholipid biosynthetic regulatory mutants. In this system, the *INO1* 5' flanking region and part of its coding sequence were fused in frame to the *E. coli lacZ* gene.[59] This fusion was integrated in single copy at the *ura3* genomic locus (Fig. 3). The *INO1-lacZ* fusion is regulated in response to the soluble phospholipid precursors inositol and choline in a fashion identical to the regulation of native *INO1* gene. Wild-type cells containing the fusion construct are blue when grown on X-Gal medium lacking inositol and choline (derepressing condition) and white on X-Gal medium containing inositol and choline (repressing condition). Colonies that exhibit altered regulation of the *INO1-lacZ* fusion can be readily detected by visual screening for altered colony color on X-Gal plates. Using this strategy, mutants that are unable to repress the fusion construct have been isolated. The mutants isolated remain blue on X-Gal medium even when inositol and choline are present in the medium. The advantage of this screening procedure is that it relies on a sensitive phenotype directly related to *INO1* expression. Previous screening procedures such as the search for inositol auxotrophs and Opi⁻ mutants, required gross under- or overexpression of the *INO1* gene. For example, *opi1* mutants overexpress I1PS and the *INO1* gene 2- to 3-fold. In contrast, the screening employing the gene fusion allows isolation of a class of mutants that simply fail to repress the *INO1* gene but do not overexpress the

---

[56] B. S. Loewy, Ph.D. Thesis, Albert Einstein College of Medicine, Bronx, NY (1985).

[57] R. J. Rothstein, this series, Vol. 101, p. 202.

[58] M. Dean-Johnson and S. A. Henry, *J. Biol. Chem.* **264,** 1274 (1989).

[59] J. M. Lopes, J. P. Hirsch, P.A. Chorgo, K. L. Schulze, and S. A. Henry, *Nuc. Acid Res.* **19,** 1687 (1991).

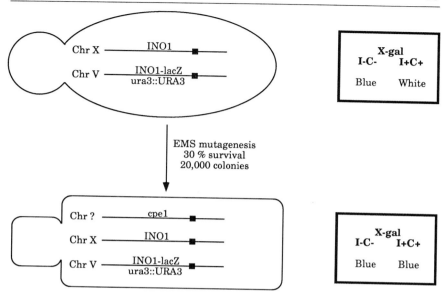

FIG. 3. Schematic representation of a mutagenic strategy used to isolate *cpe1* mutant strains. A strain harboring a wild-type copy of an *INO1-lacZ* fusion gene integrated at the *ura3* locus was mutagenized with ethylmethane sulfonate (EMS) to 30% survival. The parental strain grows as a blue colony on X-Gal media (derepressed conditions; lacking inositol and choline; I−C−) unless supplemented with 75 μM inositol and 1 mM choline (repressing conditions; I+C+), in which case it grows as white colonies. Mutants were identified that grow as blue colonies on X-Gal I+C+ media and designated *cpe1* (constitutive phospholipid expression).

gene compared to wild-type derepressed levels. Three recessive mutations (*cpe1* for constitutive phospholipid expression) possessing this phenotype have been identified (Table I). The three mutants, belonging to a single complementation group, are currently under investigation (J. Lopes and K. Hudak, unpublished data).

Media

The growth media used are as follows.

Synthetic Defined Medium

2% Glucose (w/v)
0.67% Vitamin-free yeast base (w/v) (Difco)[60]
Adenine 10 mg/liter, uracil 10 mg/liter

[60] Difco Vitamin-free yeast base has been discontinued. All other Difco yeast base formulas contain inositol. In the future, it will be necessary to make inositol-free medium completely from scratch using the above vitamins plus ammonium sulfate, salts and trace elements following the recipe for reconstituting Difco medium as previously described.[8]

Amino acids (lysine 20 mg/liter, arginine 10 mg/liter, leucine 10 mg/
  liter, methionine 10 mg/liter)
Vitamins (biotin 2 μg/liter, calcium pantothenate 400 μg/liter, folic
  acid 2 μg/liter, niacin 400 μg/liter, p-aminobenzoic acid 200 μg/
  liter, pyridoxine hydrochloride 400 μg/liter)
Inositol 50–75 μM
Choline 1 mM as needed

YEPD Medium

1% Yeast extract (w/v)
2% Peptone (w/v)
2% Glucose (w/v)

Conclusion

The isolation of mutants defective in various aspects of phospholipid
biosynthesis has provided some important insights into the regulation of
phospholipid metabolism in yeast. In particular, the pleiotropic phenotypes of regulatory mutants have revealed that many of the enzymes
of phospholipid biosynthesis are under common genetic control in *S.
cerevisiae*. In addition, many mutants possess phenotypes such as overproduction of inositol that would never have been predicted in advance.
Furthermore, it is possible to produce mutant yeast cells completely lacking major phospholipids that were assumed to be essential. These results
illustrate the value of mutant analysis in the study of major metabolic
pathways.

[4] Strategies for Isolating Somatic Cell Mutants Defective in Lipid Biosynthesis

By RAPHAEL A. ZOELLER and CHRISTIAN R. H. RAETZ

Introduction

Animal cells contain a wide variety of lipid molecular species. Little is
known about the functions of specific lipids or how their levels are regulated within particular membranes. Most of the enzymes involved in phospholipid biosynthesis in animal cells have not been isolated or characterized, because of their relatively low abundance and membrane association
(making them difficult to work with). The availability of animal cell mutants

that are deficient in the biosynthesis of a particular phospholipid species enables one to ask questions concerning the function(s) of that lipid. Mutants can also reveal a great deal about the regulation of, and interrelationships between, various biosynthetic pathways. A prime example is the discovery of the unexpected interdependence of mammalian phosphatidylserine, phosphatidylethanolamine, and phosphatidylcholine biosynthesis.[1-4] In *Escherichia coli* and yeast, lipid biosynthetic mutants have also been useful as tools to isolate relevant genes and enzymes.[5,6] The purification of most enzymes involved in phospholipid biosynthesis in *Escherichia coli* began with the isolation of a structural gene by complementation of a mutant, followed by overexpression of that gene and gene product.[5] Although this has not yet been achieved in the area of animal cell phospholipid biosynthesis, examples of this approach in other areas of animal cell physiology do exist.[7]

In this chapter, we describe three techniques that have been used to isolate somatic cell mutants defective in phospholipid biosynthesis: (1) resistance to photosensitized killing; (2) selection by tritium suicide; and (3) enzyme-targeted screening by colony autoradiography. As an example, we describe specific techniques used to isolate animal cell mutants that are deficient in plasmalogens, but variations of these methods have been used to isolate many other mutants in lipid metabolism.[1,2,8-13]

Plasmalogens

To understand the rationale of our selections, it is important to describe the structure and biosynthesis of plasmalogens. Figure 1 illustrates the structures of the three forms in which ethanolamine phospholipids can occur in animal cells: 1,2-diacyl-*sn*-glycero-3-phosphoethanolamine (phosphatidylethanolamine), 1-alkyl-2-acyl-*sn*-glycero-3-phosphoethanolamine (plasmanylethanolamine), and 1-alk-1'-enyl-2-acyl-*sn*-glycero-3-phospho-

[1] O. Kuge, M. Nishijima, and Y. Akamatsu, *Proc. Natl. Acad. Sci. U.S.A.* **82,** 1926 (1985).
[2] O. Kuge, M. Nishijima, and Y. Akamatsu, *J. Biol. Chem.* **261,** 5790 (1986).
[3] O. Kuge, M. Nishijima, and Y. Akamatsu, *J. Biol. Chem.* **261,** 5795 (1986).
[4] D. R. Voelker and J. L. Frazier, *J. Biol. Chem.* **261,** 1002 (1988).
[5] C. R. H. Raetz, *Annu. Rev. Genet.* **20,** 253 (1986).
[6] G. M. Carman and S. A. Henry, *Annu. Rev. Biochem.* **58,** 635 (1989).
[7] R. C. Mulligan and P. Berg, *Proc. Natl. Acad. Sci. U.S.A.* **78,** 2072 (1981).
[8] J. D. Esko and C. R. H. Raetz, *Proc. Natl. Acad. Sci. U.S.A.* **77,** 5192 (1980).
[9] J. D. Esko and C. R. H. Raetz, *Proc. Natl. Acad. Sci. U.S.A.* **75,** 1190 (1978).
[10] E. J. Neufeld, T. E. Bross, and P. W. Majerus, *J. Biol. Chem.* **259,** 1986 (1984).
[11] M. A. Polokoff, D. C. Wing, and C. R. H. Raetz, *J. Biol. Chem.* **256,** 7687 (1981).
[12] K. Hanada, M. Nishijima, and Y. Akamatsu, *J. Biol. Chem.* **265,** 22137 (1990).
[13] H. M. Rath, G. A. R. Doyle, and D. F. Silbert, *J. Biol. Chem.* **264,** 13387 (1989).

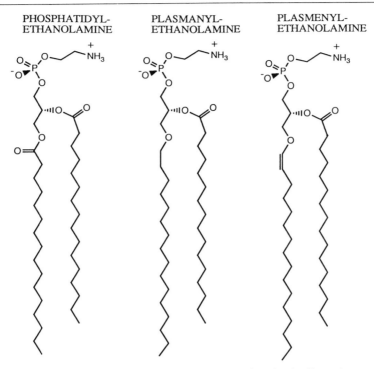

PHOSPHATIDYL- PLASMANYL- PLASMENYL-
ETHANOLAMINE ETHANOLAMINE ETHANOLAMINE

FIG. 1. Ethanolamine head-group species found in animal cell membranes.

ethanolamine (plasmenylethanolamine; the plasmalogen form). Phosphatidylethanolamine is esterified with long-chain fatty acids at the *sn*-1 and
sn-2 positions. The other two ethanolamine-linked lipids are characterized
by an ether-linked long-chain fatty alcohol at the *sn*-1 position. Plasmalogens (e.g., plasmenylethanolamine) always have a cis double bond between the first and second carbons of the fatty alcohol moiety. Depending
on the tissue and organism, up to 75% of the ethanolamine- and/or choline-
linked phospholipid of an animal cell may be recovered as the plasmenyl
variant.[14]

The biosynthesis of plasmenylethanolamine is shown in Fig. 2. The
first two steps, catalyzed by peroxisomal dihydroxyacetone phosphate
(DHAP) acyltransferase (glycerone-phosphate acyltransferase, EC
2.3.1.42) and alkyl-DHAP synthase (alkylglycerone-phosphate synthase,
EC 2.5.1.26), have been localized to the peroxisomal membrane. The

[14] L. A. Horrocks and M. Sharma, *in* "Phospholipids" (J. N. Hawthorne and G. B. Ansell,
eds.), p. 51. Elsevier, Amsterdam, 1982.

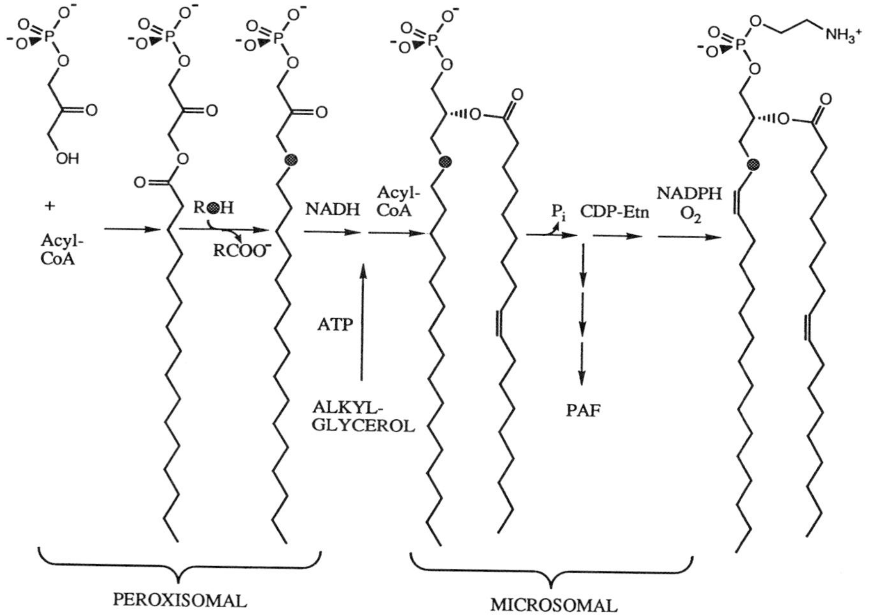

FIG. 2. Biosynthesis of plasmenylethanolamine in animal cells. PAF, Platelet activating factor.

remaining steps are catalyzed by enzymes that can be found in the endoplasmic reticulum. All of the remaining reactions, with exception of the final step, are also common to the biosynthesis of phosphatidylethanolamine. The final step in the biosynthesis of plasmenylethanolamine is the introduction of the α,β-cis double bond by the $\Delta1'$-desaturase (plasmanylethanolamine desaturase, EC 1.14.99.19). The steady-state level of the plasmanyl intermediate is very low in most cell types. The biosynthesis of plasmenylcholine has not been elucidated, although evidence points to head-group remodeling, using plasmenylethanolamine.[15] We concentrate on plasmenylethanolamine in this chapter.

General Considerations for Mutant Isolation

Choice of Cell Lines

For most mutant isolations, we have used two somatic cell lines, CHO-K1 and RAW 264.7. The primary consideration is that both cell lines contain plasmenylethanolamine. Approximately 40% of the ethanolamine-

[15] D. A. Ford and R. W. Gross, J. Biol. Chem. 263, 2644 (1988).

containing phospholipids are plasmenylethanolamine in both of these cell lines, while neither cell line contains any plasmenylcholine (although the RAW cells do contain a small amount of plasmanylcholine). Also, both of these cell lines are immortal, grow rapidly (12–15 hr doubling time), adhere to tissue culture plastic during growth, and can be grown clonally. All of these characteristics are important when trying to isolate mutant cell lines.

CHO-K1 (CHO-K1; American Type Culture Collection, Rockville, MD, CCL 61) is a fibroblast-like cell line obtained from Chinese hamster ovarian tissue.[16] It may be viewed as the "*E. coli*" of animal cell genetics in that it is biochemically and genetically well described, it is phenotypically stable over many generations, and many of the molecular biological techniques for animal cells have been developed using this cell line.[17] The RAW cell line (RAW 264.7; American Type Culture Collection, TIB 71) is a murine macrophage-like cell.[18] It is less well characterized, genetically, but responds to a variety of stimuli to perform functions in which plasmalogens have been proposed to play a role. The protocols described below have been optimized for CHO cells. The techniques are similar for RAW cells, but they must be altered somewhat due to certain differences, such as growth rate, sensitivity to trypsin, and adherence characteristics. Optimal conditions for all of the selection techniques must be determined whenever a new cell line is used for mutant isolation.

Mutagenesis

Spontaneous mutants occur on an average of 1 in 10^6 cells and can be isolated using the selection protocols described below. Mutagenesis increases this frequency 100-fold, making it possible to isolate a wider variety of genetic lesions. Also, mutagenesis is crucial for the success of nonselective techniques, such as colony autoradiography, in which only 10,000–20,000 cells can be screened.

Ethylmethane sulfonate (EMS; Sigma Chemical Co., St. Louis, MO) is a convenient mutagen which typically yields point mutations. Cells are plated out in 100-mm-diameter tissue culture plates (Corning, Corning, NY) at a density of 5×10^5 cells/dish and allowed to attach and adjust overnight in 10 ml medium [Ham's F12 medium supplemented with 1 mM glutamine, 1.176 g/liter $NaHCO_3$ (JRH Biosciences, Lenexa, KS), and 10% fetal calf serum (Hyclone Laboratories, Eugene, OR)]. The following day, 5 ml of medium containing 1.05 mg/ml EMS is added (350 μg/ml final concentration). EMS is an oil at room temperature and does not

---

[16] F.-T. Kao and T. T. Puck, *Proc. Natl. Acad. Sci. U.S.A.* **60**, 1275 (1968).

[17] L. H. Thompson, this series, Vol. 58, p. 308.

[18] W. C. Rascke, S. Baird, P. Ralph, and I. Nakoinz, *Cell (Cambridge, Mass.)* **15**, 261 (1978).

immediately dissolve in medium. The oil will sink to the bottom, and gentle swirling over a 10-min period results in its complete dispersion. The fetal calf serum which is added to the medium aids in its solubilization. Exposure of cells to 350 μg EMS/ml for 16–20 hr reduces survival to approximately 20–40%, while ouabain-resistant cells are generated at a fairly high frequency (10^{-4}). Ouabain resistance is a dominant marker used to judge the efficiency of mutagenesis. Optimal conditions may vary, depending on the cell line used. After mutagenesis, the cells are usually grown for 5–7 days to allow the cells bearing lethal mutations to die and for the mutant alleles to segregate. During this time the cells may become confluent and must be passaged, in order to reduce their density to 5×10^5 cells/dish.

Methods

Method 1: Resistance to Photosensitized Killing (P9OH/UV Selection)

Principle. First described by Moseley *et al.*[19] to isolate animal cell mutants defective in the binding and internalization of low-density lipoproteins (LDL), the photosensitized killing method is based on the fact that cells which take up and accumulate pyrene, or compounds bearing a pyrene tag, become hypersensitive to long-wavelength (>300 nm) ultraviolet irradiation. This is due to the generation of radical oxygen species following excitation of pyrene within the cell. The toxicity is directly related to the amount of pyrene in the cells as well as the duration and intensity of the UV exposure. Therefore, a cell that takes up more pyrene will be killed by exposure to a smaller dose of UV light.

In selecting plasmalogen-deficient mutants, we use a pyrene-labeled fatty alcohol analog, 9-(1'-pyrene)nonanol (P9OH), in which a pyrene moiety is covalently attached to the terminal methyl group. When long-chain fatty alcohols are taken up by most animal cells there are two major metabolic fates. They can be oxidized to the fatty acid by long-chain fatty alcohol : NAD$^+$ oxidoreductase (long-chain-alcohol dehydrogenase, EC 1.1.1.192) and incorporated, by esterification, into complex lipids. Alternatively, they can be incorporated directly into ether lipids (e.g., plasmenyl-ethanolamine) as the fatty alcohol (Fig. 2). Cells that do not synthesize plasmalogens incorporate significantly less fatty alcohol,[20] such as P9OH, accumulate much less pyrene, and, therefore, are less sensitive to long-wavelength UV light.[20]

[19] S. T. Moseley, J. L. Goldstein, M. S. Brown, J. R. Falck, and R. G. W. Anderson, *Proc. Natl. Acad. Sci. U.S.A.* **78**, 5717 (1981).

[20] O. H. Morand, L.-A. H. Allen, R. A. Zoeller, and C. R. H. Raetz, *Biochim. Biophys. Acta* **1034**, 132 (1990).

Reagents and Materials

9-(1'-Pyrene)nonanol (P9OH; Molecular Probes, Eugene, OR) comes as a powder, and 20 mM stock solutions are made up in dimethyl sulfoxide (DMSO). It is important to store these stocks at $-20°$ in the dark, and under argon or nitrogen, as pyrene is sensitive to light and oxygen.

Blak-Ray UV lamp (Model XX-15L, UVP, Inc., San Gabriel, CA; equipped with two 15-W Sylvania F1ST8 Blacklight Blue bulbs)

Blak-Ray UV intensity meter (Model J-221, UVP, Inc.)

Glass plates (1.5–2.0 mm thick); we use the glass backings from 20-cm silica gel plates (Merck, Darmstadt, FRG)

Tissue culture dishes; 100 mm diameter and 24 well (Corning)

P9OH/UV Selection. Mutagen-treated cells are plated out at 1–2 × 10^5 cells/100-mm tissue culture dish in 10 ml medium and allowed to attach overnight. The following day, 5 ml of medium containing 15 μM P9OH (final concentration 5 μM) is added, and the cells are incubated for 3 hr at 37°. The medium is removed, and the cells are washed once with 5 ml medium. Next, 15 ml of fresh medium (without P9OH) is added, and the cells are incubated for 1 hr. During this crucial "wash cycle," any unmetabolized P9OH which may have accumulated is removed from the cells. The dishes are then placed on a glass plate suspended approximately 10–15 cm above the UV source and irradiated at an intensity of approximately 2000 $\mu W/cm^2$ for a period of time which has been determined to be lethal to the cells. The glass plate serves two functions. It serves as a support for the dish containing the cells, while also serving to filter out any short-wavelength UV light which may, nonselectively, kill cells owing to its effects on DNA. The UV intensity can be adjusted by altering the height of the glass over the UV source as measured through both the glass and the bottom of a plastic tissue culture dish using the UV meter.

Once the population has been exposed to selective conditions, surviving cells are allowed to grow out for several days and develop into macroscopic colonies (3–5 mm diameter). We typically get 100–1000 colonies from 5 × 10^5 cells. These cells are then harvested, replated at a density of 3 × 10^3 cells/100-mm tissue culture dish, taken through a second round of selection, and allowed to grow out once again. The second round of selection is used to remove any wild-type like cells which, for some reason, survived the first round. If the initial round of selection is stringent enough (few surviving colonies), a second round may not be required. Surviving cells are once again allowed to grow out, and clonal lines are developed from each P9OH/UV-resistant population using limiting dilution.

Determining Conditions for P9OH/UV Selection. Prior to the selection, conditions that are lethal to wild-type cells must be determined. The conditions required to kill cells may vary, depending on the cell line used, the purity of the P9OH, and the strength of the UV source. Because the P9OH in the stock solutions slowly degrades on storage, the selection conditions must be recalibrated occasionally. To determine optimal conditions, it is most convenient to keep the P9OH concentration, P9OH exposure scheme, and the UV intensity constant, while varying the length of time that the cells are exposed to UV light. This calibration can be performed by plating out wild-type cells in a row of 24-well tissue culture dishes at a fairly low density (500 cells/well) in 0.5 ml medium. The cells are allowed to attach overnight. The following day, 0.25 ml of medium containing 15 μM P9OH is added to all of the wells. Cells are treated with P9OH as described above (including the 1-hr "wash cycle") and exposed to UV light for varying times. This UV exposure time is varied by placing a thin sheet of aluminum foil between the dish and the glass plate to block the UV light. By pulling the foil back and exposing successive wells, a temporal gradient can be achieved. The cells are then allowed to grow out for 7 days, after which the medium is removed, the cells are gently washed with phosphate-buffered saline (PBS), and the colonies resulting from surviving cells are stained for 20 min with 1 ml of 0.1% (w/v) Coomassie Brilliant Blue in methanol–water–acetic acid (45 : 45 : 10, v/v). The stain is removed, and the wells are rinsed gently with water.

Method 2: Tritium Suicide

Principle. Tritium suicide is a well-established selection method, used to isolate mutant bacterial and animal cell lines.[21,22] Neufeld et al.[10] were the first to use this technique to isolate animal cell mutants defective in lipid metabolism. Tritium suicide is a selection that involves the uptake of a conditionally toxic precursor which is normally incorporated into phospholipid(s). To select for plasmalogen-deficient mutants, the conditionally toxic compound is [9,10-^{3}H]hexadecanol. Cells which take up even a small amount of tritium [0.5–10.0 disintegrations/min (dpm)/cell], become susceptible to damage through radiolyis while stored in a frozen state. The energy of the tritium radiation is low, and neighboring cells are not affected. This effect is dependent on the dose (dpm/cell) and exposure time while in frozen storage. As in the case of P9OH/UV selection, cells that are less able to synthesize, and accumulate, plasmalogens take up

[21] J. E. Cronan, T. K. Ray, and P. R. Vagelos, *Proc. Natl. Acad. Sci. U.S.A.* **65,** 737 (1970).
[22] J. Pouyssegur, A. Franchi, J. C. Salomon, and P. Silveste, *Proc. Natl. Acad. Sci. U.S.A.* **77,** 2698 (1980).

less tritiated hexadecanol and can, therefore, withstand longer periods of storage.

Reagents and Equipment

[9,10-³H]Hexadecanol is not commercially available, but it can be easily synthesized from [9,10-³H]hexadecanoic acid (Du Pont/ NEN, Boston, MA) using the method of Davis and Hajra.[23]

Freezing vials (Fisher Scientific, Pittsburgh, PA)

Tritium Suicide Selection. A mutagen-treated population of cells is plated out at a density of 5×10^6 in 10 ml medium in two 100-mm-diameter tissue culture dishes. The cells are allowed to attach overnight at 37°. The following morning, 5 ml of medium containing 6 μM tritiated hexadecanol (60 μCi/ml) is added to achieve a final concentration of 2 μM tritiated hexadecanol at 20 μCi/ml. The cells are incubated for 3 hr at 37°, and then the medium is removed. The cells are washed once with 10 ml of medium and incubated for 1 hr, at 37°, in unlabeled medium (the "wash cycle"). The wash medium is removed, and the cells are harvested (either by trypsin or by scraping with a rubber policeman), pelleted by centrifugation, and resuspended in cryogenic medium (10%, w/v, glycerol in serum-supplemented medium) at a concentration of 10^6 cells/ml. The suspension is then aliquoted into freezing vials (1 ml/vial), and the cells are frozen in liquid nitrogen using standard procedures. Each week, after the cells are frozen, one vial is thawed, plated out in 15 ml medium in a 100-mm tissue culture dish, and the surviving cells are allowed to grow out for 2 weeks. When only 50–200 colonies arise from a plate seeded with 5×10^5 tritium-treated cells, the cells in those colonies can be harvested for clonal purification.

Method 3: Colony Autoradiography

Principle. First developed by Raetz to isolate phospholipid biosynthetic mutants in *E. coli*,[24] and later adapted for use with animal cells by Esko and Raetz,[8] colony autoradiography is a more labor-intensive technique than the first two methods of mutant isolation because it is not a selection but, rather, a visual screening method. The procedure involves the growth of colonies (generated from single cells) into polyester films in which they are immobilized.[25] Each colony is then used as a miniature reactor with which to detect the presence or absence of a specific biochemi-

[23] P. A. Davis and A. K. Hajra, *Arch. Biochem. Biophys.* **211**, 20 (1981).
[24] C. R. H. Raetz, *Proc. Natl. Acad. Sci. U.S.A.* **72**, 2274 (1975).
[25] C. R. H. Raetz, M. M. Wermuth, T. McIntyre, J. D. Esko, and D. C. Wing, *Proc. Natl. Acad. Sci. U.S.A.* **79**, 3223 (1982).

cal event. Since many colonies (each representing an individual mutant strain) can be grown into a relatively small area of polyester cloth (5–6 colonies/cm^2), 10,000–20,000 mutant candidates can be screened rather easily. Many of the finer points of colony autoradiography have been previously discussed.[26] We will deal, primarily, with the specific aspects of using this technique for the detection and isolation of CHO mutants that are deficient in the first step of plasmenylethanolamine biosynthesis, the peroxisomal DHAP acyltransferase (Fig. 2).

Reagents and Materials

^{32}P-Labeled DHAP is not commercially available but can be synthesized by the technique of Schlossman and Bell,[27] using dihydroxyacetone, glycerokinase (Sigma), and [γ-^{32}P]ATP (Du Pont/NEN).

Polyester cloth disks (17 μm mesh; Tekto, Elmsford, NY); these are cut to fit into a 100-mm-diameter tissue culture dish and autoclaved

Borosilicate glass beads, 4 mm

Kodak XAR-5 film and film cassettes (Sigma)

Trypsin solution (GIBCO, Grand Island, NY)

Growth of Colonies into Polyesters. The mutagen-treated cells are plated into forty to sixty 100-mm-diameter tissue culture dishes, in 15 ml medium, at a concentration of 500 cells/dish, and allowed to attach overnight. As CHO cells usually display a 50% plating efficiency, this represents 250 cells/dish or 4 cells/cm^2. The following day, the cells are overlaid with sterile polyester cloth. Each polyester disk should be labeled with a No. 2 pencil, along the edge, prior to sterilization for later identification. Each dish should also be labeled with the number of that polyester disk. The polyester is then weighted down with a monolayer of sterile glass beads, ensuring a close contact between the polyester cloth and the growing cells. The cells are allowed to grow for 10 days, at 37°. During this time, the cells form colonies, not only on the dish, but also in the polyester cloth. If the cells are grown at 33° (to look for temperature-sensitive mutants), it will take longer (21 days) for the cells to grow and form these colonies. The medium should be changed at least once during this period (once, at day 7 for 37°; twice, at days 10 and 18 for 33° growth).

Harvesting of Polyester Disks. After the colonies have formed, the medium is removed by aspiration, and the beads are carefully poured off into a large (500–1000 ml volume) sterile beaker (the polyester cloth will remain attached to the bottom of the dish). A sterile Kleenex or Kimwipe tissue can be used to wipe any medium from the rim of the tissue culture

[26] J. D. Esko, this series, Vol. 129, p. 237.

[27] D. M. Schlossman and R. M. Bell, *Arch. Biochem. Biophys.* **182,** 732 (1977).

dish after the beads are removed. With a magic marker, a small mark is made on the outside, bottom, edge of the tissue culture dish where the penciled identification number was previously made on the polyester disk. This will help in later localization of any putative mutant colonies. With a sterile pair of tweezers, the polyester disk is peeled off and placed into a 500-ml bath of ice-cold PBS. Next, 5 ml of medium is immediately pipetted into the tissue culture dish (master dish), being careful to add the medium gently to avoid dislodging any colonies. The medium is gently swirled and removed by aspiration. This wash removes dislodged cells and reduces cross-contamination of the colonies. Another 10 ml of medium is added, and the master dish, bearing viable colonies, is stored in a CO_2 incubator without an overlay at 28°. This will keep the cells viable, while preventing appreciable growth.

Each polyester disk containing the immobilized colonies is washed 2 times by dipping it in successive ice-cold PBS baths. The disk is then briefly blotted onto a paper towel (identification number facing up) and placed into a small bath of ice-cold homogenization buffer [0.25 M sucrose, 25 mM Tris-HCl, pH 7.4, 0.5 mM dithiothreitol (DTT), 1.0 mM EDTA, and 0.02% (w/v) sodium azide]. Each disk is once again blotted (identification number up) and placed into a 100-mm-diameter petri dish which is frozen at −20° in order to lyse the cells.

Colony Autoradiography for Detection of Peroxisomal DHAP Acyltransferase Activity. For each polyester disk, 0.8 ml of an assay mixture is added to a 100-mm-diameter petri dish, and this is placed at 37°. The assay mixture consists of 100 mM N-tris(hydroxymethyl)methyl-2-aminoethanesulfonic acid (TES) and 100 mM 4-morpholinoethanesulfonic acid (MES) (pH 5.5), 100 μM palmitoyl-CoA, 1.0 mM [^{32}P]DHAP (4–6 μCi/μmol), 8 mM NaF, 5 mM MgCl$_2$, 50 mM KCl, 2 mM KCN, 5 mM N-ethylmaleimide (NEM), and 2 mg/ml bovine serum albumin (fraction V, Sigma). There are two forms of DHAP acyltransferase, a peroxisomal and a microsomal isozyme, but only the peroxisomal form is active at pH 5.5 in the presence of NEM.[27] All of the components, except for palmitoyl-CoA and NEM, can be assembled as a 2-fold concentrated solution and stored for at least 1 year at 4°. A white precipitate may develop on prolonged storage which can be removed by filtration. Palmitoyl-CoA and NEM must be added on the day of the assay from a stock solution. The 100 mM NEM stock is prepared on the day of the assay by dissolving 12.5 mg NEM in 100 μl absolute ethanol followed by the addition of 900 μl water.

After the dishes containing the assay mixture have equilibrated for at least 20 min in a 37° incubator, the polyesters bearing the colonies are thawed and placed (identification number down) into the dishes. The assay mixture spreads as a thin film across the bottom of the polyester disk. It is important not to allow bubbles to develop, since these may result in apparent mutants. If bubbles are observed under the polyester, they can be removed by tilting the dish at a slight angle and gently swirling or tapping. After a 15-min incubation, 2 ml of 20% trichloroacetic acid (TCA) is added to stop the reaction and precipitate the TCA-insoluble product, 1-acyl-[^{32}P]DHAP, which remains associated with the TCA-fixed colony. The TCA-soluble substrate, [^{32}P]DHAP, is removed by placing each polyester on a Büchner funnel on which has been placed a Whatman (Clifton, NJ) No. 1 filter disk for support. The polyester is then rinsed with three 50-ml portions of 3% TCA, under reduced pressure. As an alternative, instead of washing each polyester individually, all of the polyesters can be transferred to a bath of ice-cold 3% TCA (about 500 ml TCA for 50 cloths) and allowed to soak, with occasional swirling, for about 30 min. The TCA solution is removed, and the polyesters are washed in 3 more changes of 3% TCA. The disks are allowed to soak for 30 min with an occasional, gentle, swirl during each TCA wash. Each polyester is then removed, blotted dry, and allowed to air dry completely for 30 min. The latter washing procedure is less tedious, uses less TCA solution, and works as effectively as the first method. The polyesters should be stirred only occasionally, and gently, as friction between the polyesters may cause some of the colonies to detach.

The polyesters are allowed to air dry for at least 1 hr and then are mounted on paper, using a small piece of tape attached to the edge of the polyester disk. The disks bearing the radioactive colonies are placed against a piece of X-ray film. After exposure to the film for 3–4 days at −80°, the film is developed to reveal signals from the radioactive colonies. Each region on the developed X-ray film is marked to denote the number of the polyester disk from which it was obtained. The colonies are then stained by placing the polyesters in a 0.1% (w/v) solution of Coomassie Brilliant Blue in methanol–water–acetic acid (45 : 45 : 10, v/v) for about 30 min. The polyesters are destained using ethanol and then air dried. The blue colony pattern on the stained polyester disk can be lined up with the autoradiographic pattern from that disk (Fig. 3). This is best done by taping a polyester disk to the surface of a light box and overlaying it with the X-ray film. Each blue-stained colony should correspond to a signal on X-ray film. A colony that fails to display a signal is interpreted as being deficient in peroxisomal DHAP acyltransferase (DHAPAT).

Once a putative DHAPAT-deficient mutant has been identified, the viable colony can be located by lining up the polyester disk under the bottom of the master dish. The colony is marked by circling it, from the bottom of the dish, with a magic marker. The colony can then be recovered by removing the medium and placing a sterile, steel pennicylinder (Fisher Scientific) so that it surrounds the colony to be recovered. A trypsin

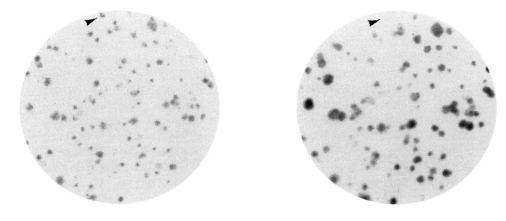

polyester cloth   x-ray film

FIG. 3. Autoradiographic detection of peroxisomal DHAP acyltransferase activity in CHO colonies immobilized in polyester. The arrowhead indicates an enzyme-deficient mutant.

solution (200 μl) is added into the cylinder, and the dish is left undisturbed for 5 min in the hood. The weight of the steel cylinder prevents the trypsin solution from leaking out, if left undisturbed, and the released cells are easily recovered with a Pasteur pipette. The cells are then transferred to a well from a 24-well tissue culture dish containing 1 ml medium and allowed to grow to confluence. If the colony was relatively well defined on the master dish, at least one-half of the cells from the recovered colony should be DHAPAT-deficient mutants. Purified cell lines can now be generated using limiting dilution, and, usually, only 2–3 isolates need to be tested for DHAPAT activity.

Preliminary Characterization of the Isolates

Plasmenylethanolamine Content

Once the isolates are obtained, the first thing that must be done is to determine if the mutant strains contain plasmenylethanolamine. The majority, but not every isolate, that we have obtained with these techniques has a reduced plasmenylethanolamine content. Also, in the isolates that are plasmalogen-deficient, the severity of the deficiency varies. To obtain an estimate of the plasmenylethanolamine content in a number of isolates, we have employed long-term labeling of the cells with [1-³H]ethanolamine, followed by separation of the phospholipid species with a two-stage, single-dimensional thin-layer chromatography (TLC) system. In this way, 10–20 isolates can be screened for their plasmalogen content on one TLC plate.

Cells are plated out in 24-well dishes at 10^4 cells/well in 1 ml medium containing 1 μCi [1-³H]ethanolamine, and the cells are allowed to grow for 3 days (5–6 generations). Virtually all of the chloroform-soluble label recovered from CHO cells (and RAW cells) is found in the ethanolamine-linked phospholipid (i.e., there is little or no label in the choline or serine head-group species). The medium is removed, the cells are washed with 2 ml PBS, and 1 ml methanol is added. After 5 min, the methanol (containing the cellular phospholipids) is transferred to a screw-capped test tube, the well is washed once with 1 ml methanol, and the methanol extracts are combined. Next, 2 ml chloroform and 1.8 ml PBS are added, the mixture is vortexed, and the organic and aqueous phases are separated by centrifugation (1000 g for 5 min). The lower (organic) phase is removed and transferred to a clean test tube. The solvent is evaporated with a stream of N_2. The lipids are dissolved in 50 μl chloroform spotted in a 1-cm lane on a silica gel 60 TLC plate (EM, Merck), and developed in chloroform–methanol–acetic acid–water (25 : 15 : 3 : 1.5, v/v) for 7 cm. This first development separates the endogenous lysophosphatidyletha-nolamine from the intact ethanolamine phospholipids.

The plate is allowed to dry and then sprayed with 10 mM HgCl$_2$ in acetic acid.[28] The double bond, adjacent to the ether linkage in the plasmen-ylethanolamine, is preferentially cleaved, generating a 2-acylglycero-phosphoethanolamine. The plate is dried for 30 min in a fume hood to re-move the acetic acid and is redeveloped in the same solvent system. The newly generated 2-acylglycerophosphoethanolamine (from plasmenyl-ethanolamine) is separated from the unaffected plasmanyl- and phosphati-dylethanolamine species. The final result is the separation, in one dimen-sion, of endogenous 1-acylglycerophosphoethanolamine, the HgCl$_2$-generated 2-acyl compound (which represents plasmenylethanolamine), and a band consisting of plasmanyl- and phosphatidylethanolamine. The plate is allowed to dry, sprayed with EN³HANCE (Du Pont/NEN), and exposed to X-ray film for 3 days at −80°. The plate generates the pattern on the X-ray film shown in Fig. 4. The bands can be scraped from the TLC plate into scintillation vials and quantitated by liquid scintillation spectrometry. The ratio of radioactivity found in the plasmenylethanol-

[28] K. Owens, *Biochem. J.* **100**, 354 (1966).

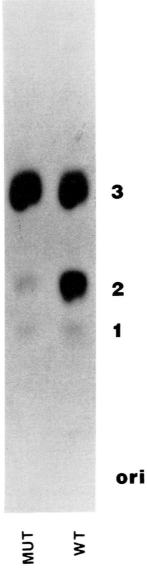

FIG. 4. Separation of plasmenylethanolamine from plasmanyl- and phosphatidyletha-nolamine using one-dimensional thin-layer chromatography. Autoradiography of TLC plate from [1-³H]ethanolamine-labeled lipids from wild-type (WT) and plasmalogen-deficient mutant (MUT) cells. Band 1, lysophosphatidylethanolamine; band 2, plasmenylethanolamine; band 3, phosphatidyl- and plasmanylethanolamine; ori, origin. Band 1 is usually very faint.

amine band to that found in the plasmanyl-/phosphatidylethanolamine band (band 2 : band 3) obtained from putative mutant strains is then compared to that obtained from wild-type cells. This ratio is much lower in the plasmalogen-deficient mutants. Other, more quantitative techniques are available.[28,29]

Rapid Assay of Peroxisomal DHAP Acyltransferase Activity in Replica-Plated Pseudocolonies

Almost every plasmalogen-deficient mutant examined to date has displayed reduced levels of peroxisomal DHAP acyltransferase. Once a set of putative mutants has been isolated, the peroxisomal DHAPAT activity can be rapidly assayed by the following procedure. Instead of measuring peroxisomal DHAPAT activity in whole-cell homogenates from each mutant strain, the enzyme is assayed in replica-plated pseudocolonies using autoradiography (Method 3). Pseudocolonies are generated by spotting a concentrated suspension of cells from each strain in a small volume (5 μl) on a tissue culture dish. In this way, macroscopic colonies can be grown into a polyester cloth in 2–3 days, and the DHAPAT activity of 40–50 strains can be determined qualitatively using one polyester cloth.

Clonal isolates are grown to near confluence in 24-well plates. The cells are then harvested with trypsin, collected by centrifugation, and resuspended in 0.05–0.3 ml (depending on the degree of confluence as judged by microscopic examination). Next, 5 μl of each cell suspension is pipetted onto the surface of a single 100-mm-diameter tissue culture dish, and the cells are allowed to attach for 2 hr at 37°. The concentrated cell suspensions form immediate macroscopic patches (3–5 mm diameter) composed of genetically homogeneous cells. The dish is marked on the bottom prior to spotting to identify each patch. After the cells have attached, 15 ml of medium is gently added. The pseudocolonies are overlayed with polyester cloth the next day and need only to be grown for 3 days (at 37°) prior to removal and freezing of the polyester cloth. A more quantitative measure of activity can later be performed on chosen isolates.

Comparison of Techniques

All of the methods described above have been used to isolate plasmalogen-deficient CHO mutants. Each technique has certain advantages and disadvantages.

Resistance to Photosensitized Killing. The easiest of the three methods

[29] M. L. Blank, M. Robinson, V. Fitzgerald, and F. Snyder, *J. Chromatogr.* **298**, 473 (1984).

is P9OH/UV selection. It is rapid, requires little specialized equipment, and is inexpensive. All of the P9OH/UV-resistant CHO cell lines that have been characterized by ourselves[20] and others[30] are deficient in both peroxisomal DHAPAT and alkyl-DHAP synthase, as the primary lesion is the loss of functional peroxisomes. However, P9OH/UV-resistant mutants derived from the murine, macrophage-like RAW 264.7 cell line contain peroxisomes and are deficient only in selected enzymatic activities such as peroxisomal DHAPAT.[31] It appears that different cell lines can yield different mutants using the same selection technique.

A problem associated with fluorescent analogs is that the presence of a bulky group, such as pyrene, often alters the metabolism of the analog from that of the compound it was intended to mimic. For instance, it appears that P9OH is not readily oxidized into fatty acid in intact cells.[31] Also, pyrene-labeled fatty acyl-CoA is not so readily degraded by mitochondrial β-oxidation.[32]

Tritium Suicide. Tritium suicide is also a relatively easy method to use, but, in contrast to P9OH/UV selection, it is based on natural precursors. Using [9,10-³H]hexadecanol, we have been able to select not only plasmalogen-deficient cell lines,[31] but also cell lines that are deficient in long-chain fatty alcohol : NAD$^+$ oxidoreductase (the activity responsible for the conversion of fatty alcohol to fatty acid[33]). In comparison to P9OH/UV selection, tritium suicide seems to yield a greater variety of genetic lesions.

A drawback to tritium suicide is the increased time associated with the selection process. It can take months (depending on the amount and nature of the labeled compound incorporated into the cells) for the tritium to kill the cells during storage. Also, as with pyrene-labeled molecules, one must be certain that the tritiated compound is incorporated into only one or two specific end products. Therefore, the metabolic fate(s) of the tritiated compound must be established prior to a selection attempt.

Colony Autoradiography. Although colony autoradiography is somewhat tedious, it is probably the most versatile. Using colony autoradiography, one can target specific enzymes in a pathway within lysed colonies, as in the peroxisomal DHAP acyltransferase screenings, or one can use polyesters containing viable colonies (not subjected to freeze–thawing) to monitor the uptake of lipid precursors from the medium. The latter approach has been used for the isolation of phosphatidylcholine biosynthesis mutants by screening for colonies defective in the incorporation of ¹⁴C-

[30] T. Tsukamoto, S. Yokota, and Y. Fujiki, *J. Cell Biol.* **110,** 651 (1990).

[31] R. A. Zoeller, unpublished results, 1991.

[32] S. Gatt, J. Bremer, and H. Osmundsen, *Biochim. Biophys. Acta* **958,** 130 (1988).

[33] P. F. James, W. B. Rizzo, J. Lee, and R. A. Zoeller, *Proc. Natl. Acad. Sci. U.S.A.* **87,** 6102 (1990).

labeled choline into TCA-precipitable material.[8] Any block in the pathway of phosphatidylcholine synthesis results in a reduction in choline incorporation.

Often, a selection scheme cannot be developed a priori. In these cases, colony autoradiography (a nonselective technique) may be the only method available. The initial mutants obtained in this manner can be used to develop selective methods. For instance, the plasmalogen-deficient CHO mutants, first isolated using colony autoradiography,[34] were later found to be amenable to the P9OH/UV selection procedure.[20]

Colony autoradiography can also be used in conjunction with selection techniques, since, quite often, only 1% of the survivors of a selection possess the desired genotype. For instance, cells surviving tritium suicide selection using [9,10-³H]hexadecanol (Method 2) were screened, using colony autoradiography, in order to identify those mutants which were truly less able to take up the fatty alcohol.[33] Other examples of screening a selected population using colony autoradiography have been reported.[10,13] Also, as described above, autoradiography of replica-plated pseudocolonies has been used to screen rapidly mutants obtained from P9OH/UV and tritium suicide selection for DHAP acyltransferase activity.

A restriction associated with colony autoradiography is that the cell line must be adherent. Cells which only grow in suspension cannot be analyzed. Also, the technique depends on the feasibility of separating unused substrate (or precursor) from product. This is not always possible, particularly when assaying enzyme activities in lysed cell colonies. It would be difficult to assay acyl-CoA : diacylglycerol acyltransferase using colony autoradiography since the substrates (diacylglycerol and acyl-CoA) and the product (triacylglycerol) are all TCA-insoluble.

---

[34] R. A. Zoeller and C. R. H. Raetz, *Proc. Natl. Acad. Sci. U.S.A.* **83**, 5170 (1986).

Section II

Acyltransferases

[5] Glycerophosphate Acyltransferase from *Escherichia coli*

By MARK A. SCHEIDELER and ROBERT M. BELL

Introduction

The initial step in the assembly of membrane phospholipids in *Escherichia coli* is controlled by the membrane-associated *sn*-glycerol-3-phosphate acyltransferase (EC 2.3.1.15). Two water-soluble substrates, *sn*-glycerol 3-phosphate and an acyl-coenzyme A (CoA) or acyl carrier protein (ACP) thioester, are catalytically utilized to form a 1-acylglycerol 3-phosphate (lysophosphatidic acid, LPA) lipid product.[1] The further acylation of LPA to phosphatidic acid (PA) is promoted by a distinct monoacylglycerol-3-phosphate acyltransferase.[2] Since both enzyme activities are tightly associated with the cytoplasmic side of the inner membrane,[3] it has been difficult to determine clearly the positional specificity and acyl thioester preference of the glycerophosphate acyltransferase or to establish directly its role in the thermal regulation of phospholipid acyl chain composition and membrane fluidity. For this reason, it has been necessary to first devise purification conditions which yield stable and homogeneous enzyme preparations which are devoid of the monoacylglycerophosphate acyltransferase activity.

Several key strategies have been implemented in order to purify and characterize the glycerophosphate acyltransferase. These are (1) cloning of the *plsB* structural gene encoding the glycerophosphate acyltransferase and construction of strains which overproduce the enzyme; (2) the effective use of nonionic detergents to solubilize the glycerophosphate acyltransferase from membranes and the development of chromatographic steps which resolve the solubilized, hydrophobic proteins; and (3) the design of a quantitatively efficient reconstitution methodology to restore activity to enzyme preparations which are stable, but inactive, in the presence of detergent.

Bacterial Strains

The *plsB* structural gene encoding the 83,000 M_r glycerophosphate acyltransferase polypeptide[4] was cloned by monitoring the ability of DNA

---

[1] C. R. H. Raetz and W. Dowhan, *J. Biol. Chem.* **265**, 1235 (1990).

[2] J. Coleman, *J. Biol. Chem.* **265**, 17215 (1990).

[3] C. O. Rock, S. E. Goeltz, and J. E. Cronan, Jr., *J. Biol. Chem.* **256**, 737 (1981).

[4] Apparent molecular weight estimated by polyacrylamide gel electrophoresis in the presence of sodium dodecyl sulfate. Alignment of the open reading frame of the sequenced

from a total *E. coli* library to overcome the glycerol 3-phosphate growth requirement of a *plsB⁻* mutant strain.[5] A strain (VL3/pVL1) containing hybrid plasmids constructed from this DNA expresses a level of enzyme activity[6,7] that is 10-fold higher than that present in wild-type *E. coli*[6,7]; this strain was used in the initial purification experiments. In later studies, the enzyme was purified from a bacterial strain (VL3/pLB3-4) with a higher plasmid copy number and 30-fold higher level of expression.[8] Bacterial strains containing hybrid plasmids in which the *plsB* gene was under the control of a heat- (λP_L) or isopropyl-β-D-thiogalactopyranoside-(tac) inducible promoter have permitted structural studies of the enzyme.[9,10]

Assay of Glycerophosphate Acyltransferase

Glycerophosphate acyltransferase is routinely estimated at 25° by monitoring the incorporation of [³H]glycerophosphate into chloroform-soluble material.[11] The *sn*-[³H]glycerol 3-phosphate used in the assay is prepared enzymatically with glycerol kinase from [2-³H]glycerol (200 mCi/mmol, New England Nuclear, Boston, MA).[12] The specific activity of the product resolved on Dowex AG-1X8 is then diluted with *sn*-glycerol 3-phosphate. Alternatively, the [2-³H]glycerol can be diluted with glycerol prior to phosphorylation. The palmitoyl- and oleoyl-CoA acyl donors used in the assay can be purchased from Pharmacia P-L Biochemicals (Piscataway, NJ). Palmitoyl- and *cis*-vaccenoyl-ACP are enzymatically synthesized by incubating ACP with free fatty acid and acyl-ACP synthetase.[13] The efficiency of this reaction is enhanced by using the reduced ACP species resolved from total *E. coli* ACP. Briefly, 10 ml of ACP (7.5 mg/ml) is incubated overnight at 4° with 100 m*M* dithiothreitol (DTT) to maximize formation of the reduced species. Hydroxylamine is added to a concentration of 0.2 *M*, pH 8.0, and the solution is passed over a 1.5 × 40 cm

---

DNA fragment bearing *plsB* with partial peptide sequence information obtained for the purified enzyme indicates a molecular weight of 91,260.

[5] V. A. Lightner, T. J. Larson, P. Tailleur, G. D. Kantor, C. R. H. Raetz, R. M. Bell, and P. Modrich, *J. Biol. Chem.* **255**, 9413 (1980).

[6] P. R. Green, A. H. Merrill, Jr., and R. M. Bell, *J. Biol. Chem.* **256**, 11151 (1981).

[7] M. A. Scheideler and R. M. Bell, *J. Biol. Chem.* **261**, 10990 (1986).

[8] M. A. Scheideler and R. M. Bell, *J. Biol. Chem.* **264**, 12455 (1989).

[9] W. O. Wilkison, J. P. Walsh, J. M. Corless, and R. M. Bell, *J. Biol. Chem.* **261**, 9951 (1986).

[10] W. O. Wilkison and R. M. Bell, *J. Biol. Chem.* **263**, 14505 (1988).

[11] M. D. Snider and E. P. Kennedy, *J. Bacteriol.* **130**, 1072 (1977).

[12] Y. Chang and E. P. Kennedy, *J. Lipid Res.* **8**, 447 (1967).

[13] C. O. Rock and J. L. Garwin, *J. Biol. Chem.* **254**, 7123 (1979).

Sephadex G-25 (fine) column to separate ACP and DTT. ACP elution is monitored by protein analysis using a modification of the Lowry method[14] and DTT by absorbance at A_{410}. Reduced ACP binds to Affi-Gel 501 (Bio-Rad, Richmond, CA) and is eluted with 5 mM 2-mercaptoethanol; this accounts for 25% of the total ACP.

The final 100 μl glycerophosphate acyltransferase reaction mixture contains 150 mM Tris, pH 8.2, 0.2 M NaCl, 5 mM 2-mercaptoethanol, 1 mg/ml bovine serum albumin (BSA), 0–0.3 milliunits (mU) of the enzyme activity present in membranes or reconstituted samples (see below), and an acyl donor such as palmitoyl-CoA. After these additions are made at 4° the samples are equilibrated for 10 min at 25°. After the reaction is initiated by the addition of [³H]glycerophosphate, the incorporation of radiolabel into LPA is linear for at least 7 min at 25° and is terminated by the addition of 0.6 ml of 1% HClO₄ (v/v). The labeled lipid products are extracted as follows: 3 ml of CHCl₃/MeOH (1 : 2, v/v) is mixed with each sample. Two separate phases result after the further addition of 1 ml of CHCl₃ and 1 ml of 1% HClO₄ (v/v). After brief centrifugation in a clinical centrifuge, the upper phase is aspirated and the remaining lower phase washed twice with 3 ml of 1% HClO₄ (v/v). A 1-ml sample is then added to a scintillation vial, evaporated to dryness, and mixed with 5 ml of Aquasol-2 (New England Nuclear) prior to counting. A unit of enzyme activity is defined as the incorporation of 1 μmol of glycerophosphate per minute under these conditions.

The routine measurement of maximal enzyme velocity is performed in the presence of saturating palmitoyl-CoA and [³H]glycerophosphate; final concentrations of these substrates are 25 μM and 5 mM, respectively. Adding BSA to the assay alleviates the detergent effect of palmitoyl- and oleoyl-CoA, but it is not required when using acyl-ACP derivatives or mixed micellar samples (see below). A total ionic strength of 0.2–0.4 M is required for optimal utilization of both acyl-CoA and acyl-ACP substrates; however, the presence of divalent cations in the assay has proved nonessential.[6] The glycerophosphate acyltransferase has a broad pH optimum which in homogeneous preparations is dictated by the lipids used to reconstitute activity.[8]

Purification of Glycerophosphate Acyltransferase

Membrane Preparation and Solubilization

Cells from *E. coli* strain VL3/pVL1 are harvested by centrifugation at 5000 g for 15 min at 4°, washed once in 50 mM Tris, pH 8.2, containing 10 mM MgCl₂, and resuspended (10 g wet weight/100 ml) in this same

---

[14] G. L. Petersen, *Anal. Biochem.* **83,** 346 (1977).

buffer. The cell suspension is mixed with DNase I (0.5 mg/100 ml, Sigma, St. Louis, MO) at 4° for 30 min; this greatly reduces the viscosity of the preparation, thereby facilitating cell breakage. The cell digest is disrupted by two passes through an ice-cold French pressure cell at 16,000 psi. The flow-through changes color from light to dark brown as cells are broken; stabilization of the color signals that maximal disruption has been achieved. Membranes are collected by centrifugation at 200,000 g for 1 hr at 4°, homogenized (5 mg/ml of protein) in 50 mM Tris, pH 8.2, containing 0.5 M NaCl, and stirred at 4° for 45 min. Although salt extraction leads to a loss in enzyme activity (Table I) several proteins are removed in this step which are not resolved by subsequent chromatography steps.

The salt-extracted membranes are recovered by centrifugation at 200,000 g for 1 hr at 4° and homogenized (10 mg/ml) in elution buffer [25 mM Tris, pH 8.2, containing 20% glycerol (w/v) and 5 mM 2-mercaptoethanol]. The glycerophosphate acyltransferase is extracted by adding 20% Triton X-100 (w/v) to achieve a final detergent concentration of 0.2% (w/v). After stirring for 45 min at 4°, the 83,000 M_r glycerophosphate acyltransferase polypeptide[4] is recovered in the supernatant following high-speed centrifugation. The ratio of detergent to protein is critical in this step; higher Triton X-100 concentrations than

TABLE I
RECOVERY OF GLYCEROPHOSPHATE ACYLTRANSFERASE ACTIVITY AND
83,000 M_r POLYPEPTIDE DURING PURIFICATION[a]

Purification step	Total activity[b] (units)	Total protein (mg)	Yield of 83,000 M_r polypeptide (% of [^{3}H]leucine-labeled protein)[c]	Purification (-fold) 83,000 M_r polypeptide	Specific activity
Membranes	58 (14)	650	2.6	1	1
Salt-extracted membranes	40 (10)	375	3.9	1.5	1.2
Triton extract	32	150	5.6	2.2	2.4
Matrex Gel Green A	3.2	7.5	22.8	8.8	4.8
Octyl-Sepharose CL-4B	3.2	2.0	81.0	31.2	18.0
Hydroxylapatite-HTP	1.0	0.3	100.0	38.0	38.0

[a] Chromatography steps were performed in the presence of 0.5% Triton X-100 (w/v). Reprinted from Ref. 7 with permission of the *Journal of Biological Chemistry*.
[b] Activities listed are for samples that were solubilized and reconstituted. Values in parentheses were obtained with membranes which were *not* solubilized and reconstituted prior to assay of glycerophosphate acyltransferase activity.
[c] Procedures used to label VL3/pVL1 cell protein with [^{3}H]leucine and quantitate the recovery of 83,000 M_r polypeptide are described in Ref. 7.

used here do not further enhance the extraction. Although activity in the membrane preparations deteriorates during storage at $-70°$, solubilized preparations of the enzyme are indefinitely stable at this temperature.

Chromatographic Fractionation

Purification is accomplished following three successive steps of fractionation based on hydrophobic dye interaction, combined sizing and hydrophobic interaction, and ion exchange. The efficiency of 83,000 M_r polypeptide recovery and reconstitution of activity under conditions which are quantitatively efficient is described for each step in Table I. All steps are performed at $4°$.

Matrex Gel Green A. The Triton X-100 extract (150 mg of protein) is diluted 1 : 1 (v/v) with elution buffer containing 0.2% Triton X-100 (w/v) and 0.2% deoxycholate (w/v); this is mixed batchwise with 4 ml of Matrex Gel Green A (Amicon, Danvers, MA) that had been equilibrated in elution buffer containing 0.2% Triton X-100 (w/v) and 0.1% deoxycholate (w/v). After gentle mixing for 1 hr the dye resin is pelleted by centrifugation at 3000 g for 5 min, washed twice with 90 ml of elution buffer containing 0.2% Triton X-100 (w/v) and 0.1% deoxycholate (w/v), and then washed twice with this same buffer containing 0.5 M NaCl. The glycerophosphate acyltransferase is then eluted from the resin by washing twice with 90 ml of this same buffer containing 3 M NaCl. The inclusion of deoxycholate reduces irreversible, nonspecific interactions between enzyme and the dye resin by adding a negative charge to the Triton X-100 micelles.[15] Binding of the glycerophosphate acyltransferase to Matrex Gel Red A is also high, but recoveries are lower. The enzyme does not bind strongly to either Blue or Orange Matrex Gel A resins.

Octyl-Sepharose CL-4B. The pooled elution fractions from the previous step are directly loaded onto a 1.5 × 70 cm octyl-Sepharose CL-4B column (Pharmacia) that had been equilibrated in elution buffer containing 0.2% Triton X-100 (w/v). The column is eluted with a 900-ml linear gradient of 3 to 0.5 M NaCl, in elution buffer containing 0.2% Triton X-100 (w/v). Resolution of the glycerophosphate acyltransferase at this step is highly dependent on both the column geometry, indicating a partial fractionation based on size, and ionic strength of the eluting buffer, as predicted for the disruption of a hydrophobic interaction.

Hydroxylapatite. The hydroxylapatite step serves to resolve inactive

---

[15] J. B. Robinson, Jr., J. M. Strottman, D. G. Wick, and E. Stellwagen, *Proc. Natl. Acad. Sci. U.S.A.* **77**, 5874 (1980).

83,000 M_r polypeptide generated during the Matrex Gel Green A step and concentrate homogeneous enzyme[7]; this step can be additionally employed to exchange Triton X-100 with another detergent.[8] The pooled fractions from the previous step are gently mixed for 1 hr with 4 g of hydroxylapatite-HTP (Bio-Rad) that had been equilibrated with elution buffer. The slurry is then loaded onto a 1.5-cm column, washed with 60 ml of elution buffer containing 0.2% Triton X-100 (w/v), and eluted with a 100-ml gradient of 0.05 to 0.5 M potassium phosphate, pH 7.0, in elution buffer containing 0.2% Triton X-100 (w/v). Preparations are homogeneous after this step; only a single 83,000 M_r band is apparent when eluting fractions are resolved electrophoretically on polyacrylamide gels run in the presence of sodium dodecyl sulfate. However, most of the protein eluting in the early (low ionic strength) fractions is inactive. Fractions eluting at high ionic strength contain high levels of glycerophosphate acyltransferase activity. These are pooled and stored at $-70°$.

Comment. Use of the *E. coli* strain VL3/pLB3-4 shortens the purification procedure considerably by obviating the need for an octyl-Sepharose CL-4B step.[8] Elution fractions from the Matrex Gel Green A step are diluted 3-fold in elution buffer containing 0.2% Triton X-100 to lower the salt concentration and then directly mixed with hydroxylapatite. In addition, enzyme inactivation during the Matrex Gel Green A step is mostly avoided; only active enzyme is subsequently observed eluting from hydroxylapatite. The large proportion of overproduced enzyme present in membranes from VL3/pLB3-4 is a dimeric species[9] which may be resistant to inactivation.

Reconstitution of Glycerophosphate Acyltransferase

Incorporation into Vesicles

Glycerophosphate acyltransferse activities of solubilized membrane samples, and of samples taken at each step of the purification, are reconstituted by addition to phospholipid.[7,11] In this procedure, a 5-μl sample (0–0.3 mU of enzyme activity) in elution buffer containing 0.2% Triton X-100 (w/v) is diluted below the critical micellar concentration (CMC)[16] of the detergent, namely, 0.05% (w/v), by addition of 30 μl of a concentrated solution of preformed phospholipid vesicles. The reconstituted sample is then added to the assay reaction mixture previously described. The recovery of activity is strictly dependent on

[16] A. Helenius, D. R. McCaslin, E. Fries, and C. Tanford, this series, Vol. 56, p. 734.

both the order of mixing and the phospholipid concentration. Quantitative enzyme reactivation observed by this procedure closely parallels the recovery of 83,000 M_r polypeptide during the purification (Table I).

The total lipid extract used to make phospholipid vesicles is prepared from *E. coli* membranes[17] and stored in chloroform at $-20°$. The major lipids present are phosphatidylethanolamine (PE), phosphatidylglycerol (PG), and diphosphatidylglycerol (cardiolipin, CL) in the approximate molar ratio of 6:1:1.[18] Following chloroform removal under nitrogen, phospholipids are suspended in 150 mM Tris, pH 8.2, containing 10 mM 2-mercaptoethanol at a concentration of 15 mg/ml. Phospholipid vesicles are formed by direct probe sonication of the suspension at room temperature; this should be continued until clearing of the suspension ceases and an orange-blue tint typical of vesicle formation is evident. Titaninum shed by the probe is pelleted by centrifugation in a clinical centrifuge for 5 min.

Mixed Micellar Reconstitution

Mixed micellar reconstitution is a soluble assay system in which the glycerophosphate acyltransferase is activated by supplementing detergent micelles containing the enzyme with limiting amounts of specific cofactors. The goal of assaying the glycerophosphate acyltransferase in mixed micelles is to permit *in vitro* investigation of the lipid–protein stoichiometry required for the reconstitution of *in vivo* enzyme kinetics. $C_{12}E_8$ is a chemically pure, nonionic detergent which forms a stable micelle size of 123 mol/micelle above its 110 μM CMC value,[19] and it does not inhibit glycerophosphate acyltransferase activity when added to assays at concentrations exceeding its CMC.[8] During the glycerophosphate acyltransferase purification, 0.1% $C_{12}E_8$ (w/v) can be substituted for Triton X-100 in the wash and elution steps of hydroxylapatite chromatography. However, $C_{12}E_8$ is too mild to be used in the initial membrane solubilization and fractionation steps.

Maximal levels of activity, equal to those measured following vesicle reconstitution of homogeneous enzyme, are achieved by mixing homogeneous glycerophosphate acyltransferase (0–0.3 mU) in 0.1% $C_{12}E_8$ (w/v) with 450 μg of *E. coli* phospholipid vesicles, or with 150 μg of vesicles formed from pure *E. coli* phosphatidylglycerol or cardiolipin

---

[17] E. A. Bligh and W. J. Dyer, *Can. J. Biochem. Physiol.* **37**, 911 (1959).
[18] J. E. Cronan, Jr., and P. R. Vagelos, *Biochim. Biophys. Acta* **265**, 25 (1972).
[19] C. Tanford, Y. Nozaki, and M. F. Rhode, *J. Phys. Chem.* **81**, 1555 (1977).

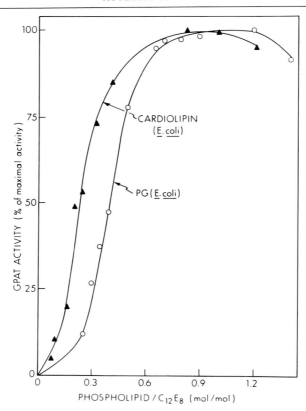

FIG. 1. Specific phospholipid requirement for enzyme activation. Data are given as the final ratio (mol/mol) in the assay of phospholipid to detergent. (Reprinted from Ref. 8 with permission of the *Journal of Biological Chemistry*.)

(Fig. 1). It is important to note that the size of a nonionic detergent micelle increases predictably as phospholipid is added,[20] up to 50 mol%. Thus, the final phospholipid/$C_{12}E_8$ ratio after mixing and addition to the assay reaction mixture should not exceed 2:1 (mol/mol). Unilamellar phospholipid vesicles are employed as the lipid source during mixing with detergent owing to their ease of solubilization. Reconstitution of the glycerophosphate acyltransferase at a low phospholipid/$C_{12}E_8$ ratio in cardiolipin/$C_{12}E_8$ mixed micelles eliminates the stringent order of enzyme–phospholipid–detergent addition required when using *E. coli*

[20] R. J. Robson and E. A. Dennis, *Biochim. Biophys. Acta* **508,** 513 (1978).

lipid preparations containing a low percentage of activating cofactor lipid.

Characterization of Glycerophosphate Acyltransferase

Reconstituted samples of homogeneous glycerophosphate acyltransferase acylate *sn*-glycerol 3-phosphate at the *sn*-1 position. Reaction products migrate with authentic 1-acylglyerol phosphate on thin-layer chromatography plates, and with 1-acylglycerol following phosphatase treatment.[6] Further, *sn*-glycerol 3-phosphate analogs which lack a secondary hydroxyl group (dihydroxyacetone phosphate, ethylene glycol phosphate, and 1,3-propanediol phosphate) work in its place as substrates.[21] Taken together, these results suggest that the primary acylation product of the homogeneous/reconstituted enzyme is 1-acylglycerol 3-phosphate.

The reconstituted enzyme preparations utilize the palmitoyl and *cis*-vaccenoyl thioester analogs of ACP as acyl donor in the assay, in addition to the palmitoyl and oleoyl analogs of coenzyme A.[6,7] However, a detailed kinetic investigation to establish an *in vitro* acyl donor preference has not yet been done.

Large amounts of *E. coli* phospholipid are required to activate fully the homogeneous glycerophosphate acyltransferase[6–8]; this result suggests that minor lipids present in the total lipid extract are responsible for activation. Enzyme reconstituted in mixed phospholipid–$C_{12}E_8$ micelles is activated by approximately 50 mol of phosphatidylglycerol or cardiolipin per mole of enzyme.[8] Phosphatidylethanolamine, the major lipid component of the *E. coli* extracts, reconstitutes activity inefficiently and in a pH-dependent manner.

Much of the membrane-associated glycerophosphate acyltransferase activity present in membranes from *E. coli* strains which overproduce the enzyme is latent until solubilized and reconstituted into an excess of activating phospholipid. The latent activity is organized as crystalline arrays of dimeric enzyme,[9,10] whereas the active form of the enzyme reconstituted into mixed cardiolipin–$C_{12}E_8$ micelles is monomeric.[8] Recent kinetic analyses of monomeric and dimeric preparations of homogeneous enzyme have revealed profound differences in their relative efficiencies of glycerol phosphate and palmitoyl-CoA utilization.[22]

[21] P. R. Green and R. M. Bell, *Biochim. Biophys. Acta* **795**, 348 (1984).
[22] M. A. Scheideler and R. M. Bell, *J. Biol. Chem.* **266**, 14321 (1991).

[6] Glycerophosphate Acyltransferase from Liver

By Dipak Haldar and Ales Vancura

Introduction

Glycerol 3-phosphate acyltransferase (acyl-CoA : sn-glycerol 3-phosphate O-acyltransferase, EC 2.3.1.15) catalyzes the following reaction:

sn-Glycerol 3-phosphate + acyl-CoA → 1-acyl-sn-glycerol 3-phosphate + CoA

The enzyme converts water-soluble glycerophosphate to a lipid product and catalyzes the committed step in the biosynthesis of phosphoglycerides and triacylglycerols from glycerophosphate. The enzyme can divert fatty acids from β-oxidation to esterification. Thus, the reaction represents a probable regulatory site in fatty acid metabolism.

In mammalian organs, glycerophosphate acyltranferase is present in both the endoplasmic reticulum (microsomes) and mitochondrial outer membrane.[1,2] In liver the glycerophosphate acyltransferase activity is nearly equal in these two organelles, whereas in other organs the microsomal acyltransferase is approximately 10 times more active than the mitochondrial enzyme.[3]

Assay

Principle. The enzyme activity can be measured by following the conversion of sn-[2-³H]glycerol 3-phosphate to 1-butanol-extractable lipid.[2,3]

Reagents

0.20 M MES–TES–glycylglycine buffer, pH 7.5 [4-morpho-line-ethanesulfonic acid, N-tris(hydroxymethyl)methyl-2-aminoethane-sulfonic acid, and glycylglycine]

0.10 M MgCl$_2$

0.72 mM Palmitoyl-CoA

0.10 M KCN

50 mg/ml Bovine serum albumin (BSA; fatty acid free)

20 mg/ml Asolectin (sonicated suspension)

[1] L. N. W. Daae and J. Bremer, *Biochim. Biophys. Acta* **210,** 92 (1970).
[2] G. Monroy, F. H. Rola, and M. E. Pullman, *J. Biol. Chem.* **247,** 6884 (1972).
[3] D. Haldar, W.-W. Tso, and M. E. Pullman, *J. Biol. Chem.* **254,** 4502 (1979).

7.5 mM sn-[2-^{3}H]Glycerol 3-phosphate [~10,000 counts/min (cpm)/
nmol]

Scintillation cocktail

MES–TES–glycylglycine buffer[2] is prepared by dissolving appropriate
amounts of each compound in water to a concentration of 0.4 M, adjusting
the pH with KOH, and finally diluting the solution to 0.2 M. The concentration of palmitoyl-CoA is determined from the optical density of the solution
at 232 and 260 nm using the millimolar extinction coefficients of 8.7 and
16.8,[2,4] respectively. An unsaturated fatty acyl-CoA can also be used as
an acyl group donor in the glycerophosphate acyltransferase assay, but it
should be freshly prepared to avoid oxidation. KCN is used to prevent
oxidation of acyl-CoAs by mitochondria and should also be freshly prepared. It has no effect on glycerophosphate acyltransferase. Asolectin, a
soybean phospholipid preparation, is suspended in 10 mM Tris-HCl, pH
7.5, and the mixture is exposed at 0° to sonic irradiation until a translucent
suspension is obtained. Fifteen-second exposures with 15-sec intervals
between exposures for a total exposure of 2 to 3 min with a Megason
Ultrasonic Disintegrator (Ultrasonic Instruments International, Farmingdale, NY) at nine-tenths of its maximum output is adequate. The suspension is then centrifuged at 105,000 g for 10 min and the supernatant taken.
The suspension can be stored at 4° for a period not exceeding 2 weeks,
after which the phospholipid starts sedimenting. Radioactive glycerophosphate is commercially available but can also be prepared enzymatically
from [2-^{3}H]glycerol by phosphorylation with ATP in the presence of glycerokinase[5] followed by purification.[6] The specific activity of the radioactive
glycerophosphate is expressed in cpm/nmol. Although the assay is carried
out with 1.5 mM glycerol 3-phosphate for economic reasons, maximal
activity of the acyltransferase is achieved at higher concentrations (7.5 to
15 mM glycerol 3-phosphate for mitochondria). Unless mentioned otherwise, all of the above reagents can be stored at $-20°$.

Preparation of Subcellular Fractions

Rat liver mitochondrial and microsomal fractions can be prepared
following any standard procedure such as the one previously described.[2,7]
An 8000 to 12,000 g intermediate fraction is discarded. The washed mitochondrial and microsomal fractions are suspended in 0.25 M sucrose (5–10
mg subcellular protein/ml) and are either immediately used or divided into
0.5- to 1.0-ml aliquots and stored at $-70°$. Storage under these conditions

[4] E. Stadtman, this series, Vol. 3, p. 931.
[5] E. P. Kennedy, this series, Vol. 5, p. 476.
[6] E. E. Hill, D. R. Husbands, and W. E. M. Lands, *J. Biol. Chem.* **243**, 4440 (1968).
[7] E. C. Weinbach, *Anal. Biochem.* **2**, 335 (1961).

results in only a little loss of activity over the course of 6 months. Protein is estimated by the method of Lowry et al.[8] using BSA as a standard.

Procedure. Reaction mixtures contain the following in a final volume of 0.5 ml: 40 mM MES–TES–glycylglycine buffer; 2 mM $MgCl_2$; an optimal concentration (20–100 μM) of palmitoyl-CoA; 2 mM KCN; an optimal concentration (1–4 mg) of BSA; 0.2 mg asolectin; 1.5 mM sn-[2-^{3}H]glycerol 3-phosphate; and 0.05 to 0.2 mg subcellular protein. The reaction mixture is preincubated for 3 min at 37° before the subcellular fraction is added, and the incubation is continued for another 3 min. The reaction is stopped by adding 0.5 ml of 1-butanol and vortexing the mixture. One milliliter of water is added, mixed well, and the mixture centrifuged for 5 min at room temperature in a tabletop centrifuge. The bottom water layer is removed and may be stored and used later for recycling of the radioactive glycerophosphate. The butanol layer (upper) is washed with 1.5 ml of butanol-saturated water. To an aliquot (usually 0.1–0.2 ml) of the butanol is added a scintillation cocktail, and the mixture is counted. The specific activity of the enzyme is calculated on the basis of nanomoles of glycerophosphate acylated per minute of incubation in the presence of the enzyme per milligram of protein (nmol/min/mg). An endogenous control, incubated without palmitoyl-CoA, is subtracted from the experimental value. Typical values for freshly prepared rat liver mitochondria and microsomes, using palmitoyl-CoA as the acyl donor, are 3–5 and 4–6 nmol/min/mg, respectively.

The reaction rate is rectilinear with time (up to 9 min) and subcellular protein (up to 2 mg/ml). The optimal concentration of palmitoyl-CoA presumably depends on the amount of total protein—subcellular fraction and BSA—and is lower for the mitochondria than for microsomes. The optimal concentration of BSA varied in our hands according to the source of the protein. Asolectin, although stimulatory to the glycerophosphate acyltransferase, is not essential for the assay and is usually omitted from the assay because of the complexities involved in the preparation and stability of the suspension of this soybean phospholipid mixture.

Variations of the assay include the use of palmitoylcarnitine and CoA to generate palmitoyl-CoA.[1] This assay works, but when an activator or inhibitor is used, one is not sure if the effect is on glycerophosphate acyltransferase or palmitoylcarnitine acyltransferase. In another variation,[9] chloroform–methanol (2 : 1) is used instead of 1-butanol to extract the lipids. However, chloroform–methanol shows some bias against lysophosphatidic acid,[2] which is the main product of mitochondrial incubation

[8] O. H. Lowry, N. J. Rosebrough, A. L. Farr, and R. J. Randall, *J. Biol. Chem.* **193,** 265 (1951).
[9] H. Eibl, E. F. Hill, and W. E. M. Lands, *Eur. J. Biochem.* **9,** 250 (1969).

under the above conditions.[2,3] Also, the butanol extraction procedure is somewhat more efficient.[1,2]

The products of acylation of glycerophosphate under the above conditions are mainly lysophosphatidic acid and phosphatidic acid along with a small amount, if any, of mono- and diglycerides. The lipids can be separated by thin-layer chromatography on silica gel plates with a solvent system containing chloroform, methanol, acetic acid, and water (65 : 25 : 1 : 4).[2,10] An input of approximately 2000 cpm is adequate. The radioactivity of the lipids is determined by scraping 0.5 × 2 cm sections of each lane into mini-scintillation vials. The scrapings are soaked for 15 min in 0.2-ml portions of 10 mM EGTA, pH 4.6. Three milliliters of scintillation fluid is added and the mixture counted. Recovery of radioactivity from the plates is between 80 and 110%.

Properties

pH Optimum and K_m. Both the microsomal[11] and mitochondrial glycerophosphate acyltransferase exhibit a broad pH optimum between 6.6 and 9.0. The K_m for *sn*-glycerol 3-phosphate for the microsomal enzyme is lower (0.1–0.2 mM) than that for the mitochondrial acyltransferase (~1 mM).[2,11]

Substrate and Position Specificity. In the glycerophosphate acyltransferase assay, the microsomes from liver as well as from other mammalian organs can use both saturated and unsaturated fatty acyl-CoAs with almost equal efficiency. Mitochondria, on the other hand, prefer saturated acyl-CoA thioesters.[2] However, regardless of whether saturated or unsaturated acyl-CoA is used as acyl donor, 1-acyl-*sn*-glycerol 3-phosphate is the major if not the sole product of *sn*-glycerol 3-phosphate acylation. During the processing of the samples for determining positional specificity, approximately 6% isomerization of 1- to 2-monoacylglycerol and 20% conversion of the 2- to the 1-isomer have occurred.[3]

The above properties of the mitochondrial glycerophosphate acyltransferase provide an excellent mechanism for the observed preferential positioning of saturated fatty acids in the *sn*-1 position and unsaturated fatty acids in the *sn*-2 position in naturally occurring acylglycerols.[2] This possibility is further supported by the facts that (1) in Ehrlich cells, which exhibit an abnormally high proportion of unsaturated fatty acids at the *sn*-1 position of some phosphoglycerides, a mitochondrial glycerophosphate acyltransferase is undectable[3] and (2) in cell cultures, there is an age-

---

[10] V. P. Skipski and M. Barclay, this series, Vol. 14, p. 530.
[11] S. Yamashita and S. Numa, this series, Vol. 71, p. 550.

dependent decline of mitochondrial glycerophosphate acyltransferase activity and a parallel increase in the amount of unsaturated fatty acids at the *sn*-1 position in choline phosphoglycerides.[12]

Recently, experimental support of the above hypothesis has come from the observation that in the presence of BSA, mitochondrially produced monoacylglycerophosphate can exit the organelles, be translocated to the microsomes, and converted to diacylglycerophosphate.[13] Also, a mono-acylglycerophosphate-binding liver cytosolic protein has been purified, which may be involved in the transport of mitochondrially made monoacyl-glycerophosphate to the endoplasmic reticulum.[14]

Activators and Inhibitors. Some divalent cations, such as Mg^{2+} and Ca^{2+}, stimulate both the mitochondrial[15] and microsomal[11] glycerophosphate acyltranferase, especially when the enzyme is partially purified (see later). Acetone[3] and polymyxin B[16] stimulate the mitochondrial but inhibit the microsomal acyltranferase. Sulfhydryl group reagents, such as *N*-ethylmaleimide and iodoacetamide, strongly inhibit the microsomal acyltransferase; the mitochondrial activity is unaffected by these reagents.[2,3] This distinguishing property of mitochondrial and microsomal glycerophosphate acyltransferase has been used to determine the micro-somal contamination of the mitochondrial fraction[3,17] or to assay the mitochondrial enzyme present in a whole cell homogenate.[12,18]

Phospholipids modulate liver mitochondrial and microsomal glycero-phosphate acyltransferase. Phosphatidylserine, asolectin, and phosphati-dylcholine stimulate, whereas mono- and diacylglycerophosphate and cardiolipin inhibit the mitochondrial activity.[15] Microsomes are stimulated by phosphatidylcholine and phosphatidylethanolamine but are inhibited by phosphatidylserine and phosphatidylinositol.[11] These results have been obtained from experiments performed with partially purified subcellular glycerophosphate acyltransferase. Among other lipids, monoacylglycerols and some of their analogs inhibit both the mitochondrial and microsomal glycerophosphate acyltransferase.[18]

The liver microsomal glycerophosphate acyltransferase is inhibited by

[12] W. Stern and M. E. Pullman, *J. Biol. Chem.* **253,** 8047 (1978).
[13] D. Haldar and L. Lipfert, *J. Biol. Chem.* **265,** 11014 (1990).
[14] A. Vancura, M. A. Carroll, and D. Haldar, *Biochem. Biophys. Res. Commun.* **175,** 339 (1991).
[15] G. Monroy, H. C. Kelker, and M. E. Pullman, *J. Biol. Chem.* **248,** 2845 (1973).
[16] M. A. Carroll, P. E. Morris, C. D. Grosjean, T. Anzalone, and D. Haldar, *Arch. Biochem. Biophys.* **214,** 17 (1982).
[17] R. J. Pavlica, C. B. Hesler, L. Lipfert, I. N. Hirshfield, and D. Haldar, *Biochim. Biophys. Acta* **1022,** 115 (1990).
[18] R. A. Coleman, *Biochim. Biophys. Acta* **963,** 367 (1988).

all proteases.[19,20] The mitochondrial enzyme, on the other hand, is not inhibited by trypsin and chymotrypsin. Non-site-specific proteases, such as proteinase K and subtilisin, inhibit the mitochondrial enzyme. The degree of inhibition increases if the ionic strength of the incubation medium is lowered. The protease sensitivity of the microsomal enzyme does not respond to any such changes in the ionic strength of the incubation medium. Exposure of trypsin- or chymotrypsin-sensitive domains of glycerophosphate acyltransferase on the inner surface of the mitochondrial outer membrane can be directly demonstrated by incubating trypsin-loaded outer membrane vesicles.[20] These results, taken together, suggest that the mitochondrial glycerophosphate acyltransferase spans the transverse plane of the outer membrane.

Purification

Both microsomal and mitochondrial glycerophosphate acyltransferase have been only partially purified. In contrast, the *Escherichia coli* glycerophosphate acyltranferase has been completely purified, its properties studied,[21] and the gene cloned and sequenced.[22] The successful purification of the prokaryotic enzyme has been facilitated by massive overproduction of the enzyme owing to molecular cloning.[21]

Microsomal Glycerophosphate Acyltransferase

All operations are conducted at $0°$ to $5°$. Rat liver microsomes are treated for 2 hr with 6 mM Triton X-100 at pH 8.6 at a protein concentration of 10 mg/ml and then passed through a Sepharose 2B column. The first few fractions of the eluate which contain protein and the enzyme activity are combined and centrifuged through a sucrose gradient (0.5 to 1.5 M layered on 2 M). This treatment resolves the two enzyme activities: glycerophosphate acyltransferase and 1-acylglycerophosphate acyltransferase. However, the enrichment of glycerophosphate acyltransferase is only 3.6 times, with approximately 90% loss of activity. For a detailed account of this method, the reader is referred to an earlier volume in this series.[11] It is interesting to note that, unlike the microsomes, the partially purified enzyme prefers saturated over unsaturated fatty acyl-CoA thioesters as substrates.

[19] R. A. Coleman and R. M. Bell, *J. Cell Biol.* **76,** 245 (1978).
[20] C. B. Hesler, M. A. Carroll, and D. Haldar, *J. Biol. Chem.* **260,** 7452 (1985).
[21] M. A. Scheideler and R. M. Bell, *J. Biol. Chem.* **264,** 12455 (1989).
[22] P. R. Green, T. C. Vanaman, P. Mordich, and R. M. Bell, *J. Biol. Chem.* **258,** 10862 (1983).

Mitochondrial Glycerophosphate Acyltransferase

There are two methods available for the partial purification of the mitochondrial acyltransferase. All operations are carried out at 0° to 4°. In the first method,[15] submitochondrial particles are prepared by sonic irradiation of an aqueous suspension of mitochondria (10 mg/ml) for 10 min with a Branson Sonifier operating at maximum output. The particles are centrifuged down and finally suspended (10 mg/ml) for 10 min in a mixture containing 0.25 M sucrose, 10 mM Tris-HCl, pH 8.0, 0.5% potassium cholate, pH 7.8, and 1.0 M KCl. The mixture is centrifuged at 150,000 g, and to the supernatant is added 0.35% asolectin suspension followed by the addition of a saturated solution of ammonium sulfate to a final concentration of 15%. After 10 min the mixture is centrifuged at 105,000 g for 15 min. The supernatant fluid is adjusted to 45% saturation with ammonium sulfate and centrifuged. The precipitate is dissolved in Tris-HCl, pH 8.0, containing 0.35% asolectin and dialyzed thoroughly against the same solution. This preparation represents 6-fold purification with no monoacylglycerophosphate acyltransferase activity.

In the other method,[23] submitochondrial particles are prepared by suspending the mitochondria (10 mg/ml) in 20 mM Tris-HCl buffer, pH 8.4, containing 1 mM phenylmethylsulfonyl fluoride. The mixture is exposed to sonic irradiation for 4 min (30-sec exposures with 15-sec intervals) using a Megason Ultrasonic Disintegrator working at nine-tenths of its maximum output. The mixture is then centrifuged at 170,000 g for 90 min. The sediment containing the submitochondrial particles is suspended in a solution containing 0.25 M sucrose, 1 M KCl, 20 mM Tris-HCl buffer, pH 8.4, 1 mM phenylmethylsulfonyl fluoride, and 0.5% (ultrapure) Lubrol PX. The detergent-to-protein ratio is maintained between 2 and 4.

After 30 min, the mixture is centrifuged at 170,000 g for 90 min. The supernatant fraction, containing most of the enzyme activity, is desalted on a Pharmacia (Piscataway, NJ) PD-10 column equilibrated with 20 mM Tris-HCl, pH 8.4. The mixture is then loaded on Sepharose Q column (1.5 × 30 cm), equilibrated with 20 mM Tris-HCl, pH 8.4, containing 0.5% Lubrol PX and 10% glycerol. The column is eluted with the same buffer, and fractions containing the glycerophosphate acyltransferase activity are combined and loaded on a BioGel HT (Bio-Rad, Richmond, CA) column (1 × 10 cm) equilibrated with 20 mM Tris-HCl, pH 8.4, containing 0.5% Lubrol PX and 10% glycerol. The column is washed with 100 ml of the same buffer. Delipidated glycerophosphate acyltransferase is eluted from the column with 50 ml of a gradient of 20 to 500 mM potassium phosphate

---

[23] A. Vancura and D. Haldar, manuscript in preparation.

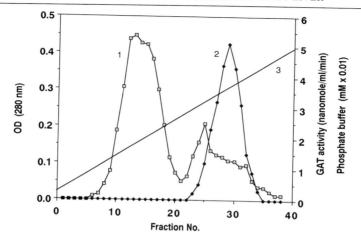

FIG. 1. Purification of mitochondrial glycerophosphate acyltransferase on a BioGel HT column. The column was eluted at 14 ml/hr with a linear gradient of phosphate buffer (20–500 mM) (line 3), and 1.3-ml fractions were collected for determining absorbance at 280 nm (line 1) and glycerophosphate acyltransferase activity (line 2).

buffer, pH 7.4, containing 0.2% Lubrol PX and 10% glycerol (Fig. 1). From this step, determination of the glycerophosphate acyltransferase activity is completely dependent on reconstitution of the enzyme with asolectin suspension. Fractions exhibiting glycerophosphate acyltransferase activity are combined, supplemented with NaCl to 0.5 M concentration, and loaded on an octyl-Sepharose CL-4B column (1 × 5 cm), equilibrated with 20 mM Tris-HCl buffer, pH 7.4, containing 0.2% Lubrol PX, 10% glycerol, and 0.5 M NaCl. The column is washed with 10 ml of the same buffer, and then the enzyme is eluted with 20 mM Tris-HCl, pH 8.4, containing 0.5% Lubrol PX and 10% glycerol. Results of a typical purification procedure are summarized in Table I.

TABLE I

PURIFICATION OF MITOCHONDRIAL GLYCEROPHOSPHATE ACYLTRANSFERASE

Step	Protein (mg)	Total activity (nmol/min)	Specific activity (nmol/min/mg)	Purification (-fold)	Yield (%)
Submitochondrial particles	119.2	122.1	1.03	1.00	100
Lubrol PX extract	84.9	73.1	0.86	0.83	60
Sepharose Q fast flow	12.1	41.3	3.42	3.32	34
BioGel HT	2.2	33.8	15.36	14.91	28
Octyl-Sepharose CL-4B	1.5	31.7	21.20	20.58	26

Additional improvement in purification of mitochondrial glycerophosphate acyltranferase was achieved using affinity chromatography on palmityl-CoA-agarose or glycerol 3-phosphate-agarose. When a glycerol 3-phosphate agarose column was used instead of octyl-Sepharose CL-4B column, the overall purification of glycerophosphate acyltransferase was over 40-fold.

The molecular weight of the native glycerophosphate acyltransferase was 60–85 kDa as determined by gel filtration on Sephacryl S-300 HR in 0.2% CHAPS. Comparison of this result with electrophoretic data strongly suggests the mitochondrial glycerophosphate acyltransferase to be a monomeric enzyme. SDS–PAGE of the purified glycerophosphate acyltransferase followed by Coomassie blue staining exhibited a single band with a molecular weight of 80–85 kDa.

Acknowledgment

This work was supported by a grant from the National Science Foundation (DCB-8801535).

[7] Coenzyme A-Independent Acyltransferase

By TAKAYUKI SUGIURA and KEIZO WAKU

Introduction

The coenzyme A (CoA)-independent transacylation system was first described by Kramer and Deykin[1,2] for human platelets. Similar enzyme activity was also found for several mammalian tissues and cells[3–6] including rabbit alveolar macrophages.[7–9] This system catalyzes the transfer of fatty

[1] R. M. Kramer and D. Deykin, *J. Biol. Chem.* **258**, 13806 (1983).
[2] R. M. Kramer, G. M. Patton, C. R. Pritzker, and D. Deykin, *J. Biol. Chem.* **259**, 13316 (1984).
[3] P. V. Reddy and H. H. O. Schmid, *Biochem. Biophys. Res. Commun.* **129**, 381 (1985).
[4] Y. Masuzawa, S. Okano, Y. Nakagawa, A. Ojima, and K. Waku, *Biochim. Biophys. Acta* **876**, 80 (1986).
[5] A. Ojima, Y. Nakagawa, T. Sugiura, Y. Masuzawa, and K. Waku, *J. Neurochem.* **48**, 1403 (1987).
[6] O. V. Reddy and H. H. O. Schmid, *Biochim. Biophys. Acta* **879**, 369 (1986).
[7] T. Sugiura and K. Waku, *Biochem. Biophys. Res. Commun.* **127**, 384 (1985).
[8] M. Robinson, M. L. Blank, and F. Snyder, *J. Biol. Chem.* **260**, 7889 (1985).
[9] T. Sugiura, Y. Masuzawa, Y. Nakagawa, and K. Waku, *J. Biol. Chem.* **262**, 1199 (1987).

acids from diradyl phospholipids to various lysophospholipids in the absence of any cofactors, differing from the CoA-dependent transacylation reaction which requires the presence of CoA.[10–14] Free fatty acids cannot be introduced into phospholipids via this system. The types of fatty acids transferred by the CoA-independent system are restricted to C_{20} and C_{22} polyunsaturated fatty acids esterified at the 2-position of diradyl phospholipids, especially diacylglycerophosphocholine (diacyl-GPC). This is quite different from either the CoA-dependent transacylation system or the lysophospholipase-mediated transacylation reaction. Furthermore, the distribution among tissues and the subcellular localization of CoA-independent transacylation activity are also considerably different from those of CoA-dependent transacylation and lysophospholipase-mediated transacylation activities. Thus, CoA-independent transacylation is presumed to be a novel enzyme reaction which may play important roles in the remodeling of phospholipids in mammalian cells.

It has been demonstrated[15–21] that C_{20} and C_{22} polyunsaturated fatty acids are gradually transferred from diacyl-GPC to ether-containing phospholipids in several inflammatory cells, endothelial cells, and testis under various conditions. It appears that such transfers can be responsible for the accumulation of C_{20} and C_{22} polyunsaturated fatty acids in ether phospholipids in several cell types. Although the precise mechanism for this transfer observed in living cells is not fully understood, it has been postulated that the CoA-independent transacylation system could be involved, at least in part, in the gradual transfer of polyunsaturated fatty acids between phospholipids. In fact, several lines of evidence suggest that the CoA-independent transacylation system is important in the reacylation of ether-containing lysophospholipids in mammalian tissues to provide polyunsaturated fatty acid-containing ether phospholipids as follows: (1)

[10] R. F. Irvine and R. M. C. Dawson, *Biochem. Biophys. Res. Commun.* **91**, 1399 (1979).
[11] R. M. Kramer, C. R. Pritzker, and D. Deykin, *J. Biol. Chem.* **259**, 2403 (1984).
[12] O. Colard, M. Breton, and G. Bereziat, *Biochim. Biophys. Acta* **793**, 42 (1984).
[13] J. Trotter, I. Flesch, B. Schmidt, and E. Ferber, *J. Biol. Chem.* **257**, 1816 (1982).
[14] T. Sugiura, Y. Masuzawa, and K. Waku, *J. Biol. Chem.* **263**, 17490 (1988).
[15] M. L. Blank, R. L. Wykle, and F. Snyder, *Biochim. Biophys. Acta* **316**, 28 (1973).
[16] S. Rittenhouse-Simmons, F. A. Russell, and D. Deykin, *Biochim. Biophys. Acta* **488**, 370 (1977).
[17] T. Sugiura, O. Katayama, J. Fukui, Y. Nakagawa, and K. Waku, *FEBS Lett.* **165**, 273 (1984).
[18] T. Sugiura, Y. Masuzawa, and K. Waku, *Biochem. Biophys. Res. Commun.* **133**, 574 (1985).
[19] O. Colard, M. Breton, and G. Bereziat, *Biochem. J.* **222**, 657 (1984).
[20] F. H. Chilton and R. C. Murphy, *J. Biol. Chem.* **261**, 7771 (1986).
[21] F. H. Chilton, J. S. Hadley, and R. C. Murphy, *Biochim. Biophys. Acta* **917**, 48 (1987).

the enzyme activities of acyl-CoA : 1-alkyl-GPC and acyl-CoA : 1-alkenyl-GPC acyltransferases are usually low, and (2) fatty acid specificities of the acyl-CoA-mediated acylation of ether-containing lysophospholipids *in vitro* are different from those observed for the acylation profiles of ether-containing lysophospholipids in living cells. In contrast, (3) the acylation profiles of the CoA-independent transacylation reaction observed *in vitro* closely resemble the acylation pattern of ether-containing lysophospholipids in living cells (see below), and (4) CoA-independent transacylation activity was shown to be present in microsomal fractions of various mammalian tissues and cells except for liver, in which ether-containing phospholipids are known to be almost absent.[14] We suppose that the CoA-independent transacylation system will be effective in promptly disposing of ether-containing lysophospholipids generated within the cell to form nontoxic diradyl phospholipids. This might be different from the case of acyl-containing lysophospholipids which can be rapidly metabolized by lysophospholipase or by acyl-CoA : lysophospholipid acyltransferase.

The properties of the CoA-independent transacylation reaction have been studied by several investigators.[1-9] However, a detailed mechanism of the reaction remains unclear. Purification and characterization of enzyme protein have not yet been successful. Thus, further efforts are required to investigate the properties of the enzyme involved in CoA-independent transacylation system and understand the precise mechanism of the reaction. In particular, detailed comparative studies of CoA-independent transacylation activity and phospholipase A_2 activity seem to be necessary in order to clarify the mechanism of the transacylation reaction, since the transacylation system involves the degradation of diradyl phospholipids in its initial step.

In a previous study, we have shown that CoA-independent transacylation activity is almost negligible in macrophage cytosolic fractions in which a remarkably high activity of phospholipase A_2 was noted,[9] indicating that cytosolic phospholipase A_2 is incapable of catalyzing the transacylation reaction. Furthermore, we found that the microsomal fraction obtained from the slug *Incilaria bilineata* does not contain CoA-independent transacylation activity, despite the presence of phospholipase A_2 activity in this fraction and high amounts of ether phospholipids in this animal.[22] The absence of CoA-independent transacylation activity was also observed for the microsomal fraction from the earthworm *Eisenia foetida*,[22a] suggesting that the acylation system for ether-containing lysophospholipids may be somewhat different between lower animals and mammals. In any event, it is obvious that phospholipase A_2 lacks the ability to catalyze the

[22] T. Sugiura, T. Ojima, T. Fukuda, K. Satouchi, K. Saito, and K. Waku, *J. Lipid Res.*, in press.
[22a] T. Sugiura, manuscript in preparation.

transacylation reaction in these cases. Recent progress in the structural analysis of phospholipase A_2 in mammalian tissues provided the evidence for the occurrence of several types of phospholipase A_2. There remains the possibility, therefore, that some forms can catalyze the transfer of polyunsaturated fatty acids between phospholipids. Alternatively, intrinsic enzyme protein other than phospholipase A_2 may catalyze the reaction. Further detailed studies on this issue may answer the question and contribute to a better understanding of the regulation of polyunsaturated fatty acid metabolism in mammalian tissues.

Procedures

Dual-assay systems are possible, one containing radiolabeled exogenous or endogenous phospholipids as acyl donors and nonradiolabeled lysophospholipids as acyl acceptors, the other containing endogenous membrane phospholipids as acyl donors and radiolabeled lysophospholipids as acyl acceptors. Both systems have advantages and disadvantages.

Transacylation from Exogenous Phospholipids

CoA-independent transacylation activity is mainly found in the microsomal fraction (105,000 g pellet fraction) in various mammalian tissues. The mitochondrial fraction (7000 g pellet fraction) has also been shown to contain some activity, whereas the cytosolic fraction (105,000 g supernatant fraction) has not. We used, therefore, the microsomal fraction or membrane fraction as the enzyme source.[4,5,7,9,18,23]

The assay system for the transacylation of 1-alkyl-GPC by the macrophage microsomal fraction consists of a microsomal fraction (0.2 mg of protein), 5 nmol of 1-alkyl-GPC [lyso-platelet-activating factor (lyso-PAF)] (final 20 μM), ^{14}C-labeled fatty acid-containing phospholipids [20,000 disintegrations/min (dpm)], and 250 μl of 0.1 M Tris-HCl buffer (pH 7.4) containing 5 mM EGTA.[9] ^{14}C-Labeled fatty acid-containing phospholipids are sonicated in advance in distilled water with a Branson Sonifier (Danbury, CT) (40 W, 4 periods of 10 sec each). Alternatively, labeled phospholipids can be previously dissolved in ethanol and then directly added to the incubation mixture. Similar results are obtained in either case. If the ethanolamine-containing 1-alkyl(alkenyl)-GPE or 1-acyl-GPE is used as acceptor instead of choline-containing lysophospholipids, it is also necessary to sonicate these lysophospholipids in distilled water.

The incubation is carried out at 37° for various periods of time. The

---

[23] T. Sugiura, T. Fukuda, Y. Masuzawa, and K. Waku, *Biochim. Biophys. Acta* **1047**, 223 (1990).

reaction is linear at least up to 30–60 min, which may be dependent on the concentration of the microsomal protein. If higher concentrations of 1-alkyl-GPC are employed, the reaction is rather reduced, probably owing to the detergent effect of 1-alkyl-GPC. In the case of ethanolamine-containing lysophospholipids such as 1-alkenyl(alkyl)-GPE, we could not find strong inhibition. The incubation is stopped by adding chloroform–methanol, and total lipids are extracted by the method of Bligh and Dyer. Individual phospholipids are separated by two-dimensional thin-layer chromatography (TLC), first with chloroform–methanol–28% NH$_4$OH (65:35:5, v/v) and second with chloroform–acetone–methanol–acetic acid–water (5:2:1:1.5:0.5, v/v).

To measure the distribution of radioactivities among subclasses of individual phospholipids (especially for CGP and EGP), they are hydrolyzed with phospholipase C and subsequently acetylated with acetic anhydride and pyridine. The resultant 1,2-diradyl-3-acetylglycerols are separated by TLC developed first with petroleum ether–diethyl ether–acetic acid (90:10:1, v/v) and then with toluene. This technique enables the resolution of individual phospholipids into alkenylacyl, alkylacyl, and diacyl subclasses. Such separation is essential for the analysis of the transacylation reaction where donor phospholipids and acceptor lysophospholipids have the same polar head groups. To estimate the transfer of ^{14}C-labeled fatty acids between the same phospholipid class (e.g., between diacyl-GPC and 1-acyl-GPC), different fatty acyl moieties at the 1-position are selected for donor and acceptor phospholipids. The reaction product (transacylated phospholipids) can be separated from the original material (donor phospholipids) by high-performance liquid chromatography (HPLC) on the basis of differences of the fatty acids at the 1-position after conversion to 1,2-diradyl-3-acetylglycerol derivatives as described below.

The assay method with exogenous donor phospholipids has the advantage of using chemically defined phospholipids with various types of fatty acids at the 1- and 2-positions. However, the estimation by this method might be influenced to some extent by the amount and composition of endogenous phospholipids in the microsomal (membrane) fraction used as the enzyme sources. Furthermore, physicochemical properties of donor phospholipids in the incubation mixture must be taken into account. In order to exclude such possibilities, we mix the donor phospholipids containing different fatty acids and then used this mixture in some experiments. Similar fatty acid specificities are observed even in the case of the mixed donor phospholipids. Moreover, the fatty acid specificity determined using the exogenous phospholipids coincides with that obtained with endogenous membrane phospholipids.[5,9]

Transacylation from Endogenous Phospholipids

The assay system contains a microsomal fraction (0.2 mg of protein) and 5 nmol of various radiolabeled lysophospholipids such as 1-[^{3}H]alkyl-GPC (40,000 dpm, final 20 μM) in 250 μl of 0.1 M Tris-HCl buffer (pH 7.4) containing 5 mM EGTA.[9] The incubation is carried out at 37° for various periods of time. The incubation is stopped by adding chloroform–methanol, and total lipids are then extracted by the method of Bligh and Dyer. The individual phospholipids are separated by two-dimensional TLC, and the subclasses are separated from each other as described above. The enzyme activity is calculated from the radioactivity found in the product (the acylated lysophospholipid) and that in the original lysophospholipid.

To examine the fatty acid specificity of the reaction, the purified 1,2-diradyl-3-acetylglycerols are further fractionated into molecular species by reversed-phase HPLC according to the method of Nakagawa and Horrocks.[24] Briefly, the sample, dissolved in 20 μl of methanol, is injected into a HPLC system (Shimadzu, Kyoto, Japan, LC-6A) equipped with a 4.6 mm × 25 cm Zorbax ODS column (Du Pont Co., Wilmington, DE) and an ultraviolet detector which is operated at 205 nm. Alkenylacyl and alkylacyl compounds are eluted with acetonitrile–2-propanol–methyl *tert*-butyl ether–water (63 : 28 : 7 : 2, v/v), and diacyl compounds are separated with a solvent system of acetonitrile–2-propanol–methyl *tert*-butyl ether–water (72 : 18 : 8 : 2, v/v). Figure 1 shows typical acylation profiles of 1-[^{3}H]alkyl(18 : 0)-GPC by intact cells (Fig. 1a) and by the microsomal fraction in the absence of cofactors (Fig. 1b). The values are the radioactivities from individual fractions (tube/40 sec) collected between 12 and 52 min after injection of the samples.

In this method, the enzyme utilizes endogenous phospholipids as acyl donors, making it unnecessary to prepare various types of phospholipids having different radiolabeled fatty acids. Furthermore, the physicochemical effects on the enzyme activity of dispersing donor phospholipids into water can be neglected. On the other hand, the disadvantages of this method may be that the fatty acid composition of the membrane phospholipids affect the fatty acid specificity of the reaction. In addition, it seems important to take special care in the interpretation of the results, especially in experiments where acyl-containing lysophospholipids are used as acyl acceptors, since acyl-containing lysophospholipids can be acylated either via the CoA-independent transacylation reaction or via the lysophospholipase-mediated transacylation reaction. In addition to the above types of

[24] Y. Nakagawa and L. A. Horrocks, *J. Lipid Res.* **24**, 1268 (1983).

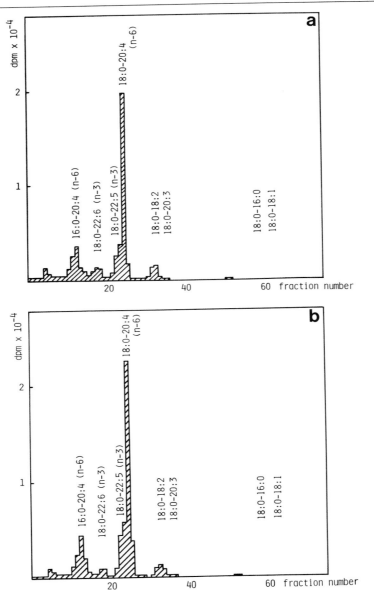

FIG. 1. Chromatographic analysis of the acylation profiles of 1-[³-H]alkyl(18 : 0)-GPC by intact macrophages (a) and by macrophage microsomes (b).⁹ 1-[³H]-Alkyl-GPC was incubated with intact macrophages (a) and macrophage microsomes alone (b) at 37° for 60 min. The acylated product was purified and then analyzed following suitable modifications.

experiments, it is also possible to use prelabeled endogenous phospholipids as acyl donors. Kramer and Deykin[1] studied the CoA-independent transacylation reaction using ^{3}H-prelabeled platelet membranes. This method is also very useful, although the donor phospholipid may not be accurately specified in this case compared with the experiments where chemically defined donor phospholipids are employed.

Properties

The CoA-independent transacylation system was shown to catalyze the transfer of various C_{20} and C_{22} polyunsaturated fatty acids either from exogenously added phospholipids or from endogenous membrane phospholipids.[1-9,23] Both n-6 and n-3 acids can be transferred. On the other hand, the transfer rates for C_{16} and C_{18} acids such as 16:0, 18:0, 18:1, and 18:2 were low, if any. Thus, the CoA-independent transacylation system appears to have high specificity with respect to the transferable fatty acyl residues. Concerning the types of donor phospholipids, CGP, especially diacyl-GPC, was shown to be the most preferred substrate.[1,9] Although diacyl-GPE also serves as the donor phospholipid to some extent, diacylglycerophosphoinositol (diacyl-GPI) does not act at least in the case where 1-alkyl-GPC is used as an acceptor. As for the acceptor lysophospholipids, three subclasses of choline-containing lysophospholipids (1-alkenyl-GPC, 1-alkyl-GPC, and 1-acyl-GPC) were shown to be effective acceptors, 1-alkyl-GPC being most rapidly acylated among them.[9] 1-Alkenyl-GPE, 1-alkyl-GPE, and 1-acyl-GPE were also acylated with 20:4 transferred from diacyl-GPC. On the other hand, 1-acyl-GPI and 1-acylglycerophosphoserine (1-acyl-GPS), as well as 1-acylglycerophosphate, do not serve as effective acceptors.[1,9]

Free fatty acids once liberated from donor phospholipids could not be involved in the transacylation reaction, since exogenously added free ^{3}H-labeled 20:4 acids failed to be incorporated into phospholipids in the absence of cofactors.[7] Furthermore, Kramer and Deykin[1] demonstrated that the addition of unlabeled 20:4 acids did not affect the arachidonoyl transacylation reaction at least at lower concentrations, though a slight inhibition was observed at higher concentrations. We concluded that such inhibition could be attributed to the nonspecific physicochemical effects of free fatty acids, since the addition of 18:2 acids also gave a similar inhibition curve.[9] It has already been suggested that the possible formation of the arachidonoyl enzyme intermediate may take place before the reesterification of 20:4 acids into lysophospholipids,[1] although the precise mechanism is still unknown.

The CoA-independent transacylation has a broad pH optimum, ranging from 7 to 8 for platelet membranes[1] and from 7.5 to 8.5 for the heart microsomal fraction.[6] The enzyme activity was shown to be sensitive to detergents. Triton X-100 at concentrations above 0.1 mg/ml in platelet membranes[1] and above 0.2 mg/ml in macrophage microsomes[9] totally inhibited the transacylation activity. A similar but weaker inhibition was observed for cholate. Several sulfhydryl agents such as N-ethylmaleimide, 5,5'-dithiobis(2-nitrobenzoic acid) (DTNB), and p-chloromercuribenzene-sulfonic acid (pCMBS) also inhibited the enzyme reaction,[1,6,9] suggesting that the SH group(s) in the enzyme protein is important for maintaining the catalytic activity. On the other hand, the reaction was not influenced by the presence of EDTA or EGTA, indicating that the presence of divalent cations such as Ca^{2+} and Mg^{2+} is not required for enzyme activity. The addition of Ca^{2+} to the incubation mixture did not markedly affect the transacylation from endogenous membrane phospholipids, whereas the transfer from exogenous phospholipids was somewhat enhanced by Ca^{2+}.[25] The reason for this discrepancy is as yet unknown. A possible explanation is that Ca^{2+} may alter the affinity of exogenous substrates for the membrane-bound enzyme.

Kinetic constants have also been reported by several investigators. Robinson et al.[8] compared the activities of three acylation systems toward 1-alkyl-GPC using relatively low concentrations of 1-alkyl-GPC and macrophage microsomes. They concluded that the CoA-independent transacylation system has the highest affinity for 1-alkyl-GPC in comparison with the CoA-dependent transacylation system and acyl-CoA : 1-alkyl-GPC acyltransferase. The apparent K_m value for 1-alkyl-GPC was 1.1 μM, and the apparent V_{max} was 3.2 nmol/min/mg. In the case of platelet membranes, the apparent K_m value for 1-alkyl-GPC was calculated to be 12 μM, and V_{max} was 0.87 nmol/min/mg.[2]

[25] T. Sugiura, unpublished results, 1990.

[8] Lysophosphatidylcholine Acyltransferase

By PATRICK C. CHOY, PAUL G. TARDI, and J. J. MUKHERJEE

Introduction

Structural analysis of phosphatidylcholine (PC) indicates that saturated fatty acids are usually esterified at the C-1 position and unsaturated fatty acids at the C-2 position.[1] The observed fatty acid distribution does not seem to result from the limited selectivity of CDP-choline : 1,2-diacylglycerol phosphocholinetransferase (EC 2.7.8.2, cholinephosphotransferase) for diacylglycerol species during the *de novo* biosynthesis of PC.[2] The remodeling of newly synthesized PC via deacylation–reacylation has been regarded as an important mechanism for the selection of PC acyl groups in mammalian tissues.

The pathway for the remodeling of PC was first identified by Lands.[3] In this pathway, PC is first deacylated, and the resultant lyso-PC is acylated back to PC by the action of lyso-PC : acyl-CoA acyltransferase as follows:

$$\text{Lyso-PC} + \text{acyl-CoA} \rightarrow \text{PC} + \text{CoA-SH}$$

Lyso-PC : acyl-CoA acyltransferase (EC 2.3.1.23, 1-acylglycerophosphocholine acyltransferase) activity has been detected in various mammalian tissues,[1] bacteria, protozoa, and plants.[4] The transfer of acyl groups from acyl-CoA to 1-acylglycerophosphocholine (1-acyl-GPC) is catalyzed by 1-acyl-GPC : acyl-CoA acyltransferase, whereas transfer of fatty acids to 2-acyl-GPC is catalyzed by 2-acyl-GPC : acyl-CoA acyltransferase. These two reactions appear to be carried out by separate enzymes.[5,6] Most of the studies on acyltransferase have focused on the acylation of 1-acyl-GPC, and only limited information is available on the acylation of 2-acyl-GPC. The procedures described here are for the assay of 1-acyl-GPC : acyl-CoA acyltransferase.

Assay Methods

Assay with Radioactive Substrate

Principle. Enzyme activity is determined by the incorporation of a radiolabeled acyl group into 1-acyl-GPC to form 1,2-diacyl-GPC. The radiolabeled product is isolated from the substrate by phase separation followed by thin-layer chromatography.[7]

Preparation of Substrate. 1-Acyl-GPC is available from commercial

[1] P. C. Choy and G. Arthur, *in* "Phosphatidylcholine Metabolism" (D. E. Vance, ed.), p. 87. CRC Press, Boca Raton, Florida, 1989.
[2] G. Arthur and P. C. Choy, *Biochim. Biophys. Acta* **795,** 221 (1984).
[3] W. E. M. Lands, *J. Biol. Chem.* **235,** 2233 (1960).
[4] K. A. Devor and J. B. Mudd, *J. Lipid Res.* **12,** 412 (1971).
[5] W. E. M. Lands and P. Hart, *J. Biol. Chem.* **240,** 1905 (1965).
[6] G. Arthur, *Biochem. J.* **261,** 575 (1989).
[7] G. Arthur and P. C. Choy, *Biochem. J.* **236,** 481 (1986).

sources as lysophosphatidylcholine. Usually, the preparation consists of 95–99% 1-acyl-GPC with the remainder as 2-acyl-GPC. Synthetic 1-acyl-GPC with specific acyl groups are also commercially available. Both labeled and unlabeled oleoyl-CoA are available in sufficiently pure form (over 95% purity) from commercial sources. However, the purity of arachidonoyl-CoA must be carefully assessed prior to use.

Procedure. The reaction mixture contains 80 mM Tris-HCl (pH 7.4), 100 μM [1-^{14}C]oleoyl-CoA [1000 disintegrations/min (dpm)/nmol], 150 μM 1-acyl-GPC, and enzyme source in a final volume of 0.7 ml. The mixture is preincubated at 25° for 5 min, and the reaction is initiated by the addition of enzyme. The reaction mixture is incubated in a shaking water bath at 25° for 10 min. The reaction is terminated by the addition of 3 ml of chloroform–methanol (2 : 1, v/v). Phase separation is caused by the addition of 0.8 ml of 0.9% KCl to the mixture. After brief centrifugation, the upper phase is removed, and aliquots of the lower phase are applied onto a 20 × 20 cm thin-layer chromatographic plate (Silica Gel G, Fisher Scientific, Ottawa, Ontario) etched with 2-cm lanes. Unlabeled PC (pig liver) is used as a carrier. The plate is developed in chloroform–methanol–water–acetic acid (70 : 30 : 4 : 2, v/v). After development, the PC fraction on the plate (R_f 0.43) is visualized by exposure of the plate to iodine vapor. The silica gel containing the PC fraction is removed from the plate and placed in a scintillation vial for radioactivity determination. The specific activity of the enzyme in rat liver[8] and heart[9] have been reported to be 58 and 13 nmol/min/mg protein, respectively.

Spectrophotometric Assay

Principle. The spectrophotometric assay is dependent on the release of CoA-SH from the enzyme reaction, which reacts with 5,5′-dithiobis(2-nitrobenzoic acid) (DTNB) to form thionitrobenzoic acid (TNB). The amount of TNB produced is monitored spectrophotometrically at 412 nm.[5]

Procedure. The complete assay mixture contains 40 mM Tris-HCl (pH 7.4), 50 μM oleoyl-CoA, 100 μM 1-acyl-GPC, 0.1 mM DTNB, and 10–20 μl of enzyme preparation in a final volume of 1 ml. The DTNB stock solution (1 mM) is prepared fresh prior to the assay. The reaction is carried out at 25° in a spectrophotometric cuvette (1-cm path length), and the change of absorbance is continuously monitored at 412 nm. A molar extinction coefficient of 13,600 is used to quantitate the amount of DTNB reduced. Since DTNB might also react slowly with the thiol groups of

---

[8] Y. Kawashima, T. Matsunaga, A. Hirose, T. Ogata, and H. Kozuka, *Biochim. Biophys. Acta* **1006,** 214 (1989).

[9] P. C. Choy, K. O, R. Y. K. Man, and A. C. Chan, *Biochim. Biophys. Acta* **1005,** 225 (1989).

proteins and with CoA-SH formed by some specific or unspecific hydrolysis of acyl-CoA, a control that contains all the assay components except 1-acyl-GPC is usually employed.[5] The control mixture is placed into the reference cuvette (in a double-beam spectrophotometer), and the absorbance due to nonspecific reduction of DTNB is automatically subtracted from the reaction mixture. Owing to the nonspecific reduction of DTNB, the use of crude tissue homogenate for this assay is not recommended. Any enzyme preparation containing large particles may interfere with the spectrophotometric assay.

Other Assays

Assays for acyltransferase activity with 1-[1-^{14}C]palmitoyl-GPC or a 1-[1-^{14}C]stearyl-GPC and unlabeled acyl-CoA have been reported.[10,11] The use of these labeled substrates is not recommended when there is a substantial amount of lyso-PC: lyso-PC acyltransferase activity in the enzyme preparation.

Specificity of Assays

The assay with radioactive acyl-CoA is very specific for the enzyme reaction. The spectrophotometric assay is more rapid, but the assay is based on the determination of a thiol group, and therefore is less specific. The data obtained from the spectrophotometric assay must be confirmed by a more specific assay.

Subcellular Localization and Purification

1-Acyl-GPC: acyl-CoA acyltransferase has been found in all mammalian organs. Most of the enzyme activity is located in the microsomal fraction.[12] The enzyme appears to have a transmembrane orientation in the microsomal membrane vesicle.[13] Acyltransferase activity has also been reported in the mitochondrial and plasma membrane fractions.[14,15] Low enzyme activity has been detected in the cytosolic fraction of the rabbit heart.[16]

[10] B. J. Holub, J. A. MacNaughton, and J. Piekarski, *Biochim. Biophys. Acta* **572**, 413 (1979).
[11] N. Deka, G. Y. Sun, and R. MacQuarrie, *Arch. Biochem. Biophys.* **246**, 554 (1986).
[12] H. Eibl, E. E. Hill, and W. E. M. Lands, *Eur. J. Biochem.* **9**, 250 (1969).
[13] W. Renooij and F. Snyder, *Biochim. Biophys. Acta* **666**, 468 (1981).
[14] G. Arthur, L. L. Page, C. L. Zaborniak, and P. C. Choy, *Biochem. J.* **242**, 171 (1987).
[15] O. Colard, D. Bard, G. Bereziat, and J. Polonovski, *Biochim. Biophys. Acta* **618**, 88 (1980).
[16] P. Needleman, A. Wyche, H. Sprecher, W. J. Elliott, and A. Evers, *Biochim. Biophys. Acta* **836**, 267 (1985).

Attempts to purify the 1-acyl-GPC : acyl-CoA acyltransferase were hampered by difficulties in its solubilization from the membrane domain. The use of detergent for this purpose has not been very effective since enzyme activity is inevitably inhibited by the presence of detergents. For example, when microsomes were treated with 1% cholate, 0.25% deoxycholate, or 0.05% Triton X-100, over 90% of enzyme activity was inhibited. The solubilization of the enzyme with 1-acyl-GPC and its analogs appears to be a more viable approach.[17] The purification of the enzyme from bovine brain and bovine heart has been reported.[11,18]

Acyl Specificity

Since dipalmitoyl-PC is the major form of PC in the lung, the selectivity of the acyltransferase toward palmitoyl-CoA was expected to be higher than that for oleoyl-CoA. However, no difference in specificity of the enzyme toward palmitoyl-CoA and oleoyl-CoA has been detected.[1] Interestingly, 1-palmitoyl-GPC appears to be a better acyl acceptor than 1-stearyl-GPC.[19] In rat liver, the enzyme has a definite specificity toward acyl-CoA with unsaturated acyl groups,[15,20] and 1-palmitoyl-GPC is more effective than 1-stearyl-GPC as an acyl acceptor.[10]

In guinea pig heart, the microsomal enzyme is more active toward unsaturated acyl-CoAs, but displays little selectivity with respect to the degree of unsaturation.[7] Interestingly, the mitochondrial enzyme appears to have a very high specificity toward linoleoyl-CoA.[14]

The acyl specificity of the enzyme appears to be affected by the presence of detergent.[21] The acyl specificity is also affected during enzyme purification. For example, the purified acyltransferase from bovine brain displays a higher degree of selectivity toward arachidonate than the microsomal enzyme.[11] Stimulation of smooth muscle cells with phorbol myristate results in the enhancement of arachidonate incorporation into 1-acyl-GPC.[22]

[17] H. U. Weltzien, G. Richter, and E. Ferber, *J. Biol. Chem.* **254**, 3652 (1979).
[18] M. Sanjanwala, G. Y. Sun, M. A. Cutrera, and R. A. MacQuarrie, *Arch. Biochem. Biophys.* **265**, 476 (1988).
[19] G. P. H. van Heusden, H. P. J. M. Noteborn, and H. van den Bosch, *Biochim. Biophys. Acta* **664**, 49 (1981).
[20] H. van den Bosch, L. M. G. van Golde, H. Eibl, and L. L. M. van Deenen, *Biochim. Biophys. Acta* **144**, 613 (1967).
[21] H. Okuyama, K. Yamada, and H. Ikezawa, *J. Biol. Chem.* **250**, 1710 (1975).
[22] T. Kanzaki, N. Morisaki, Y. Saito, and S. Yoshida, *Lipids* **24**, 1024 (1989).

Characteristics and Kinetics

The acyltransferase reaction is reversible and is dependent on the concentrations of the substrates. The pH optimum for the mammalian enzyme is between 7 and 8 and may be dependent on the transfer of a specific acyl group. Using the partially purified enzyme from rat liver, the transfer of oleoyl-CoA was found to be more active at pH 7, whereas arachidonoyl-CoA transfer was more rapid at a higher pH.[23] Recently, the molecular weight of the purified enzyme was determined to be 43,000 for bovine brain[11] and 64,000 for bovine heart.[18] The partially purified enzyme from rabbit lung displayed an iso-Ping-Pong mechanism, and the K_m values for palmitoyl-CoA and 1-acyl-GPC were found to be 8.5 and 61 μM, respectively.[24] In human platelets, the K_m values for saturated and unsaturated acyl-CoA vary from 1.05 to 5.70 μM in the presence of 100 μM 1-acyl-GPC.[25] In most mammalian tissues, the K_m values for acyl-CoA and 1-acyl-GPC vary between 1–15 μM and 50–100 μM, respectively.

Regulation of Enzyme Activity

1-Acyl-GPC acyltransferase activity in rat lung was inhibited by high concentrations of 1-acyl-GPC, but the inhibition by high concentrations of acyl-CoAs was not prominent.[26] Enzyme activity in the liver microsomes was relatively unaffected by sulfhydryl-binding reagents such as iodoacetate, N-ethylmaleimide, and p-chloromercuriphenylsulfonic acid, but the activity of the partially purified enzyme was inhibited by these reagents.[23] The compounds WY-14643 and clofibric acid were found to inhibit mitochondrial 1-acyl-GPC : acyl-CoA acyltransferase.[27] Methyllidocaine, a local anesthetic, caused the inhibition of enzyme activity in mammalian heart.[28] In addition, divalent cations such as Mg^{2+} and Ca^{2+} were inhibitory to enzyme activity.[7,29] Acyltransferase activity *in vivo* was affected by long-term administration of clofibric acid and chronic administration of isoproterenol.[8,29] Enzyme activity in rat liver, rat heart, and rabbit gastric mucosa microsomes was inhibited by detergents and

[23] H. Hasegawa-Sasaki and K. Ohno, *Biochim. Biophys. Acta* **617**, 205 (1980).

[24] R. Arche, P. Estrada, and C. Acebal, *Arch. Biochem. Biophys.* **257**, 131 (1987).

[25] M. L. McKean, J. B. Smith, and M. J. Silver, *J. Biol. Chem.* **257**, 11278 (1982).

[26] H. Hasegawa-Sasaki and K. Ohno, *Biochim. Biophys. Acta* **380**, 486 (1975).

[27] W. W. Riley and D. R. Pfeiffer, *J. Biol. Chem.* **261**, 14018 (1986).

[28] P. G. Tardi, R. Y. K. Man, C. R. McMaster, and P. C. Choy, *Biochem. Cell Biol.* **68**, 745 (1990).

[29] K. Yashiro, Y. Kameyama, M. Mizuno, S. Hayashi, Y. Sakashita, and Y. Yokota, *Biochim. Biophys. Acta* **1005**, 56 (1989).

inhibitors of cyclic nucleotide phosphodiesterase.[30] The regulation of enzyme activity by a phosphorylation–dephosphorylation cycle has been postulated.[31]

Concluding Remarks

Since the discovery of acyltransferases it has been postulated that there may be separate enzymes for each acyl group. Indirect evidence with Triton X-100-treated microsomes indicates that there may be a specific acyltransferase for oleoyl-CoA.[21] The identification of an acyltransferase with an absolute specificity for lineoleoyl-CoA lends support to this hypothesis.[14] Purification of the enzyme from bovine brain causes an enrichment of specificity toward arachidonoyl-CoA.[11] Thus, acyl specificities previously determined in subcellular fractions could simply reflect the quantitative distribution of different acyltransferases in these fractions. Direct evidence for the existence of multiple forms of the acyltransferase is still lacking.

[30] W. T. Shier, *Biochem. Biophys. Res. Commun.* **75**, 186 (1977).
[31] S. L. Reinhold, G. A. Zimmerman, S. M. Prescott, and T. M. McIntyre, *J. Biol. Chem.* **264**, 21652 (1989).

[9] 1-Alkyl- and 1-Alkenylglycerophosphocholine Acyltransferases

By PATRICK C. CHOY and CHRISTOPHER R. McMASTER

Introduction

Three forms of choline-containing phospholipids are found in mammalian tissues.[1] The 1,2-diacyl form (phosphatidylcholine) is the most abundant form and is present as the major phospholipid in all mammalian tissues. The 1-alkyl-2-acyl form (plasmanylcholine) is found in significant amounts in circulating cells such as neutrophils and macrophages but in low amounts in other tissues. The 1-alkenyl-2-acyl form (plasmenylcholine) comprises up to 40% of the choline-containing phospholipids in electrically active tissues such as the heart.

The major catabolic pathway for 1-alkyl-2-acylglycerophosphocholine (1-alkyl-2-acyl-GPC) or 1-alkenyl-2-acylglycerophosphocholine

[1] L. A. Horrocks and M. Sharma, *in* "Phospholipids" (J. N. Hawthorne and G. B. Ansell, eds.), p. 51. Elsevier Biomedical Press, Amsterdam, New York, and Oxford, 1982.

(1-alkenyl-2-acyl-GPC) is via the deacylation reaction catalyzed by phospholipase A_2.[2,3] 1-Alkyl-GPC and 1-alkenyl-GPC are reacylated back into their parent phospholipids by the actions of 1-alkyl-GPC : acyl-CoA acyltransferase[4] (EC 2.3.1.63, 1-alkylglycerophosphocholine acyltransferase) or 1-alkenyl-GPC : acyl-COA acyltransferase[5] (EC 2.3.1.25, plasmalogen synthase) as follows:

1-Alkyl-GPC + acyl-CoA → 1-alkyl-2-acyl-GPC + CoA-SH
1-Alkenyl-GPC + acyl-CoA → 1-alkenyl-2-acyl-GPC + CoA-SH

The acylation of these lysophospholipids serves a number of functions including the selection of the appropriate acyl group at the C-2 position of the phospholipid. The acylation reaction may also be important for preventing the accumulation of cytotoxic lysophospholipids in the cells. 1-Alkylglycerophosphocholine acyltransferase for long-chain acyl-CoA has been found in mammalian muscle,[4] Ehrlich ascites cells,[6] and several hemopoietic cells.[4] This enzyme is different from the 1-alkyl-GPC : acetyl-CoA acyltransferase[2] (for the formation of platelet-activating factor) which is described in Section IX of this volume. 1-Alkenylglycerophosphocholine acyltransferase activity is most abundant in mammalian heart[5] and skeletal muscles,[7] although activity has also been located in erythrocytes, testis, and Ehrlich ascites cells.[6,7]

Assay Methods

Assay of 1-Alkylglycerophosphocholine Acyltransferase with Radiolabeled Substrate

Principle. Enzyme activity is determined by the incorporation of a radiolabeled acyl group into 1-alkyl-GPC to form 1-alkyl-2-acyl-GPC. The radiolabeled product is isolated from the substrate by phase separation followed by thin-layer chromatography.[4]

Preparation of Substrate. 1-Alkyl-GPC is available from commercial sources, either as 1-alkyl-GPC or lyso-PAF (lyso-platelet-activating factor). Most of the top-grade commercial preparations can be used directly without further purification. However, the purity of these preparations

[2] F. Snyder, T.-C. Lee, and M. L. Blank, *in* "Phosphatidylcholine Metabolism" (D. E. Vance, ed.), p. 143. CRC Press, Boca Raton, Florida, 1989.
[3] G. Arthur, L. Page, T. Mock, and P. C. Choy, *Biochem. J.* **236,** 475 (1986).
[4] K. Waku and Y. Nakazawa, *J. Biochem. (Tokyo)* **68,** 459 (1970).
[5] G. Arthur and P. C. Choy, *Biochem. J.* **236,** 481 (1986).
[6] K. Waku and Y. Nakazawa, *J. Biochem. (Tokyo)* **72,** 495 (1972).
[7] K. Waku and W. E. M. Lands, *J. Biol. Chem.* **243,** 2654 (1968).

should be assessed. The usual contaminants found in these preparations are 1-alkenyl- and 1-acyl-GPC. These impurities can be removed by treating the sample with 1 N methanolic NaOH.[4]

1-Alkyl-GPC can also be prepared from bovine heart choline phosphoglycerides according to the procedure of Waku and Nakazawa.[4] Briefly, choline phosphoglycerides are hydrogenated with Pd–charcoal, and the acyl and alkenyl groups remaining in the choline phosphoglycerides are hydrolyzed by treatment with 1 N methanolic NaOH. Subsequently, 1-alkyl-GPC is purified by silicic acid chromatography. Alternatively, the chemical synthesis of 1-alkyl-GPC has been reported.[8]

Both labeled and unlabeled linoleoyl-CoA are usually available in sufficiently pure form (over 95% purity) from commercial sources. However, their purity should be carefully assessed.

Procedure. The reaction mixture contains 100 mM Tris-HCl (pH 7.4), 75 μM [1-^{14}C]linoleoyi-CoA [5000 disintegrations/min (dpm)/nmol], and 150 μM 1-alkyl-GPC in a total volume of 0.7 ml. The reaction is initiated by the addition of enzyme preparation. The mixture is incubated in a shaking water bath at 25° for 30 min, and the reaction is terminated by the addition of 3 ml chloroform–methanol (2 : 1, v/v). Phase separation is achieved by the addition of 0.8 ml of 0.9% KCl to the mixture. After brief centrifugation, the upper phase is removed from the mixture. The solvent in the lower phase is removed by evaporation, and the solute is dissolved in 50 μl of chloroform. An aliquot of this solution (25 μl) is applied onto a 20 × 20 cm thin-layer chromatographic plate (Silica Gel G, Fisher Scientific, Ottawa, Ontario) etched with 2-cm lanes. Unlabeled phosphatidylcholine (pig liver) is used as a carrier. The plate is developed in chloroform–methanol–water–acetic acid (70 : 30 : 4 : 2, v–v). After development, the phosphatidylcholine fraction on the plate is visualized by exposure of the plate to iodine vapor. The silica gel containing the phosphatidylcholine fraction is removed from the plate and placed in a scintillation vial for radioactivity determination. Enzyme activities determined without the addition of 1-alkyl-GPC are used as controls. The specific activity of the enzyme in rabbit sarcoplasmic reticulum has been reported to be 0.46 nmol/min/mg protein.[4]

Confirmation of Product. Product confirmation is important especially when small amounts of 1-acyl-GPC and/or 1-alkenyl-GPC are present in the assay mixture. These lipids may originate from the substrate source (as impurities to 1-alkyl-GPC) or from the enzyme preparation. In order to confirm quantitatively the formation of labeled 1-alkyl-2-linoleoyl-GPC from the reaction, the lipid material in the phosphatidylcholine fraction

[8] M. P. Murari, R. Murari, S. Parthasarathy, C. A. Guy, V. V. Kumar, B. Malewicz, and W. J. Baumann, *Lipids* **25,** 606 (1990).

after thin-layer chromatography is eluted from the silica gel and treated with phospholipase C (*Clostridium perfringens*). The diradyl glycerols are converted to their benzoate derivatives and subsequently analyzed by thin-layer chromatography in a solvent system containing benzene–hexane–diethyl ether (50 : 45 : 4).[9] The purpose of the experiment is to convert the choline phosphoglycerides to diradyl glycerol benzoate to facilitate the separation of the 1-alkyl-2-acyl-moiety (R_f 0.45) from the 1,2-diacyl (R_f 0.33) and the 1-alkenyl-2-acyl (R_f 0.56) moieties by thin-layer chromatography. After separation, the radioactivity in the 1-alkyl-2-acylglycerol benzoate fraction is determined.

Assay of 1-Alkenylglycerophosphocholine Acyltransferase with Radiolabeled Substrate

Principle. Enzyme activity is determined by the incorporation of a radiolabeled acyl group into 1-alkenyl-GPC to form 1-alkenyl-2-acyl-GPC. The radiolabeled product is isolated by phase separation followed by thin-layer chromatography.[5]

Preparation of Substrate. There is no reliable 1-alkenyl-GPC reagent available from commercial sources, despite claims of over 90% purity. The best methods to produce large amounts of 1-alkenyl-GPC is to purify it from an enriched choline phosphoglyceride source, for example, bovine heart.[10] Briefly, choline phosphoglycerides were isolated from bovine heart total lipid extracts by silicic acid column chromatography. The lipid fraction, containing 100 μmol of choline phosphoglyceride, is evaporated to dryness and redissolved in 120 ml chloroform–methanol (1 : 1, v–v) and 20 ml of 0.35 N NaOH in 96% methanol (v/v). The mixture is incubated at 25° for 40 min in order to hydrolyze the acyl groups. After incubation, 100 ml chlorofom and 51 ml water are added to cause phase separation. The upper phase is removed, and the solvent in the lower phase is evaporated under reduced pressure. The sample is redissolved in chloroform and applied to a silicic acid column equilibrated with chloroform. Lipids are eluted from the column by a stepwise methanol gradient, with the elution of 1-alkenyl-GPC at 60–70% methanol. Alternatively, 1-alkenyl-GPC can be prepared by chemical synthesis according to the methods described by Paltauf.[11]

Procedure. The assay mixture contains 75 mM Tris–succinate, pH 6.5, 150 μM 1-alkenyl-GPC, 75 μM [1-^{14}C]linoleoyl-CoA (2000 dpm/nmol), and the appropriate amount of enzyme protein in a total volume of 0.7 ml. The

[9] M. L. Blank, M. Robinson, V. Fitzgerald, and F. Snyder, *J. Chromatogr.* **298**, 473 (1984).
[10] O. Renkonen, *Acta Chem. Scand.* **17**, 634 (1963).
[11] F. Paltauf, *in* "Ether Lipids" (H. K. Mangold and F. Paltauf, eds.), p. 49. Academic Press, New York and London, 1983.

reactants are mixed, and the reaction is initiated by the addition of the enzyme. The mixture is incubated in a shaking water bath at 25° for 30 min. The reaction is stopped by the addition of 3 ml chloroform–methanol (2 : 1, v/v), followed by the addition of 0.8 ml of 0.9% KCl. After phase separation, the upper phase is removed, and the labeled 1-alkenyl-2-acyl-GPC in the lower phase is isolated by thin-layer chromatography as described in the assay for 1-alkylglycerophosphocholine acyltransferase. Enzyme activities determined without the addition of 1-alkenyl-GPC are used as controls. In guinea pig heart microsomes, the specific activity of the enzyme has been reported to be 6 nmol/min/mg protein.[5]

Confirmation of Product. Radiolabeled 1-alkenyl-2-acyl-GPC can be confirmed by the same procedure as described in the 1-alkylglycerophosphocholine acyltransferase assay. Alternatively, treatment of the product with 0.35 N HCl for 20 min results in the hydrolysis of the alkenyl group without affecting the acyl or alkyl groups. The 2-acyl-GPC, which originates from 1-alkenyl-2-acyl-GPC, is separated from other choline phosphoglycerides by thin-layer chromatography. The radioactivity in the 2-acyl-GPC fraction is determined.

Spectrophotometric Assay of 1-Alkenylglycerophosphocholine Acyltransferase

Principle. The CoA-SH produced by the enzyme reaction reacts with 5,5'-dithiobis(2-nitrobenzoic acid) (DTNB), resulting in the formation of thionitrobenzoic acid (TNB). The amount of TNB produced is continuously monitored by a spectrophotometer at 412 nm.[7]

Procedure. The complete assay mixture contains 10 mM Tris-HCl, pH 7.4, 75 μM 1-alkenyl-GPC, 75 μM linoleoyl-CoA, 0.1 mM DTNB, and enzyme preparation in a total volume of 0.9 ml. The stock DTNB solution (1 mM) is prepared fresh prior to the assay. The reaction is carried out at 25° in a spectrophotometric cuvette (1-cm path length), and the change of absorbance is continuously monitored at 412 nm. A molar extinction coefficient of 13,600 is used to quantitate the amount of DTNB reduced. Since DTNB also reacts slowly with the thiol groups of proteins and with CoA-SH formed by unspecific hydrolysis of acyl-CoA, a control that contains all the assay components except 1-alkenyl-GPC is usually employed. The control mixture is placed into the reference cuvette (in a double-beam spectrophotometer), and the absorbance due to nonspecific reduction of DTNB is automatically subtracted from the reaction mixture. Owing to the nonspecific reduction of DTNB, the use of crude tissue homogenates for this assay is not recommended. Any enzyme preparation containing large particles may interfere with the spectrophotometric assay.

Other Assays

An assay for 1-alkenylglycerophosphocholine acyltransferase activity with labeled alkenyl-GPC has been reported.[5]

Specificity of Assays

The assays with radioactive acyl-CoA are very specific for both enzyme reactions. The spectrophotometric assay for 1-alkenylglycerophosphocholine acyltranferase is not specific, and the results must be confirmed by a more specific assay. Owing to the relatively low activity of 1-alkylglycerophosphocholine acyltransferase, use of the spectrophotometric assay is not recommended.

Properties of 1-Alkylglycerophosphocholine Acyltransferase

Enzyme activity has been found in the sarcoplasmic reticulum of rabbit muscle. It has an optimal pH between 7 and 8 and does not seem to require cations for maximal activity.[4] The enzyme appears to have specificity toward linoleoyl-CoA and linolenoyl-CoA but not the saturated acyl-CoA. The specificity of the enzyme toward other acyl-CoA species have not been examined. In Ehrlich ascites cells, the enzyme is active toward linoleoyl-CoA, linolenoyl-CoA, and arachidonoyl-CoA. The enzyme does not show any activity toward the saturated acyl-CoA.[6]

Properties of 1-Alkenylglycerophosphocholine Acyltransferase

Enzyme activities have been found in the microsomal and mitochondrial fractions from mammalian tissues. The microsomal enzyme[5] has an optimal pH range between 6 and 7 and has no absolute cation requirement. However, enzyme activity was enhanced by the presence of Ca^{2+} or Mg^{2+}. The enzyme exhibits a broad specificity toward various molecular species of acyl-CoA but with a distinct preference for the unsaturated species. Over 90% of enzyme activity was lost when the preparation was incubated at 55° for 1 min. The enzyme activity was inhibited by 1-acyl-GPC in a noncompetitive manner.

A mitochondrial enzyme has been identified in mammalian heart.[12] The heart enzyme displays a broad pH range with an optimum at 7.0, and it is specific for linoleoyl-CoA. Interestingly, the enzyme has no absolute requirement for cations. Enzyme activity does not appear to be affected by Mg^{2+} but is slightly inhibited by Ca^{2+}.

[12] G. Arthur, L. L. Page, C. L. Zaborniak, and P. C. Choy, *Biochem. J.* **242**, 171 (1987).

Concluding Remarks

Current evidence suggests that the activities of 1-alkylglycero-phosphocholine and 1-alkenylglycerophosphocholine acyltransferases are located only in selected tissues. Since small amounts of 1-alkyl-2-acyl-GPC and 1-alkenyl-2-acyl-GPC are always found in major mammalian organs, and these phospholipids undergo acyl chain remodeling, the enzymes for such remodeling should also be present. It is likely that the acyltransferases are present in these tissues, but their low activities have eluded detection.

[10] Dihydroxyacetone Phosphate Acyltransferase

By Keith O. Webber and Amiya K. Hajra

Introduction

Dihydroxyacetone phosphate acyltransferase (DHAPAT; glycerone-3-phosphate acyltransferase, EC 2.3.1.42) catalyzes the transfer of the fatty acid moiety from long-chain acyl-coenzyme A (acyl-CoA) to the free hydroxyl group of dihydroxyacetone phosphate:

Dihydroxyacetone phosphate + acyl-CoA → acyldihydroxyacetone phosphate + CoASH

This reaction initiates the synthesis of the ether-linked glycerolipids as well as the more common glycerol ester lipids.[1,2] DHAPAT is an integral membrane-bound protein located on the luminal side of animal cell peroxisomes.[3–5] This acyltransferase has been solubilized and partially purified from guinea pig liver.[6,7] The present method is based on that described for the preparation of highly purified enzyme from guinea pig liver peroxisomes.[7]

---

[1] A. K. Hajra, *Biochem. Soc. Trans.* **5,** 34 (1977).

[2] R. Manning and D. N. Brindley, *Biochem. J.* **130,** 1003 (1972).

[3] A. K. Hajra, C. L. Burke, and C. L. Jones, *J. Biol. Chem.* **244,** 8289 (1989).

[4] A. K. Hajra and J. E. Bishop, *Ann. N.Y. Acad. Sci.* **386,** 170 (1982).

[5] C. L. Jones and A. K. Hajra, *J. Biol. Chem.* **255,** 8289 (1980).

[6] C. L. Jones and A. K. Hajra, *Arch. Biochem. Biophys.* **226,** 155 (1983).

[7] K. O. Webber, Ph.D. Dissertation, University of Michigan, Ann Arbor (1988).

METHODS IN ENZYMOLOGY, VOL. 209

Assay Method

Principle. Enzyme activity is measured as the amount of lipophilic (at low pH) radioactivity formed from [^{32}P]DHAP in the presence of palmitoyl-CoA and enzyme.[8]

Reagents

5 mM Dihydroxyacetone [^{32}P]phosphate: [^{32}P]DHAP is prepared by enzymatic phosphorylation of dihydroxyacetone with [γ-^{32}P]ATP.[8,9] The radioactive DHAP is diluted with nonradioactive DHAP (Li$^+$ salt) to the desired specific activity [5000–10,000 counts/min (cpm)/nmol]

1 mM Palmitoyl-CoA, lithium salt

0.3 M Morpholinoethanesulfonic acid (MES), pH 5.7

0.3 M Tris-HCl buffer, pH 7.5

0.1 M Sodium fluoride

0.1 M Magnesium chloride

20 mg/ml Bovine serum albumin (BSA), fatty acid poor, fraction V

Asolectin suspension (liposomes) in 10 mM Tris-HCl, 1 mM EDTA, pH 7.5: Suspend 2 g asolectin (Associated Concentrates, Woodside, NY) in 100 ml of 10 mM Tris-HCl (pH 7.5)–1 mM EDTA by sonicating with an ultrasonic probe and then centrifuge the suspension at 35,000 g for 30 min. Use the supernatant (15–20 μmol lipid phosphate/ml) for the assay. When stored at 4° under N$_2$, this suspension is stable for 1 month.

2 M potassium chloride in 0.2 M phosphoric acid

Chloroform

Chloroform–methanol (1 : 2, v/v)

Chloroform–methanol–0.5 N aqueous H$_3$PO$_4$ (1 : 12 : 12, v/v)

Procedure. Add the following reagents to 16 × 125 mm screw-topped tubes: buffer[10] (0.3 M Tris-HCl (or 0.3 M MES) 0.15 ml, BSA 0.05 ml, palmitoyl-CoA 0.05 ml, NaF 0.1 ml, MgCl$_2$ 0.05 ml, asolectin suspension 0.1 ml,[11] enzyme sample, and water to make the final volume 0.6 ml. Incubate the mixture at 37° for 15 min in a reciprocating shaker water bath. Stop the reaction by adding 2.25 ml chloroform–methanol (1 : 2, v–v) and mix (vortex). Add 0.75 ml methanol and 0.75 ml KCl–H$_3$PO$_4$ (2–0.2 M), vortex, and centrifuge for 5 min at 1000 g. Aspirate off the upper layers

---

[8] A. K. Hajra, *J. Biol. Chem.* **243**, 3458 (1968).

[9] A. K. Hajra and C. L. Burke, *J. Neurochem.* **31**, 125 (1978).

[10] Use MES buffer for the membrane-bound enzyme and Tris buffer for the solubilized enzyme.

[11] Addition of asolectin is not necessary for the assay of membrane-bound enzyme.

and add 2.5 ml of chloroform–methanol–0.5 N aqueous H_3PO_4 (1 : 12 : 12, v/v), mix, and centrifuge. Aspirate off the upper layers and wash the lower layers again with chloroform–methanol–aqueous H_3PO_4 as described above. Transfer aliquots of the washed lower layers (generally 1 ml, i.e., two-thirds of the total) to counting vials, evaporate off the solvents with a stream of air or N_2, add scintillation solvent mixture, and determine the radioactivity in a liquid scintillation counter.

Units. One unit (U) of enzyme activity is defined as 1 nmol of product formed per minute at 37°. The specific activity is expressed as nanomoles per minute per milligram protein.

Purification of Enzyme

The purification of DHAPAT can be divided into three processes: (1) isolation of peroxisomes, (2) solubilization of the peroxisomal membranes, and (3) chromatographic purification of DHAPAT.

Purification of Peroxisomes from Guinea Pig Liver

Subcellular Fractionation. Livers are homogenized according to the protocol of deDuve *et al.*[12] in a buffer containing 0.25 M sucrose, 10 mM N-tris(hydroxymethyl)methyl-2-aminoethanesulfonic acid (TES; pH 7.5), 1 mM EDTA, 0.1% ethanol (v/v), 0.2 mM phenylmethylsulfonyl fluoride (PMSF), and 1 μM leupeptin. Fractionate the liver by differential centrifugation to produce the "light mitochondrial" fraction (L-fraction, sedimenting between 33,000 and 250,000 g.min),[12] which is enriched in peroxisomes. The method used in our laboratory for subcellular fractionation of guinea pig liver is essentially that described by deDuve *et al.*[12] with minor modifications.[3] Suspend the L-fraction in the homogenization buffer to a volume of 0.25 ml/g of original liver weight.

Density Gradient Centrifugation. The peroxisomes are isolated from the L-fraction by centrifugation through 30% (w/v) Nycodenz as described in Ref. 13 with some modifications.[7]

1. Into 25-ml polycarbonate centrifuge bottles (Beckman, Fullerton, CA, #340382), place 15 ml of a solution containing 30% Nycodenz (w/v), 10 mM TES, pH 7.5, and 1 mM EDTA.
2. Overlay the 15 ml of Nycodenz solution with 2 ml of L-fraction.
3. Centrifuge at 75,000 g for 45 min in a Ti-55 fixed-angle rotor (Beck-

[12] C. deDuve, B. C. Pressman, R. Gianetto, R. Wattiaux, and F. Appelmans, *Biochem. J.* **60,** 604 (1955).
[13] M. K. Ghosh and A. K. Hajra, *Anal. Biochem.* **159,** 169 (1986).

man). Peroxisomes will sediment through the Nycodenz solution while mitochondria, microsomes, and lysosomes will not.

4. Carefully aspirate away the entire supernatant.
5. Resuspend the remaining pellet in homogenization buffer to a volume equivalent to 20% of the original liver weight. Store at $-20°$.

Solubilization of Dihydroxyacetone Phosphate

Membrane Preparation. Initial purification of DHAPAT entails separation of the peroxisomal membrane from the soluble, matrix proteins. The peroxisomes are osmotically ruptured by 10-fold dilution with 10 m*M* sodium pyrophosphate, pH 9.0, containing 1 μ*M* leupeptin, 1 μ*M* pepstatin, 0.2 m*M* PMSF, and 1 m*M* EDTA. Stir the mixture for 1 hr on ice. Centrifuge at 100,000 *g* for 30 min to separate membranes (pellet) from the soluble matrix proteins (supernatant). Resuspend the pellet in 25 m*M* TES (pH 7) to a protein concentration of approximately 10 mg/ml. This membrane suspension may be stored at $-20°$.

Solubilization. Mix equal volumes of peroxisomal membrane suspension and a solution containing 300 m*M* NaCl, 30 m*M* 3-[(3-cholamidopropyl)dimethylammonio]-1-propanesulfonate (CHAPS), 20 m*M* MES (pH 6.5), 2 m*M* dithiothreitol (DTT), 2 μ*M* leupeptin, 2 μ*M* pepstatin, 2 m*M* EDTA, and 0.4 m*M* PMSF. Stir this mixture gently for 15 min on ice. Centrifuge the mixture at 100,000 *g* for 60 min. The supernatant (a clear brown liquid) contains the solubilized DHAPAT with a specific activity of approximately 80–90 U/mg protein.

Column Chromatography

The solubilized DHAPAT is purified to near homogeneity by a multistep regimen of both low-pressure and high-pressure column chromatography.

Low-Pressure Size-Exclusion Chromatography. Pass the solubilized peroxisomal membrane proteins over a column of Sephacryl S-200 (1.6 × 95 cm; Pharmacia, Piscataway, NJ) at 4°. The mobile phase is 150 m*M* NaCl, 5 mg/ml CHAPS, 10 m*M* MES (pH 6.5), 1 m*M* DTT, 0.02% NaN_3 flowing at 6.0 ml/hr. Collect 3-ml fractions. Pool the fractions (fractions 12–15) comprising the peak DHAPAT activity.

Cation-Exchange Chromatography. Cation-exchange chromatography is performed on a 0.5 × 5 cm Mono S column (Pharmacia) attached to a suitable high-performance liquid chromatography (HPLC) system allowing binary gradient elution and monitoring of effluent absorbance at 280 nm. Equilibrate the column with 25 m*M* MES, pH 6.5, 5 mg/ml CHAPS, 1 m*M* DTT (buffer A) and then load the sample onto the column

TABLE I
PURIFICATION OF DIHYDROXYACETONE PHOSPHATE ACYLTRANSFERASE

Fraction	Protein (mg)	Total activity (U)	Specific activity (U/mg)	Enrichment (-fold)	Yield (%)
Liver homogenate	15,150	7900	0.52	1	100
Peroxisomes	276	2820	10.2	19.6	36
Solubilized membranes	94	6020	64	123	76
Sephacryl S-200	31	3640	118	227	46
Mono S	1.0	1400	1400	2700	18
Hydroxylapatite	0.18	450	2400	4600	6
TSK-3000	0.08	260	3350	6440	3

at 0.5 ml/min. Wash the unbound proteins off the column with buffer A until the 280 nm absorbance of the effluent returns to baseline. While collecting 1.0-ml fractions, elute the adherent proteins from the column with a linear salt gradient from 0 to 400 mM NaCl in buffer A at a flow rate of 0.5 ml/min (total gradient volume 40 ml). Pool the fractions (fractions 11–15) comprising the peak DHAPAT activity.

Hydroxylapatite Chromatography. Equilibrate a high-pressure hydroxylapatite column (0.5 × 4.8 cm; Bio-Rad Laboratories, Richmond, CA) with 10 mM potassium phosphate, pH 6.8, 0.3 mM CaCl$_2$, 5 mg/ml CHAPS, 50 mM NaCl, 1 mM DTT, 0.05% NaN$_3$. Load the pooled DHAPAT-containing fractions from the cation-exchange column at 0.5 ml/min. Wash the unbound proteins from the column with equilibration buffer until the absorbance of the effluent at 280 nm returns to baseline. Elute the bound proteins with a 25-ml linear phosphate gradient starting with equilibration buffer and ending with 0.3 M potassium phosphate, pH 6.8, 10 μM CaCl$_2$, 5 mg/ml CHAPS, 1 mM DTT, 0.05% NaN$_3$. Collect 1-ml fractions.

High-Pressure Size-Exclusion Chromatography. Pool the DHAPAT-containing fractions (fractions 12–14) eluted from the hydroxylapatite column and concentrate the combination to approximately 1.0 ml in a centrifugal microconcentrator (Bio-Rad). Inject the concentrated enzyme solution onto the HPLC gel-filtration column (TSK-3000; Toyo-Soda, Japan) and elute with 150 mM NaCl, 5 mg/ml CHAPS, 10 mM Bis–Tris–propane, pH 7.5, 1 mM DTT, 0.02% NaN$_3$ at 0.5 ml/min. Collect 1.0-ml fractions while monitoring the absorbance of the effluent at 280 nm. The enzyme generally elutes in fractions 4 to 7. Combine the fractions, concentrate the enzyme by ultrafiltration (centrifugal microconcentrator), and store at 4°.

The enzyme could be purified further by chromatofocusing but at such a low yield that the specific activity could not be accurately determined.[7] A typical purification procedure is summarized in Table I.

Properties

Physical Characteristics. DHAPAT activity copurifies with a protein which has an apparent M_r of 69,000 as determined by polyacrylamide gel electrophoresis in the presence of sodium dodecyl sulfate (SDS–PAGE). Gel filtration of trace amounts of solubilized DHAPAT gives a M_r determination of approximately 90,000, indicating that the active enzyme is probably not multimeric.

Thermostability. Membrane-bound peroxisomal DHAP acyltransferase is remarkably heat stable. In fact, heating membranes to 50° prior to assay increases the measurable enzyme activity. The solubilized enzyme does not possess this thermostability and loses significant activity at temperatures as low as 40°.

Kinetics. The guinea pig liver enzyme is most active at pH 5.5 in the membrane-bound state; however, on solubilization with either ionic or nonionic detergents, the pH optimum shifts to pH 7.4 and becomes somewhat less stringent. At pH 7.4, the membrane-bound enzyme gives biphasic kinetics with K_m (DHAP) values of 40 and 200 μM which increase to 100 and 500 μM on solubilization. When the solubilized enzyme is assayed in the presence of asolectin (soybean lipids) the kinetics are monophasic with a K_m (DHAP) of 100 μM. After extensive purification (as above), the K_m (DHAP) was determined to be about 60 μM; however, at subsaturating levels of palmitoyl-CoA the value drops to 35 μM.

Activators and Inhibitors. The crude membrane-bound enzyme activity is stimulated by Mg^{2+} and F^-; however, after solubilization these ions have no effects on the enzyme activity. The solubilized enzyme is stimulated by phospholipid vesicles, such as asolectin or phosphatidylcholine.[6] Palmitoyl-CoA inhibits the enzyme activity, and the presence of BSA in the reaction mixture prevents this inhibition. Both the membrane-bound and solubilized enzymes are relatively unaffected by thiol-modifying agents such as N-ethylmaleimide and iodoacetamide; however, bulkier, charged sulfhydryl-reactive compounds such as p-chloromercuriphenylsulfonic acid or 5,5′-dithiobis(2-nitrobenzoic acid) produce moderate levels of inhibition.

Diagnostic Use of Enzyme

A number of genetic diseases involving peroxisomal disorders are prenatally or postnatally diagnosed by the decreased activity of DHAPAT in aminocytes, chorionic villi, leukocytes, or in cultured skin fibroblasts obtained from the patients.[14–16]

[14] N. S. Datta, G. N. Wilson, and A. K. Hajra, *N. Engl. J. Med.* **311,** 1080 (1984).

[15] A. K. Hajra, N. S. Datta, G. L. Jackson, A. B. Moser, H. W. Moser, J. W. Larsen, and J. Powers, *N. Engl. J. Med.* **312,** 455 (1985).

[16] R. B. Schutgens, H. S. Heyman, R. J. Wanders, H. van den Bosch, and J. M. Tager, *Eur. J. Pediatr.* **144,** 430 (1986).

[11] Diacylglycerol Acyltransferase and Monoacylglycerol Acyltransferase from Liver and Intestine

By Rosalind A. Coleman

Introduction

The acylation of diacylglycerol catalyzed by the diacylglycerol acyltransferase is the only enzyme reaction unique to triacylglycerol synthesis. This reaction lies at the diacylglycerol branch point of phosphatidylcholine, phosphatidylethanolamine, and triacylglycerol synthesis.[1] In most tissues the major pathway for the biosynthesis of the diacylglycerol substrate proceeds from the acylation of glycerol 3-phosphate. In intestine and liver, however, the monoacylglycerol pathway provides an alternate route for diacylglycerol synthesis. 2-Monoacylglycerol, produced by the action of gastric and pancreatic lipases on dietary triacylglycerol, enters the intestinal mucosa where monoacylglycerol acyltransferase plays a major role in resynthesizing triacylglycerol. In the liver of certain species, monoacylglycerol acyltransferase activity is high during early development; however, neither its specific role in liver nor the source of the monoacylglycerol substrate is known.[2–4]

Both diacylglycerol acyltransferase and monoacylglycerol acyltransferase are intrinsic membrane proteins whose active sites face the cytosolic surface of the endoplasmic reticulum.[5] Neither activity has been substantially purified.

[1] R. M. Bell and R. A. Coleman, *in* "The Enzymes" (P. D. Boyer, ed.), 3rd Ed., Vol. 16, p. 87. Academic Press, New York, 1983.

[2] R. A. Coleman and E. B. Haynes, *J. Biol. Chem.* **259,** 8934 (1984).

[3] K. Sansbury, D. S. Millington, and R. A. Coleman, *J. Lipid Res.* **30,** 1251 (1989).

[4] R. A. Coleman, E. B. Haynes, and C. D. Coates, *J. Lipid Res.* **28,** 320 (1987).

[5] R. A. Coleman and R. M. Bell, *in* "The Enzymes" (P. D. Boyer, ed.), 3rd Ed., Vol. 16, p. 605. Academic Press, New York, 1983.

Preparation of Microsomes from Rat Liver or Intestinal Mucosa

Rat liver is minced with a pair of scissors and then homogenized in 9 volumes of STE buffer (0.25 M sucrose, 10 mM Tris-HCl, pH 7.4, 1.0 mM EDTA) with a Teflon–glass homogenizer by 10 up-and-down strokes at medium speed. The homogenate is centrifuged at 1000 g for 10 min at 4°. The supernatant is removed and centrifuged at 25,000 g for 15 min at 4°. The second supernatant is then centrifuged at 100,000 g for 1 hr at 4° to obtain a microsomal pellet. Microsomes are resuspended in STE buffer by 5 up-and-down strokes in a Teflon–glass homogenizer, and aliquots are stored at −70°. Aliquots are thawed for use once and then discarded.

Sections (6 cm) of rat intestine are rinsed by forcing STE buffer into the lumen with a needleless 10-ml syringe. The sections are then cut open with a pair of scissors and placed with the mucosal side up on a piece of glass which rests on a bed of ice. The mucosa·is scraped off with a glass slide and put into a glass homogenizing vessel with STE buffer. The preparation is then treated as described above for liver.

Diacylglycerol Acyltransferase Assay Procedure

sn-1,2-Diacylglycerol is acylated with [^{3}H]palmitoyl-CoA. The [^{3}H]triacylglycerol product is separated from the water-soluble [^{3}H]palmitoyl-CoA substrate by extraction into heptane. Initial rates can be measured in total particulate preparations or in microsomes.

Reagents

Tris-HCl, 1.0 M, pH 8.0
MgCl$_2$, 160 mM
Bovine serum albumin (essentially fatty acid free), 10 mg/ml
[^{3}H]Palmitoyl-CoA, 0.2 mM, 50–100 mCi/mmol
sn-1,2-Dioleoylglycerol, 2.0 mM; dried under N$_2$ and then dispersed in cold acetone
Stop solution: 2-propanol–heptane–water (80 : 20 : 2, v/v) in a 1-liter dispenser
Heptane in a 1-liter dispenser
Water in a dispenser
Alkaline ethanol: ethanol–0.5 N NaOH–water (50 : 10 : 40, v/v) in a 1-liter dispenser

The albumin, [^{3}H]palmitoyl-CoA, and dioleoylglycerol solutions are stored at −20°. Only small amounts of 1,2-dioleoylglycerol in acetone are prepared since, over time, the acyl chains migrate to form increasing amounts of the 1,3-isomer. A mixture of 3.5 ml Tris buffer, 1.0 ml MgCl$_2$, and 2.0 ml albumin (enough for 100 assays) is made and stored frozen. Tris buffer is refrigerated, and the $MgCl_2$ is kept at room temperature.

Reagent Sources. Serdary Research Laboratory (London, ON, Canada) is suggested for the *sn*-1,2-diacylglycerol substrate since diacylglycerol from other suppliers may have isomerized to the 1,3-stereoisomer. If many assays are planned, it is cost-effective to synthesize [³H]palmitoyl-CoA enzymatically.[6] In this synthesis [³H]palmitate product with a very low background can be obtained by solvent extraction of fresh [³H]palmitic acid. The radiolabeled palmitic acid is dried under a stream of N_2, and resuspended in 1.0 ml Dole's reagent (2-propanol–heptane–2 N H_2SO_4, 80 : 20 : 2, v/v). Then 0.5 ml water and 2 ml heptane are added. The upper heptane phase is transferred to a clean tube, and the aqueous phase is rinsed with 2 ml of heptane. The heptane phases are pooled and rinsed with 4 ml of alkaline ethanol. The upper heptane phase is discarded. The alkaline ethanol phase is acidified with 75 μl of concentrated HCl and extracted twice with 4 ml heptane. The heptane is transferred to a clean tube, and the purity of the [³H]palmitic acid is verified by thin-layer chromatography as described below. More than 98.5% of the counts per minute (cpm) should be in fatty acid before the material is used to synthesize palmitoyl-CoA.

Method. The assays are performed in 13 × 100 mm glass tubes. Each individual reaction mixture contains the following, in a total volume of 200 μl: 65 ml of the Tris/MgCl$_2$/albumin mixture, water to make up to 0.2 ml, 1.0 to 6.0 μg of microsomal protein, 20 μl dioleoylglycerol, and 30 μl [³H]palmitoyl-CoA. The first three reagents are added with a Pipetman, and the last two are added with a Hamilton syringe directly into the reaction mixture. The order of addition is important: the reaction should be started with the palmitoyl-CoA, followed by vortexing gently and briefly. After 10 min at room temperature, the reaction is terminated by adding 1.5 ml of the stop solution and vortexing briefly. One milliliter of heptane and 0.5 ml of water are added, and the mixture is vortexed. Using a 6-inch Pasteur pipette, as much as possible of the top phase is transferred to a clean 13 × 100 mm glass tube, 2.0 ml of the alkaline ethanol solution is added, and the mixture is vortexed. A 0.65-ml aliquot of the top heptane phase is pipetted into a scintillation vial, 4.0 ml of CytoScint (ICN Biomedicals, Inc., Costa Mesa, CA) is added, and the amount of radioactivity is determined in a liquid scintillation counter.

With microsomes from adult rat liver and adipose tissue, more than 93% of the labeled product is triacylglycerol. However, microsomal lipases may hydrolyze some of the diacylglycerol substrate to form *sn*-2-monoacylglycerol, a substrate for the monoacylglycerol acyltransferase activity

[6] A. H. Merrill, S. Gidwitz, and R. M. Bell, *J. Lipid Res.* **23**, 1368 (1982).

which is present in intestinal mucosa, in perinatal liver, and in embryonic chicken liver. Thus, since the monoacylglycerol formed by action of diacylglycerol lipase may be reacylated to produce labeled diacylglycerol, it is wise to check the product distribution for each new tissue that is analyzed. This is most conveniently done by thin-layer chromatography on 10-cm silica gel G plates developed in heptane–isopropyl ether–acetic acid (30 : 20 : 2, v/v) using known standards as carrier. After assay products have been visualized by exposure to I_2 vapor, the silica gel is scraped into scintillation vials and counted.

Comments

1. The rat intestine activity has been solubilized in taurocholate and purified an estimated 145-fold from the homogenate after phenyl-Sepharose chromatography.[7] This partially purified preparation also contained fatty-acid-CoA ligase and monoacylglycerol acyltransferase activities. Rat liver diacylglycerol acyltransferase was solubilized in sodium cholate,[8] but chromatography on Sepharose 4B and a sucrose gradient achieved only a 9-fold purification. Additional purification steps resulted in complete loss of activity. Solubilization in CHAPS and 2978-fold partial purification from soybean cotyledons has also been reported.[9]

2. Diacylglycerol dispersed in ethanol was used to measure diacylglycerol acyltransferase in adipose tissue.[10,11] Since liver, unlike adipocytes, contains high levels of an enzyme activity that can acylate ethanol, acetone is the preferred vehicle for dispersing the diacylglycerol substrate in the latter tissue.

3. The specific activity of diacylglycerol acyltransferase varies considerably in microsomes from different tissues. Highest activities are found in isolated fat cells, intestinal mucosa, lactating mammary gland, and differentiated 3T3-L1 adipocytes (20–100 nmol/min/mg microsomal protein); much lower activities have been reported in liver and other tissues.[11–13] Seemingly conflicting data have been presented that phosphorylation activates[14] and inactivates[15] diacylglycerol acyltransferase.

[7] F. Manganaro and A. Kuksis, *Can. J. Biochem. Cell Biol.* **63,** 107 (1985).

[8] M. A. Polokoff and R. M. Bell, *Biochim. Biophys. Acta* **618,** 129 (1980).

[9] P. Kwanyuen and R. F. Wilson, *Biochim. Biophys. Acta* **877,** 238 (1986).

[10] P. Goldman and P. R. Vagelos, *J. Biol. Chem.* **236,** 2620 (1961).

[11] R. A. Coleman and R. M. Bell, *J. Biol. Chem.* **251,** 4537 (1976).

[12] M. R. Grigor and R. M. Bell, *Biochim. Biophys. Acta* **712,** 464 (1982).

[13] R. A. Coleman, B. C. Reed, J. C. Mackall, A. K. Student, M. D. Lane, and R. M. Bell, *J. Biol. Chem.* **253,** 7256 (1978).

[14] H.-D. Soling, W. Fest, T. Schmidt, H. Esselmann, and V. Bachmann, *J. Biol. Chem.* **264,** 10643 (1989).

[15] H. P. Haagsman, C. G. M. de Haas, M. J. H. Geelen, and L. M. G. van Golde, *J. Biol. Chem.* **257,** 10593 (1982).

4. Unlike other acyltransferases which have been investigated, the diacylglycerol acyltranferase can use fatty acyl-CoA substrates that are 8 to 12 carbons in length.[16–18] Acylation of *sn*-2,3-diacylglycerol occurs at 20% of the rate of the *sn*-1,2-stereoisomer.[18,19]

Monoacylglycerol Acyltransferase Assay Procedure

sn-2-Monoacylglycerol is stereospecifically acylated with [³H]palmitoyl-CoA. Since the labeled *sn*-1,2-diacylglycerol product is a substrate for diacylglycerol acyltransferase (which is always present in microsomes), the assay products must be identified by thin-layer chromatography. Initial rates can be measured in total particulate preparations or in microsomes.

Reagents

Tris-HCl, 1.0 *M*, pH 7.0
Bovine serum albumin (essentially fatty acid free), 10 mg/ml
Phosphatidylcholine, 1.5 mg/ml, and phosphatidylserine, 1.5 mg/ml, sonicated in three 10-sec bursts in a cuphorn sonicator in 10 m*M* Tris-HCl, pH 7.4
sn-2-Monooleoylglycerol, 2.0 m*M*; dried under N_2 and then dispersed in acetone
[³H]Palmitoyl-CoA, 0.2 m*M*, 50–100 mCi/mmol
Stop solution: 2-propanol–heptane–water (80 : 20 : 2, v/v) in a 1-liter dispenser
Heptane in a 1-liter dispenser
Water in a dispenser
Alkaline ethanol: ethanol–0.5 *N* NaOH–water (50 : 10 : 40, v/v) in a 1-liter dispenser

Reagents are stored as noted above for diacylglycerol acyltransferase. Only small amounts of monooleoylglycerol in acetone are prepared since, over time, the acyl chains may migrate to form increasing amounts of the 1(3)-isomer. A mixture of 2.0 ml Tris buffer and 2.0 ml albumin (enough for 100 assays) is made and stored frozen. The phospholipid mixture is stored frozen and sonicated before each use.

Reagent Sources. Serdary Research Laboratory is suggested for the *sn*-2-monoacylglycerol substrate since monoacylglycerol from other suppliers may have partially isomerized to the *sn*-1(3)-position. See the note above concerning the [³H]palmitoyl-CoA.

---

[16] N. Mayorek and J. Bar-Tana, *J. Biol. Chem.* **258,** 6789 (1983).
[17] C. Y. Lin, S. Smith, and S. Abraham, *J. Lipid Res.* **17,** 647 (1976).
[18] M. O. Marshall and J. Knudsen, *Eur. J. Biochem.* **81,** 259 (1977).
[19] S. B. Weiss, E. P. Kennedy, and J. Y. Kiyasu, *J. Biol. Chem.* **235,** 40 (1960).

Method. The assays are performed in 13 × 100 mm glass tubes. Each individual reaction mixture contains the following, in 200 μl: 40 ml of the Tris/albumin mixture, water to make up to 0.2 ml, 10 μl of the phosphatidylcholine/phosphatidylserine mixture, 0.5 to 2.0 μg of microsomal protein, 5 μl of 2-monooleoylglycerol, and 25 μl [³H]palmitoyl-CoA. The first four reagents are added with a Pipetman and the last two with a Hamilton syringe directly into the reaction mixture. The reaction can be started with either the microsomes or the [³H]palmitoyl-CoA, followed by vortexing gently and briefly. After 5 min at room temperature, the reaction is terminated by adding 1.5 ml of the stop solution and vortexing. The remainder of the extraction procedure is carried out as described above for the diacylglycerol acyltransferase.

Typically, in intestinal mucosa and suckling rat liver, 80% of the labeled product is diacylglycerol and 20% is triacylglycerol. However, the relative proportions may vary considerably. In chick embryo liver, for example, the amount of triacylglycerol can be as high as 60% of the labeled product.[2] Thus, in order to correct the specific activity, it is essential to identify the products by thin-layer chromatography. After removing 0.65 ml for counting, the rest of the heptane phase is removed, dried (most conveniently in a Speed-Vac concentrator), and chromatographed with authentic standards as described for the diacylglycerol acyltransferase product. After standards have been visualized with I_2 vapor, the silica gel which contains triacylglycerol and diacylglycerol is scraped into separate scintillation vials, 4.0 ml of CytoScint is added, and the vials are counted. The counts are corrected by subtracting one-half the counts in the triacylglycerol fraction, and the specific activity is expressed as nanomoles diacylglycerol synthesized per minute per milligram of microsomal protein. To determine how much endogenous monoacylglycerol in the microsomal preparation is available for acylation, all reagents are added including microsomes and acetone, but the 2-monooleoylglycerol is omitted. Typically, the activity with endogenous substrate is less than 1 and 6% of the total activity in liver and intestine, respectively.

Comments

1. Analysis of double-reciprocal plots indicates that the apparent K_m for palmitoyl-CoA is higher than the concentration used in the assay; however, higher concentrations of palmitoyl-CoA inhibit the reaction.

2. The monoacylglycerol acyltransferase activities from rat intestine and from suckling rat liver have different properties and appear to be isoenzymes.[20] The liver activity is highly specific for *sn*-2-monoacylglycerol; activity with 1(3)-monooleylglycerol, *sn*-2-monooleylglycerol

[20] R. A. Coleman and E. B. Haynes, *J. Biol. Chem.* **261**, 224 (1986).

ether, or 1(3)-monooleylglycerol ether is less than 10% of the activity obtained with 2-monooleoylglycerol. In contrast, the enzyme from intestinal mucosa uses each of the alternate substrates at rates that are about 50% of that observed with 2-monooleoylglycerol. Other differences between the liver and intestinal activities include greater susceptibility to inhibition of the liver activity by diethyl pyrocarbonate (which can be reversed by NH_2OH) and to trinitrobenzenesulfonic acid (but not to sulfhydryl reagents), suggesting that critical histidine and lysine residues are present within the active site.

3. Monoacylglycerol acyltransferase activity is 12 times higher with palmitoyl-CoA than with octanoyl-CoA. Longer chain monoacylglycerols seem to be preferred.[2]

4. Little activity is present in rat tissues other than intestinal mucosa and neonatal liver. Activity in the latter is 700-fold higher than present in adult rat liver.[2]

5. Rat intestinal monoacylglycerol acyltransferase has been solubilized in sodium taurocholate and partially purified, but the activity was not separated from fatty-acid-CoA ligase or diacylglycerol acyltransferase.[7]

Acknowledgments

Research was supported by Grant HD19068 from the National Institutes of Health.

[12] Biosynthesis of Bis(monoacylglycero)phosphate in Liver and Macrophage Lysosomes

By Karl Y. Hostetler, S. Julia Huterer, and John R. Wherrett

Introduction

Bis(monoacylglycero)phosphate is a polyglycerophosphatide which is usually a minor constituent of mammalian cell phospholipids[1] and is concentrated exclusively in secondary lysosomes.[2,3] In contrast to the related

[1] K. Y. Hostetler, in "Phospholipids" (J. N. Hawthorne and G. B. Ansell, eds.), p. 215. Elsevier, New York, 1982.
[2] R. J. Mason, T. P. Stossel, and M. Vaughan, J. Clin. Invest. 51, 2399 (1972).
[3] J. R. Wherrett and S. Huterer, J. Biol. Chem. 247, 4114 (1972).

phospholipid, phosphatidylglycerol [(1,2-diacyl)glycero-sn-3-phospho-sn-1'-glycerol], bis(monoacylglycero)phosphate may have the unconventional stereoconfiguration of glycero-sn-1-phospho-sn-1'-glycerol.[4] Degradation of bis(monoacylglycero)phosphate by phospholipases A of defined specificity[2] suggested that the single fatty acid esterified to each glycerol may be linked through either of the two available hydroxyl groups, whereas a nuclear magnetic resonance (NMR) study[5] suggested esterification to terminal hydroxyls only. The fatty acid composition of bis (monoacylglycero)phosphate is characterized by a high proportion of polyunsaturated fatty acids.[5] With liver lysosomal fractions, either 1-acyllysophosphatidylglycerol or 2-acyllysophosphatidylglycerol may be converted to bis(monoacylglycero)phosphate.[6] Although usually maintained at a low concentration in the lysosome, bis(monoacylglycero) phosphate may increase greatly in genetic lipidoses[7] and drug-induced lysosomal phospholipid storage.[8] High levels of bis(monoacylglycero) phosphate have been found in pulmonary alveolar macrophages,[2] and turnover of bis(monoacylglycero)phosphate in alveolar macrophages has been linked to arachidonic acid release and leukotriene synthesis.[9]

Studies using rat liver established that the glycero–phospho–glycerol backbone of bis(monoacylglycero)phosphate is derived from phosphatidylglycerol[10,11] or cardiolipin[12] and that the transacylations required for its biosynthesis occur in lysosomes.[6] Acylation of the free (nonacylated) glycerol of lysophosphatidylglycerol was shown to occur by transacylation from a lysosomal phospholipid donor in both liver[11] and macrophages.[13] The enzyme(s) which catalyzes transacylation of the free glycerol of lysophosphatidylglycerol can be directly assayed but has not been extensively purified or characterized.

[4] J. Brotherus, O. Renkonen, J. Herrmann, and W. Fischer, *Chem. Phys. Lipids* **13**, 178 (1974).
[5] J. R. Wherrett and S. Huterer, *Lipids* **8**, 531 (1973).
[6] B. J. H. M. Poorthuis and K. Y. Hostetler, *J. Biol. Chem.* **250**, 3297 (1975).
[7] G. Rouser, G. Kritchevsky, A. Yamamoto, A. G. Knudson, and G. Simon, *Lipids* **3**, 287 (1968).
[8] A. Yamamoto, S. Adachi, K. Ishikawa, T. Yokomura, T. Kitani, T. Nasu, T. Imoto, and M. Nishikawa, *J. Biochem. (Tokyo)* **70**, 775 (1971).
[9] M. Waite, V. Roddick, T. Thornburg, L. King, and F. Cochran, *FASEB J.* **1**, 318 (1987).
[10] B. J. H. M. Poorthuis and K. Y. Hostetler, *J. Biol. Chem.* **251**, 4596 (1976).
[11] Y. Matsuzawa, B. J. H. M. Poorthuis, and K. Y. Hostetler, *J. Biol. Chem.* **253**, 6650 (1978).
[12] B. J. H. M. Poorthuis and K. Y. Hostetler, *J. Lipid Res.* **19**, 309 (1978).
[13] S. J. Huterer and J. R. Wherrett, *Biochim. Biophys. Acta* **1001**, 68 (1989).

Methods

Bis(monoacylglycero)phosphate Formation from
Lysophosphatidylglycerol and Phospholipid Acyl Donor

Principle. De novo synthesis of the glycero–phospho–glycerol back-bone of bis(monoacylglycero)phosphate is determined by measuring the CDPdiacylglycerol-stimulated incorporation of labeled glycero-3-phosphate into bis(monoacylglycero)phosphate in homogenates and crude mitochondria/lysosome fractions.[10,14] Conversion of lysophosphatidylglycerol to bis(monoacylglycero)phosphate is measured using substrate prepared from phosphatidylglycerol containing a radioactive label introduced biosynthetically.[13]

Preparation of Liver Lysosome Fractions. Purified lysosomes, essentially free of other subcellular membranes, are isolated from the liver of rats treated with Triton WR-1339 as described by Trouet.[15] The lysosomal fraction in 0.125 M sucrose is subjected to 10 cycles of freezing and thawing. This fraction is treated with 1 volume of ice-cold, water-saturated n-butanol and centrifuged at 100,000 g for 1 hr at 4°. The supernatant fraction containing the enzyme activity is desalted by passage over a Sephadex G-25 column (1.5 × 30 cm) eluted with 20 μM Tris-HCl (pH 7.2). After determination of protein, the fraction may be frozen in aliquots and used for enzyme assay.

Preparation of Macrophage Fractions. New Zealand White rabbits weighing 2–3 kg and judged free of infection are used. They are sacrificed by intravenous injection of T-61 (Hoechst), the trachea is exposed and canulated, and the lungs are lavaged four times with chilled sterile 0.9% saline delivered and removed with a 50-ml syringe. Cells are collected by centrifugation, washed twice with saline, and suspended in 0.1 M Tris-HCl (pH 7.4) containing 0.25 M sucrose. Contaminating red cells are removed during the washing by suspension of cells in sterile 0.2% saline for 90 sec followed by addition of an equal volume of 1.6% saline and centrifugation. The final suspension is examined under the microscope with trypan blue staining to assess contamination and viability.

Cells suspended in Tris–sucrose are disrupted by sonication for 1–2 min using the microtip of the Sonic Dismembrator (Fisher Scientific, Model 300, Toronto, ON, Canada), and completeness is assessed by phase-contrast microscopy. A fraction containing mitochondria and lysosomes is obtained by centrifuging the homogenate at 2000 rpm (Sorvall RC-5, SS-34 rotor, Dupont, Wilmington, DE) for 10 min to remove nuclei

---

[14] S. Huterer and J. R. Wherrett, *J. Lipid Res.* **20**, 966 (1979).
[15] A. Trouet, this series, Vol. 31, p. 323.

and unbroken cells followed by centrifuging at 12,000 rpm for 20 min. The pellet obtained is suspended in Tris–sucrose, and the enzyme is solubilized by delipidation. An equal volume of ice-cold, water-saturated n-butanol is added to the mitochondria/lysosome suspension and the two phases mixed by vortexing just sufficiently to mix the phases. Following centrifugation at 45,000 rpm (IEC B-60, A321 rotor, Fisher Scientific, Tustin, CA) for 1 hr, the clear lower phase is carefully removed and used for incubations. Butanol retained in the aqueous phase does not affect the reaction. The clear upper butanol phase is removed, dried down under N_2, and taken up in chloroform–methanol (2 : 1, v/v) for addition to incubations.

Preparation of Labeled Lysophosphatidylglycerol. Phosphatidyl[U-¹⁴C]glycerol is prepared biosynthetically by incubating rat liver mitochondria with sn-[U-¹⁴C]glycerol 3-phosphate in the presence of CDPdiacylglycerol following the procedure of Poorthuis and Hostetler.[10] The reaction is stopped with the addition of chloroform–methanol (2 : 1, v/v), and the washed radioactive lipids are loaded on a DEAE-Sephadex column and fractionated into acidic and nonacidic fractions.[16] Phosphatidylglycerol is eluted in the acidic fraction with methanol–chloroform–0.2 M sodium acetate (60 : 30 : 8, v/v/v) and further purified by thin-layer chromatography (TLC) using a Radiomatic Scanner (Model RTLC, Radiomatic, Meriden, CT) for detection of radioactivity. Purified phosphatidyl[U-¹⁴C]glycerol is converted to 1-acyllysophosphatidyl[U-¹⁴C]glycerol using *Crotalus adamanteus* venom phospholipase A₂ (Sigma Chemical Co., St. Louis, MO), isolated by lipid extraction, and purified by TLC.[17]

Reagents

Sodium acetate buffer, 50 mM, pH 4.5
2-Mercaptoethanol, 10 mM
Lysophosphatidyl[U-¹⁴C]glycerol, 20 μM, 2 mCi/mmol
Washed butanol-extracted lipids from the liver lysosomal or macrophage mitochondrial/lysosomal fractions equivalent to 100 μM lipid phosphorus
Phospholipid standards, unlabeled: lysophosphatidylglycerol (Sigma), bis(monoacylglycero)phosphate purified from macrophages[14] or liver lysosomes[6]

Assay Procedure. Butanol-extracted lipids and substrate are dried in incubation tubes under N_2 and dispersed in 50 μl of distilled water by sonicating the tubes in a bath sonicator (Mettler, Hightstown, NJ, Electronics Sonic Cleaner, Model ME 2.1, set at 6) for 10 min. Mercaptoetha-

---

[16] R. K. Yu and R. W. Ledeen, *J. Lipid Res.* **13,** 680 (1972).
[17] R. Haverkate and L. L. M. van Deenen, *Biochim. Biophys. Acta* **106,** 78 (1965).

nol and buffer are then added, and the reaction is started by addition of enzyme. The final volume is 0.2 ml, and the reaction is carried out at 37° for 30 min.

   Extraction and Measurement of Reaction Products. The reaction is stopped by addition of 2 ml chloroform–methanol (2 : 1, v/v) and the extracted lipids washed by partitioning with 0.2 ml of 0.1 M KCl and collected in the lower phase. Unreacted lysophosphatidylglycerol remaining in the aqueous phase is recovered by adding 10 μl concentrated HCl to the aqueous phase and then extracting twice with 1.5 ml chloroform and combining these extracts with the original chloroform extract. The radioactive lipids are separated by TLC along with added pure unlabeled lipid standards, using silica gel H plates and the solvent system chloroform–methanol–concentrated ammonia (65 : 25 : 4, v/v/v). Plates are dried under a gentle nitrogen or air stream for 5 min. Unreacted substrate and the bis(monoacylglycero)phosphate product are detected by exposure to I_2 vapor, and the spots are aspirated into Pasteur pipettes plugged with glass wool. The labeled lipids are eluted with 2 ml chloroform–methanol (2 : 1, v/v) followed by 4 ml methanol into counting vials, dried under nitrogen, dissolved in 0.2 ml methanol and 10 ml aqueous counting scintillant (Amersham, Arlington Heights, IL), and then counted. Bis(monoacylglycero)phosphate synthase activity is taken as the amount of labeled lysophosphatidylglycerol converted to bis(monoacylglycero)phosphate. Alternatively, phosphatidyl[U-^{14}C]glycerol may be used as substrate[10] and the products analyzed as described above.

Results

Properties of Liver Lysosomal Enzyme
(Phosphoglyceride:Lysophosphatidylglycerol Acyltransferase)

   Liver Lysosomes. When a crude lysosomal soluble fraction from rat liver was incubated with phosphatidylglycerol, a pH optimum for bis (monoacylglycero)phosphate formation was noted at 4.0–4.8. In addition, either 1-acyl- or 2-acyllysophosphatidylglycerol served as substrate. Bis-(monoacylglycero)phosphate formation was not affected by calcium, magnesium, or EDTA, but mercuric chloride, *p*-chloromercuribenzoate, and detergents inhibited the enzyme activity.[6] Acyl-CoA and magnesium were not effective in catalyzing acylation of the free glycerol of phosphatidylglycerol or its lyso compound.

   The delipidated lysosomal soluble preparation required a phospholipid acyl donor; bis(monoacylglycero)phosphate formation was stimulated 10-fold by the addition of total lipids isolated from the lysosomal fraction.[11]

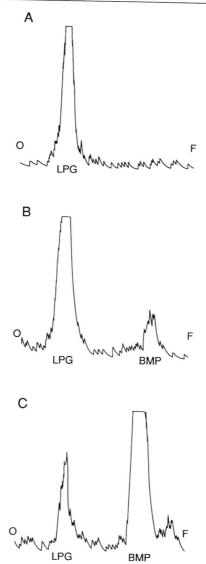

FIG. 1. Thin-layer chromatogram scan of lysophosphatidyl[U-^{14}C]glycerol conversion to bis(monoacylglycero)phosphate by the soluble, delipidated enzyme from a rabbit alveolar macrophage lysosomal fraction. The enzyme isolation and incubations were done as described in the text. (A) Incubation without added enzyme; (B) incubation with enzyme and without added lipids; (C) incubation with enzyme and added lipids. The scans were made with a Packard Model 7230 radiochromatogram scanner. LPG, Lysophosphatidyl[U-^{14}C]glycerol; BMP, bis(monoacylglycero)phosphate; O, origin; F, front.

Lysosomal phosphatidylinositol was identified as the most active phospholipid in catalyzing the acyl transfer to phosphatidylglycerol. Recently, purified lysosomal phospholipase A_1[18] was found to catalyze transesterification of an acyl group from phosphatidylcholine to bis(monoacylglycero) phosphate to form acylphosphatidylglycerol.[19]

Properties of Macrophage Enzyme (Phosphoglyceride:Lysophosphatidylglycerol Acyltransferase)

Macrophage Lysosomes. Figure 1 shows thin-layer scanner tracings of an experiment with the macrophage lysosomal enzyme incubated without protein (Fig. 1A), without added lipids (Fig. 1B), and with added lipids (Fig. 1C). The tracings demonstrate clearly the lack of bis(monoacylglycero)phosphate formation without added enzyme protein and the marked stimulatory effect of adding unlabeled phospholipids which serve as acyl donors in the reaction.

Conversion of lysophosphatidylglycerol to bis(monoacylglycero)phosphate was maximal at pH 4.5, with activity falling to one-third at pH 7.5 (50 mM Tris-HCl). Activity in the butanol-solubilized preparation was dependent on the presence, nature, and amount of lipids added. Optimal rates were measured with the total fraction of lipids extracted with butanol. Rate continued to increase with concentrations up to 250 μM lipid phosphorus and were linear for up to 1 hr with concentrations less than 100 μM. Lipid fractions containing phosphatidylcholine or bis(monoacylglycero)phosphate or the purified lipids per se as well as commercial phosphatidylglycerol also stimulated but to a much lesser extent than the total lipid fraction. Activity was inhibited by Ca^{2+} and the detergents Triton X-100, Triton WR-1339, and sodium taurocholate. CoA and oleoyl-CoA also inhibited, indicating that a CoA-mediated acylation is unlikely to be involved. The enzyme preparation catalyzed the transfer of labeled oleic acid from either di[1-^{14}C]oleoyl or 1-acyl-2-[1-^{14}C]oleoylphosphatidylcholine to form bis(monoacylglycero)phosphate but at activities much less than observed with the glycerol-labeled substrate. In the presence of bis(monoacylglycero)phosphate the enzyme preparation also catalyzed the formation of acylphosphatidylglycerol.[19]

[18] K. Y. Hostetler, P. Yazaki, and H. van den Bosch, *J. Biol. Chem.* **257**, 13367 (1982).

[19] S. J. Huterer and J. R. Wherrett, *Biochem. Cell Biol.* **68**, 366 (1990).

[13] 2-Acylglycerophosphoethanolamine Acyltransferase/ Acyl-[Acyl-Carrier-Protein] Synthetase from *Escherichia coli*

By SUZANNE JACKOWSKI, LI HSU, and CHARLES O. ROCK

Introduction

2-Acylglycerophosphoethanolamine (2-acyl-GPE) acyltransferase is a membrane-bound enzyme that either activates fatty acids for acyl transfer in the presence of ATP and Mg^{2+} or transfers fatty acids from acyl-acyl carrier protein (ACP) to the 1-position of lysophospholipids.[1-3] 2-Acyl-GPE acyltransferase is a heterodimer composed of a hydrophobic membrane-bound subunit and ACP.[4] The catalytic cycle of the acyltransferase/synthetase is shown in Fig. 1. The first step is the ligation of a fatty acid to the 4'-phosphopantetheine sulfhydryl group of the ACP subunit. This reaction requires ATP and Mg^{2+}, and AMP and pyrophosphate are the products.[5] The bound acyl-ACP remains associated with the complex *in vivo*, since exogenous fatty acids activated by this enzyme are not made available to other intracellular enzymes that utilize acyl-ACP.[3,4] However, *in vitro*, the presence of high salt concentrations in the assay mixture lowers the affinity of the acyltransferase subunit for ACP, leading to the dissociation and accumulation of acyl-ACP.[4] This property forms the basis of the acyl-ACP synthetase activity measurement,[6] and it is clear that the acyltransferase and synthetase are dual catalytic activities of the same protein. The second step in the acyltransferase reaction is the transfer of the acyl moiety from ACP to the 2-acyl-GPE substrate to form phosphatidylethanolamine (PtdEtn).

The 2-acyl-GPE acyltransferase is responsible for a number of physiological processes. The primary function of the enzyme system is to acylate 2-acyl-GPE that arises from the transfer of 1-position fatty acids to outer membrane lipoproteins or from the degradation of membrane phospholipids by phospholipase A_1.[3,7] The acyltransferase is also responsible for the

[1] S. S. Taylor and E. C. Heath, *J. Biol. Chem.* **244**, 6605 (1969).
[2] H. Homma, M. Nishijima, T. Kobayashi, H. Okuyama, and S. Nojima, *Biochim. Biophys. Acta* **508**, 165 (1981).
[3] C. O. Rock, *J. Biol. Chem.* **259**, 6188 (1984).
[4] C. L. Cooper, L. Hsu, S. Jackowski, and C. O. Rock, *J. Biol. Chem.* **264**, 7384 (1989).
[5] T. K. Ray and J. E. Cronan, Jr., *Proc. Natl. Acad. Sci. U.S.A.* **73**, 4374 (1976).
[6] C. O. Rock and J. E. Cronan, Jr., *J. Biol. Chem.* **254**, 7116 (1979).
[7] S. Jackowski and C. O. Rock, *J. Biol. Chem.* **261**, 11328 (1986).

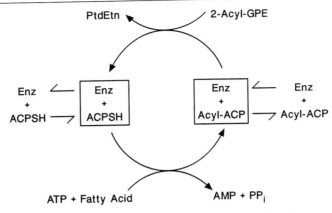

FIG. 1. Catalytic cycle of 2-acyl-GPE acyltransferase/acyl-ACP synthetase.

acyl-CoA-independent uptake and incorporation of exogenous fatty acids and lysophospholipids into the membrane.[8,9]

Assay Method

Principle. The assay method for 2-acyl-GPE acyltransferase is based on measuring the formation of [^{14}C]PtdEtn from 2-acyl-GPE and [^{14}C]palmitic acid in the presence of ATP, Mg^{2+}, and ACP. The reaction is terminated by applying an aliquot of the incubation mixture to the preadsorbant layer of a thin-layer chromatography plate, and, following development to separate unreacted fatty acid from PtdEtn, the amount of [^{14}C]PtdEtn formed is quantitated by scintillation counting. The acyltransferase/synthetase may also be detected using the acyl-ACP synthetase assay.[5,6,10] The acyl-ACP synthetase assay technique is easier to perform and faster than the acyltransferase assay described in this chapter, but it measures a nonphysiological, low specific activity side reaction of the acyltransferase rather than its normal catalytic activity.[4]

Reagents. [^{14}C]Palmitic acid (specific activity ~55 Ci/mol) from any supplier is suitable. Purified (oxidant-free) Triton X-100, ATP, octyl-β-D-glucoside, and *Rhizopus arrhizus* lipase are purchased from Boehringer Mannheim (Indianapolis, IN). *Escherichia coli* PtdEtn is purchased from Serdary Research Laboratories (London, ON, Canada). ACP is purified

[8] C. O. Rock and S. Jackowski, *J. Biol. Chem.* **260,** 12720 (1985).
[9] L. Hsu, S. Jackowski, and C. O. Rock, *J. Bacteriol.* **171,** 1203 (1989).
[10] See this series, Vol. 71 [21].

TABLE I
REAGENTS AND STOCK SOLUTIONS FOR
2-ACYLGLYCEROPHOSPHOETHANOLAMINE ACYLTRANSFERASE ASSAY

Reagent	Microliters/assay	Final concentration
Tris-HCl, 1 M, pH 8.0	4	0.1 M
ATP, 0.1 M, pH 7.0	2	5 mM
MgCl$_2$, 0.1 M	2	5 mM
ACP, 100 μM	4	10 μM
Dithiothreitol, 40 mM	2	2 mM
2-Acyl-GPE,[a] 1 mM	4	100 μM
[^{14}C]Palmitic acid,[a] 1 mM	2	50 μM
Protein in 2% Triton X-100	20	0–10 mU

[a] These solutions are prepared in 2% Triton X-100.

as described in an earlier volume in this series.[11] The stock solutions required for the assay are listed in Table I and can be stored at −20° for months. The exception is dithiothreitol, which should be prepared daily.

Preparation of 2-Acyl-GPE Substrate. 2-Acyl-GPE is prepared by the digestion of *E. coli* PtdEtn essentially as described by Homma and Nojima.[12] PtdEtn (4 μmol) is added to a screw-top tube and the solvent evaporated. Next, 1.6 ml of 50 mM Bis-Tris, pH 5.6, is added, and the PtdEtn is suspended by two 10-sec rounds of sonication. Then, 0.2 ml of 100 mM CaCl$_2$, 0.2 ml of 2 mg/ml *R. arrhizus* lipase, and 0.2 ml of diethyl ether are added. The mixture is incubated at 25° for 3 hr with shaking. Methanol (2 ml) is added to stop the reaction and the mixture extracted twice with 4 ml of hexane to remove the fatty acid. Next, 2.2 ml of chloroform, 0.5 ml methanol, and 0.2 ml of 0.5 M citric acid are added. The 2-acyl-GPE partitions into the bottom phase, which is collected, the solvent removed, and the 2-acyl-GPE suspended in chloroform–methanol (1 : 1, v/v). 2-Acyl-GPE concentrations are determined by the spectrophotometric assay of Stewart[13] using *E. coli* PtdEtn as a standard. 2-Acyl-GPE is aliquoted and stored under nitrogen at −20°. Stocks of 2-acyl-GPE should not be stored for longer than 1 week because by this time a significant fraction of the acyl groups have migrated to the 1-position and 2-acyl-GPE acyltransferase will not acylate 1-acyl-GPE.

Method. The number of assays to be performed is determined, and the appropriate volumes of the assay regents (Table I) are combined. To each

---

[11] See this series, Vol. 71 [41]; see also C. O. Rock and J. E. Cronan, Jr., *Anal. Biochem.* **102**, 362 (1980).

[12] H. Homma and S. Nojima, *J. Biochem.* (*Tokyo*) **91**, 1103 (1982).

[13] J. C. M. Stewart, *Anal. Biochem.* **104**, 10 (1980).

10×75 mm glass assay tube, 20 μl of the reaction mixture is added, and the reaction is initiated by the addition of 20 μl of the protein solution in 2% Triton X-100. The tubes are briefly vortex mixed and placed in a 37° water bath for 10 min. Incubations are terminated by the addition of 0.2 ml of ethanol. The mixture is evaporated to dryness under a stream of nitrogen, resuspended in chloroform–methanol (1 : 1, v/v), and the entire sample applied to the preadsorbant layer of a silica gel G plate. The thin-layer plate is developed with chloroform–methanol–acetic acid (85 : 15 : 10, v/v), and the areas corresponding to PtdEtn and fatty acid are located by comparison with standards chromatographed in one lane of the same plate. These two areas are removed, and the amount of radioactivity incorporated into PtdEtn is determined by scintillation counting. Counting the fatty acid area is a useful control to confirm that the amount of radioactivity recovered from the thin-layer plate corresponds to the amount of carbon-14 initially added to the assay. Results are expressed as nanomoles per minute per milligram protein.

Purification

Initial Purification Steps. The initial steps in the purification of 2-acyl-GPE acyltransferase/acyl-ACP synthetase are identical to the procedure outlined by Rock and Cronan for the purification of acyl-ACP synthetase activity.[6,10] All procedures are performed at 4° unless otherwise indicated. The cells are disrupted by passage through a French pressure cell at 16,000 psi. The homogenate is then centrifuged at 15,000 g for 20 min to remove debris and unbroken cells, and the pellet is discarded. The supernatant is adjusted to 10 mM in $MgCl_2$ by the addition of the appropriate volume of 1 M $MgCl_2$. The suspension then is sedimented at 80,000 g for 90 min, and the supernatant is discarded. The resulting pellets are homogenized thoroughly in 150 ml of 50 mM Tris-HCl, pH 8.0, and then 150 ml of 50 mM Tris-HCl, pH 8.0, containing 1 M NaCl, and 20 mM $MgCl_2$ is added to the membrane suspension. The solution is gently stirred for 15 min, and the membranes are sedimented at 80,000 g for 90 min. The membrane pellets from the salt wash are homogenized thoroughly in 100 ml of 50 mM Tris-HCl, pH 8.0, and to this suspension 100 ml of 50 mM Tris-HCl, pH 8.0, containing 4% Triton X-100 and 20 mM $MgCl_2$ is added, and the solution is gently stirred for 30 min. The suspension then is centrifuged at 80,000 g for 90 min to remove unsolubilized material, and the Triton X-100 supernatant is saved.

The Triton X-100 extract (200 ml) is then adjusted to 5 mM ATP by the addition of 20 ml of 0.1 M ATP in 50 mM Tris-HCl, pH 8.0. The protein concentration should be between 2 and 5 mg–ml. The Triton X-100 extract is then placed in a 37° water bath, stirred until the temperature reaches 33°, and then placed in a 55° water bath and stirred on ice and centrifuged at 15,000 *g* for 20 min to remove denatured proteins.

Blue-Sepharose Chromatography. A column (1.2 × 35 cm) is packed with Blue-Sepharose CL-6B and equilibrated with 50 m*M* Tris-HCl, pH 8.0, containing 2% Triton X-100 at 4°. The supernatant from the heat step is applied to the column. The column is then rinsed with 6 column volumes of 50 m*M* Tris-HCl, pH 8.0, containing 2% Triton X-100, followed by 6 column volumes of the same buffer containing 0.6 *M* NaCl. The majority of the protein adsorbed to the Blue-Sepharose column is eluted with 0.6 *M* NaCl, and acyltransferase/synthetase activity is desorbed from the column by elution with 0.5 *M* KSCN in 50 m*M* Tris-HCl, pH 8.0, 2% Triton X-100.

ATP-Charged DEAE-Cellulose Chromatography. The final step in the procedure is chromatography on ATP-charged DEAE-cellulose. A 17-ml column of Whatman (Clifton, NJ) DE-52 is packed and equilibrated in 50 m*M* Tris-HCl, pH 8.0, 2% Triton X-100. The column is then washed with 50 ml of 5 m*M* ATP in the same buffer and excess nucleotide removed by washing with buffer without ATP. The dialyzed KSCN eluate from the Blue-Sepharose step is loaded on the ATP-charged Whatman DE-52 column and the column washed with the column buffer. In the case where the samples are to be analyzed by sodium dodecyl sulfate (SDS) gel electrophoresis, the column is exchanged in 50 m*M* Tris-HCl, pH 8.0, 30 m*M* octyl-β-D-glucoside. Activity is eluted with 0.5 *M* NaCl in either octyl-β-D-glucoside or Triton X-100 buffer.

Comments on Purification. The results of the purification procedure are presented in Table II. Throughout the purification procedure, the ratio of 2-acyl-GPE acyltransferase to acyl-ACP synthetase activity remains

TABLE II
PURIFICATION OF 2-ACYLGLYCEROPHOSPHOETHANOLAMINE ACYLTRANSFERASE/
ACYL-[ACYL-CARRIER-PROTEIN] SYNTHETASE

Purification step	2-Acyl-GPE acyltransferase (nmol/min/mg)	Acyl-ACP synthetase (nmol/min/mg)	Ratio	Recovery (%)
Membranes	0.300	0.092	3.26	100
Washed membranes	0.437	0.117	3.74	67.8
Triton extract	0.958	0.273	3.51	48.5
Heat supernatant	1.73	0.490	3.53	45.8
Blue-Sepharose	50.6	16.24	3.12	9.9
DEAE-cellulose	102.4	28.4	3.61	7.7

constant. The specific activities of the preparations for 2-acyl-GPE acyltransferase are consistently 10-fold higher than the specific activity of acyl-ACP synthetase. At stages of purification prior to the Blue-Sepharose chromatography, 2-acyl-GPE acyltransferase activity is not dependent on the addition of ACP to the assay. However, the conditions of the Blue-Sepharose step dissociate the ACP subunit from the acyltransferase, and, after this step, acyltransferase activity becomes completely dependent on the addition of ACP to the assay. This purification procedures results in the 3000-fold purification of 2-acyl-GPE acyltransferase/acyl-ACP synthetase in 7.7% yield. SDS gel electrophoresis of the final product shows a single band with an apparent molecular weight of 27,000.

Properties

2-Acyl-GPE acyltransferase/acyl-ACP synthetase is an inner membrane enzyme that is not subject to catabolite repression, and it is genetically and biochemically distinct from acyl-CoA synthetase and glycerolphosphate and 1-acylglycerol-phosphate acyltransferases. The enzyme specifically catalyzes the transfer of fatty acids to the 1-position; 1-acyl-GPE is not a substrate. The enzyme acylates 2-acylglycerophosphocholine as readily as 2-acyl-GPE but does not acylate 2-acylglycerophosphoglycerol. The enzyme prefers saturated fatty acids (palmitate and myristate) as substrates and is less active on shorter (decanoate) and unsaturated (oleate) fatty acids. Poor substrates have both a higher K_m and lower V_{max}. The acyltransferase binds ACP with high affinity (K_d ~60 nM) and catalyzes the acyltransferase reaction without the dissociation of ACP or acyl-ACP from the enzyme. ACP is dissociated from the protein complex by exposure to high salt. This property accounts for the accumulation of acyl-ACP in assays performed in the presence of 0.4 M LiCl and explains the acyl-ACP synthetase activity expressed by the acyltransferase. There is little doubt that the protein functions only as an acyltranferase *in vivo* and does not catalyze the formation of acyl-ACP that can be made available to other proteins.

Genetics and Molecular Biology

A replica print procedure was used to isolate mutants that lacked both 2-acyl-GPE acyltransferase and acyl-ACP synthetase activities. This mutation maps to the 61-min region of the *E. coli* chromosome, and the allele is designated *aas*.[14] The *aas* mutants were then employed to isolate

[14] L. Hsu, S. Jackowski, and C. O. Rock, *J. Biol. Chem.* **266,** 13783 (1991).

plasmids that carry the *aas* gene. One such plasmid, pLCH3, contains the full *aas* coding sequence, and strains that harbor this plasmid possess 10-fold higher specific activities of both 2-acyl-GPE acyltransferase and acyl-ACP synthetase.[14] Thus, strains harboring this plasmid are an excellent source for purifying the acyltransferase/synthetase either for biochemical characterization or in routine preparation of the enzyme for the synthesis of acyl-ACPs.

Synthesis of Acyl-ACP

Although the acyltransferase/synthetase functions only as an acyltransferase *in vivo*, its ability to synthesize acyl-ACP *in vitro* has made it a useful tool for the preparation of defined acyl-ACPs to be used as substrates for other enzymes.[15] Our laboratory still uses the techniques outlined in a previous volume in this series[15] to synthesize acyl-ACP using the acyltransferase/synthetase, but there is a useful modification introduced by others[16,17] that warrants consideration. These investigators[16,17] have taken advantage of the tight binding of the acyltransferase to Blue-Sepharose and the fact that the synthetase remains active when coupled to the resin.[6,16,17] The enzymatically active column is used directly to catalyze the formation of acyl-ACP by the continuous cycling of a reaction mixture (~60 ml) over the column for 24 hr with a peristaltic pump at 4°.[17] The reaction mixture contains 0.6 mg/ml ACP, 0.1 M Tris-HCl, pH 8.0, 0.4 M LiCl, 10 mM ATP, 10 mM $MgCl_2$, 2 mM dithiothreitol, 0.07% Triton X-100, and 100 μM of the desired fatty acid.[17] Yields of acyl-ACP up to 90% or greater have been reported, and the acyl-ACP synthetase/Blue Sepharose column remains stable for at least 4 months at 4°C.[17]

Acknowledgments

This work was supported by National Institutes of Health Grant GM 28035, Cancer Center (CORE) Support Grant CA 21765, and the American Lebanese Syrian Associated Charities. We thank Pam Jackson for excellent technical assistance.

---

[15] See this series, Vol. 72 [27]; see also C. O. Rock and J. L. Garwin, *J. Biol. Chem.* **254,** 7123 (1979).

[16] P. R. Green, A. H. Merrill, Jr., and R. M. Bell, *J. Biol. Chem.* **256,** 11151 (1981).

[17] M. S. Anderson, C. E. Bulawa, and C. R. H. Raetz, *J. Biol. Chem.* **260,** 15536 (1985).

Section III

Kinases

[14] Choline/Ethanolamine Kinase from Rat Kidney

By KOZO ISHIDATE and YASUO NAKAZAWA

Introduction

Choline kinase (ATP:choline phosphotransferase, EC 2.7.1.32) and ethanolamine kinase (ATP:ethanolamine O-phosphotransferase, EC 2.7.1.82) catalyze the phosphorylation of choline/ethanolamine by ATP in the presence of Mg^{2+}, yielding phosphocholine/phosphoethanolamine and ADP:

$$(CH_3)_3N^+CH_2CH_2OH \ + \ ATP \quad \xrightarrow[\text{choline kinase}]{Mg^{2+}} \quad (CH_3)_3N^+CH_2CH_2OPO_3^{2-} \ + \ ADP$$

$$H_2NCH_2CH_2OH \ + \ ATP \quad \xrightarrow[\text{ethanolamine kinase}]{Mg^{2+}} \quad H_2NCH_2CH_2OPO_3^{2-} \ + \ ADP$$

This enzyme step commits choline (ethanolamine) to the CDPcholine (CDPethanolamine) pathway for the biosynthesis of phosphatidylcholine (phosphatidylethanolamine) in all animal cells.

Although it has long been a controversy as to whether the same enzyme catalyzes the phosphorylation of choline and ethanolamine in various tissue sources, recent purification and immunological studies clearly demonstrated that both reactions are catalyzed by the same enzyme(s) in the rat.[1,2] It is also strongly suggested that the enzyme does not exist in one particular active form, but exists in several isoforms (probably isozymes) in rat tissues.[3,4]

Assay for Choline Kinase

Reagents

Tris-HCl, 0.2 M, pH 8.75
Disodium ATP, 100 mM
$MgCl_2$, 100 mM
[*methyl*-^{14}C]Choline chloride, 5 mM; specific radioactivity, 0.7 Ci/mol (stored at $-20°$)

[1] K. Ishidate, K. Furusawa, and Y. Nakazawa, *Biochim. Biophys. Acta* **836,** 119 (1985).
[2] T. J. Porter and C. Kent, *J. Biol. Chem.* **265,** 414 (1990).
[3] K. Ishidate, K. Iida, K. Tadokoro, and Y. Nakazawa, *Biochim. Biophys. Acta* **833,** 1 (1985).
[4] K. Tadokoro, K. Ishidate, and Y. Nakazawa, *Biochim. Biophys. Acta* **835,** 501 (1985).

Enzyme; a high-speed supernatant (postmicrosomal) fraction from
20% rat tissue homogenates, 10–15 mg protein/ml

Dowex 1-X8 (OH⁻ form) resin; 200–400 mesh, 50% suspension in
distilled water, stored in a refrigerator

Procedure. Aliquots of the following reagents are added to a small
centrifuge tube (1.3 × 10 cm) placed in an ice bath: 150 μl of Tris-HCl
(pH 8.75); 30 μl of disodium ATP; 36 μl of MgCl$_2$; 15 μl of [*methyl-*
^{14}*C*]choline chloride [~110,000 disintegrations/min (dpm) per incubation];
water; and, finally, less than 35 μl of enzyme to give a final volume of 0.3
ml. The mixture is incubated with moderate shaking for 15 min at 37°.
The reaction is started immediately with the addition of enzyme but is
conducted more conveniently by placing the tubes containing the whole
mixture in a shaking water bath. The reaction is stopped by placing the
tubes for 2 min in a boiling water bath. When necessary, blanks are
prepared by stopping the reaction immediately after the addition of
enzyme.

After brief centrifugation, the entire content of each tube is applied
carefully to a Dowex 1-X8 (OH⁻) minicolumn (see below). The column is
washed first with 1 ml of 5 mM choline chloride, then with 3 ml water.
Any radioactive choline is eluted by this washing. The column is then
placed into a clean test tube of appropriate size so that the column can be
kept in place for the later centrifugation step. Phospho[^{14}C]choline is
completely eluted with 0.3 ml of 1 M NaOH followed by 0.9 ml of 0.1 M
NaOH. Finally, any NaOH solution retained in the column is collected
into the tube by brief centrifugation. The eluate is transferred to a counting
vial and neutralized with 5 M HCl, then the radioactivity is measured with
10 ml ACS-II (Amersham, Arlington Heights, IL).

The time required for the assay runs to about 2 hr for 50 samples. The
reaction rate is constant for at least 30 min and proportional to the amount
of enzyme within the range of 0.5 mg protein per incubation, when crude
rat tissue high-speed supernatant is the enzyme source assayed. The enzyme activity is calculated from the specific radioactivity of [^{14}C]choline
chloride (~1500 dpm/nmol), and 1 unit of activity represents 1 μmol of
phosphocholine formed per minute of incubation. The specific activities
achieved with a high-speed supernatant fraction of several rat tissue homogenates are in the following ranges: 2.5–3.5 mU/mg for liver, 2.0–3.0
mU/mg for kidney, 0.6–0.8 mU/mg for lung, and 3.5–4.5 mU/mg for whole
intestine.

Comments. This method was first applied for the choline kinase assay
by McCaman,[5] and later some modifications were made by Weinhold and

---

[5] R. E. McCaman, *J. Biol. Chem.* **237**, 672 (1962).

Rethy.[6] We have further simplified the procedure by utilizing a Gilson-type (P-1000) pipette tip with a cotton plug at the bottom as a column, in which 0.5 ml of the Dowex resin is packed.

Other methods to assay choline kinase activity have been reported: (1) removal of unreacted free [^{14}C]choline by precipitation as an iodine complex[7] or as a reineckate salt[8] when the reaction is to be stopped (the phospho[^{14}C]choline, which is not precipitated by this reaction, is then measured); (2) extraction of unreacted [^{14}C]choline by an organic cation exchanger, sodium tetraphenylboron in allyl cyanide[9] or 3-heptanone,[10] with an aliquot of the remaining aqueous phase containing phospho[^{14}C]-choline counted; (3) an enzymatic coupling method in which one of the reaction products, ADP, is coupled to pyruvate kinase and lactate dehydrogenase reactions in the presence of phosphoenolpyruvate and NADH; with the disappearance of NADH being monitored spectrophotometrically[11]; and (4) separation of phospho[^{14}C]choline from unreacted [^{14}C]choline by paper chromatography,[12] thin-layer chromatography,[13] or high-performance liquid chromatography (HPLC) equipped with a cation-exchange column.[14]

Assay for Ethanolamine Kinase

Reagents

Glycylglycine, sodium salt, 0.3 M, pH 8.5
Disodium ATP, 100 mM
MgCl$_2$, 100 mM
[1,2-^{14}C]Ethanolamine-HCl, 5 mM; specific radioactivity, 1 Ci/mol (stored at $-20°$)
KCl, 1 M
Enzyme: a high-speed supernatant (postmicrosomal) fraction from 20% rat tissue homogenates, 10–15 mg protein/ml
Dowex 50W-X8 (H$^+$ form) resin: 100–200 mesh, 50% suspension in distilled water, stored in a refrigerator

[6] P. A. Weinhold and V. B. Rethy, *Biochemistry* **13**, 5135 (1974).
[7] M. A. Brostrom and E. T. Browning, *J. Biol. Chem.* **248**, 2364 (1973).
[8] R. E. McCaman, S. A. Dewhurst, and A. M. Goldberg, *Anal. Biochem.* **42**, 171 (1971).
[9] A. M. Burt and S. A. Brody, *Anal. Biochem.* **65**, 215 (1975).
[10] S. A. Dewhurst, *J. Neurochem.* **19**, 2217 (1972).
[11] J. Wittenberg and A. Kornberg, *J. Biol. Chem.* **202**, 431 (1953).
[12] R. E. Ulane, L. L. Stephenson, and P. F. Farrell, *Anal. Biochem.* **79**, 526 (1977).
[13] S. L. Pelech, E. Power, and D. E. Vance, *Can. J. Biochem. Cell Biol.* **61**, 1147 (1983).
[14] L. D. Nelson, N. D. Brown, and W. P. Wiesmann, *J. Chromatogr.* **324**, 203 (1985).

Procedure. Aliquots of the following reagents are added to a small centrifuge tube (1.3 × 10 cm) placed in an ice bath: 60 μl of glycylglycine (pH 8.5); 30 μl of disodium ATP; 36 μl of MgCl$_2$; 30 μl of [1,2-^{14}C]ethanol-amine-HCl (~300,000 dpm per incubation); 30 μl of KCl; water; and, finally, less than 35 μl of enzyme to give a final volume of 0.3 ml. The mixture is incubated with moderate shaking for 20 min at 37°. The reaction is started immediately with the addition of the enzyme but is performed more conveniently by placing the tubes containing the whole mixture in a shaking water bath. The reaction is stopped by placing the tubes for 2 min in a boiling water bath. Blanks are prepared by stopping the reaction immediately after the addition of enzyme.

After brief centrifugation, the entire content of each tube is applied carefully onto the Dowex 50W-X8 (H$^+$ form) minicolumn (0.75 ml resin) prepared as described above. The column is eluted with 3 ml of water, and the eluate is collected in a clean test tube of appropriate size that the column can be retained on top, affording the next centrifugation step. Phospho[^{14}C]ethanolamine is not retained on the column and is recovered completely into the tube by brief centrifugation. The eluate is transferred to a counting vial, then the radioactivity measured with 10 ml ACS-II in a gel state. The reaction rate is constant for at least 30 min and proportional to the amount of enzyme within the range of 0.5 mg protein per incubation, when crude rat tissue high-speed supernatant is the enzyme source assayed. The enzyme activity is calculated from the specific radioactivity of [^{14}C]ethanolamine (~2000 dpm/nmol), and 1 unit of activity represents 1 μmol of phosphoethanolamine formed per minute of incubation. The specific activity with rat tissues is about one-third that of choline kinase.

Preparation of Cytosol from Kidney

Rats fasted overnight are sacrificed by decapitation, and the kidneys are quickly excised and rinsed with ice-cold saline. The kidneys are first cut into small pieces and then homogenized in 4 volumes of 0.154 *M* KCl/20 m*M* Tris-HCl (pH 7.5)/2 m*M* 2-mercaptoethanol (2-ME)/1 m*M* EDTA/0.5 m*M* phenylmethylsulfonyl fluoride (PMSF), using a motor-driven Teflon pestle homogenizer. The homogenates are centrifuged at 9000 *g* for 20 min, and the supernatants are recentrifuged at 105,000 *g* for 60 min. An aliquot of the resulting high-speed supernatant is used as the crude enzyme preparation.

Purification of Choline/Ethanolamine Kinase from Rat Kidney

Purification is started with the high-speed supernatant fraction from fresh kidneys (~200 g wet weight). All procedures are carried out below 5°.

Step 1: Acid Precipitation. The pH of the supernatant is adjusted to 5.1 with 1 M acetic acid while stirring at 0°, and the resulting precipitates are quickly removed by centrifugation at 20,000 g for 15 min.

Step 2: Ammonium Sulfate Precipitation. Tris-HCl (1 M, pH 7.5) is added to the clear supernatant to a final Tris concentration of 0.1 M, then solid ammonium sulfate is added to 30% saturation. The mixture is incubated at 0° for 30 min with moderate stirring, then centrifuged at 20,000 g for 20 min. To the resulting supernatant, solid ammonium sulfate is added to 45% saturation, and the mixture is incubated and centrifuged in the same manner. The activities of both choline and ethanolamine kinases can be efficiently concentrated into the 30–45% saturated ammonium sulfate-precipitable fraction. The precipitate is dissolved in an appropriate amount (~50 ml) of 20 mM Tris-HCl (pH 7.5), 2 mM 2-ME, then dialyzed overnight against the same buffer. Undissolved materials are removed by centrifugation at 20,000 g for 10 min.

Step 3: DEAE-Cellulose Column. The supernatant is applied on a DEAE-cellulose column (2.6 × 45 cm) previously equilibrated with 20 mM Tris-HCl (pH 7.5) and 2 mM 2-ME. The column is eluted first with 500 ml of 0.1 M KCl in 20 mM Tris-HCl (pH 7.5) and 2 mM 2-ME followed by a 1000-ml linear gradient of 0.1–0.4 M KCl in the same buffer. Fractions of 10 ml are collected and assayed for choline/ethanolamine kinase activity. Fractions with high enzyme activity are pooled and treated again with solid ammonium sulfate to 50% saturation. The resulting precipitates are collected by centrifugation, redissolved in a small amount (~12 ml) of 20 mM Tris-HCl (pH 7.5), 2 mM 2-ME containing 0.1 M KCl.

Step 4: Sephadex G-150 Gel Chromatography. A portion (usually 6 ml, ~230 mg protein) of the above fraction is applied onto Sephadex G-150 column (2.6 × 75 cm) previously equilibrated with 0.1 M KCl, 20 mM Tris-HCl (pH 7.5), 2 mM 2-ME and eluted with the same buffer at a flow rate of 8 ml/hr. Fractions of 5 ml are collected and assayed for enzyme activity. The fractions with high kinase activity are pooled and concentrated through a membrane filter [Type A, 10,000 MWCO (molecular weight cutoff) from Spectrum Sciences, Santa Clara, CA] to give a final protein concentration of approximately 20 mg/ml, then dialyzed against 50 mM Tris-HCl (pH 7.5) and 2 mM 2-ME to remove most of the KCl from the sample.

Step 5: Choline-Sepharose Affinity Column. An aliquot of the fraction (1–2 ml) is applied on choline-Sepharose column (1.3 × 6 cm, see below) at a flow rate of 3 ml/hr. After washing with 10 volumes of the equilibration buffer, the column is eluted with a linear gradient of 0–0.3 M NaCl in the same buffer at a flow rate of 6 ml/hr. Fractions of 2 ml are collected and assayed for the kinase activity. A typical elution pattern of choline kinase activity through the affinity column is shown in Fig. 1.

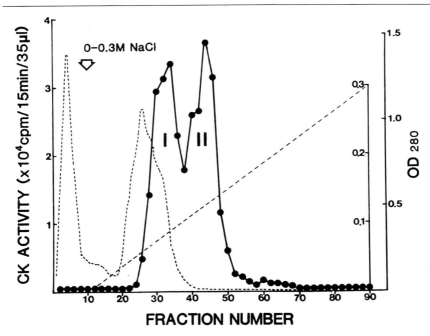

FIG. 1. Choline-Sepharose 6B affinity chromatography of rat kidney choline kinase preparation. A portion of the Sephadex G-150-purified fraction of kidney choline kinase (1 ml, ~20 mg protein) was applied on the affinity column (1.3 × 6 cm). Fractions of 2 ml each were collected, and a 35-μl portion was assayed for choline kinase activity. [From K. Ishidate, K. Nakagomi, and Y. Nakazawa, *J. Biol. Chem.* **259**, 14706 (1984), with permission.] (●——●), Choline kinase activity; (---), absorbance at 280 nm; (---), NaCl concentration (estimated).

Although a considerable amount of nonspecific binding occurred, probably owing to the ionic interaction between the ligand (choline) and proteins, these low affinity materials could easily be eluted with a relatively low NaCl concentration. The activity was eluted thereafter in two major fractions, I and II, with approximately 80% recovery of the total applied activity. The activity in fraction I was not clearly separated from the materials absorbing at 280 nm. On the other hand, the activity in fraction II was eluted with almost no detectable protein absorbance at 280 nm, which indicated that a form of kinase protein in fraction II could be highly purified through this affinity column. As expected, only one major protein band was detected in fraction II, which migrated at a relative molecular size of 42 kDa on polyacrylamide gel electrophoresis in the presence of sodium dodecyl sulfate (SDS–PAGE).[15] Fraction I appeared to have a

[15] K. Ishidate, K. Nakagomi, and Y. Nakazawa, *J. Biol. Chem.* **259**, 14706 (1984).

TABLE I
COPURIFICATION OF CHOLINE KINASE AND ETHANOLAMINE KINASE ACTIVITIES
FROM RAT KIDNEY CYTOSOL[a,b]

Preparation step	Total protein (mg)	CK activity		EK activity		Activity ratio EK/CK
		Specific activity (mU/mg protein)	Total activity (mU)	Specific activity (mU/mg protein)	Total activity (mU)	
1. 105,000 g Supernatant	12,457	2.2	27,280	0.9	11,460	0.42
2. pH 5.1 Supernatant	8720	2.5	21,364	1.0	8720	0.41
3. 30–45% Ammonium sulfate	2413	9.8	23,599	3.3	8059	0.34
4. DEAE-cellulose	462	32.7	15,126	9.5	4371	0.29
5. Sephadex G-150	194	49.3	9572	16.2	3135	0.33
6. Choline-Sepharose I	79	41.4	3263	13.9	1100	0.34
7. Choline-Sepharose II	0.7	3300	2343	1258	893	0.38
Steps 1–7 purification (-fold):		1500		1398		
Total recovery (%):			8.6		8.0	

[a] The purification was initiated with newly prepared kidney cytosol from 98 Wistar rats. CK, Choline kinase; EK, ethanolamine kinase.

[b] From K. Ishidate, K. Furusawa, and Y. Nakazawa, *Biochim. Biophys. Acta* **836,** 119 (1985), with permission.

42-kDa band also, though a considerable amount of other proteins were present as contaminants. To evaluate whether the 42-kDa band in fraction II is a form of rat kidney choline/ethanolamine kinase protein, we examined the patterns obtained with native PAGE, then compared the location of protein band with choline kinase activities which were determined in 3-mm gel slices after electrophoresis. The result showed that the only one detectable band after staining with Coomassie Brilliant blue was located at exactly the same R_f position as that with the highest kinase activity.[15] Thus, we concluded that the 42-kDa polypeptide in fraction II is actually a rat kidney choline kinase protein.

A typical purification index is shown in Table I. The molecular size of the intact form of rat kidney choline/ethanolamine kinase was estimated as 75–80 kDa through a standardized Sephadex G-150 column (0.88 × 68 cm), which indicates that the enzyme in rat kidney exists most likely in a dimeric form.[15] The purified kidney enzyme is stable when sterilized and stored at −20° without any stabilizing additives such as bovine serum albumin (BSA) or glycerol.

Preparation of Choline-Sepharose Column. The choline-Sepharose affinity gel is prepared as follows: 3 g of epoxy-activated Sepharose 6B (Pharmacia, Uppsala, Sweden) is swollen and washed with 300 ml distilled water on a G-3 glass filter. The coupling reaction is conducted by incubating the gel in 20 ml of 10 mM NaOH containing 10 mM choline chloride at 45° for 16 hr. The gel is transferred onto the glass filter and washed successively with 10 mM NaOH, 0.1 M NaHCO$_3$, and 0.1 M acetate buffer (pH 4.0). After the gel is neutralized by washing with distilled water, excess activated epoxy groups are blocked by incubating with 1 M ethanolamine for 4 hr at room temperature. Then the gel is washed several times with 0.1 M Tris-HCl (pH 8.75) containing 0.5 M NaCl and 0.1 M acetate buffer (pH 4.0) containing 0.5 M NaCl, alternately. The gel is finally equilibrated in 50 mM Tris-HCl (pH 7.5) and 2 mM 2-ME, deaerated under aspiration, and packed in a small column.

Subcellular Localization

Except for a few instances, both kinase activities are recovered quantitatively in a high-speed supernatant fraction of tissue homogenates, indicating that their locations in the cell are cytosolic. Essentially no activity is detectable in blood plasma, although significant choline kinase activity can be detected in the plasma from carbon tetrachloride-intoxicated rats, probably owing to release from damaged parenchymal hepatocytes.[16] In brain, some activity of choline/ethanolamine kinase has been reported to exist in a membrane-associated form.[17–20] Other exceptions include choline kinase activity associating to plasma membranes in the anaerobic protozoan *Endodinium caudatum*,[21] and to mitochondrial fractions in the filaments of *Cuscuta reflexa*.[22]

General Properties

Optimal pH. There is general agreement for a rather alkaline pH optimum between pH 8.0 and 9.5 for both kinase reactions. In most instances, the activity is considerably reduced at neutral pH, and essentially no activity is detected in acidic pH ranges.

[16] K. Ishidate, S. Enosawa, and Y. Nakazawa, *Biochem. Biophys. Res. Commun.* **111,** 683 (1983).
[17] R. R. Reinhardt and L. Wecker, *J. Neurochem.* **41,** 623 (1983).
[18] R. R. Reinhardt, L. Wecker, and P. F. Cook, *J. Biol. Chem.* **259,** 7446 (1984).
[19] T. Kunishita, K. K. Vaswani, C. R. Morrow, and R. W. Ledeen, *Neurochem. Res.* **12,** 351 (1987).
[20] T. Kunishita, K. K. Vaswani, C. R. Morrow, G. P. Novak, and R. W. Ledeen, *J. Neurochem.* **48,** 1 (1987).
[21] F. L. Bygrave and R. M. C. Dawson, *Biochem. J.* **160,** 481 (1976).
[22] P. N. Setty and P. S. Krishnan, *Biochem. J.* **126,** 313 (1972).

Kinetic Parameters. A large number of reports now exist on the kinetic characterization of both choline and ethanolamine kinase reactions from various enzyme sources (reviewed in Ref. 23), and typical parameters from several purification studies are summarized in Table II. There have been wide differences in the values reported for the apparent K_m for choline (ethanolamine) ranging from micromolar to millimolar. It was reported that double-reciprocal plots of the initial velocity versus choline concentration were nonlinear for choline kinase from brewer's yeast at relatively high choline concentrations.[7] This nonlinearity was apparently not recognized in early investigations and is possibly the cause of conflicting reports for the apparent K_m for choline.

More than one K_m has been reported even in highly purified choline kinase preparations from rabbit brain,[24] rooster liver,[25] and primate lung[26] cytosols. These observations could be accounted for either by the presence of two catalytic species (or sites) with different affinities for choline or by a mechanism of negative cooperativity between choline binding sites; choline bound to one site would increase the K_m of the enzyme for choline at another site on the same protein.[7] In the case of the first possibility, there is evidence for the existence of multiple forms of the enzyme,[3,4] most likely physiologically relevant isozymes or, less likely, artifacts of proteolysis during preparation. For the second possibility, a homogeneous choline/ethanolamine kinase preparation from rat kidney has been shown to have two binding sites for choline (ethanolamine),[1] probably one catalytic and the other regulatory. Thus, more precise kinetic characterization with homogeneous preparations of each isozyme will be required to distinguish the two possibilities.

The reported apparent K_m values for ATP–Mg^{2+} also have wide fluctuation from tissue to tissue.[23] Several investigations indicate that the highest activity is obtained with an ATP/Mg^{2+} ratio of 1.0 and that excess ATP is inhibitory. In other studies, excess Mg^{2+} besides ATP–Mg^{2+} is required for maximal activity. We have observed that the apparent K_m value of a homogeneous rat kidney choline/ethanolamine kinase for ATP–Mg^{2+} is extremely high (10 mM) when estimated in the presence of equivalent amounts of ATP and Mg^{2+}, whereas when the value is estimated in a 1.5-fold higher concentration of Mg^{2+} than ATP, an apparent K_m for ATP of 1.5 mM is obtained in either of the kinase reactions.[1] Thus, the

[23] K. Ishidate, *in* "Phospholipid Metabolism" (D. E. Vance, ed.), p. 9. CRC Press, Boca Raton, Florida, 1989.

[24] D. R. Haubrich, *J. Neurochem.* **21,** 315 (1973).

[25] H. B. Paddon, C. Vigo, and D. E. Vance, *Biochim. Biophys. Acta* **710,** 112 (1982).

[26] R. E. Ulane, *in* "Lung Development: Biological and Clinical Perspectives" (P. M. Farrell, ed.), Vol. 1, p. 295. Academic Press, New York and London, 1982.

TABLE II
KINETIC PARAMETERS OF PARTIALLY OR HIGHLY PURIFIED CHOLINE KINASE
FROM VARIOUS TISSUES

Source	Purification index (% recovery)	Estimated molecular size (Da)	Kinetic constants and optimal pH
Cuscuta reflexa[a] (mitochondria)	283-fold (33%)		K_m (choline) 2.5 mM (allosteric nature), K_m (ATP) 5 mM, optimal pH 8.5
Brewer's yeast[b,c] (cell extract)	308-fold (5%)	67,000 (gel filtration and sedimentation constant)	K_m (choline) 15 μM, K_m (ATP) 0.14 mM, optimal pH 8.0–9.5
Rabbit brain[d] (aceton powder)	203-fold (10%)		K_m (choline) 32 μM (low), 0.31 mM (high), K_m (ATP) 1.1 mM, optimal pH 9.0–10.5
Rat liver (cytosol)	68-fold[e] (20%)	166,000 (gel filtration)	K_m (choline) 30 μM, K_m (ATP) 3.7 mM, optimal pH 8.0
	550-fold[f,g] (19%)		K_m (choline) 33 μM
	26,000-fold[h] (8%)	47,000 (SDS–PAGE)[i] 160,000 (gel filtration)	K_m (choline) 13 μM, K_m (ethanolamine) 1.2 mM
Monkey lung[j,k,l] (cytosol)	1000-fold (10%)	80,000 (gel filtration and sedimentation constant)	K_m (choline) 30 μM, K_m (ATP) 5 mM, optimal pH 8.5
Chicken liver (cytosol)	780-fold[m] (36%)		K_m (choline) 40 μM (low), 0.2–0.5 mM (high), K_m (ATP) 2.1 mM
	601-fold[n] (17%)	36,000 (gel filtration)	K_m (choline) 60 μM, K_m (ATP) 13.3 mM, optimal pH 9.0
Rat kidney[o,p] (cytosol)	1500-fold (9%)	42,000 (SDS–PAGE) 75,000–80,000 (gel filtration)	K_m (choline) 0.1 mM, K_m (ethanolamine) 0.6 mM, K_m (ATP) 1.5 mM, optimal pH 8.5–9.0
Rat brain[q] (cytosol)	15,000-fold (3%)	44,000 (SDS–PAGE) 90,000 (gel filtration)	K_m (choline) 14 μM, K_m (ATP) 1.0 mM, optimal pH >8.5

[a] P. N. Setty and P. S. Krishnan, *Biochem. J.* **126**, 313 (1972).

[b] J. Wittenberg and A. Kornberg, *J. Biol. Chem.* **202**, 431 (1953).

[c] M. A. Brostrom and E. T. Browning, *J. Biol. Chem.* **248**, 2364 (1973).

[d] D. R. Haubrich, *J. Neurochem.* **21**, 315 (1973).

[e] P. A. Weinhold and V. B. Rethy, *Biochemistry* **13**, 5135 (1974).

[f] P. J. Brophy and D. E. Vance, *FEBS Lett.* **62**, 123 (1976).

[g] P. J. Brophy, P. C. Choy, J. R. Toone, and D. E. Vance, *Eur. J. Biochem.* **78**, 491 (1977).

[h] T. J. Porter and C. Kent, *J. Biol. Chem.* **265**, 414 (1990).

[i] SDS–PAGE, Sodium dodecyl sulfate-polyacrylamide gel electrophoresis.

[j] R. E. Ulane, *in* "Lung Development: Biological and Clinical Perspectives" (P. M. Farrell, ed.), Vol. 1, p. 295. Academic Press, New York and London, 1982.

[k] R. E. Ulane, L. L. Stephenson, and P. M. Farrell, *Anal. Biochem.* **79**, 526 (1977).

[l] R. E. Ulane, L. L. Stephenson, and P. M. Farrell, *Biochim. Biophys. Acta* **531**, 295 (1978).

[m] H. B. Paddon, C. Vigo, and D. E. Vance, *Biochim. Biophys. Acta* **710**, 112 (1982).

[n] G. R. Kulkarni and S. K. Murthy, *Ind. J. Biochem. Biophys.* **23**, 90 (1986).

[o] K. Ishidate, K. Nakagomi, and Y. Nakazawa, *J. Biol. Chem.* **259**, 14706 (1984).

[p] K. Ishidate, K. Furusawa, and Y. Nakazawa, *Biochim. Biophys. Acta* **836**, 119 (1985).

[q] T. Uchida and S. Yamashita, *Biochim. Biophys. Acta* **1043**, 281 (1990).

affinity of rat kidney choline/ethanolamine kinase for ATP is highly dependent on the free Mg^{2+} concentration. Several other investigators have reported similar results for both choline kinase[27,28] and ethanolamine kinase[6] reactions. For crude enzyme preparations such as tissue homogenates, account must be taken of the endogenous hydrolytic activities toward ATP, which have been shown to be particularly high in certain membrane fractions.

Phosphate Donor and Mg^{2+} Requirement.[23] ATP is specifically required as a phosphate donor for the reaction. No detectable phosphocholine was formed in the presence of CTP, GTP, ITP, or UTP, with the exception of one report in which GTP gave 50% of the activity with ATP and UTP 20%. Like other kinase reactions, choline kinase activity is totally dependent on Mg^{2+} ion. Other divalent cations so far examined (Mn^{2+}, Hg^{2+}, Co^{2+}, Ca^{2+}, Zn^{2+}, Ni^{2+}, Cd^{2+}, Cs^{2+}, Cu^{2+}, Ba^{2+}, and Fe^{2+}) could not effectively substitute for Mg^{2+}.

Reaction Mechanism. The early initial velocity, product, and inhibitor studies suggested a random equilibrium mechanism rather than an ordered one.[6,7] Subsequent studies, however, indicate that the forward reaction follows a sequentially ordered mechanism with $ATP-Mg^{2+}$ (or choline) binding to the enzyme first, followed by choline ($ATP-Mg^{2+}$), and then activation of the ternary complex by free Mg^{2+}. The release of phosphocholine occurs prior to that of $ADP-Mg^{2+}$. Thus, the overall rate of the reaction is probably limited by the release of $ADP-Mg^{2+}$ from the complex.[23]

Substrate Specificity and Structure–Activity Relationships. The initial studies by Wittenberg and Kornberg[11] demonstrated that the yeast choline kinase preparation could phosphorylate dimethyl- and monomethylethanolamine, diethyl- and monoethylethanolamine, and, though very weakly, ethanolamine. Similar results were reported with partially purified choline kinase preparations from rabbit brain,[24] *Phormina regina* larvae,[29] and rape seeds.[30] Choline kinase preparations lacking ethanolamine kinase activity have also been obtained from soybean seeds,[31] *E. caudatum*,[32] spinach leaves,[33] *Culex pipiens fatigans*,[34] and rat liver.[35] On the other

[27] S. Spanner and G. B. Ansell, *Biochem. J.* **178**, 753 (1979).

[28] J. P. Infante and J. E. Kinsella, *Int. J. Biochem.* **7**, 483 (1976).

[29] R. M. Shelley and E. Hodgson, *Insect Biochem.* **1**, 149 (1971).

[30] T. Ramasarma and L. R. Wetter, *Can. J. Biochem. Physiol.* **35**, 853 (1957).

[31] J. Wharfe and J. L. Horwood, *Biochim. Biophys. Acta* **575**, 102 (1979).

[32] T. E. Broad and R. M. C. Dawson, *Biochem. Soc. Trans.* **2**, 1272 (1974).

[33] B. A. Macher and J. B. Mudd, *Arch. Biochem. Biophys.* **177**, 24 (1976).

[34] P. Ramabrahmam and D. Subrahmanyam, *Arch. Biochem. Biophys.* **207**, 55 (1981).

[35] P. J. Brophy, P. C. Choy, J. R. Toone, and D. E. Vance, *Eur. J. Biochem.* **78**, 491 (1977).

hand, the highly purified choline kinase preparations from primate lung,[36] rat liver,[2,37] rat kidney,[1] rat brain,[38] chicken liver,[39] and human liver[40] have all been shown to have a considerable catalytic activity toward ethanolamine phosphorylation. Recently, the relationship between substrate structure and activity has been investigated precisely with yeast choline kinase preparations and more than 50 structural analogs of choline.[41]

Inhibitors. There is general agreement that choline kinase is inhibited weakly by ethanolamine and strongly by hemicholinium-3 (HC-3). The inhibition by ethanolamine is competitive in most instances, whereas the inhibition by HC-3 has been reported to be competitive, noncompetitive, uncompetitive, or of a mixed-type manner versus choline. That ethanolamine was not an effective inhibitor for choline kinase reaction has also been reported.[23]

There is little information on the inhibition of choline kinase by structural analogs of ATP. ADP, a product of the reaction, has been reported to be a competitive, noncompetitive, or uncompetitive inhibitor with respect to ATP. 5'-AMP and AMP-PNP, an ATP analog, were found to be competitive versus ATP in some systems.[23]

Recently, various analogs of ATP were examined in our laboratory with a partially purified rat kidney choline kinase preparation. Interestingly, the most efficient inhibition was obtained by adenosine, followed by 3',5'-cAMP > 5'-ADP > 3'-AMP > 2',3'-cAMP, and no significant inhibition was observed by 5'-AMP, 2',5'-ADP, 3',5'-ADP, and 2'-AMP when they were added to the choline kinase assay mixture up to 5 mM. The inhibition by either adenosine or 3',5'-cAMP appeared to be competitive versus ATP, and the estimated K_i value was 0.26 or 4 mM, respectively.[23]

Activators. Polyamines stimulated choline kinase activity severalfold in a partially purified enzyme preparation from rat liver cytosol by increasing the affinity of the enzyme for ATP–Mg^{2+}. This effect was strongest

[36] R. E. Ulane, L. L. Stephenson, and P. M. Farrell, *Biochim. Biophys. Acta* **531,** 295 (1978).
[37] P. J. Brophy and D. E. Vance, *FEBS Lett.* **62,** 123 (1976).
[38] T. Uchida and S. Yamashita, *Biochim. Biophys. Acta* **1043,** 281 (1990).
[39] G. R. Kulkarni and S. K. Murthy, *Ind. J. Biochem. Biophys.* **23,** 90 (1986).
[40] E. Draus, J. Niefind, K. Vietor, and B. Havsteen, *Biochim. Biophys. Acta* **1045,** 195 (1990).
[41] G. L. Clary, C.-F. Tsai, and R. W. Guynn, *Arch. Biochem. Biophys.* **254,** 214 (1987).

TABLE III
INDUCTION OF CHOLINE/ETHANOLAMINE KINASE ACTIVITY
BY VARIOUS EXPERIMENTAL MANIPULATIONS

Tissue source	Factor	Proposed mechanism	Degree of induction (%)	Ref.
Liver (rat)	Essential fatty acid deficiency	?	350	a
	3-Methylcholanthrene	Enzyme induction	200	b, c
	Carbon tetrachloride	Enzyme induction	300–400	d, e
Liver (chicken)	Estrogens	Enzyme induction	300–400	f, g, h
3A2 liver cells	Insulin	Enzyme induction	220	i
Mammary gland (mouse)	Insulin + cortisol + prolactin	Activation by polyamines	370–560	j
Nb2 lymphoma cells	Prolactin	Enzyme induction	210–440	k
Soybean	*Rhizobium* infection	Enzyme induction	350	l, m
3T3 and C3H 10T fibroblasts	10% serum	?	200–300	n, o
	Ha-*ras* tranformation	?	200	p, q

[a] J. P. Infante and J. E. Kinsella, *Biochem. J.* **176**, 631 (1978).
[b] K. Ishidate, M. Tsuruoka, and Y. Nakazawa, *Biochem. Biophys. Res. Commun.* **96**, 946 (1980).
[c] K. Ishidate, M. Tsuruoka, and Y. Nakazawa, *Biochim. Biophys. Acta* **620**, 49 (1980).
[d] K. Ishidate, S. Enosawa, and Y. Nakazawa, *Biochem. Biophys. Res. Commun.* **111**, 683 (1983).
[e] K. Tadokoro, K. Ishidate, and Y. Nakazawa, *Biochim. Biophys. Acta* **835**, 501 (1985).
[f] H. B. Paddon, C. Vigo, and D. E. Vance, *Biochim. Biophys. Acta* **710**, 112 (1982).
[g] C. Vigo, H. B. Paddon, F. C. Millard, P. H. Pritchard, and D. E. Vance, *Biochim. Biophys. Acta* **665**, 546 (1981).
[h] G. R. Kulkarni and S. K. Murthy, *Ind. J. Biochem. Biophys.* **23**, 254 (1986).
[i] R. E. Ulane and M. M. Ulane, *Life Sci.* **26**, 2143 (1980).
[j] T. Oka and J. W. Perry, *Dev. Biol.* **68**, 311 (1979).
[k] K. W. S. Ko, H. W. Cook, and D. E. Vance, *J. Biol. Chem.* **261**, 7846 (1986).
[l] R. B. Mellor, T. M. I. E. Christensen, and D. Werner, *Proc. Natl. Acad. Sci. U.S.A.* **83**, 659 (1986).
[m] R. B. Mellor, H. Thierfelder, G. Pausch, and D. Werner, *J. Plant Physiol.* **128**, 169 (1987).
[n] C. H. Warden and M. Friedkin, *Biochim. Biophys. Acta* **792**, 279 (1984).
[o] C. H. Warden and M. Friedkin, *J. Biol. Chem.* **260**, 6006 (1985).
[p] I. G. Macara, *Mol. Cell. Biol.* **9**, 325 (1989).
[q] D. Teegarden, E. J. Taparowsky, and C. Kent, *J. Biol. Chem.* **265**, 6042 (1990).

with spermine followed by spermidine, cadaverine, and putrescine.[42] Similar *in vitro* results have been reported with other enzyme systems. The possible physiological role of polyamines in modulating choline kinase activity has been discussed.[23]

[42] H. Fukuyama and S. Yamashita, *FEBS Lett.* **71**, 33 (1976).

Specific Features of Choline/Ethanolamine Kinase

One of the most important features of choline/ethanolamine kinase is its inducibility in various experimental systems (Table III). Although the physiological significance of the inducibility has not yet been fully understood, the induction of choline/ethanolamine kinase by several means suggests that more than one mechanism could be involved in this process. Actually, distinct mechanisms of choline kinase induction have been demonstrated between rooster liver by estrogens and rat liver by certain hepatotoxins. A new isozyme of choline/ethanolamine kinase appears to be preferentially induced in the latter case. For more information on the induction of choline/ethanolamine kinase, characterization of isozymes, and possible involvement of the phosphorylation–dephosphorylation mechanism in the regulation of the kinase reaction, the reader is referred to a recent review[23] on choline transport and choline kinase.

[15] Choline/Ethanolamine Kinase from Rat Liver

By THOMAS J. PORTER and CLAUDIA KENT

Introduction

Choline kinase (EC 2.7.1.32) is the initial enzyme in the CDPcholine pathway for phosphatidylcholine biosynthesis in mammalian cells. It catalyzes the Mg^{2+}–ATP-dependent phosphorylation of choline. Choline kinase is rate-limiting for phosphatidylcholine formation in several cells and tissues (see citations in Ref. 1). The enzyme exists as at least two isoelectric forms in all tissues examined,[1-3] and a third form is inducible with polycyclic aromatic hydrocarbons[2] and hepatoxic agents[4] in liver. Both constitutive isoforms in liver have ethanolamine kinase activity, and kinetic and immunological data show that both kinase activities are catalyzed by the same enzymes.[1] These results on the regulation and enzymatic activities of choline kinase indicate that the enzyme plays an important role in the long-term regulation of the CDPcholine pathway and may be involved in the coordinate regulation of phosphatidylethanolamine biosynthesis.

[1] T. J. Porter and C. Kent, *J. Biol. Chem.* **265**, 414 (1990).
[2] K. Tadokoro, K. Ishidate, and Y. Nakazawa, *Biochim. Biophys. Acta* **835**, 501 (1985).
[3] K. Ishidate, K. Nakagomi, and Y. Nakazawa, *J. Biol. Chem.* **259**, 14706 (1984).
[4] K. Ishidate, M. Tsuruoka, and Y. Nakazawa, *Biochem. Biophys. Res. Commun.* **96**, 946 (1980).

To study choline and ethanolamine phosphorylation with a purified kinase preparation, and to investigate the nature of the isoforms, we have developed a purification protocol for rat liver choline–ethanolamine kinase.[1] With the use of size-exclusion chromatography as the final step, a preparation of very high specific activity can be obtained. To obtain homogeneous, though less active, enzyme, chromatofocusing is used as the final step.

Assay Methods

Choline Kinase

Principle. We have combined the assay procedures of McCaman and Dewhurst[5] and Ulane[6] for the assay of choline kinase activity in column fractions. Radiolabeled choline is used as the substrate, and the choline remaining after the enzymatic reaction is removed by precipitation as choline reineckate. The radioactivity of the supernatant fraction, which contains the soluble phosphocholine, is determined.

Procedure. The reaction mixture contains the following (final concentrations): [*methyl*-^{3}H]choline (1–2 mCi/mmol, Research Products International, Inc., Mount Prospect, IL), 5 mM; ATP, 10 mM; MgCl$_2$, 11 mM; dithiothreitol (DTT), 1.3 mM; Tris-HCl, pH 8.5, 67 mM; and 5 μl of enzyme preparation in a final volume of 60 μl in 0.75-ml microcentrifuge tubes. For convenience, a 5× concentrated cocktail of all the assay components minus the enzyme can be made, aliquoted, and stored at −20° for at least 3 months. After a 5-min incubation in a 37° water bath, the reactions are stopped with 20 μl of 10 mM choline chloride in 0.5 N NaOH, followed by 50 μl of freshly prepared satured ammonium reineckate. The solutions are mixed and the choline reineckate pelleted by centrifugation in a microcentrifuge at 10,000 rpm for 1 min. One hundred microliters of the supernatant solution is transferred to another set of microcentrifuge tubes that each contain 10 μl of 40 mM choline chloride and 20 μl of saturated ammonium reineckate solution. The solutions are mixed and the choline reineckate pelleted as above.

One hundred microliters of the supernatant solution is transferred to a 6-ml scintillation vial; 5 ml of Budget Solve scintillation cocktail (Research Products International, Inc.) is added, and the radioactivity of the sample is determined. The counts per minute (cpm) of the blank (assay without

---

[5] R. E. McCaman and S. A. Dewhurst, *Anal. Biochem.* **42,** 171 (1971).

[6] R. E. Ulane, *in* "Lung Development: Biological and Clinical Perspectives" (P. M. Farrell and M. Philip, eds.), Vol. 1, p. 295. Academic Press, New York, 1982.

enzyme) are subtracted from the counts per minute of the sample, and the resulting number is multiplied by a correction factor of 1.7 to allow for the portion of the assay not counted. This assay has a high blank radioactivity which is reduced by ion-exchange purification of the [³H]choline on Bio-Rad (Richmond, CA) AG1-X8, hydroxide form. The choline does not bind and can be eluted with water; the impurities remain bound to the support.

This assay is rapid and is used for all purification steps. Only a 5-min incubation is required for column fractions. The high blank values preclude routine use of this assay with samples of low specific activity, such as crude extracts from certain cells. In addition, polyethylene glycol interferes. For samples with low specific activity or that contain polyethylene glycol, we use a modification of the assay procedure of Ulane.[1,6]

Ethanolamine Kinase

Principle. Ethanolamine kinase activity is determined by the method of Ulane[6] with some modifications. Paper chromatography is used to separate ethanolamine from phosphoethanolamine, which remains at the origin.

Procedure. The final volume of the assay is 100 μl, and [1,2-¹⁴C]ethanolamine hydrochloride (0.5–1 mCi/mmol, New England Nuclear, Boston, MA), 5 mM, is used. The [1,2-¹⁴C[ethanolamine is purified by ion exchange on AG-50 H⁺ resin.[7] Other components of the assay mixture are the same as those indicated above for the choline kinase assay. The assay is stopped with 10 μl glacial acetic acid, and 10 μl of the mixture is then spotted on the paper chromatogram. To facilitate the assay of a large number of samples, 1.5-cm-wide lanes are marked lightly in pencil on a 9 × 24 cm (height × width) piece of Whatman (Clifton, NJ) 3MM paper, and chromatography is performed in a rectangular developing tank in a solvent system of ethanol–ammonium hydroxide–n-butanol (3 : 5. 2 : 6, v/v). After the paper is developed and dried; a 1-cm piece containing the origin is cut out and counted.

Purification of Choline Kinase

Principle. After some initial precipitation steps, choline kinase is highly purified by affinity chromatography on choline-Sepharose. The few contaminating proteins that remain are removed by subsequent chromatographic steps, namely, strong anion exchange, hydrophobic interaction, and chromatofocusing. Some activity is lost at the chromatofocusing step,

[7] P. A. Weinhold and V. B. Rethy, *Biochemistry* **13**, 5135 (1974).

however. If a highly active preparation is desirable, size-exclusion chromatography is used instead of chromatofocusing, but the enzyme prepared by size exclusion is not homogeneous.

Materials

Buffer A: 20 mM Tris-HCl, 154 mM KCl, 2 mM 2-mercaptoethanol, 1 mM [ethylenebis(oxyethylenenitrilo)]tetraacetic acid (EGTA), 1 mM phenylmethylsulfonyl fluoride, 1 mg/liter each of leupeptin, pepstatin, and chymostatin, 2 mg/liter antipain, and 10 mg/liter benzamidine, pH 7.5

Buffer B: 50 mM Tris-HCl, 0.5% Tween 20, 2 mM 2-mercaptoethanol, 1 mM EGTA, 0.2 mM phenylmethylsulfonyl fluoride, 0.5 mg/liter each of leupeptin, pepstatin, and chymostatin, 1 mg/liter antipain, and 5 mg/liter benzamidine, pH 7.5

Buffer C: 20 mM Tris-HCl, 0.01% Tween 20, 2 mM 2-mercaptoethanol, 1 mM EGTA, and protease inhibitors as in buffer B, pH 7.5

Buffer D: 25 mM Bis–Tris–HCl, 0.01% Tween 20, 2 mM 2-mercaptoethanol, and protease inhibitors as in buffer B, without EGTA, pH 6.3

Buffer E: 1 : 10 dilution of Polybuffer 74 (Pharmacia-LKB, Piscataway, NJ), 0.01% Tween 20, 2 mM 2-mercaptoethanol, and protease inibitors as in buffer B without EGTA, pH 4.0

Protease inhibitors, stored as stock solutions as follows[8]: 100 mM phenylmethylsulfonyl fluoride in 2-propanol at room temperature; leupeptin (1 mg/ml), antipain (2 mg/ml), and benzamidine (10 mg/ml) together in water at $-20°$; pepstatin and chymostatin together at 1 mg/ml in dimethyl sulfoxide at $-20°$.

Choline-Sepharose: Sepharose 6B is activated according to Sundberg and Porath.[9] To obtain a choline ligand density of approximately 50 μmol/ml, a 30% solution of butanediol diglycidyl ether is used during the activation. Sepharose 6B is washed with water and suction dried on a sintered glass funnel. Dried Sepharose (40 g) is then suspended in a 500-ml Erlenmeyer flask in 40 ml of 0.6 N NaOH that contains 2 mg/ml sodium borohydride. Forty milliliters of butanediol diglycidyl ether (Aldrich, Milwaukee, WI, 95%) is added, and the flask is stoppered, secured to an orbital shaker, and mixed for 16 hr at 25°. The activated Sepharose is washed with 3 liters of

[8] G. V. Ronnet, V. P. Knutson, R. A. Kohanski, T. L. Simpson, and M. D. Lane, *J. Biol. Chem.* **259**, 4566 (1984).

[9] L. Sundberg and J. Porath, *J. Chromatogr.* **90**, 87 (1974).

water. Ligand is coupled by suspension of epoxy-activated Sepharose in 80 ml of a solution of 1.5 M choline chloride in 10 mM NaOH. The suspension is degassed and placed in a 45° shaking water bath for 16 hr. The coupled support is then washed successively with 1 liter of each of the following per 40 g support: 10 mM NaOH, 0.1 M NaHCO$_3$, and 0.1 M sodium acetate, pH 4.0. The support is washed with water until the pH is neutral, then is blocked with 100 ml of 1 M ethanolamine hydrochloride for 4 hr at room temperature. Final washing steps are with 500 ml each of 0.1 M Tris-HCl–0.5 M NaCl, pH 8.75, then 0.1 M sodium acetate–0.5 M NaCl, pH 4.0; the Tris–NaCl and sodium acetate–NaCl washes are repeated. The support is washed with water until neutral, then stored in 20% ethanol at 4°. The choline ligand density is determined in a parallel reaction with [*methyl*-^{3}H]choline chloride. The optimum salt and choline chloride concentrations used in the column buffers during purification must be determined for each batch of affinity support synthesized.

Procedure

Step 1: Preparation of Cytosol. Twenty female weanling Wistar rats are anesthetized with 0.5 ml chloral hydrate (200 mg/ml). Livers are perfused with buffer A without protease inhibitors or 2-mercaptoethanol, then excised and weighed. (The average liver weight is 12 g/rat.) The livers are homogenized in 4 volumes (ml/g tissue) buffer A with a motor-driven, Teflon pestle–glass homogenizer. All procedures from the homogenization through the affinity column are performed at 4°. The homogenate is subjected to centrifugation at 10,000 g for 10 min. The resulting supernatant fraction is subjected to centrifugation at 105,000 g for 60 min to obtain the cytosol.

Step 2: Preparation of Acid Supernatant Fraction. Cytosol is acidified to pH 5.0 with 1 M acetic acid over a period of 15 min with rapid stirring. The acid is added to the preparation with a peristaltic pump. This suspension is immediately subjected to centrifugation at 10,000 g for 25 min. The supernatant fraction is neutralized with 0.1 volume of 1 M Tris-HCl, pH 7.5.

Step 3: Ammonium Sulfate Fractionation. The neutralized acid supernatant fraction is brought to 30% saturated ammonium sulfate with saturated ammonium sulfate solution (4°, pH 7.0) over a period of 5 min with rapid stirring. This is stirred for an additional 5 min and subjected to centrifugation at 10,000 g for 25 min. The pellet is discarded. Similarly, this supernatant fraction is brought to 45% saturated ammonium sulfate.

After centrifugation, the supernatant fraction is discarded, and the pellet is dissolved in approximately 20 ml of buffer B. This solution is clarified by centrifugation at 20,000 g for 5 min.

Step 4: Desalting. The clarified 30–45% ammonium sulfate cut is loaded onto a Sephadex G-25 (medium) desalting column (2.5 × 80 cm), previously equilibrated with 600 ml of buffer B. Protein is eluted with 400 ml buffer B without protease inhibitors or 2-mercaptoethanol, at a flow rate of 200 ml/hr. Fractions of 8 ml are collected. The void peak, which contains material that absorbs at 280 nm, is pooled.

Step 5: Affinity Chromatography. The desalted protein is loaded at a flow rate of 80 ml/hr onto a choline-Sepharose column (2.5 × 7 cm, 50 μmol/ml, Fig. 1) equilibrated in buffer C. The column is washed at 80 ml/hr with 750 ml buffer C containing 50 mM KCl. After a wash with 75 ml buffer C, the column is connected to a DE-52 column (1.5 × 2.3 cm). Choline kinase is eluted from the affinity column onto the DE-52 column at 50 ml/hr with 100 ml buffer C containing 45 mM choline chloride. After another 75 ml wash with buffer C, the columns are disconnected. Enzyme is eluted from the DE-52 column at 25 ml/hr with a 50-ml linear salt gradient from 0 to 0.5 M KCl in buffer C. Fractions of 2 ml are collected.

Step 6: Strong Anion-Exchange Chromatography. The strong anion-

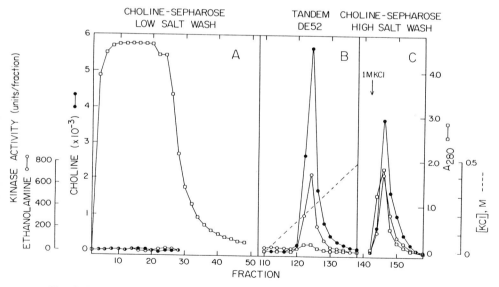

FIG. 1. Tandem choline-Sepharose/DE-52 chromatography. (A) Wash of choline-Sepharose with 50 mM KCl. (B) Elution of enzyme from the tanden DE-52 column. (C) High salt wash of choline-Sepharose. (From Ref. 1.)

exchange chromatography and all other steps are carried out at a flow rate of 1 ml/min at room temperature on a Rainin (Woburn, MA), gradient high-performance liquid chromatography (HPLC) system and monitored at 280 nm unless otherwise noted. All buffers are filtered through 0.22-μm nylon filters. Samples greater than 2 ml are loaded with a 50-ml superloop (Pharmacia-LKB).

The activity peak from the DE-52 column is pooled, concentrated to 2 ml with an Amicon (Danvers, MA) concentrator with a YM30 membrane, diluted 5-fold with buffer C, and loaded onto a Mono Q HR 5/5 column (Pharmacia-LKB, Fig. 2) equilibrated in buffer C. After a 10-min buffer C/0.1 M KCl wash, a 40-min linear gradient is run from 0.1 to 0.3 M KCl.

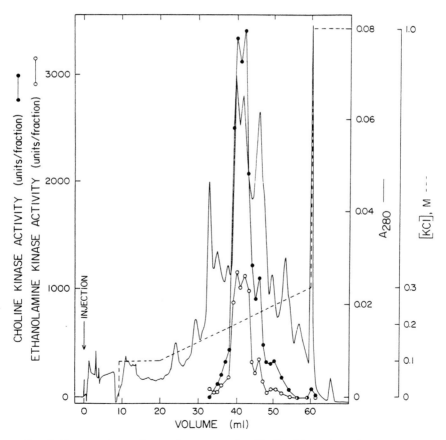

FIG. 2. Strong anion-exchange chromatography. Choline kinase from the DE-52 column was chromatographed on a Mono Q column as described in Step 6. (From Ref. 1.)

Fractions of 1 ml are collected. Two major peaks of activity are observed, with a minor peak that elutes at a slightly higher salt concentration. The two major peaks are pooled.

Step 7: Hydrophobic Interaction Chromatography. The pooled peaks of activity from the Mono Q column are made 0.5 *M* in ammonium sulfate with 3 *M* ammonium sulfate in buffer C without Tween 20. This is loaded onto a TSK gel phenyl-5PW column (TosoHaas, Philadelphia, PA, 7.5 × 0.75 cm, Fig. 3) equilibrated in 0.5 *M* ammonium sulfate in buffer C without Tween 20. The column is washed with a 15-ml step of 0.3 *M* ammonium sulfate in buffer C without Tween 20, and activity is eluted with a 25-min

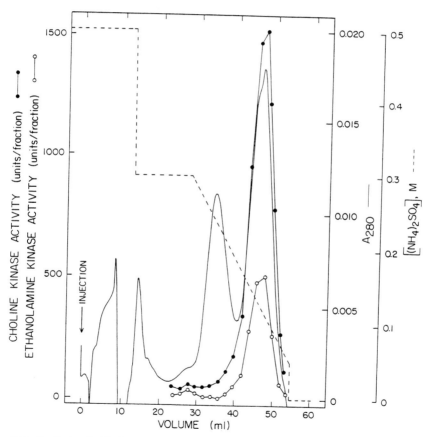

FIG. 3. Hydrophobic interaction chromatography. Choline kinase from the Mono Q column was chromatographed on a TSK gel phenyl-5PW hydrophobic interaction column as described in Step 7. (From Ref. 1.)

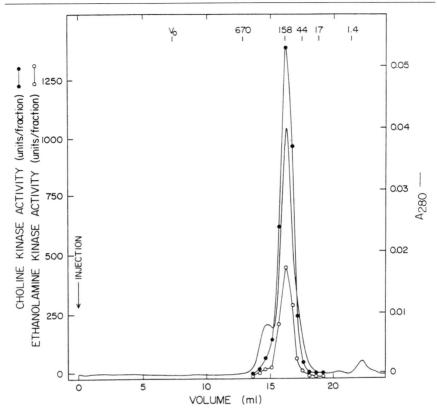

FIG. 4. Size exclusion. The activity peak from the hydrophobic interaction column was chromatographed on a Superose 6 column as described in Step 8. (From Ref. 1.)

linear gradient to 0.05 M ammonium sulfate in buffer C. Fractions of 2.0 ml are collected.

Step 8: Size Exclusion. The activity peak from the hydrophobic interaction column is concentrated to approximately 150 μl with Centricon 30 microconcentrators (Amicon) and loaded onto a Superose 6 HR 10/30 (Pharmacia-LKB) column equilibrated in buffer C with 0.1 M KCl (Fig. 4). Enzyme is eluted at a flow rate of 0.5 ml/min. Fractions of 1 ml are collected. The specific activity of the enzyme purified through the size-exclusion step is 143,000 nmol/min/mg, with a purification of 26,000-fold and an 8% yield (Table I).

Step 8a: Chromatofocusing. To obtain homogeneous enzyme, chromatofocusing is used as an alternative to size-exclusion chromatography (Fig.

TABLE I
PURIFICATION OF RAT LIVER CHOLINE KINASE[a]

Step	Protein (mg)[c]	Activity (nmol/min)	Specific activity (nmol/min/mg)	Purification (-fold)	Yield (%)	Activity ratio, (CK/EK)
Unperfused cytosol[b]	—	—	5.5	1.0	—	—
Perfused cytosol	8,000	74,700	9.6	1.7	100	3.8
pH 5 Supernatant	5,630	74,400	13.5	2.5	100	3.8
30–45% ammonium sulfate	978	55,000	58.8	10.7	73	4.3
Sephadex G-25	885	54,200	63.9	12.8	72	4.3
Choline-Sepharose	1.03	25,800	25,100	4,600	35	3.8
Mono Q	0.26	15,100	57,700	10,500	20	3.0
Phenyl-5PW	0.096	10,100	105,000	19,100	14	3.3
Superose 6	0.041	5,860	143,000	26,000	8	3.0

[a] From Ref. 1.

[b] Data are included for comparison.

[c] Total protein was determined by a modification of the method of Bradford with bovine serum albumin as a standard (G. L. Peterson, this series, Vol. 91, p. 95). When the samples contained Tween 20, all samples and standards were brought to equivalent concentrations of detergent.

5). The hydrophobic interaction peak is concentrated to about 200 μl with Centricon 30 microconcentrators. This is diluted 10-fold with buffer D and loaded onto a Mono P HR 5/20 column (Pharmacia-LKB) equilibrated in buffer D. After a 5-min wash with buffer D, the column is developed with buffer E. Fractions of 1 ml are collected. Two peaks of activity elute from the column, at pH 4.7 and 4.5. The isoforms eluting at pH 4.7 and 4.5 are referred to as CKI and CKII, respectively.

Properties

Physical Properties. The pH 4.5 peak from chromatofocusing is homogeneous for CKII. The subunit molecular mass of CKII, as estimated by sodium dodecyl sulfate (SDS) gel electrophoresis, is 47,000 Da (Fig. 6). Western blot analysis indicates that CKI has the same subunit mass.[1] The native molecular mass of both species, as estimated by high-performance size-exclusion chromatography, is 160,000 Da, which suggests a tetrameric structure. The brain enzyme, which we have found to have a subunit molecular mass identical to that of the liver enzyme,[1] has native mass of 90,000 Da by size exclusion and 87,600 Da by sucrose density gradient centrifugation, and it appears to be a dimer.[10] The native isoelectric points

[10] T. Uchida and S. Yamashita, *Biochim. Biophys. Acta* **1043**, 281 (1990).

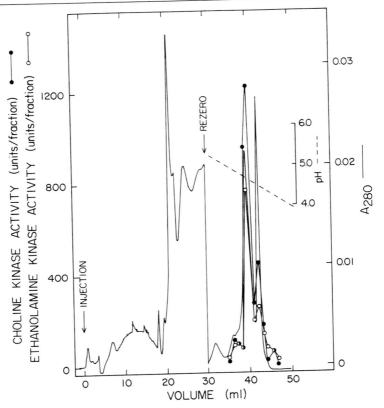

FIG. 5. Chromatofocusing. The activity peak from the hydrophobic interaction column was chromatographed on a Mono P column as described in Step 8a. (From Ref. 1.)

of CKI and CKII are 4.8 and 4.7, respectively, as determined by flat-bed isoelectric focusing.[11]

Kinetic Properties. The K_m and K_i values for choline and ethanolamine were determined with an enzyme preparation purified through the hydrophobic interaction step.[1] Choline and ethanolamine are mutually competitive inhibitors. The respective K_m values, 0.013 and 1.2 mM, are similar to the K_i values of 0.014 and 2 mM. These results suggest that binding of both substrates occurs at the same active site. Precise K_m values for Mg^{2+}–ATP were not obtained for the phosphorylation of choline, as the plots of $1/v$ versus $1/S$ were not linear over an ATP concentration range of 0.75 to 10 mM. The data can be fit to a

[11] T. J. Porter, Ph.D. Thesis, Purdue University, West Lafayette, Indiana (1989).

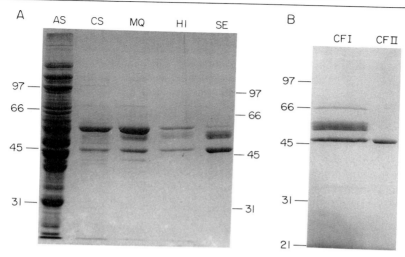

FIG. 6. Results of 10% SDS-polyacrylamide gel electrophoresis. (A) Coomassie blue staining pattern of the purification: ammonium sulfate cut (AS); choline-Sepharose (CS); Mono Q (MQ); hydrophobic interaction chromatography (HI); size exclusion (SE). (B) Peaks from chromatofocusing: peak eluting at pH 4.7 (CFI); peak eluting at pH 4.5 (CFII). Numbers refer to molecular weight markers. (From Ref. 1.)

Michaelis–Menten model that assumes two Mg^{2+}–ATP binding sites. The K_m values obtained at 0.04 and 8 mM at saturating choline concentrations. Multiple binding sites with distinct affinities for choline are not detected over the choline concentration range of 7.5 to 150 μM, as shown by the linear double-reciprocal plot. In the case of ethanolamine kinase activity, multiple ATP binding sites with different affinities are not observed within the Mg^{2+}–ATP concentration range of 0.125 to 3 mM, and the reciprocal plot of $1/v$ versus $1/S$ is linear and gives a K_m value of 0.7 mM for Mg^{2+}–ATP.

Notes on Procedure

This procedure takes about 3 days to complete, and the chromatofocusing step yields about 20 μg of homogeneous choline kinase (Fig. 6). The procedure is limited by the amount of material that can be conveniently processed in the ultracentrifuge. The ultracentrifugation step could possibly be replaced by a longer low-speed spin which would allow considerably more homogenate to be processed. Fresh tissue is always used for the purification. When frozen liver ($-80°$ for 1 month) was used, we noticed additional peaks of activity appearing during anion-exchange chromatography.

We have frequently used the same liver homogenates for preparation of both choline kinase and choline-phosphate cytidylyltransferase according to the procedure of Weinhold *et al.*[12] In such a case the steps up through acid precipitation are performed as prescribed for cytidylyltransferase purification.[12] The two enzymes are separated at the acid precipitation step, where choline kinase is soluble and cytidylyltransferase insoluble. The presence of lipids in the acid precipitation step has no effect on the purification of choline kinase.

For affinity chromatography we prepare our own epoxy-activated Sepharose because commercial preparations do not yield a sufficiently high ligand density, and choline kinase is not bound tightly enough to the resulting choline-Sepharose. The use of the DE-52 column after choline-Sepharose serves to concentrate the enzyme and enables the assay of the affinity step without the interference of unlabeled choline. Approximately 50% of the recovered activity remains on the choline-Sepharose column after the substrate elution. This activity can be eluted from choline-Sepharose with high salt (0.2 *M* or higher) and, after concentration and desalting, can be reloaded onto the affinity column. The same elution procedure (i.e., substrate, then high salt) again yields two peaks, one which elutes with substrate and one which elutes with high salt. Rechromatography of the material that elutes with choline chloride from the first choline-Sepharose column also results in two peaks of activity on a second choline-Sepharose column. This behavior may indicate an equilibrium between two modes of binding, ion exchange and an affinity interaction.

The specific activities of the isoforms separated by the chromatofocusing column (CKI and CKII) are only 70 and 20%, respectively, of the specific activity of the preparation that was loaded onto the column. This loss of activity is not prevented by elution into 1 *M* Tris-HCl (100 μl per fraction). When equal volumes of the activities are combined, the resulting mixture gives an average of the two activities, which suggests that the two activities are not interdependent. At all stages of purification, the only protein band that reacts specifically on a Western blot with antibody raised to CKII has a molecular mass of 47,000 Da.[1]

We purified rat brain enzyme, through the choline-Sepharose step, to a specific activity of 19,500 nmol/min/mg. Chromatofocusing of this material gives two peaks of activity that elute at approximately the same positions as the liver isoforms. This and immunological data[1] indicate that the isoforms from different tissues are quite similar, and that this purification protocol should be generally applicable to other choline kinases. A distinctly different protocol has also been used to purify an isozyme of rat brain choline kinase to homogeneity.[10]

[12] P. A. Weinhold, M. E. Rounsifer, and D. A. Feldman, *J. Biol. Chem.* **261**, 5104 (1986).

[16] Choline/Ethanolamine Kinase from Rat Brain

By Tsutomu Uchida and Satoshi Yamashita

Introduction

Choline (ethanolamine) + ATP → phosphorylcholine (phosphorylethanolamine) + ADP

Choline kinase or ATP : choline phosphotransferase (EC 2.7.1.32) catalyzes the phosphorylation of choline with ATP yielding phosphorylcholine and ADP.[1] This is the initial enzyme of the CDPcholine pathway[2] which serves as the principal pathway of phosphatidylcholine synthesis in animal tissues. Choline kinase has been highly purified from lung,[3] kidney,[4] liver,[5] and brain.[6] The brain is one of the richest sources of choline kinase. The enzyme of the soluble fraction from brain has been purified to homogeneity and shown to mediate the phosphorylation of choline, N,N-dimethylethanolamine, N-monomethylethanolamine, and ethanolamine.[6]

Assay Methods

Principle. Choline kinase activity is assayed by determining the rate of formation of either phosphorylcholine isotopically or ADP spectrophotometrically. In method I, [14C]choline is used as substrate, and the phosphorylcholine formed is separated from unreacted choline by solvent extraction and then counted. In method II, choline kinase is coupled with the pyruvate kinase–lactate dehydrogenase system [reactions (1) and (2)], so that the ADP formed can be determined spectrophotometrically by measuring the decrease in NADH.

$$ADP + phosphoenolpyruvate \rightarrow pyruvate + ATP \qquad (1)$$
$$Pyruvate + NADH + H^+ \rightarrow lactate + NAD^+ \qquad (2)$$

[1] J. Wittenberg and A. Kornberg, *J. Biol. Chem.* **202,** 431 (1953).
[2] E. P. Kennedy, *Fed. Proc.* **20,** 934 (1961).
[3] R. E. Ulane, L. L. Stephenson, and P. M. Farrell, *Biochim. Biophys. Acta* **531,** 295 (1978).
[4] K. Ishidata, K. Nakagomi, and Y. Nakazawa, *J. Biol. Chem.* **259,** 14706 (1984).
[5] T. J. Porter and C. Kent, *J. Biol. Chem.* **265,** 414 (1990).
[6] T. Uchida and S. Yamashita, *Biochim. Biophys. Acta* **1043,** 281 (1990).

Method I

Reagents. All aqueous solutions are made with water which has been deionized and then glass-distilled.

 Glycine–NaOH, 0.5 M, pH 9.0
 ATP, 0.1 M, pH 7
 $MgCl_2$, 0.2 M
 [*methyl*-^{14}C]Choline chloride, 20 mM, approximately 10^{12} disintegrations/min (dpm)/mol
 KCl, 1 M
 Bovine serum albumin, 10 mg/ml
 Tetraphenylboron, 30 mg/ml in butyronitrile

Procedure. The assay mixture contains 10 μmol glycine–NaOH, pH 9.0, 1 μmol ATP, 1.2 μmol $MgCl_2$, 15 μmol KCl, 0.2 μmol [*methyl*-^{14}C]choline chloride, and the enzyme in a total volume of 100 μl. The reaction is started by the addition of the enzyme and carried out at 30°. After 5 min the reaction is stopped by the addition of 250 μl of the tetraphenylboron solution, followed by vigorous mixing on a vortex mixer. The tube is centrifuged at 1000 g for 2 min, and the upper phase is aspirated off. The lower phase is washed 3 times with the tetraphenylboron solution to remove unreacted [^{14}C]choline, transferred to a counting vial, and then counted in the presence of a toluene/Triton X-100 scintillant.

Comments. In this method solvent extraction[7] is utilized for separation of the phosphorylcholine produced from unreacted choline. This makes the assay much easier to perform when compared to the previous assay using a Dowex I column.[8] When crude enzyme, for example, a 100,000 g supernatant, is used, the reaction proceeds linearly for only a short time. Thus a 5-min incubation is routinely used. When the assay is performed at low protein concentrations as in the assay of the purified enzyme, the mixture is supplemented with 0.5 mg/ml bovine serum albumin.

Method II

Reagents

 Glycine–NaOH, 0.5 M, pH 9.0
 ATP, 0.1 M, pH 7
 $MgCl_2$, 0.2 M
 Choline chloride, 20 mM
 KCl, 1 M
 Bovine serum albumin, 10 mg/ml

[7] A. M. Burt and S. A. Brody, *Anal. Biochem.* **65,** 215 (1975).
[8] H. Fukuyama and S. Yamashita, *FEBS Lett.* **71,** 33 (1976).

Pyruvate kinase (EC 2.7.1.40) (Boehringer-Mannheim, Germany), 100
    units/ml, dialyzed against 10 mM glycine–NaOH, pH 9.0
Lactate dehydrogenase (EC 1.1.1.27) (Boehringer-Mannheim), 400
    units/ml, dialyzed against 10 mM glycine–NaOH, pH 9.0
Phosphoenolpyruvate, 10 mM
NADH, 4 mM, prepared on the day of use

Procedure. The assay mixture contains 100 μmol glycine–NaOH, 2
μmol choline chloride, 10 μmol ATP, 12 μmol MgCl$_2$, 150 μmol KCl, 0.5
mg bovine serum albumin, 1 μmol phosphoenolpyruvate, 0.4 μmol
NADH, 7 units of pyruvate kinase, and 32 units of lactate dehydrogenase
in a total volume of 1.0 ml. The reaction is started by the addition of the
enzyme, and then the decrease in absorbance at 340 nm is followed at 30°
in a 2-ml quartz cuvette (1-cm light path). The control lacking choline is
run in the same manner. Choline kinase activity (in units) is calculated
according to the following formula:

$$\text{Activity} = [A_{340} \text{ (sample)} - A_{340} \text{ (control)}]/6.22t$$

where A_{340} is the absorbance at 340 nm and t is time in minutes.

Comments. Assay method II is seriously compromised if the enzyme
preparation is contaminated by other ADP-producing or NADH-oxidizing
activities, and hence this method is not suitable for assay of crude enzyme.
However, it is a convenient means of assay for column fractions from 1,6-
diaminohexane-agarose. This method is also suitable for the assay of the
ethanolamine, *N*-monomethylethanolamine, and *N,N*-dimethylethanol-
amine kinase activities of the purified enzyme because labeled substrates
are not required.

Definition of Unit. The formation of 1 μmol of phosphorylcholine or
ADP per minute at 30° is defined as 1 unit of enzyme. Specific activity is
expressed as units per milligram of protein.

Identification of Phosphorylcholine as a Reaction Product. The
method I mixture is incubated and applied onto a silica gel 60 plate (Merck,
Darmstadt, Germany) together with authentic phosphorylcholine and developed with 95% ethanol/2% ammonia (1 : 1, v/v). The product is located
by autoradiography.

Purification Procedure

All operations are performed at 2°–4°. A summary of the purification
is given in Table I.

Step 1: Preparation of Crude Extract. Wistar rat brains (130 g wet
weight) are cut into small pieces in 3 volumes of a solution containing 240
mM sucrose, 20 mM Tris-HCl, pH 7.5, 5 mM EDTA, 5 mM 2-mer-

TABLE I
PURIFICATION OF RAT BRAIN CHOLINE KINASE[a]

Fraction	Protein[b] (mg)	Total activity (units)	Specific activity (units/mg)	Yield (%)	Purification (-fold)
100,000 g supernatant	2616	7016	0.00268	100	1
Q-Sepharose eluate	531	6907	0.0130	98	4.9
$(NH_4)_2SO_4$ precipitate	319	6590	0.0206	94	7.7
AF-Blue-Toyopearl flow-through	204	4610	0.0226	66	8.4
1,6-Diaminohexane-agarose eluate	0.0053	214	40.4	3	15,000

[a] From T. Uchida and S. Yamashita, *Biochim. Biophys. Acta* **1043**, 281 (1990).
[b] Protein was determined by the method of M. M. Bradford, *Anal. Biochem.* **72**, 248 (1976), using bovine serum albumin as the standard.

captoethanol, and 0.2 mM phenylmethylsulfonyl fluoride using a Waring blender and then homogenized with a glass–Teflon homogenizer. The homogenate is centrifuged at 10,000 g for 15 min, and the resulting supernatant is further centrifuged at 100,000 g for 60 min. The 100,000 g supernatant is used as the crude extract.

Step 2: Q-Sepharose Column Chromatography. Solid NaCl is added to the crude extract to 75 mM. The mixture is applied onto a 1.6 × 21.5 cm column of Q-Sepharose (Pharmacia LKB, Uppsala, Sweden) equilibrated with buffer A (20 mM Tris-HCl, pH 7.5, 1 mM EDTA, 5 mM 2-mercaptoethanol, and 0.2 mM phenylmethylsulfonyl fluoride) containing 100 mM NaCl. The column is washed with 150 ml of buffer A containing 100 mM NaCl and then eluted with a linear gradient of NaCl in buffer A (0.1 to 0.5 M, 320 ml). The flow rate is adjusted to 150 ml/hr. Fractions containing the enzyme activity are pooled.

Step 3: Ammonium Sulfate Fractionation. To the pooled eluate (~100 ml) is added solid ammonium sulfate. The fraction precipitating between 25 and 50% saturation is collected by centrifugation and then dialyzed against four changes of 1 liter of buffer A.

Step 4: AF-Blue-Toyopearl 650ML Column Chromatography. The dialysate is centrifuged to remove insoluble materials and then loaded onto a 1 × 11.5 cm column of AF-Blue Toyopearl 650ML (Toso, Tokyo, Japan) which has been pretreated with 4 M urea and then equilibrated with buffer

A. The column is eluted with the same buffer at a flow rate of 60 ml/hr. Flow-through fractions containing the activity are collected.

Step 5: 1,6-Diaminohexane-Agarose Affinity Chromatography. The eluate is loaded onto a 1 × 7.5 cm 1,6-diaminohexane-agarose column equilibrated with buffer A. The column is sequentially washed with 5 ml of buffer A, 40 ml of a linear gradient of NaCl (0–500 mM) in buffer A, 200 ml of 500 mM NaCl in buffer A, and 20 ml of 100 mM NaCl in buffer A. Then the enzyme is eluted with buffer A containing 100 mM NaCl and 200 mM choline chloride at the flow rate of 20 ml/hr.

Comments. We usually use rat brains stored at $-60°$ as the enzyme source, but the above purification procedure can also be used for fresh brains. Most of the choline kinase activity of brain (88%) passes through the Blue-dye column. The retained activity (12%), probably representing a different isozyme, can be eluted from the column with a linear gradient of NaCl (0–1 M). Kidney and liver contain more of the Blue-dye-retained activity than brain.

1,6-Diaminohexane-agarose chromatography is the most efficient step of the purification. For reproducible results, 1,6-diaminohexane-agarose is prepared as follows. A suction-packed cake (4 g) of Sepharose 6B (Pharmacia LKB) is suspended in 5 ml of 2 M Na_2CO_3 and incubated with 75–150 μl of acetonitrile containing CNBr (1 g/ml) at 20° for 2 min, followed by successive washing with 100 ml of water and 100 ml of 0.1 M $NaHCO_3$, pH 9.0, at 0° on a sintered glass funnel. The activated gel is resuspended in 10 ml of 1 M 1,6-diaminohexane, pH 9.0, incubated at 25° overnight with gentle shaking, and then thoroughly washed with water. We recommend that the affinity adsorbent be prepared using different amounts of CNBr and then tested. The adsorbent used can be regenerated by washing with 5 M NaOH and reused at least 10 times.

Properties

Purity and Stability. The final preparation has a very high specific activity and gives a single protein band on electrophoresis in the presence of sodium dodecyl sulfate. The purified enzyme is very stable, even at a low protein concentration (1 μg/ml). The activity of the 1,6-diaminohexane-agarose eluate did not decrease significantly on storage at 4° for at least 1 month.

Molecular Properties. The M_r of the sodium dodecyl sulfate-denatured enzyme was determined to be 44,000 on electrophoresis. The M_r and Stokes radius of the native enzyme determined by Superose-12 column chromatography were estimated to be 90,000 and 4.2 nm, respectively. The sedimentation coefficient $s_{20,w}$ was 4.8 S. Assuming the partial specific

TABLE II
ACTIVITIES FOR VARIOUS CHOLINE-RELATED AMINO ALCOHOLS[a]

Compound	Apparent K_m (μM)	Relative activity[b]
Choline	15.1	1.00
N,N-Dimethylethanolamine	22.2	1.17
N-Monomethylethanolamine	118	0.81
Ethanolamine	787	0.83
N,N-Diethylethanolamine	25.0	2.66
N,N-Diisopropylethanolamine	400	0.97
N,N-Dibutylethanolamine	—[c]	0
N,N-Dimethylaminopropanol	208	0.69
N,N-Dimethylaminobutanol	—[c]	0
β-Methylcholine	444	0.79

[a] From T. Uchida and S. Yamashita, *Biochim. Biophys. Acta* **1043**, 281 (1990).

[b] The V_{max} value for choline was taken as 1.00.

[c] Not measurable.

volume of the enzyme to be 0.74, the M_r of the native enzyme was calculated to be 87,600. Hence, brain choline kinase enzyme is a dimer.

pH Dependence. Brain choline kinase is active in an alkaline pH range, but the enzyme does not exhibit a distinct pH optimum. The activity increases progressively when the pH of the reaction mixture is raised up to pH 10.5.

Substrate Specificity. The enzyme exhibits significant specificity to ATP. The second best nucleotide triphosphate substrate is CTP, giving 12% activity. dATP, ITP, GTP, and UTP are poor substrates. The enzyme is relatively specific for a group of amino alcohols related to choline. Table II summarizes the apparent K_m values and relative V_{max} values. Among the naturally occurring amino alcohols examined, choline and N,N-dimethylethanolamine are good substrates. Although the apparent K_m values are high, the enzyme exhibits considerable activity with N-monomethylethanolamine and ethanolamine.

Activators. Among various divalent cations examined, Mg^{2+} stimulates the enzyme most efficiently. Mn^{2+} is one-sixth as effective as Mg^{2+}. Co^{2+}, Cd^{2+}, Ca^{2+}, Zn^{2+}, and Ni^{2+} are ineffective. The Mg^{2+} concentration required varies with the ATP concentration used. Maximal activity is obtained when Mg^{2+} and ATP are present in nearly equal amounts. NH_4^+, Na^+, K^+, and Li^+ at concentrations of 0.1 to 0.2 M in the presence of Mg^{2+} stimulate the activity. NH_4^+ is the most effective, with 2-fold stimulation.

Polyamines are also known to activate the enzyme.[8] In the presence of a sufficient amount of Mg^{2+}, 3 mM spermine increase the activity nearly 4-fold. Spermidine at 3 mM stimulates the enzyme 3-fold. Putrescine is only slightly effective. Spermine and spermidine decrease the K_m for ATP and increase V_{max}.

Acetylcholine and its analogs also activate brain choline kinase.[9,10] At a concentration of 10 mM, acetylcholine chloride, carpronium chloride, and chlorocholine chloride stimulate the activity approximately 2-fold. Introduction of one negative charge completely abolishes the stimulatory effect (betaine and carnitine).

[9] D. R. Haubrich, *J. Neurochem.* **21**, 315 (1973).
[10] C.-P. Sung and R. M. Johnstone, *Biochem. J.* **105**, 497 (1967).

[17] Diacylglycerol Kinase from *Escherichia coli*

By James P. Walsh *and* Robert M. Bell

Introduction

Diacylglycerol kinases (EC 2.7.1.107) catalyze the ATP-dependent phosphorylation of *sn*-1,2-diacylglycerols. In *Escherichia coli*, diacylglyc-erol kinase (DG kinase) activity was first described by Pieringer and Kunnes.[1] The activity is tightly associated with the inner membrane fraction.[2] *Escherichia coli* DG kinase functions to recycle diacylglycerol (DAG), which is generated largely as a by-product of membrane-derived oligosac-charide biosynthesis.[3] Recent observations with mutants defective in the regulation of DG kinase expression suggest that DAG may play a regula-tory role in *E. coli* analogous to its role in eukaryotic signal transduction.[4] Diacylglycerol kinase from *E. coli* has found use in a radiochemical assay of DAG levels in crude lipid extracts of eukaryotic cells.[5,6]

The structural gene for *E. coli* DG kinase has been cloned and its DNA sequence determined.[7] Bacterial strains which express high levels of

[1] R. A. Pieringer and R. S. Kunnes, *J. Biol. Chem.* **240**, 2833 (1965).
[2] E. G. Schneider and E. P. Kennedy, *Biochim. Biophys. Acta* **441**, 201 (1976).
[3] H. Rotering and C. R. H. Raetz, *J. Biol. Chem.* **258**, 8068 (1983).
[4] J. P. Walsh, C. R. Loomis, and R. M. Bell, *J. Biol. Chem.* **261**, 11021 (1986).
[5] J. Preiss, C. R. Loomis, W. R. Bishop, R. Stein, J. E. Niedel, and R. M. Bell, *J. Biol. Chem.* **261**, 8597 (1986).
[6] J. E. Preiss, C. R. Loomis, R. M. Bell, and J. E. Niedel, this series, Vol. 141, p. 294.
[7] V. A. Lightner, R. M. Bell, and P. Modrich, *J. Biol. Chem.* **258**, 10856 (1983).

enzyme activity have been constructed and employed to purify the enzyme to homogeneity.[8] Sequence analysis of the purified kinase allowed alignment of the protein with the structural gene.[8] With 121 amino acids and a molecular weight of 13,114, *E. coli* DG kinase is remarkably small compared to other known kinases.[8,9] Primary sequence analyses suggest three membrane-spanning α helixes and an 18-amino acid amphipathic helix, consistent with the enzyme's being an integral membrane protein.[8] With 70% nonpolar amino acids, *E. coli* DG kinase is one of the most hydrophobic proteins known.[8] The enzyme can be quantitatively extracted from membranes into acidified organic solvent, from which the activity may be subsequently recovered.[8,10] The enzyme exhibits a broad specificity with regard to its lipid substrate. Diacylglycerols and monoacylglycerols ranging in acyl chain length from 4 to 18 carbons are readily phosphorylated.[11] Kinetic analyses indicate that in addition to MgATP and diacylglycerol, *E. coli* DG kinase requires a second divalent metal cation and a lipid cofactor.[12,13] Several lines of kinetic evidence suggest that the lipid cofactor induces a conformational change in the enzyme prior to its interaction with DAG, MgATP, and the divalent metal activator.[13]

Assay of *Escherichia coli* Diacylglycerol Kinase

Principle. The most convenient assay employs $[\gamma^{-32}P]$ATP as substrate. The $[^{32}P]$phosphatidic acid product is readily extracted into acidified organic solvent and quantitated by scintillation counting. Delivery of water-insoluble DAG and lipid activators to DG kinase is accomplished by means of mixed micelles.[14] Octylglucoside will readily cosolubilize enzyme, DAG, and lipid activator in a single homogeneous phase.[12] Under these assay conditions, activity is not limited by the rate of delivery of DAG to the enzyme.[12] Specific conditions for the radiochemical assay of total diacylglycerols in cell lipid extracts using *E. coli* DG kinase are detailed elsewhere.[5,6]

Materials: Disodium ATP is from P-L Biochemicals (Madison, WI). $[\gamma^{-32}P]$ATP is from New England Nuclear (Boston, MA) or Amersham (Arlington Heights, IL). Octyl-β-D-glucopyranoside (octylglucoside) is from Calbiochem (La Jolla, CA) or Sigma (St. Louis, MO) and is recrystal-

[8] C. R. Loomis, J. P. Walsh, and R. M. Bell, *J. Biol. Chem.* **260**, 4091 (1985).
[9] C. M. Anderson, F. H. Zucker, and T. A. Steitz, *Science* **204**, 375 (1979).
[10] E. Bohnenberger and H. Sandermann, *Eur. J. Biochem.* **94**, 401 (1979).
[11] J. P. Walsh, L. Fahrner, and R. M. Bell, *J. Biol. Chem.* **265**, 4374 (1990).
[12] J. P. Walsh and R. M. Bell, *J. Biol. Chem.* **261**, 6239 (1986).
[13] J. P. Walsh and R. M. Bell, *J. Biol. Chem.* **261**, 15062 (1986).
[14] D. Lichtenberg, R. J. Robson, and E. A. Dennis, *Biochim. Biophys. Acta* **737**, 285 (1983).

lized as described below. Imidazole base (thrice recrystallized, fluorescent grade), ethylenediaminetetraacetic acid (EDTA), [ethylenebis(oxyethylenenitrilo)]tetraacetic acid (EGTA), diethylenetriaminepentaacetic acid (DTPA), and cholesteryl acetate are from Sigma. *Escherichia coli* cardiolipin, dioleoylphosphatidylglycerol, and phosphatidylcholines are from Avanti Polar Lipids (Birmingham, AL). *Bacillus cereus* phospholipase C (2500 units/mg protein) is from Sigma. The assay and organic extractions are performed in 13 × 100 mm borosilicate glass test tubes with Teflon-lined screw caps. We have noted some batches of commercial test tubes to inhibit DG kinase activity up to 99%.[11] This inhibition can be prevented by washing the tubes with 2 N HNO$_3$ prior to the assay.

Stock Solutions

[γ-^{32}P]ATP: Commercial [γ-^{32}P]ATP is added to a solution of 50 mM Na$_2$ATP, 100 mM imidazole base, 0.1 mM DTPA. Small aliquots (150 μl) are dispensed into 1.5-ml conical polypropylene centrifuge tubes, rapidly frozen on acetone/dry ice, and stored at −20°. The concentration of the stock [γ-^{32}P]ATP solution is verified by measuring the absorbance of appropriate dilutions (in 50 mM NaH$_2$PO$_4$, 1 mM EDTA, pH 7.0) at 259 nm and using an extinction coefficient for ATP of 15.4 cm^{-1} mM^{-1}.[15] The specific radioactivity of the [γ-^{32}P]ATP is determined by thin-layer chromatography (TLC) on polyethyleneimine-cellulose plates (J. T. Baker, Phillipsburg, NJ) developed in aqueous 1 M formic acid, 1 M LiCl.[16] The radiochemical purity of the stock ATP, prepared and stored in this manner, does not change over periods of up to 4 weeks.

Diacylglycerol: *sn*-1,2-Dioleoylglycerol is prepared by digestion of dioleoylphosphatidylcholine with *Bacillus cereus* phospholipase C.[17] Dioleoylphosphatidylcholine, 200 mg in chloroform, is evaporated under a stream of N$_2$ and the lipid residue dissolved in 5 ml of diethyl ether. To this is added 2 ml of 50 mM potassium phosphate, pH 7.0, and 50 units of *B. cereus* phospholipase C. The emulsion is kept at room temperature and vortexed intermittently. The pH of the aqueous phase of the reaction is followed with pH paper and maintained at approximately 7.0 by addition of aqueous 2.0 M Tris base. The reaction is complete when an emulsion no longer forms, usually within 2 hr. The ether is then evaporated under a stream of N$_2$, leaving a drop of dioleoylglycerol floating on

---

[15] R. M. Bock, N.-S. Ling, S. A. Morell, and S. H. Lipton, *Arch. Biochem. Biophys.* **62**, 253 (1956).

[16] Randerath K. and E. Randerath, this series, Vol. 12, p. 323.

[17] B. R. Ganong and R. M. Bell, *Biochemistry* **23**, 4977 (1984).

the aqueous buffer. This mixture is extracted 3 times with 3 ml of ethanol-free chloroform, and the chloroform extracts are passed over a 1.0-ml bed of silicic acid in a Pasteur pipette. The chloroform solution is stored under argon at $-20°$ in a glass test tube with a Teflon-lined screw cap. The threads of the test tube are wrapped with Teflon tape to prevent evaporation. The concentration of the stock dioleoylglycerol solution is determined by the ester assay method of Stern and Shapiro using cholesteryl acetate as a standard.[18] The purity of diacylglycerols is readily checked by TLC on 0.25-mm silica gel H plates (E. Merck, Darmstadt, Germany) developed in heptane–diethyl ether–acetic acid (25 : 75 : 1, v/v).[12] Only traces (<1%) of 1,3-dioleoylglycerol are detectible and acyl migration is not observed on storage for periods of up to 2 months.[12] Other diacylglycerols (dioctanoylglycerol, dibutyrylglycerol) can be prepared in an identical fashion starting with appropriate phosphatidylcholines.[11,12]

Octylglucoside/cardiolipin: *Escherichia coli* cardiolipin, 50 mg in chloroform, is evaporated under N_2 and the last traces of chloroform removed under reduced pressure. Octylglucoside, 505 mg (recrystallized as described below), 6.6 ml of water, and 6.8 μl of a 100 mM stock solution of DTPA (pH 7.0 in water) are then added. The cardiolipin slowly dissolves with intermittent vortexing. The resulting solution is 255 mM in octylglucoside, 5.0 mM in cardiolipin, and 0.1 mM in DTPA. It is stored under argon at $-20°$ and is stable over numerous freeze–thaw cycles. Other lipids may be substituted for the cardiolipin.[12,13] Commercially purchased octylglucoside is recrystallized from acetone–diethyl ether.[19] Octylglucoside, 12.5 g, is dissolved by gentle warming in 50 ml of acetone and the solution filtered. Diethyl ether, 250 ml, is then added and the solution kept at $-20°$ for several days. The octylglucoside is collected by filtration and washed with a small volume of ice-cold diethyl ether. The crystals are dried under reduced pressure and stored desiccated at $-20°$.

$2\times$ DG kinase assay buffer stock solution: 100 mM imidazole, 100 mM LiCl, 25 mM $MgCl_2$, 2 mM EGTA, pH 6.6; this solution is stable for several months at room temperature

Dithiothreitol (DTT) solution, prepared fresh daily as a 100 mM solution in 1.0 mM DTPA, pH 7.0, and kept on ice

Enzyme: For routine kinetic assays, membranes of *E. coli* strain

---

[18] I. Stern and B. Shapiro, *J. Clin. Pathol.* **6,** 158 (1953).
[19] W. J. de Grip and P. H. M. Bovee-Geurts, *Chem. Phys. Lipids* **23,** 321 (1979).

N4830/pJW10 are used.[8,11-13] These cells express high levels of DG kinase under the control of an inducible promoter.[8] Specific conditions for growth of bacteria, induction of DG kinase synthesis, and preparation of membranes are detailed elsewhere.[4,8] The membranes are stored at $-20°$ in 20% (w/v) glycerol, 25 mM NaH$_2$PO$_4$, 5 mM 2-mercaptoethanol, 1.0 mM DTPA, pH 7.0. They are diluted prior to use in 10 mM imidazole, 0.1 mM DTPA, pH 6.6. The DG kinase activity of membranes stored in this way is stable over numerous freeze–thaw cycles.

Assay Procedures. The assay[11-13] contains, in a total volume of 100 μl, 51 mM (1.5 g/dl) octylglucoside, 1.0 mM (3.5 mol %) cardiolipin, 60 mM imidazole, pH 6.6, 50 mM LiCl, 12.5 mM MgCl$_2$, 1.0 mM EGTA, 50 μM DTPA, 2.0 mM DTT, 1.0 mM (3.5 mol %) dioleolyglycerol, and 5.0 mM Na$_2$ATP. Calculation of lipid mole fractions is described below (Surface Dilution Kinetics). For a standard assay, an appropriate volume of the stock diacylglycerol solution is transferred to a glass test tube and the chloroform evaporated under a stream of dry N$_2$. The DAG is then solubilized by adding 20 μl of the octylglucoside/cardiolipin solution, 50 μl of 2× assay buffer, and 2 μl of the DTT solution. When the DAG has dissolved, *E. coli* membranes and water are added to bring the total volume to 90 μl. Marked inactivation of DG kinase occurs if the DAG is not completely dissolved prior to enzyme addition. From 0.01 to 10 μg of membrane protein is used, depending on the source.[4,12] The mixture is allowed to sit for a few minutes at 25° to allow the enzyme to solubilize, and the reaction is initiated by adding 10 μl of the stock ATP solution. After 10 min at 25°, the reaction is terminated by addition of 0.7 ml of 1% (w/v) HClO$_4$ and 3 ml of methanol–chloroform (2:1, v/v). Phases are then broken by addition of 1.0 ml of 1% HClO$_4$ and 1.0 ml of chloroform and separated by brief centrifugation at 5000 g for 2 min. The upper aqueous phase is discarded and the lower chloroform phase washed twice with 2.0 ml of 1% aqueous HClO$_4$–methanol (7:1, v/v). Methanol is included in the wash solution to prevent the formation of interfacial emulsions, in which the product could be lost.[13] The volume of the remaining chloroform phase is 1.80 ml.[13] A 1.0-ml aliquot of the chloroform phase is pipetted into a scintillation vial and the radioactivity conveniently determined by direct Cerenkov counting. The chloroform-soluble, radioactive product consists exclusively of phosphatidic acid.[12] Other diacylglycerols (dioctanoylglycerol, palmitoyloleoylglycerol) can be substituted for the di-C$_{18:1}$. However, fully saturated long-chain diacylglycerols (such as dipalmitoylglycerol) are poorly solubilized by the octylglucoside and give low activities in this system. Diacylglycerols do not undergo detectable acyl migration under these assay conditions.[12]

Diacylglycerols cause rapid inactivation of DG kinase.[12] Lipid activators and MgATP protect the enzyme from this effect.[12,13] If the assay is to be performed with a weak lipid activator, an alternative procedure is used to prevent inactivation which would occur during the preincubation period.[13] The final assay conditions are identical to those above except that the volume is 0.2 ml. The enzyme is solubilized in the reaction mixture as described above, omitting the DAG and bringing the final volume to 100 μl with water. The reaction is initiated by addition of 100 μl of reaction mixture containing 2 mM dioleoylglycerol, completely solubilized in the octylglucoside/lipid activator, and 10 mM [γ-^{32}P]ATP. After 10 min at 25° the reaction is terminated by addition of 0.6 ml of 1% HClO$_4$ and 3.0 ml of methanol–chloroform (2 : 1, v/v) and the PA extracted into chloroform as described above.

With short-chain diacylglycerols, the phosphatidic acid product cannot be extracted into organic solvent.[11] With dihexanoylglycerol, 20% of the product is lost in the aqueous washings. With dibutyrylglycerol, the product cannot be extracted into chloroform, and an alternative procedure is used.[11] Dibutyrylglycerol kinase activity is assayed as described above, but using unlabeled ATP and sn-1,2-[^{3}H]dibutyrylglycerol. The reaction volume is scaled up to 250 μl. Immediately after initiating the reaction, 100 μl of the mixture is loaded onto a 1.0-ml column of AG 1-X8 (Bio-Rad, Richmond, CA) for background counts and washed onto the column with 0.5 ml of 2-propanol–water (1 : 1, v/v). After 10 min another 100 μl is loaded onto a second AG 1-X8 column in an identical fashion. The columns are washed with 4 ml of 2-propanol–water (1 : 1, v/v) and then eluted into scintillation vials with 4 ml of 2-propanol–2 N HCl (1 : 1, v : v). Quenching of the ^{3}H counts by HCl is corrected by means of an internal standard. [^{3}H]Dibutyrylglycerol is prepared from commercial [^{3}H]glycerol by phosphorylation with glycerol kinase, acylation with butyric anhydride, and dephosphorylation with calf intestinal alkaline phosphatase, as detailed elsewhere.[20] Radiochemical purity of the [^{3}H]dibutyrylglycerol is ascertained by TLC performed as described for dioleoylglycerol above.

Dibutyrylglycerol is soluble in water up to 40 mM and can be used to assay DG kinase in native $E. coli$ membranes as well as in mixed micelles.[11] This can be done following the procedure described above, but omitting the octylglucoside and lipid activator.[11] In the native membranes the lipid cofactor requirement is apparently fulfilled by the endogenous lipids.[11] As seen in Fig. 1, the activity of native membrane DG kinase closely parallels that observed in mixed micelles. The small differences seen are attribut-

---

[20] W. R. Bishop and R. M. Bell, *J. Biol. Chem.* **261,** 12513 (1986).

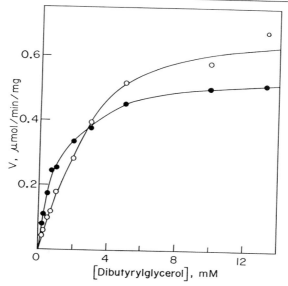

FIG. 1. Dependence of diacylglycerol kinase reaction velocity on dibutyrylglycerol. Activities were determined in native membranes (●) and mixed micelles (○) at the concentrations of *sn*-1,2-dibutyrylglycerol indicated. (From Walsh *et al.*[11])

able to surface dilution of dibutyrylglycerol by the octylglucoside and the presence of latent DG kinase activity in the unsolubilized membranes.[11]

Properties of Diacylglycerol Kinase

Surface Dilution Kinetics. If excess octylglucoside is added to the DG kinase assay, the activity will be decreased unless additional DAG and lipid activator are added to maintain a constant mixed micelle composition (Fig. 2). This dependence of activity on micelle composition and not on bulk solution concentrations has been termed "surface dilution kinetics."[21] For this reason, the DAG and lipid activator concentrations in are expressed as mole fractions of the micellar pseudophase of the reaction mixture. In calculating these mole fractions the octylglucoside concentration must be corrected by subtracting its aqueous monomer concentration, which is assumed to always be equal to 25 m*M*, the critical micelle concen-

---

[21] E. A. Dennis, *in* "The Enzymes" (P. D. Boyer, ed.), 3rd Ed., Vol. 16, p. 307. Academic Press, New York, 1983.

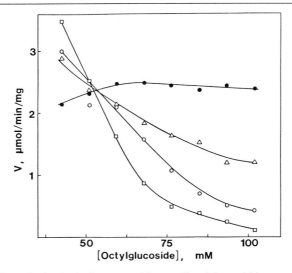

F<small>IG</small>. 2. Effect of mixed micelle composition on diacylglycerol kinase activity. The total concentration of octylglucoside in the reaction mixture was varied as shown, and the concentrations of dioleoylglycerol and cardiolipin were either varied to maintain constant mole fractions in the micellar pseudophase of 0.37 and 3.7 mol %, respectively, or maintained at constant bulk solution concentrations of 0.1 and 1.0 mM, respectively. ●, Constant mole fractions of dioleoylglycerol and cardiolipin; △, constant mole fraction of cardiolipin and constant bulk dioleoylglycerol; ○, constant mole fraction of dioleoylglycerol and constant bulk cardiolipin; □, constant bulk concentrations of dioleoylglycerol and cardiolipin. (From Walsh and Bell.[12])

tration of octylglucoside.[22,23] The DAG and lipid activator are assumed to be entirely in the micelles.

Divalent Metal Interactions. Diacylglycerol kinase requires a free divalent metal cation in addition to MgATP.[12] Several metals can fulfill this requirement, including Mg^{2+}, Mn^{2+}, Co^{2+}, Cd^{2+}, and Zn^{2+}.[12] Of these, Mg^{2+} gives the highest activity. Free Mg^{2+} activates DG kinase with an apparent K_a of 3.4 mM.[12] Other divalent cations, especially Ca^{2+}, will strongly inhibit DG kinase activity.[12] It is for this reason that EGTA is included in the assay. When performing DG kinase assays with divalent metal cations other than Mg^{2+}, other chelators must be substituted for the EGTA and DTPA.[12] Free divalent metal ion concentrations can be calculated from the apparent dissociation constants of the ATP and chela-

---

[22] M. L. Jackson, D. Schmidt, D. Lichtenberg, B. J. Litman, and A. D. Albert, *Biochemistry* **21**, 4576 (1982).

[23] A. Helenius, D. R. McCaslin, E. Fries, and C. Tanford, this series, Vol. 56, p. 734.

tor complexes of the divalent metal cations used.[12] The apparent dissociation constants for the Mg^{2+} complexes occurring in this assay system, calculated for pH 6.6, are as follows: MgATP, 158 μM; MgDTPA, 100 μM; MgEGTA, 44 mM.[12]

Diacylglycerol Substrate Specificity. The DAG substrate specificity of DG kinase has been evaluated in detail.[11] Under the standard assay conditions the apparent K_m for dioleoylglycerol is 0.92 mol %.[12] A large number of diacylglycerols, monoacylglycerols, alkylglycerols, and DAG analogs are readily phosphorylated by the enzyme.[2,11,24] The degree of saturation of the acyl chains does not seem to influence activity.[25] In a large series of compounds examined, the most critical determinants for activity were the *sn*-3-hydroxyl and *sn*-1 ester moieties.[11] *sn*-2,3-Diacylglycerol is phosphorylated at 0.1% the rate of the *sn*-1,2 isomer, and *sn*-1,3-diacylglycerols are not substrates.[26] Ceramide is readily phosphorylated by DG kinase.[27] Phosphorylation of a contaminant present in Triton X-100 is also observed.[2,11] When using DG kinase to assay DAG in cell extracts, the phosphatidic acid product must be separated from contaminating lysophosphatidic acid and ceramide phosphate by TLC.[5,6] As noted above, diacylglycerols with two long, saturated acyl chains have limited solubility in octylglucoside. These compounds are readily phosphorylated by DG kinase in a Triton X-100-based assay performed at 40°.[10] Solubility of diacylglycerols is not an impediment to using the octylglucoside-based assay to measure DAG levels in crude cell extracts. The trace amounts of fully saturated diacylglycerols present under these conditions are readily solubilized and phosphorylated.

Thermal Stability. In native membranes, DG kinase exhibits remarkable heat stability.[1] In one study, the DG kinase activity of membranes heated at 100° decayed with a $t_{1/2}$ of 20 min.[28] Solubilization by detergents results in loss of this heat stability. A Triton X-100-solubilized preparation lost activity with a $t_{1/2}$ of 12 min at 57°.[2] In octylglucoside at 25° the $t_{1/2}$ is 230 min.[12] Addition of phospholipids to solubilized DG kinase provides some protection from thermal inactivation.[2,12] Lowering the pH is also protective.[12] As discussed below, addition of DAG markedly accelerates the loss of DG kinase activity.

[24] D. A. Ford and R. A. Gross, *J. Biol. Chem.* **265**, 12280 (1990).

[25] M. L. MacDonald, K. F. Mack, B. L. Williams, W. C. King, and J. A. Glomset, *J. Biol. Chem.* **263**, 1584 (1988).

[26] R. A. Coleman, J. P. Walsh, D. S. Millington, and D. A. Maltby, *J. Lipid Res.* **27**, 158 (1986).

[27] E. G. Schneider and E. P. Kennedy, *J. Biol. Chem.* **248**, 3739 (1973).

[28] C. R. H. Raetz, G. D. Kantor, M. Nishijima, and M. L. Jones, *J. Biol. Chem.* **256**, 2109 (1981).

Inactivation by Diacylglycerols. When cosolubilized with DAG in octylglucoside micelles, DG kinase undergoes rapid irreversible inactivation.[12] This inactivation exhibits first-order kinetics with respect to time
and DAG concentration.[12] Lipid cofactors and MgATP protect the enzyme
from this effect.[12,13] For this reason, care must be taken to avoid prolonged
exposure of DG kinase to diacylglycerols after solubilization in octylglucoside. If the lipid activator to be used is a weak one, it is best to add the
DAG already solubilized in octylglucoside at the time the reaction is
initiated.[13] At concentrations greater than 3.4 mol %, diacylglycerols cause
a substrate inhibition of DG kinase which is distinct from this irreversible
inactivation.[11,12]

Lipid Cofactor Requirement. Detailed structure–function and kinetic
analyses of the DG kinase lipid cofactor requirement have been published.[13] The best activators are anionic lipids with long (C_{16}–C_{18}) acyl
chains.[13] However, compounds as diverse as cholesterol 3-sulfate, Triton
X-100, sodium dodecyl sulfate, and hexadecylphosphorylcholine can all
function as activators.[13] The activation is highly cooperative.[13] For routine
assays, 3.5 mol % (1.0 mM) cardiolipin is satisfactory[12]; 15.6 mol % (5.0
mM) phosphatidylglycerol is equally effective.[11,12] The ability of lipid
activators to protect DG kinase from DAG-induced inactivation is proportional to their ability to stimulate enzyme activity.[13] The degree of saturation of the acyl chains of lipid activators does not have much effect.
However, lipids in which the acyl chains have been oxidized lose their
ability to function as activators.

[18] Diacylglycerol Kinase Isozymes from Brain and Lymphoid Tissues

By Hideo Kanoh, Fumio Sakane, and Keiko Yamada

Introduction

$$sn\text{-}1,2\text{-Diacylglycerol} + ATP \rightarrow \text{phosphatidate} + ADP$$
$$sn\text{-}2\text{-Monoacylglycerol} + ATP \rightarrow \text{lysophosphatidate} + ADP$$

Diacylglycerol kinase (DGK, EC 2.7.1.107) converts diacylglycerol
(DG) to phosphatidate (PA). The same enzyme also catalyzes the phosphorylation of *sn*-2 (but not *sn*-1) acyl monoacylglycerols.[1] Because the

---

[1] H. Kanoh, T. Iwata, T. Ono, and T. Suzuki, *J. Biol. Chem.* **261**, 5597 (1986).

reaction of DGK reverses the normal flow of glycerolipid biosynthesis from PA to DG, the function of this enzyme in resting cells remains unknown. However, much attention has been focused on this enzyme since DG was established as a second messenger in signal transduction mediated by protein kinase C.[2] The enzyme may regulate the intracellular concentration of DG as has been proposed from the effect of the enzyme inhibitor R59022.[3] In view of the biological activity of PA or lyso-PA so far described for various types of cells, DGK quite likely serves as a critical modulator of the signal transduction process.[4]

A DGK with an apparent molecular weight of 80,000 was the first to be successfully purified from the cytosol of porcine brain.[5] This isozyme was later found to be more highly abundant in porcine and human lymphocytes rather than in brain.[6] The cDNA coding for the porcine thymus 80K DGK has been cloned,[7] and the existence of a new enzyme family having a remarkable structural feature has been disclosed, as will be discussed later. Immunological and enzyme purification studies showed that there are multiple DGK isozymes. The pattern of distribution of these isozymes is variable depending on the tissue and cell types,[6,8] although they can coexist in a single cell type such as human platelets[9] and murine fibroblasts.[10] The physiological significance of the multiple forms of DGK is an interesting subject of further study. The markedly different enzymological properties described for some of the DGK isozymes[4,8–12] suggest that they are operating under distinct regulatory mechanisms. In this chapter we discuss the properties of 80K DGK, the isozyme best characterized so far, and another 150K DGK, both of which have been purified from porcine thymus cytosol.

[2] Y. Nishizuka, *Nature (London)* **308**, 693 (1984).

[3] D. de Chaffoy de Courcelles, P. Roevens, and H. van Belle, *J. Biol. Chem.* **260**, 15762 (1985).

[4] H. Kanoh, K. Yamada, and F. Sakane, *Trends Biochem. Sci.* **15**, 47 (1990).

[5] H. Kanoh, H. Kondoh, and T. Ono, *J. Biol. Chem.* **258**, 1767 (1983).

[6] K. Yamada, F. Sakane, and H. Kanoh, *FEBS Lett.* **244**, 402 (1989).

[7] F. Sakane, K. Yamada, H. Kanoh, C. Yokoyama, and T. Tanabe, *Nature (London)* **344**, 345 (1990).

[8] K. Yamada and H. Kanoh, *Biochem. J.* **255**, 601 (1988).

[9] Y. Yada, T. Ozeki, H. Kanoh, and Y. Nozawa, *J. Biol. Chem.* **265**, 19237 (1990).

[10] M. L. MacDonald, K. F. Mack, B. W. Williams, W. C. King, and J. A. Glomset, *J. Biol. Chem.* **263**, 1584 (1988).

[11] F. Sakane, K. Yamada, and H. Kanoh, *FEBS Lett.* **255**, 409 (1989).

[12] R. N. Lemaitre, W. C. King, M. L. MacDonald, and J. A. Glomset, *Biochem. J.* **266**, 291 (1990).

Assay Method

To introduce DG, a water-insoluble substrate, into the reaction mixture, several detergents have been employed, among which deoxycholate and β-octylglucoside have been most often used. In this chapter, only the deoxycholate suspension assay will be described. (For details of the octylglucoside mixed micellar assay, see *E. coli* DGK[13] and arachidonoyl-specific DGK.[14]) It should be noted that the activities of DGK isozymes included in a crude enzyme preparation are affected differently by the detergents employed. The 80K DGK gives the highest activity with deoxycholate but is suppressed by octylglucoside.[11] On the other hand, arachidonoyl-specific DGK could be detected by the octylglucoside micellar assay but not with deoxycholate.[12] An assay method optimally applicable to all DGK isozymes has not yet been developed. Therefore, the relative activities of multiple DGK isozymes in a particular enzyme preparation may be variable depending on the assay conditions.

Reagents

5× Stock solution: 250 mM Tris-HCl (pH 7.4), 500 mM NaCl, 100 mM NaF, 5 mM Dithiothreitol (DTT), 5 mM EDTA, 50 mM MgCl$_2$ (store aliquoted at $-20°$)

DG (1,2-diolein from Sigma, St. Louis, MO): 10 mM in hexane (stored at $-20°$)

Deoxycholate: 10 mM (stored at room temperature)

[γ-^{32}P]ATP [10,000–50,000 counts/min (cpm)/nmol, 10 mM, stored at $-20°$]

n-butanol

Water saturated with *n*-butanol

Concentrated HCl

Procedure. The reaction mixture (final volume 100 μl) contains 50 mM Tris-HCl (pH 7.4), 100 mM NaCl, 20 mM NaF, 1 mM DTT, 1 mM EDTA, 1 mM deoxycholate, 10 mM MgCl$_2$, 0.5 mM DG, 1 mM [^{32}P]ATP, and enzyme. To prepare the DG suspension, the calculated amount of DG in hexane is dried under N$_2$ in a plastic tube, and the stock solution and deoxycholate are added. The tube is sonicated briefly (3 times for 30 sec using a Branson sonifier) in a bath-type sonicator at a maximum intensity until the turbidity becomes visibly constant. The suspension is pipetted into a round-bottomed glass tube (1.3 × 10 cm), [^{32}P]ATP is added, and the mixture is warmed at 30°. The reaction is initiated by adding enzyme (usually 10 μl) and continued for 2–3 min at 30°.

The PA formed can be analyzed by a conventional lipid extraction

[13] J. P. Walsh and R. M. Bell, this volume [17].
[14] R. N. Lemaitre and J. A. Glomset, this volume [19].

technique combined with thin-layer chromatography if necessary. Here we describe a rapid butanol extraction method. In this case, the incubations are terminated by the addition of 50 μl concentrated HCl followed by 1.5 ml of water. After vortex mixing, the lipids are extracted by vigorous mixing with 1.0 ml n-butanol. The tubes are then centrifuged at 1000 g for 5 min at room temperature, and 0.85 ml of the resulting upper phase is transferred to a tube containing 1.0 ml water saturated with butanol. The tubes are again mixed well, and, after centrifugation, 0.5 ml of the washed butanol phase is transferred to scintillation vials. The radioactivity can be measured directly using an aqueous scintillation mixture like toluene–Triton X-100 scintillator. After background subtraction the radioactivity measured represents one-half of the formed PA.

Throughout these procedures, adequate mixing at each step is critical to obtain reproducible data and low background radioactivity from the enzyme-free incubations. Further purification of PA by thin-layer chromatography is usually not necessary. When required, lipids extracted with butanol or other solvents are dried under N_2 with 5 μg of carrier PA (Sigma) and applied to silica gel 60 plates (Merck, Darmstadt, Germany). As the developing solvent we recommend the use of the upper phase of ethyl acetate–2,2,4-trimethylpentane–acetic acid–water (90 : 50 : 20 : 1, v/v)[10] mixed in the order mentioned. PA spots are detected by iodine vapor and scraped into scintillation vials, and 0.5 ml of methanol–water (1 : 1, v/v) is added to deactivate the silica gel. The radioactivity is measured using an aqueous scintillation mixture. One unit of enzyme activity is defined as 1 μmol PA formed per minute.

Comments. DG should be stored under anhydrous alcohol-free conditions to minimize acyl migration. Under the described storage conditions, formation of *sn*-1,3 isomer is negligible for over 1 month. The precise concentration of DG can be measured, when required, as free glycerol after saponification of the lipid extract of the reaction mixture.[15] Because the concentration of deoxycholate used is below its critical micellar concentration (4–6 mM), the physical form of the dispersed DG has not been defined. The deoxycholate assay does not require the addition of lipid activators so far described for this enzyme.

Purification of Diacylglycerol Kinase Isozymes
from Porcine Thymus Cytosol

The purification method is based on our finding that porcine thymus is highly enriched with the 80K DGK.[8] Further, we noted the presence in thymus cytosol of a minor DGK species with an apparent M_r of 150,000.

[15] E. Van Handel and D. B. Zilversmit, *J. Lab. Clin. Med.* **50,** 152 (1957).

In contrast to the 80K isozyme, the 150K DGK is stable when heat-treated at 41° for 5 min, and it does not react with anti-porcine brain 80K DGK antibodies. Because the two isozymes can be separated at the initial step of purification,[11] we modified the previous method for purifying porcine brain 80K DGK[5] to allow the purification of two DGKs from the same material.

Separation of Two Diacylglycerol Kinases. Fresh porcine thymus is collected on dry ice at a local slaughterhouse. The connective tissues and blood vessels are carefully removed, and the tissue (400 g) is homogenized in a Waring blender with 1.3 liters of buffer A (25 mM Tris-HCl, pH 7.4, 0.25 M sucrose, 1 mM EDTA, 0.5 mM DTT, 10 μM ATP, 10 μg/ml leupeptin) containing 50 mM NaCl. The cytosol obtained by ultracentrifugation is divided into two equal portions, and each portion is applied to a column (5 × 13 cm) of DE-52 (Whatman, Clifton, NJ) equilibrated with the same buffer. After washing the columns with the same buffer (2 liters each), the heat-labile 80K DGK is eluted with buffer A containing 0.1 M NaCl. After elution of the heat-labile activity, the heat-stable 150K DGK is eluted with buffer A containing 0.2 M NaCl.

Purification of 80K Diacylglycerol Kinase. Saturated ammonium sulfate is added to the combined 0.1 M NaCl eluates (usually 400 ml) from the DE-52 columns, and the fraction precipitated between 40 and 60% saturation is collected by centrifugation (10,000 rpm for 20 min). The precipitate is suspended in a small volume of buffer A/50 mM NaCl (final volume 15–20 ml), and is centrifuged to remove insoluble material. The enzyme is then subjected to gel filtration through a column (2.6 × 80 cm) of Sephacryl S-300 (Pharmacia, Piscataway, NJ) equilibrated with the same buffer. The active fractions corresponding to a molecular weight of 80,000 are combined (40–60 ml) and applied to a column (1.5 × 13 cm) of ATP-agarose (Sigma, A9264) equilibrated with the same buffer. After washing the column with buffer A/50 mM NaCl, the enzyme is eluted by a linear 50–500 mM NaCl gradient in 200 ml buffer A. The active fractions at about 0.4 M NaCl are combined (20–40 ml) and made 10 mM in phosphorus by adding 0.5 M potassium phosphate buffer (pH 7.2).

As the final step of purification, a conventional hydroxylapatite column may be used as described previously,[11,16] but we usually employ a high-performance prepacked hydroxylapatite column (HCA column, 5 × 100 mm, A-5010G, Mitsui Toh-atsu, Tokyo, Japan) connected to an FPLC (fast protein liquid chromatography) system (Pharmacia). The column equilibration buffer is 10 mM potassium phosphate buffer (pH 7.2) containing 0.25 M sucrose and 1 mM DTT. The sample from the ATP-agarose

---

[16] H. Kanoh and T. Ono, *J. Biol. Chem.* **259,** 11197 (1984).

column is pumped onto the HCA column, and the column is then washed successively with 10 ml each of 10 and 50 mM potassium phosphate buffer (pH 7.2) having the composition described above. The elution of DGK is achieved by a linear 0–2 M NaCl gradient in 20 ml of the 50 mM potassium phosphate buffer. The fraction size is 0.5 ml, and the collecting plastic tubes contain EDTA and ATP to make the final concentrations 1 mM and 10 μM, respectively. The 80K DGK forms a sharp peak at 0.5 M NaCl and is usually recovered in two or three fractions. The enzyme is aliquoted and stored at $-80°$.

Purification of 150K Diacylglycerol Kinase. The heat-stable DGK in the 0.2 M NaCl eluates from the DE-52 columns (400 ml) is stirred for 2 hr with 70 ml heparin-Sepharose (Pharmacia) equilibrated with buffer A/0.2 M NaCl. The Sepharose beads are then washed twice by decantation with 250 ml of the same buffer and packed into a short column (3 × 10 cm). After washing the column with 1 liter of the same buffer, elution is done with a linear NaCl gradient (0.2–1.0 M) in 400 ml of buffer A. The 150K DGK is eluted at about 0.7 M NaCl, and active fractions (usually 100 ml) are adjusted to 40% saturation with saturated ammonium sulfate. The precipitate is collected by centrifugation and dissolved in a small volume of buffer A/0.2 M NaCl (10–14 ml).

After insoluble material is removed by centrifugation, the sample is applied to a column (2.6 × 80 cm) of Sephacryl S-300 equilibrated with buffer A/0.2 M NaCl. The activity appears at an elution volume corresponding to a molecular weight of 150,000. The pooled fractions (40–60 ml) are diluted 2-fold with buffer A and applied to a column (1.5 × 13 cm) of ATP-agarose equilibrated with buffer A/0.1 M NaCl. When the column is eluted by a linear NaCl gradient (0.1–0.75 M) in 200 ml buffer A, the activity appears at 0.5 M NaCl. The active fractions are pooled, diluted 10-fold with buffer A to reduce the NaCl concentration, and pumped onto a Mono Q HR 5/5 column (Pharmacia) connected to an FPLC system. When eluted by a linear NaCl (50–350 mM) gradient in 20 ml of buffer A, the enzyme forms a sharp peak at 180 mM NaCl. The enzyme thus obtained is very unstable, but it could be stored at $-80°$ in the presence of bovine serum albumin (1 mg/ml).

Comments. The 80K DGK is more than 90% pure when analyzed by sodium dodecyl sulfate gel electrophoresis, whereas the purified 150K DGK gives two bands of 75K and 50K, the significance of which remains to be explored. The 150K DGK does not contain the 80K DGK protein or polypeptides reactive with anti-80K DGK antibodies.[11] The presence in thymus cytosol of two DGK isozymes is thus confirmed.

The results of enzyme purification are summarized in Table I. A large loss of the 80K DGK activity occurs at the initial DE-52 column chroma-

TABLE I
Purification of Two Diacylglycerol Kinase Isozymes from Porcine Thymus Cytosol[a]

Procedure	80K DGK				150K DGK			
	Protein (mg)	Total activity[b] (units)	Specific activity (units/mg)	Yield (%)	Protein (mg)	Total activity[b] (units)	Specific activity (units/mg)	Yield (%)
Cytosol	11,320	383.7	0.034	100	11,320	49.8	0.0044	100
DE-52								
0.1 M NaCl eluate	1,060	81.4	0.077	21.2	—	—	—	—
0.2 M NaCl eluate	—	—	—	—	897	25.3	0.028	51
Heparin-Sepharose	—	—	—	—	59.8	16.7	0.28	34
(NH₄)₂SO₄ fraction								
40–60% saturation	322	35.0	0.11	9.1	—	—	—	—
0–40% saturation	—	—	—	—	38.3	11.3	0.29	23
Sephacryl S-300	67.5	27.0	0.40	7.0	13.3	8.0	0.60	16
ATP-agarose	1.23	15.4	12.5	4.0	1.51	6.0	4.0	12
Hydroxylapatite (HCA)	0.40	6.0	15.0	1.6	—	—	—	—
Mono Q HR 5/5	—	—	—	—	0.30	2.3	7.7	4.6

[a] Fresh porcine thymus (400 g) was processed as described in the text.
[b] The activity which remained after treatment at 41° for 5 min is taken to represent the 150K DGK activity, whereas the activity lost by this treatment is assigned to 80 K DGK.[4,8,11]

tography step. This is mainly due to inactivation of the labile enzyme, although DGKs other than the two isozymes purified may also be present. Addition of various protease inhibitors failed to improve the enzyme stability. Owing to the high content of 80K DGK in the thymus, merely 400- to 500-fold purification over the cytosol results in a pure enzyme preparation.

Several DGK isozymes have been purified from rat brain[17] and human platelets.[9] The human equivalent of 80K DGK has recently been purified from white blood cells,[18] similar to the procedure described for the porcine enzyme. ATP-agarose chromatography has been used in these purifications, reflecting its effectiveness. In our experience, a new batch of the ATP-agarose beads tends to bind the enzyme more tightly than the repeatedly used matrix, and yields a purer enzyme preparation when eluted at a higher NaCl concentration. After being used several times, the beads give more reproducible enzyme elution patterns for reasons unknown.

Properties of Diacylglycerol Kinase

Primary Structure of 80K Diacylglycerol Kinase. We have cloned porcine cDNA coding for the 80K DGK.[7] The cDNA consists of 2772 base pairs (bp) and contains an open reading frame encoding 734 amino acids. The deduced structure of this enzyme revealed interesting features. The enzyme contains two Cys-rich, zinc finger-like sequences [Cys-X2-Cys-X12(14)-Cys-X2-Cys-X7-Cys-X6(7)-Cys], which are very similar to those occurring in protein kinase C.[19] Furthermore, we detected two EF hand motifs which are typical of various Ca^{2+}-binding proteins.[20] The enzyme also contains two putative ATP-binding sites, although the significance of the sequence located in the first zinc finger region remains obscure.

The structure of DGK is schematically presented in Fig. 1, and basically the same structure has been noted in the human 80K DGK.[18] In recent studies,[21] we found that the purified DGK indeed binds 2 mol of Ca^{2+} per mole of enzyme with a high affinity (apparent dissociation constant, K_d, 0.3 μM). We have thus shown that the 80K DGK is an EF hand type Ca^{2+}-binding protein. Ca^{2+} alone affects the enzyme activity to a limited extent, but Ca^{2+} plus phosphatidylserine markedly activates enzyme by reducing the K_m value for ATP. During these studies we detected in the sequence a 33-residue amphipathic α helix between the two Ca^{2+}-

[17] M. Kato and T. Takenawa, *J. Biol. Chem.* **265**, 794 (1990).
[18] D. Schaap, J. de Widt, J. van der Wal, J. Vanderkerckhove, J. van Damme, D. Gussow, H. L. Ploegh, W. J. van Blitterswijk, and R. L. van der Bend, *FEBS Lett.* **275**, 151 (1990).
[19] Y. Nishizuka, *Nature (London)* **334**, 661 (1988).
[20] R. H. Kretsinger, *Adv. Cyclic Nucleotide Res.* **11**, 1 (1979).
[21] F. Sakane, K. Yamada, S. Imai, and H. Kanoh, *J. Biol. Chem.* **266**, 7096 (1991).

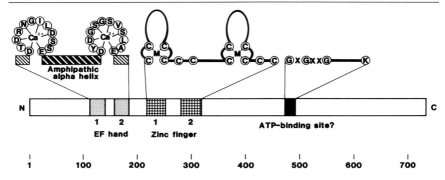

FIG. 1. Schematic presentation of the structure of porcine thymus 80K DGK. The diagram was constructed on the basis of our cDNA cloning work[7] and the Ca^{2+}-binding properties of the purified enzyme.[21] Amino acid residues at the functional regions are expressed in single-letter code. The metal coordination at the two zinc finger regions has not been proved. The putative ATP-binding site located at the first zinc finger is omitted, although this sequence is conserved in the human homolog of 80K DGK.[18] X, Any amino acid; M, metal ion; N, amino terminus; C, carboxy terminus.

binding sites, which might be involved in the Ca^{2+}-dependent interaction with phospholipids (Fig. 1).

Immunological Studies of 80K Diacylglycerol Kinase. In view of the artificial assay conditions and the presence of isozymes, it is desirable to study DGK with specific antibodies. We have prepared rabbit antibodies to porcine brain 80K DGK[1] that are also reactive with the human homolog of the isozyme.[6] By using the antibody we found that the occurrence of 80K DGK protein is quite limited, being detected only in thymus, spleen, and brain among the tissues examined.[8] Some tissues like liver and platelets lack this DGK isozyme. Three DGK isozymes recently purified from human platelets[9] are confirmed to be distinct from the 80K species. As shown in Fig. 2, immunoquantitation of the 80K enzyme protein revealed that the content of this isozyme is very high in T lymphocytes, relatively low in brain, and undetectable in platelets. Considering the DGK activity based on immunoprecipitation,[1,6,8] it now seems clear that different DGK isozymes exist in different types of cells.[6,8] An interesting question remains for future studies: Why is it necessary for resting lymphocytes to maintain such a high level of 80K DGK?

Activators and Inhibitors. Since first developed as an inhibitor of DGK from human erythrocytes and platelets,[3] R59022 has been actively tested in a variety of cells to assess the function of DGK. When tested *in vitro* with purified enzymes,[11] R59022 inhibited the 80K DGK but not the 150K DGK. Such differences for this inhibitor are also noted for DGK isozymes

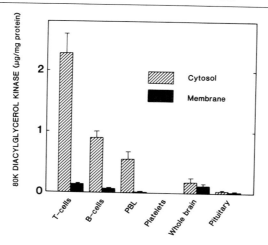

FIG. 2. Content and distribution of 80K DGK enzyme protein in various porcine cells. Soluble and membrane fractions from various porcine cells were subjected to immunoblotting using anti-porcine 80K DGK antibodies. The DGK enzyme protein was quantitated by densitometric scanning of the immunoblot using purified porcine thymus 80K DGK as the standard. Modified from Yamada et al.[6] and Kanoh et al.[4] T-cell, Splenic T lymphocytes; B-cell, splenic B lymphocytes; PBL, peripheral blood lymphocytes. In the case of platelets, no immunoreactive bands were detected.

from human platelets.[9] Interpretation of these data obtained under artificial assay conditions is difficult, but care should be taken to see whether the particular type of cells contains R59022-sensitive DGK species.

Phosphatidylserine has been used most often as an activator for DGK in octylglucoside[10,12] and DG suspension[5,11] assays. Sphingosine, which has been used as an inhibitor of protein kinase C,[22] potently activates 80K DGK both in DG suspension assays[11] and in intact cell systems.[23] On the other hand, 150K DGK is inhibited by sphingosine.[11] The mechanism of enzyme activation or inhibition by various substances has not been studied except for the effect of Ca^{2+} plus phosphatidylserine on 80K DGK.[21] Triton X-100 is extremely inhibitory to all DGK isozymes purified.

Enzyme Phosphorylation and Intracellular Translocation. We noted that 80K DGK was phosphorylated by an unidentified protein kinase contained in crude brain extracts.[24] We have further observed that the enzyme is actively phosphorylated *in vitro* by protein kinase C.[25] On the

[22] Y. A. Hannun and R. M. Bell, *Science* **243**, 500 (1989).
[23] K. Yamada, F. Sakane, S. Imai, and H. Kanoh, unpublished.
[24] H. Kanoh and T. Ono, *FEBS Lett.* **201**, 97 (1986).
[25] H. Kanoh, K. Yamada, F. Sakane, and T. Imaizumi, *Biochem. J.* **258**, 455 (1989).

other hand, membrane-bound DGK from pancreas was activated *in vitro* by cAMP or Ca^{2+}/calmodulin-dependent protein kinase but not by protein kinase C.[26] To date no reports with an intact cell system have been available to determine the significance of the enzyme phosphorylation.

In most cells and tissues the major portion of DGK activity has been detected in soluble fractions.[4,6,8,9,17] As presented in Fig. 2, the 80K enzyme protein exists principally as a soluble form, though it is also detected in membranes to variable extents. Diacylglycerol is generated within the membranes, and the phospholipid dependency of DGK action has been widely observed. Regulation of DGK activity, therefore, should involve enzyme interaction with membranes. There are several reports describing the translocation of soluble DGK activity to membranes.[27-29] These data have been obtained by assaying crude enzyme activities, and the mechanism of enzyme translocation or the type of DGK isozymes involved in this phenomenon has not been determined. We found that 80K DGK phosphorylated by protein kinase C binds to phosphatidylserine,[25] but in retrospect micromolar Ca^{2+} present in the experimental system might have contributed to this observation, since the binding of Ca^{2+} to DGK alone is sufficient to cause enzyme interaction with phospholipids or with membranes.[21]

Substrate Specificity. Recently a membrane-bound arachidonoyl-specific DGK has been found.[10,12,14] This specific isozyme is detectable by the octylglucoside (but not deoxycholate) assay. We could confirm this finding in Jurkat cells.[23] The two DGKs purified from the soluble fraction, namely, the 80K and 150K DGKs, do not possess specificity toward DG molecular species.[23] The presence of such specific DGK isozymes points to the importance of the acyl composition of DG serving as a second messenger and as a substrate of glycerolipid synthesis.

Acknowledgments

This work has been supported by the Ministry of Education, Science, and Culture, Japan, and the Uehara Memorial Foundation. We thank Professor R. Sato, Institute for Protein Research, Osaka University, for support.

[26] H.-D. Soeling, W. Fest, T. Schmidt, H. Esselmann, and V. Bachmann, *J. Biol. Chem.* **264,** 10643 (1989).

[27] J. M. Besterman, R. S. Pollenz, E. L. Booker, and P. Cuatrecasas, *Proc. Natl. Acad. Sci. U.S.A.* **83,** 9378 (1986).

[28] J. Ishitoya, A. Yamakawa, and T. Takenawa, *Biochem. Biophys. Res. Commun.* **144,** 1025 (1987).

[29] A. C. Maroney and I. G. Macara, *J. Biol. Chem.* **264,** 2537 (1989).

[19] Arachidonoyl-Specific Diacylglycerol Kinase

By ROZENN N. LEMAITRE and JOHN A. GLOMSET

Introduction

Mammalian diacylglycerol (DG) kinase appears to exist as several isoforms.[1-6] A number of cytoplasmic DG kinases have been purified,[3,5-8] and the best characterized is an 80K protein reviewed elsewhere in this volume.[9] The soluble DG kinases appear to have no preference for any DG species.[8,10] In contrast, studies with rat brain microvessels,[11] 3T3 cells,[2,12] and baboon tissues[4] revealed the existence of a DG kinase that prefers arachidonoyl-containing substrates. Whereas the soluble DG kinases can bind to membranes transiently,[13-16] the arachidonoyl-specific DG kinase (arachidonoyl-DG kinase) appears to be an integral membrane protein.[4] Although the arachidonoyl-DG kinase coexists with nonspecific DG kinases,[2,4] it has escaped detection in the past.[10] We have found that two conditions must be met if the arachidonoyl-DG kinase is to be detected: (1) the enzyme activity must be measured with an appropriate assay system, and (2) the arachidonoyl-DG kinase must be separated from

[1] K. Yamada and H. Kanoh, J. Biol. Chem. 255, 601 (1988).

[2] M. L. MacDonald, K. F. Mack, B. W. Williams, W. C. King, and J. A. Glomset, J. Biol. Chem. 263, 1584 (1988).

[3] F. Sakane, K. Yamada, and H. Kanoh, FEBS Lett. 255, 409 (1989).

[4] R. N. Lemaitre, W. C. King, M. L. MacDonald, and J. A. Glomset, Biochem. J. 266, 291 (1990).

[5] M. Kato and T. Takenawa, J. Biol. Chem. 265, 794 (1990).

[6] Y. Yada, T. Ozeki, H. Kanoh, and Y. Nozawa, J. Biol. Chem. 265, 19237 (1990).

[7] H. Kanoh, H. Kondoh, and T. Ono, J. Biol. Chem. 258, 1767 (1983).

[8] C.-H. Lin, H. Bishop, and K. P. Strickland, Lipids 21, 206 (1986).

[9] H. Kanoh, F. Sakane, and K. Yamada, this volume [18].

[10] H. H. Bishop and K. P. Strickland, Lipids 15, 285 (1980).

[11] M. Hee-Cheong, T. Fletcher, S. K. Kryski, and D. L. Severson, Biochim. Biophys. Acta 833, 59 (1985).

[12] M. L. MacDonald, K. F. Mack, C. N. Richardson, and J. A. Glomset, J. Biol. Chem. 263, 1575 (1988).

[13] J. M. Besterman, R. S. Pollenz, E. L. Booker, and P. Cuatrecasas, Proc. Natl. Acad. Sci. U.S.A. 83, 9378 (1986).

[14] J.-I. Ishitoya, A. Yamakawa, and T. Takenawa, Biochem. Biophys. Res. Commun. 144, 1025 (1987).

[15] B. Jimenez, M. M. Van Lookeren Campagne, A. Pestana, and M. Fernandez-Renart, Biochem. Biophys. Res. Commun. 150, 118 (1988).

[16] A. C. Maroney and I. G. Macara, J. Biol. Chem. 264, 2537 (1989).

the nonspecific DG kinases.[4] In this chapter we describe the assay system that we routinely use and a chromatographic technique for separating the arachidonoyl-DG kinase from the nonspecific enzymes. In addition, we discuss the importance of the choice of detergent used in the assay for detecting the substrate preference of the arachidonoyl-DG kinase.

Assay Methods

There are currently three different assay systems for measuring DG kinase activity. In each case the amount of [^{32}P]phosphatidic acid ([^{32}P]PA) formed from DG and [^{32}P]ATP is measured. The most extensively used assay system is performed in the presence of deoxycholate, which solubilizes DG and activates the soluble DG kinases.[7] In another assay system DG is suspended in buffer in the absence of detergent.[13] Both of these assays are effective for measuring the activity of the soluble DG kinases. However, the arachidonoyl-DG kinase requires a third assay system in which octylglucoside micelles containing DG and phosphatidylserine (PS) are mixed with octylglucoside micelles containing the enzyme. This assay, which we describe in the following pages, is a modification[4] of the mixed micellar assay developed for the membrane-bound DG kinase from *Escherichia coli.*[17]

Stock Solutions. The following solutions are kept at $-20°$:

sn-1-Stearoyl-2-arachidonoylglycerol (18 : 0/20 : 4 DG, Sigma, St. Louis, MO), dissolved in chloroform at 10 mg/ml

sn-1,2-Didecanoyldiacylglycerol (10 : 0/10 : 0 DG), 10 mg/ml in chloroform (Avanti Polar Lipids, Birmingham, AL)

PS, 10 mg/ml in chloroform (Avanti)

[γ-^{32}P]ATP, 3000 Ci/mmol, 10 mCi/ml (New England Nuclear, Boston, MA)

20 mM ATP in deionized water, frozen in aliquots

Stock assay buffer: 0.375 M 3-(N-Morpholino)propanesulfonic acid (MOPS; pH 7.2), 0.125 M MgCl$_2$, and 6 mM dithiothreitol (DTT), frozen in aliquots

Carrier with egg PA: 200 μl egg PA, 10 mg/ml in chloroform (Avanti), mixed with 3 ml chloroform–methanol–deionized water (75 : 25 : 2, by volume)

Carrier with both egg PA and didecanoylphosphatidic acid (PA$_{10}$): 200 μl egg PA and 200 μl PA$_{10}$, 10 mg/ml in chloroform (Avanti), mixed with 3 ml chloroform–methanol–deionized water (75 : 25 : 2, by volume)

[17] C. R. Loomis, J. P. Walsh, and R. M. Bell, *J. Biol. Chem.* **260,** 4091 (1985).

The following solutions are kept at room temperature:

Chloroform–methanol–HCl (66 : 33 : 1, by volume)

Methanol–deionized water–chloroform (48 : 47 : 3, by volume)

Chloroform–methanol–deionized water (75 : 25 : 2, by volume)

Nile Blue sulfate solution: 40 mg Nile Blue sulfate (Sigma) is dissolved in 2 liters deionized water. The solution is then acidified with 5.5 ml H_2SO_4.

Assay in Presence of 18 : 0/20 : 4 Diacylglycerol or Other Long-Chain Diacylglycerol

Solution of Mixed Micelles of Octylglucoside, Diacylglycerol, and Phosphatidylserine. A solution containing 10.15 mM 18 : 0/20 : 4 DG, 11.55 mM PS, and 6.5% (w/v) octylglucoside is prepared as follows. Appropriate amounts of 18 : 0/20 : 4 DG and PS are pipetted into a 12 × 75 mm glass tube, dried under argon or nitrogen, and immediately resuspended in a freshly prepared solution containing 6.5% octylglucoside (w/v) in deionized water. Solubilization of the lipids takes time and requires frequent vortexing over a period of 30 to 45 min. One can hasten the process by sonicating the lipid solution in an ultrasonic cleaner (Branson, Shelton, CT) for 30 sec soon after adding the octylglucoside solution. It is essential that the lipids be solubilized completely. The solution should be completely clear, and there should be no remaining lipid film at the bottom of the tube. The mixed micelles can be kept on ice for 2 to 3 hr.

Solution of [^{32}P]ATP. A solution containing 1.82 mM ATP and approximately 0.73 mCi/ml [^{32}P]ATP is prepared shortly before the assays.

Enzyme sources. Tissue extract, cell extract, or effluent from a hydroxylapatite column (see below) may be used as the enzyme source.

Assay. Activity measurements are performed for 2 min at room temperature in a 35-μl reaction volume in the presence of 55–60 mM MOPS (pH 7.2), 18 mM MgCl$_2$, 1.1 mM DTT, 2.14% (w/v) octylglucoside, 0.52 mM ATP, 2.9 mM 18 : 0/20 : 4 DG, 3.3 mM PS, 5.7% (v/v) glycerol, and 86 mM KCl or 43 mM potassium phosphate. The reaction mixture consists of 5 μl stock assay buffer [0.375 mM MOPS (pH 7.2), 0.125 mM MgCl$_2$, and 6 mM DTT] mixed with 10 μl tissue or cell extract and 10 μl mixed micelles of octylglucoside, DG, and PS (preparation described above). The reaction is initiated by the addition of 10 μl [^{32}P]ATP solution (see above). After 2 min at room temperature the reaction is quenched by the addition of 0.6 ml chloroform–methanol–HCl (66 : 33 : 1, by volume). Egg PA carrier solution is added to each assay tube (40 μl containing 25 μg egg PA). The unreacted [^{32}P]ATP is then partially removed by extraction of the quenched reaction mixture twice with 0.5 ml methanol–deionized water–chloroform (48 : 47 : 3, by volume). The aqueous and chloroform phases are separated by centrifugation at 800 g for 3 min at 4°. The upper phases are discarded into a container for radioactive waste. The lower chloroform phase contains [^{32}P]PA. At this point the lower phase can be kept at −20° for later analysis.

Isolation of [^{32}P]PA by Thin-Layer Chromatography. The chloroform phases containing [^{32}P]PA are dried down under a stream of argon or nitrogen or, if there are many samples, by centrifugation for 45 min under reduced pressure in a Speed-Vac concentrator (Forma Scientific, Marietta, OH). The samples are each resuspended in 70 μl chloroform–methanol–deionized water (75 : 25 : 2, by volume) and 25 μl aliquots are spotted on cellulose plates (EM Science, Cherry Hill, NJ). The plates are developed with the upper phase of a mixture of ethyl acetate–2,2,4-trimethylpentane–acetic acid–deionized water (90 : 50 : 20 : 30, by volume). The thin-layer chromatography solvent is prepared shortly before it is needed by stirring the above solvent mixture vigorously for 10–15 min and then decanting the two phases for 20–30 min. The [^{32}P]PA is detected by autoradiography of the air-dried cellulose plates using XAR films (Kodak, Rochester, NY) with an intensifying screen for 2 to 16 hr at −70° and by staining with a solution of Nile Blue sulfate (see Stock Solutions). The corresponding band is excised and counted for radioactivity in 5 or 10 ml scintillation liquid (Ecolume, ICN, Costa Mesa, CA).

Assay in Presence of 18 : 0/20 : 4 and 10 : 0/10 : 0 Diacylglycerols

A quick screen for the presence of arachidonoyl-DG kinase can be performed by incubating a tissue extract simultaneously with 18 : 0/20 : 4 DG, 10 : 0/10 : 0 DG, and [^{32}P]ATP, and measuring the respective amounts of long-chain [^{32}P]PA and [^{32}P]PA$_{10}$ that are formed. The procedure is identical to that of the assay with 18 : 0/20 : 4 DG except that two solutions are modified from the previous assay: (1) the solution of mixed micelles of octylglucoside, DG, and PS used in the assay contains both 10 : 0/10 : 0 DG and 18 : 0/20 : 4 DG at the same concentration, and (2) the solution of carrier PA added after the enzymatic reaction is quenched contains both egg PA and PA$_{10}$ (see Stock Solutions).

The long-chain [^{32}P]PA and the [^{32}P]PA$_{10}$ formed during the assay are separated with the thin-layer chromatography system described for the previous assay. PA$_{10}$ migrates below long-chain PA. It should be noted, however, that the effectiveness of the separation varies considerably. The following details are important for good resolution: (1) the temperature of the area where the thin-layer chromatography is performed should be no less than 24°; (2) the solvent mixture should be prepared shortly before use; (3) the chromatographic tanks should be lined with two saturation pads but should not be preequilibrated with the solvent mixture; (4) only one plate should be chromatographed in each tank; and (5) the chromatography should be performed until the solvent front reaches approximately 2.5 cm from the top of the plate, but no longer than 65 min. We also observed that different batches of cellulose plates yield different degrees of separation.

Notes About Assay Conditions

The assay described above is done at saturating substrate and ATP concentrations. Changes in the activity measured therefore represent changes in V_{max}. In addition, the activity measured is linear with time during the 2 min of the assay and with respect to protein concentration.

The activity of the arachidonoyl-DG kinase varies with the mole fraction of DG in the assay mixture rather than the absolute DG concentration. The mole fraction is defined as the ratio of the DG concentration to the concentration of octylglucoside in micelles, the latter concentration being equal to the difference between the total octylglucoside concentration and the concentration of octylglucoside monomers (critical micelle concentration). The concentration of octylglucoside monomers varies with ionic strength and the composition of the micelles. When the micellar concentration of octylglucoside is estimated to be 48 mM, the mole fraction of DG in the assay is calculated to be 6 mol%.

Variations in some of the assay conditions result in variations in the arachidonoyl-DG kinase activity. The most important factor is the nature of the detergent used in the assay (discussed under Substrate Preference of Arachidonoyl-Specific Diacylglycerol Kinase). In addition, the enzyme activity is decreased in the absence of glycerol, at both low and high ionic strength, and in the absence of PS.

Separation of Arachidonoyl-Specific Diacylglycerol Kinase from Nonspecific Diacylglycerol Kinases

In most baboon tissue homogenates, the presence of the arachidonoyl-DG kinase is masked by the contribution of nonspecific DG kinases to the overall activity.[4] It is therefore necessary to separate the specific enzyme from the nonspecific ones.

In 3T3 cells, the separation of soluble and membrane compartments allows the detection of the arachidonoyl-DG kinase in the crude membranes.[2] However, soluble DG kinases are known to bind to membranes,[13] and a significant fraction of the nonspecific DG kinases is recovered with brain particulates even after removal of most of the cytoplasmic contaminants.[4] Therefore, the DG kinase activity of cell membranes is likely to be a composite of the activities of different DG kinases. Thus, the individual species of DG kinase must be separated before one can assess which species are present in cell membranes and attribute variations in overall DG kinase activity of cell membranes to a DG kinase species. The separation is achieved by chromatography on hydroxylapatite.

Solutions. The following stocks of protease inhibitors are kept at $-20°$:

Leupeptin (Sigma), 1 mg/ml in deionized water

Pepstatin (Sigma), 1 mg/ml in methanol

Soybean trypsin inhibitor (Sigma), 10 mg/ml in deionized water

The buffers below all contain the following mixture of protease inhibitors: 1 μg/ml each pepstatin and leupeptin, and 50 μg/ml soybean trypsin inhibitor (final concentrations). They can be prepared a day in advance and kept at 4°.

Extraction buffer: 20 mM MOPS (pH 7.2), 4% (w/v) octylglucoside, 0.6 M KCl, 20% (v/v) glycerol, 1 mM DTT, and protease inhibitors

Column equilibration buffer: 20 mM MOPS (pH 7.2), 1% (w/v) octylglucoside, 0.3 M KCl, 20% (v/v) glycerol, 1 mM DTT, and protease inhibitors

Gradient buffer A: 0.15 M potassium phosphate (pH 7.2), 1% (w/v) octylglucoside, 20% (v/v) glycerol, 1 mM DTT, and protease inhibitors

Gradient buffer B: 0.60 M potassium phosphate (pH 7.2), 1% (w/v) octylglucoside, 20% (v/v) glycerol, 1 mM DTT, and protease inhibitors

Procedures

Extraction of Arachidonoyl-Specific Diacylglycerol Kinase. Extraction of the arachidonoyl-DG kinase requires both detergent and high salt. We routinely use 2% (w/v) octylglucoside and 0.3 M KCl (final concentrations) to extract the enzyme. Tissue homogenates, cell lysates, or cell membranes [typically in 20 mM MOPS (pH 7.2), 0.25 M sucrose, 1 mM DTT, 1 mM EGTA, with protease inhibitors] are extracted with an equal volume of ice-cold extraction buffer (see description above) at 7–12 mg protein/ml. The suspension is vortexed or thoroughly mixed, incubated on ice for 30 min, and centrifuged at 100,000 g for 1 hr at 4°. If the extract is cloudy owing to the presence of lipid, it should be diluted with an equal volume of buffer containing 20 mM MOPS (pH 7.2), 2% (w/v) octylglucoside, 0.3 M KCl, 20% (v/v) glycerol, 1 mM DTT, and protease inhibitors.

Chromatography on Hydroxylapatite. The following procedure was developed for tissue homogenate extracts. One may adapt it to cells or cell fractions by scaling down column size and buffer volumes proportionally. A mixture of hydroxylapatite (BioGel HTP, Bio-Rad, Richmond, CA), and Sephadex G-10 or Sephadex G-25 medium (Pharmacia, Piscataway, NJ), 80 : 20 by weight, is used to build a 2.5 × 2.5 cm column. The hydroxylapatite/Sephadex mix is hydrated for 3 hr at room temperature in buffer containing 20 mM MOPS (pH 7.2) and 0.3 M KCl, washed several times with the same buffer, mixed with glycerol to a final concentration of approximately 20% (v/v), and packed into the column. The mix is allowed to settle completely in the column with the column outlet closed (this procedure ensures good flow properties). Thereafter, all steps are performed at 4°. The column is equilibrated with 25 ml equilibration buffer (see above). The extract containing approximately 100 mg protein is loaded onto the column at 20 ml/hr. The column is washed successively with 5 ml equilibration buffer and 5 ml gradient buffer A. The DG kinases are eluted with a 120 ml gradient from 100% A to 100% B followed by 20 ml gradient buffer B.

For better recovery of DG kinase activity it is best to use each column only once. To quantitate the recovery of DG kinase activity from the hydroxylapatite accurately one must reduce the content of potassium phosphate in the hydroxylapatite fractions to approximately 0.15 M (0.6 M inhibits the activity by about 40%).

Notes. The aforementioned procedure differs from the previously published one[4] in the following respects: (1) a short and wide column is used rather than a long, narrow one, (2) the hydroxylapatite is mixed with Sephadex, and (3) all the buffers contain 20% (v/v) glycerol. Inclusion of glycerol in the chromatographic buffers maximizes recovery of DG kinase activity. The other changes are needed in order to maintain good flow properties of the hydroxylapatite column in the presence of glycerol. Very similar chromatograms are obtained with the two procedures.

We attempted to replace octylglucoside with the nonionic detergent dodecylpoly(ethylene glycol ether)[9] (sold by Boehringer, Indianapolis, IN, under the trademark Thesit) for the extraction of DG kinase and the chromatography on hydroxylapatite. Thesit extracted DG kinase activity effectively. However, the arachidonoyl-DG kinase was eluted from hydroxylapatite at a much reduced potassium phosphate concentration in the presence of Thesit.[18]

Hydroxylapatite Chromatogram. A typical profile of DG kinases separated on hydroxylapatite is shown in Fig. 1. Assays of the chromatographic fractions in the presence of 18 : 0/20 : 4 DG and 10 : 0/10 : 0 DG reveal an

---

[18] R. N. Lemaitre, unpublished results, 1990.

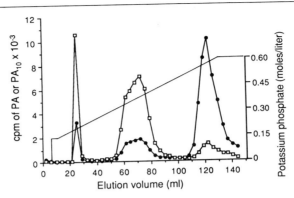

FIG. 1. Separation of DG kinase species on hydroxylapatite. An octylglucoside/KCl extract of baboon brain containing 100 mg protein was chromatographed on a 1 × 12 cm hydroxylapatite column with the buffer system described in the text except for the absence of glycerol. The DG kinase activity of 10 μl aliquots of the effluent was measured with a mixture of 18:0/20:4 DG (●) and 10:0/10:0 DG (□). The concentration of potassium phosphate in the running buffer is indicated by the solid line. (From Ref. 4, with permission.)

arachidonoyl-DG kinase that elutes at the end of the gradient, at an effluent concentration of approximately 0.50 M potassium phosphate. Two other DG kinases that show lower activities toward 18:0/20:4 DG than toward 10:0/10:0 DG elute at approximately 0.15 and 0.30 M potassium phosphate, respectively. All of the baboon tissues that we have screened (brain, muscle, liver, testis, spleen, and kidney) show similar profiles.[4] However, both the level of arachidonoyl-DG kinase activity and the relative distribution of DG kinase activities vary dramatically among the different tissues. The arachidonoyl-DG kinase is particularly abundant in brain and testis but is present only in very small amounts (100 times less) in liver. The significance of this difference is not known.

Substrate Preference of Arachidonoyl-Specific Diacylglycerol Kinase

Comparison of the activity of the arachidonoyl-DG kinase toward 18:0/20:4 DG and 10:0/10:0 DG is a convenient screening method for detecting the presence of the enzyme. However, the substrate preference is better defined by comparing the arachidonoyl-DG kinase activity toward different long-chain DGs. Table I illustrates the substrate preference of the brain arachidonoyl-DG kinase after partial purification on hydroxylapatite. The enzyme uses DGs containing arachidonic acid in the sn-2 position 5 to 10 times more efficiently than other naturally occurring DGs and 4 times more efficiently than sn-1,2-dioleoylglycerol (diolein), a DG commonly

TABLE I
SUBSTRATE SPECIFICITY OF BRAIN ARACHIDONOYL-
SPECIFIC DIACYLGLYCEROL KINASE[a]

Diacylglycerol	Enzyme activity[b] (pmol/min)
Naturally occurring	
sn-1-Stearoyl-2-arachidonoyl-G[c]	19.19 ± 0.38
sn-1-Palmitoyl-2-arachidonoyl-G	20.81 ± 0.55
sn-1-Myristoyl-2-arachidonoyl-G	19.61 ± 0.11
sn-1-Stearoyl-2-oleoyl-G	2.01 ± 0.06
sn-1-Stearoyl-2-linoleoyl-G	3.89 ± 0.38
sn-1-Stearoyl-2-docosahexaenoyl-G	1.91 ± 0.07
Synthetic	
sn-1,2-Dioleoyl-G	4.75 ± 0.09
sn-1,2-Didecanoyl-G	1.78 ± 0.09
sn-1,2-Dioctanoyl-G	1.04 ± 0.04
sn-1-Arachidonoyl-2-stearoyl-G	1.75 ± 0.03

[a] Adapted with permission from Ref. 4.
[b] Activity in a 10 μl aliquot of the brain arachidonoyl-
DG kinase eluted from hydroxylapatite as in Fig. 1 was
measured with DG species that had different fatty acid
chains.
[c] G, Glycerol.

used in DG kinase assays.[7,13] The enzyme shows least activity toward
short-chain DGs. These substrates are more soluble than long-chain DGs
and may be less accessible to the enzyme, which itself is in the mixed
micelles. Maximum activity is observed with DGs containing arachidonic
acid in the sn-2 position regardless of whether myristic acid, palmitic acid,
or stearic acid is present in the sn-1 position. A synthetic DG containing
arachidonic acid in the sn-1 position is used 10 times less efficiently,
indicating that the arachidonic acid must be in the sn-2 position. In addition, we observed 3 times less activity toward 1-hexadec-1'-enyl-2-arachi-
donoyldiradylglycerol.[4] The fatty acid chain in the sn-1 position must
therefore be linked via an ester bond.

A similar preferential utilization of arachidonoyl-DGs has been observed for the DG kinase in 3T3 cell membranes[2] and in a highly purified
preparation of the arachidonoyl-DG kinase from bovine testis membranes.[18] Thus, the substrate preference of the arachidonoyl-DG kinase is
evidently an intrinsic property of the enzyme.

The other two brain DG kinases separated on hydroxylapatite show
no preference for any DG species.[4] These nonspecific DG kinases are
predominantly soluble. Similarly, the DG kinase(s) present in 3T3 cell

cytoplasm show no substrate preference.[2] Therefore, the substrate preference of the arachidonoyl-DG kinase is not a property shared with the soluble DG kinases.

Choice of Detergent and Substrate Specificity. A peculiarity of the arachidonoyl-DG kinase is that its substrate preference is not detected in the presence of bile salts. In particular, the assay in the presence of deoxycholate initially developed by Kanoh and colleagues,[7] and most commonly used for measuring DG kinase activity,[19] results in similar activities of the arachidonoyl-DG kinase toward 18 : 0/20 : 4 DG and diolein.[4] The deoxycholate assay differs from the octylglucoside assay described herein in many respects including pH, temperature, detergent, and other buffer components. However, when octylglucoside is substituted for deoxycholate in the deoxycholate assay, the arachidonoyl-DG kinase phosphorylates 5 times more 18 : 0/20 : 4 DG than diolein, as it does in the regular octylglucoside assay. In contrast, when deoxycholate is substituted for octylglucoside in the octylglucoside assay (except for the $MgCl_2$ concentration, which has to be decreased to 10 mM to avoid precipitation with deoxycholate), no substrate preference is observed.[18] This result indicates that deoxycholate, and no other factor in the assay, abolishes the substrate specificity. We presume that the presence of deoxycholate during DG kinase assays prevented the detection of the arachidonoyl-DG kinase in the past.[10]

The octylglucoside assay is not unique in allowing detection of the arachidonoyl-DG kinase substrate preference. Assays in the presence of Triton X-100 are also effective.[4] On the other hand, assays in the presence of other bile salt derivatives such as taurocholate and CHAPS show little (CHAPS) or no (taurocholate) substrate specificity.[18]

Temperature and Substrate Specificity

The substrate specificity of the arachidonoyl-DG kinase in 3T3 cell membranes is quickly abolished when the membranes are preincubated with 18 : 0/20 : 4 DG at 37° prior to the activity measurements.[2] This result is not observed with the arachidonoyl-DG kinase in baboon tissues.[18] The basis for this difference is not known.

Acknowledgments

This work was supported by the Howard Hughes Medical Institute and by U.S. Public Health Service Grant RR.00166.

[19] M. A. Wallace and J. N. Fain, this series, Vol. 109, p. 469.

[20] Phosphatidylinositol 4-Kinase from Yeast

By George M. Carman, Charles J. Belunis,
and Joseph T. Nickels, Jr.

Introduction

Phosphatidylinositol 4-kinase (1-phosphatidylinositol kinase) catalyzes the reaction of phosphatidylinositol with ATP to form phosphatidylinositol

$$\text{Phosphatidylinositol} + \text{ATP} \rightarrow \text{phosphatidylinositol 4-phosphate} + \text{ADP}$$

4-phosphate.[1] Phosphatidylinositol 4-kinase (ATP: 1-phosphatidyl-1D-*myo*-inositol 4-phosphotransferase, EC 2.7.1.67) is the first enzyme in the phosphorylation sequence of phosphatidylinositol leading to the formation of phosphatidylinositol 4-phosphate and phosphatidylinositol 4,5-bisphosphate in *Saccharomyces cerevisiae*.[2] The synthesis and turnover of the polyphosphoinositides in *S. cerevisiae* as well as in higher eukaryotes play an important role in cell growth.[3-7] Phosphatidylinositol 4-kinase is associated with the microsomal,[8] plasma membrane,[9] and cytosolic[10] fractions of *S. cerevisiae*. Phosphatidylinositol 4-kinase has been purified to near homogeneity from the microsomal fraction of *S. cerevisiae* by standard protein purification procedures.[8] We describe here the purification and properties of the enzyme.

Assay Method

Phosphatidylinositol 4-kinase activity is measured for 10 min by following the phosphorylation of 0.4 mM phosphatidylinositol with 5 mM [γ-^{32}P]ATP [3000–5000 counts/min (cpm)/nmol] in the presence 50 mM

[1] M. Colodzin and E. P. Kennedy, *J. Biol. Chem.* **240**, 3771 (1965).

[2] S. Steiner and R. L. Lester, *Biochim. Biophys. Acta* **260**, 82 (1972).

[3] R. T. Talwalkar and R. L. Lester, *Biochim. Biophys. Acta* **306**, 412 (1973).

[4] K. Kaibuchi, A. Miyajima, K.-I. Arai, and K. Matsumoto, *Proc. Natl. Acad. Sci. U.S.A.* **83**, 8172 (1986).

[5] C. Dahl, H.-P. Biemann, and J. Dahl, *Proc. Natl. Acad. Sci. U.S.A.* **84**, 4012 (1987).

[6] I. Uno, K. Fukami, H. Kato, T. Takenawa, and T. Ishikawa, *Nature (London)* **333**, 188 (1988).

[7] M. J. Berridge, *Biochim. Biophys. Acta* **907**, 33 (1987).

[8] C. J. Belunis, M. Bae-Lee, M. J. Kelley, and G. M. Carman, *J. Biol. Chem.* **263**, 18897 (1988).

[9] A. J. Kinney and G. M. Carman, *J. Bacteriol.* **172**, 4115 (1990).

[10] R. T. Talwalker and R. L. Lester, *Biochim. Biophys. Acta* **360**, 306 (1974).

Tris–maleate buffer (pH 8.0) containing 25 mM Triton X-100, 30 mM MgCl$_2$, and enzyme protein in a total volume of 0.1 ml at 30°.[11] The reaction is terminated by the addition of 0.5 ml of 0.1 N HCl in methanol. Chloroform (1 ml) and 1 M MgCl$_2$ (1.5 ml) are added, the system is mixed, and the phases are separated by a 2-min centrifugation at 100 g at room temperature. A 0.5-ml sample of the chloroform phase is removed and taken to dryness on an 80° water bath. Betafluor (4 ml, National Diagnostics, Somerville, NJ) is added to the sample, and radioactivity is determined by scintillation counting. The chloroform-soluble phospholipid product of the reaction, phosphatidylinositol 4-phosphate, is analyzed by thin-layer chromatography with standard phosphatidylinositol 4-phosphate using the solvent system chloroform–methanol–2.5 M ammonium hydroxide (9 : 7 : 2, v–v).[11] Alternatively, activity is measured using 0.4 mM phosphatidyl[2-^{3}H]inositol (2000 cpm/nmol) and 5 mM ATP as substrates.

A unit of enzymatic activity is defined as the amount of enzyme that catalyzes the formation of 1 nmol of product per minute. Protein is determined by the method of Bradford.[12] Buffers which are identical to those containing the protein samples are used as blanks. The presence of Triton X-100 does not interfere with the protein determination, provided the blank contains a final concentration of detergent identical to that of the sample.

Growth of Yeast

Saccharomyces cerevisiae strain *ade5 MATa*[13] is used as a representative wild-type strain for enzyme purification. Cells are grown in 1% yeast extract, 2% peptone, and 2% glucose (w/v) at 28° to late exponential phase, harvested by centrifugation, and stored at −80°.[14,15] We use wild-type strain S288C for the purification of phosphatidylinositol synthase,[14,15] phosphatidylserine synthase,[16,17] and CDPdiacylglycerol synthase.[18,19] We are unable to purify phosphatidylinositol 4-kinase activity from strain S288C. The reason for this is unclear.

[11] M. A. McKenzie and G. M. Carman, *J. Food Biochem.* **6**, 77 (1982).
[12] M. M. Bradford, *Anal. Biochem.* **72**, 248 (1976).
[13] M. R. Culbertson and S. A. Henry, *Genetics* **80**, 23 (1975).
[14] A. S. Fischl and G. M. Carman, *J. Bacteriol.* **154**, 304 (1983).
[15] G. M. Carman and A. S. Fischl, this volume [36].
[16] M. Bae-Lee and G. M. Carman, *J. Biol. Chem.* **259**, 10857 (1984).
[17] G. M. Carman and M. Bae-Lee, this volume [35].
[18] M. J. Kelley and G. M. Carman, *J. Biol. Chem.* **262**, 14563 (1987).
[19] G. M. Carman and M. J. Kelley, this volume [28].

Purification Procedure

All steps are performed at 5°.

Step 1: Preparation of Cell Extract. Cells (200 g) are disrupted with glass beads (diameter, 0.5 mm) with a BioSpec Products (Bartlesville, OK) Bead-Beater in 50 mM Tris-HCl buffer (pH 7.5) containing 1 mM Na$_2$EDTA, 0.3 M sucrose, and 10 mM 2-mercaptoethanol as described previously.[14,15] Glass beads and unbroken cells are removed by centrifugation at 1500 g for 15 min to obtain the cell extract.

Step 2: Preparation of Microsomes. Microsomes are collected from the cell extract by differential centrifugation.[14,15] Microsomal pellets are washed with 50 mM Tris-HCl buffer (pH 8.0) containing 10 mM MgCl$_2$, 10 mM 2-mercaptoethanol, and 10% glycerol. Microsomes can be frozen at −80° until the enzyme is purified.

Step 3: Preparation of Triton X-100 Extract. Microsomes are suspended in 50 mM Tris-HCl buffer (pH 8.0) containing 10 mM MgCl$_2$, 10 mM 2-mercaptoethanol, 10% glycerol, and 1% Triton X-100 (v/v) at a final protein concentration of 10 mg/ml. The suspension is incubated for 1 hr on a rotary shaker at 150 rpm. After the incubation, the suspension is centrifuged at 100,000 g for 1.5 hr to obtain the Triton X-100 extract (supernatant). The Triton X-100 extract is immediately used for the next purification step since, at this stage, the enzyme is labile. The addition of protease inhibitors does not improve enzyme stability.

Step 4: DE-52 Chromatography. A DE-52 (Whatman, Clifton, NJ) column (2.5 × 5 cm) is equilibrated with 50 mM Tris-HCl buffer (pH 8.0) containing 10 mM MgCl$_2$, 10 mM 2-mercaptoethanol, 10% glycerol, and 0.5% Triton X-100. The Triton X-100 extract is applied to the column at a flow rate 20 ml/hr. The column is washed with 6 column volumes of equilibration buffer followed by elution of the enzyme with equilibration buffer containing 0.1 M NaCl. Fractions containing phosphatidylinositol 4-kinase activity are pooled and immediately used for the next step in the purification scheme. The enzyme is still labile at this stage.

Step 5: Hydroxylapatite Chromatography. A hydroxylapatite (BioGel HT, Bio-Rad, Richmond, CA) column (2.5 × 3 cm) is equilibrated with 10 mM potassium phosphate buffer (pH 8.0) containing 10 mM MgCl$_2$, 10 mM 2-mercaptoethanol, 10% glycerol, and 0.2% Triton X-100. The DE-52 purified enzyme is diluted 2 : 1 with equilibration buffer and applied to the column at a flow rate of 10 ml/hr. The column is washed with 6 column volumes of equilibration buffer followed by elution of the enzyme with 8 column volumes of a linear potassium phosphate gradient (0.01–0.3 M) in equilibration buffer at a flow rate of 10 ml/hr. The peak of phosphatidylinositol 4-kinase activity elutes from the column at a phosphate concentration of about 0.15 M. The most active fractions are pooled and used for the next step in the purification. At this stage in the purification, phosphatidylinositol 4-kinase activity is completely stable.

Step 6: Octyl-Sepharose Chromatography. An octyl-Sepharose (Pharmacia LKB Biotechnology, Piscataway, NJ) column (1.5 × 16 cm) is equilibrated with 50 mM Tris-HCl buffer (pH 8.0) containing 10 mM MgCl$_2$, 10 mM 2-mercaptoethanol, 10% glycerol, and 3 M NaCl. Solid NaCl is added to the hydroxylapatite-purified enzyme to a final concentration of 3 M, and the enzyme is applied to the octyl-Sepharose column at a flow rate of 20 ml/hr. It is necessary to add 3 M NaCl and to omit Triton X-100 from the equilibration buffer for the enzyme to bind to the octyl-Sepharose column. The column is washed with 2 column volumes of equilibration buffer followed by 6 column volumes of equilibration buffer without NaCl. Phosphatidylinositol 4-kinase activity is then eluted from the column with 8 column volumes of a linear Triton X-100 gradient (0–1%) in 50 mM Tris-HCl buffer (pH 8.0) containing 10 mM MgCl$_2$, 10 mM 2-mercaptoethanol, and 10% glycerol. The peak of activity elutes from the column at a Triton X-100 concentration of about 0.5%. The most active fractions are pooled and used for the next step in the purification.

Step 7: Mono Q Chromatography I. A anion-exchange Mono Q (Pharmacia LKB Biotechnology) column (0.5 × 5 cm) is equilibrated with 50 mM Tris-HCl buffer (pH 8.0) containing 10 mM MgCl$_2$, 10 mM 2-mercaptoethanol, 10% glycerol, and 0.5% Triton X-100. Enzyme from the previous step is applied to the column at a flow rate of 60 ml/hr. The column is washed with 4 column volumes of equilibration buffer followed by 2 column volumes of equilibration buffer containing 0.2 M NaCl. The enzyme is then eluted from the column with 10 column volumes of a linear NaCl gradient (0.2–0.5 M) in equilibration buffer. The peak of enzyme activity elutes at a NaCl concentration of about 0.22 M. Fractions containing activity are pooled and used for the next step.

Step 8: Mono Q Chromatography II. A second Mono Q column (0.5 × 5 cm) is equilibrated with 50 mM Tris-HCl buffer (pH 8.0) containing 10 mM MgCl$_2$, 10 mM 2-mercaptoethanol, 10% glycerol, 0.5% Triton X-100, and 60 mM NaCl. The enzyme from the previous Mono Q column is diluted 4 : 1 with equilibration buffer and applied to the column at a flow rate of 45 ml/hr. The column is washed with 7 column volumes of equilibration buffer followed by enzyme elution with 20 column volumes of a linear NaCl gradient (0.06–0.3 M) in equilibration buffer. Phosphatidylinositol 4-kinase activity elutes at a NaCl concentration of about 0.2 M. Fractions containing activity are pooled and stored at −80°. The purified enzyme is completely stable for at least 3 months of storage at −80°.

TABLE I
PURIFICATION OF PHOSPHATIDYLINOSITOL 4-KINASE FROM *Saccharomyces cerevisiae*[a]

Purification step	Total units (nmol–min)	Protein (mg)	Specific activity (units/mg)	Purification (-fold)	Yield (%)
1. Cell extract	5970	10,119	0.59	1	100
2. Microsomes	4418	2,559	1.72	2.9	74
3. Triton X-100 extract	3104	794	3.9	6.6	52
4. DE-52	1086	175	6.2	10.5	18
5. Hydroxylapatite	1032	57	18.1	30.6	17.2
6. Octyl-Sepharose	743	5.64	131.7	223.2	12.4
7. Mono Q I	558	0.24	2325	3940.6	9.3
8. Mono Q II	380	0.08	4750	8050.8	6.3

[a] Data from Belunis *et al.*[8]

A summary of the purification of phosphatidylinositol 4-kinase is presented in Table I. The overall purification of phosphatidylinositol 4-kinase over the cell extract is 8051-fold, with an activity yield of 6.3%.

Improved Purification Procedure. The purification scheme described above includes octyl-Sepharose chromatography. This step takes approximately 12 hr to perform and results in proteolytic degradation of the enzyme.[20] We have modified the purification scheme by omitting octyl-Sepharose chromatography and Mono Q I chromatography (Steps 6 and 7 above, respectively).[20] The elution conditions for Mono Q II chromatography (Step 8) have been changed.[20] Following hydroxylapatite chromatography (Step 5), the enzyme preparation is desalted by dialysis against 50 mM Tris-HCl buffer (pH 8.0) containing 10 mM MgCl$_2$, 10 mM 2-mercaptoethanol, 10% glycerol, and 0.5% Triton X-100 (equilibration buffer). The dialyzed enzyme is applied to a Mono Q column (0.5 × 5 cm) that is equilibrated with equilibration buffer containing 60 mM NaCl. The column is washed with 10 column volumes of equilibration buffer containing 60 mM NaCl followed by enzyme elution with 35 column volumes of a linear NaCl gradient (0.06–0.3 M) in equilibration buffer. Fractions containing phosphatidylinositol 4-kinase activity are pooled and stored at −80°. The Mono Q chromatography step results in a 370-fold increase in specific activity over the hydroxylapatite chromatography step with an activity yield of 75%. Overall the enzyme is purified 9000-fold relative to the cell extract with a final specific activity of 4400 nmol/min/mg.

[20] R. J. Buxeda, J. T. Nickels, Jr., C. J. Belunis, and G. M. Carman, *J. Biol. Chem.* **266,** 13859 (1991).

Enzyme Purity. Electrophoretic analysis using native polyacrylamide[21] and sodium dodecyl sulfate-polyacrylamide[22] gels indicates that the purified phosphatidylinositol 4-kinase preparation is nearly homogeneous. Native gels (6% polyacrylamide) contain 0.5% Triton X-100. Phosphatidylinositol 4-kinase activity is associated with the protein bands found in native polyacrylamide[8] and sodium dodecyl sulfate-polyacrylamide gels.[8,20] The subunit molecular weight of the enzyme is 45,000.[20] On storage, the enzyme is degraded to molecular weights of 35,000 and 30,000 but retains full activity.[8,20]

Product Identification

A standard phosphatidylinositol 4-kinase reaction is carried out with either [γ-^{32}P]ATP or phosphatidyl[2-^{3}H]inositol as the labeled substrate. The chloroform-soluble product of the reaction is analyzed by one-dimensional paper chromatography on EDTA-treated SG81 paper[23] with chloroform–acetone–methanol–glacial acetic acid–water (40 : 15 : 13 : 12 : 8, v/v) and chloroform–methanol–2.5 M ammonium hydroxide (9 : 7 : 2, v/v) as the solvent systems. The only phospholipid product of the reaction using either labeled substrate comigrates precisely with authentic phosphatidylinositol 4-phosphate.[8,20] The ^{3}H-labeled product of the reaction is isolated and used as the substrate for phosphoinositide-specific phospholipase C, and the inositol phosphate product is identified by high-performance liquid chromatography.[8,24] The inositol phosphate hydrolysis product of the phospholipase C reaction is identified as inositol 1,4-bisphosphate.[8] In addition, the ^{32}P-labeled product of the reaction is isolated, then deacylated with methylamine reagent, and the water-soluble product is analyzed by high-performance liquid chromatography.[25] The product is identified as glycerophosphoinositol 4-phosphate, confirming that the product of the enzyme reaction is phosphatidylinositol 4-phosphate.[20]

Properties of Phosphatidylinositol 4-Kinase

Maximum phosphatidylinositol 4-kinase activity is dependent on magnesium ions (27 mM) and Triton X-100 (25 mM) at the pH optimum of 8.0.[8] The activation energy for the reaction is 31.5 kcal/mol, and the

[21] B. Davis, *Ann. N.Y. Acad. Sci.* **121**, 404 (1964).
[22] U. K. Laemmli, *Nature (London)* **227**, 680 (1970).
[23] S. Steiner and R. L. Lester, *J. Bacteriol.* **109**, 81 (1972).
[24] C. A. Hansen, S. Mah, and J. R. Williamson, *J. Biol. Chem.* **261**, 8100 (1986).
[25] D. L. Lips, P. W. Majerus, F. R. Gorga, A. T. Young, and T. L. Benjamin, *J. Biol. Chem.* **264**, 8759 (1989).

enzyme is thermally labile above 30°.[8] Phosphatidylinositol 4-kinase activity is inhibited by calcium ions and thioreactive agents but is not affected by various nucleotides including adenosine.[8]

Kinetic experiments are conducted with a mixed micelle substrate of Triton X-100 and phosphatidylinositol.[8,20] The enzyme shows saturation kinetics with respect to the bulk (K_m of 70 μM)[8] and surface (K_m of 0.4 mol%)[20] concentrations of phosphatidylinositol. The K_m value for MgATP is 0.3–0.5 mM,[8,20] and the V_{max} is 4750 nmol/min/mg.[8] The turnover number for the enzyme is 166 min^{-1}. Results of kinetic and isotopic exchange reactions indicate that phosphatidylinositol 4-kinase catalyzes a sequential Bi–Bi reaction mechanism.[8,20] The enzyme binds to phosphatidylinositol prior to ATP, and phosphatidylinositol 4-phosphate is the first product released in the reaction.[8]

Synthetic Uses

Pure phosphatidylinositol 4-kinase can be used to synthesize radiolabeled phosphatidylinositol 4-phosphate.[8] ^{32}P-Labeled phosphatidylinositol 4-phosphate is prepared from phosphatidylinositol and [γ-^{32}P]ATP, and ^{3}H-labeled phosphatidylinositol 4-phosphate is prepared from phosphatidyl[2-^{3}H]inositol and ATP.

Acknowledgments

This work was supported by U.S. Public Health Service Grant GM-35655 from the National Institutes of Health, New Jersey State funds, and the Charles and Johanna Busch Memorial Fund.

[21] Phosphatidylinositol-4-Phosphate 5-Kinases from Human Erythrocytes

By Chantal E. Bazenet and Richard A. Anderson

Introduction

Polyphosphoinositides play a crucial role in cellular signal transduction. Phosphatidylinositol (PI) undergoes a series of phosphorylations to give phosphatidylinositol phosphate (PIP) and further phosphatidylinositol bisphosphate (PIP$_2$), which produces two second messengers, 1,2-diacylglycerol (DAG) and inositol 1,4,5-trisphosphate (IP$_3$), mediating Ca^{2+} mobilization in response to many hormones, neurotransmitters, and growth factors. It is therefore important to understand how the metabolism of PIP and PIP$_2$ is regulated. Although the breakdown of polyphosphoinositides by different hydrolytic enzymes has been studied extensively, little is yet known about the regulation of the enzymes involved in the synthesis of these phospholipids. The enzyme directly responsible for the synthesis of PIP$_2$ is the phosphatidylinositol-4-phosphate 5-kinase (PIP kinase, EC 2.7.1.68, 1-phosphatidylinositol-4-phosphate kinase), which has been studied and partially purified from a number of sources, including human erythrocytes.[1–3]

The PIP kinase activity in cells is found both in the cytosol and associated with plasma membranes.[1–10] Until now, there were few data to suggest whether membrane-bound PIP kinases were structurally related to the cytosolic kinases. If the cytosolic PIP kinases are identical to the membrane-bound PIP kinases, they could serve as a reserve pool of activity which is assembled on membranes when PIP$_2$ production is required. Recently, we have found two forms of PIP kinase in human erythrocytes: the type I PIP kinase, which is membrane bound, and the type II enzyme, which is both membrane bound and cytosolic.[11] This chapter explains in detail the preparation of two distinct forms of PIP kinases from human erythrocyte membranes and cytosol, their one-step purification on a phosphocellulose column, and the study of their structural and functional properties.

Materials

Fresh human blood is obtained from the Wisconsin Blood Bank (Madison, WI) and used within 4 days of the drawing date. [γ-^{32}P]ATP is from Du Pont–New England Nuclear (Boston, MA) or is prepared from inorganic ^{32}PO$_4$ (Du Pont–New England Nuclear) using a [γ-^{32}P]ATP synthesis kit from Promega (Madison, WI). The purity of the synthesized [γ-^{32}P]ATP

[1] A. Moritz, P. N. E. de Graan, P. F. Ekhart, W. H. Gispen, and K. W. A. Wirtz, *J. Neurochem.* **54,** 351 (1990).

[2] L. E. Ling, J. T. Schulz, and L. C. Cantley, *J. Biol. Chem.* **264,** 5080 (1989).

[3] C. Cochet and E. M. Chambaz, *Biochem. J.* **237,** 25 (1986).

[4] S. Cockcroft, J. A. Taylor, and J. D. Judah, *Biochim. Biophys. Acta* **845,** 163 (1985).

[5] E. B. Stubbs, Jr., J. A. Kelleher, and G. Y. Sun, *Biochim. Biophys. Acta* **958,** 247 (1988).

[6] C. D. Smith and W. W. Wells, *J. Biol. Chem.* **258,** 9368 (1983).

[7] E. S. Husebye and T. Flatmark, *Biochim. Biophys. Acta* **1010,** 250 (1989).

[8] M. C. Pike and C. J. Arndt, *J. Immunol.* **140,** 1967 (1988).

[9] G. A. Lundberg, R. Sundler, and B. Jergil, *Biochim. Biophys. Acta* **846,** 379 (1985).

[10] C. J. Van Dongen, H. Zwiers, and W. H. Gispen, *Biochem. J.* **223,** 197 (1984).

[11] C. E. Bazenet, A. Ruiz Ruano, J. L. Brockman, and R. A. Anderson, *J. Biol. Chem.* **265,** 18012 (1990).

is always checked by thin-layer chromatography (TLC). Phospholipids, except for PIP and PIP_2, are purchased from Avanti Polar Lipids (Birmingham, AL). PIP_2 and PIP are prepared from bovine brain by neomycin affinity chromatography.[12] Chemicals for sodium dodecyl sulfate-polyacrylamide gel electrophoresis (SDS–PAGE) are from Bethesda Research Laboratories (Gaithersburg, MD). ATP, dithiothreitol, diisopropyl fluorophosphate (DFP), phenylmethylsulfonyl fluoride (PMSF), Tris, EDTA, EGTA, pepstatin A, leupeptin, and other chemicals are from Sigma (St. Louis, MO).

Procedures

Phosphatidylinositol-4-Phosphate 5-Kinase Assays and Thin-Layer Separation of Phospholipids

A 5-μl aliquot of a column fraction is diluted to a final volume of 50 μl; the final concentrations in the reaction are 50 mM Tris, 5.0 mM $MgCl_2$, 0.5 mM EGTA, 50 μM [γ-^{32}P]ATP (2 Ci/mmol), 0.1 mM PMSF, and 80 μM PIP (pH 7.5). For quantitation of PIP kinase activity, the ionic strength of the assay is kept constant by adding KCl or water. The reaction is stopped after a 3-min incubation (20°–23°) by the addition of 50 μl of 12 N HCl and then 500 μl of chloroform–methanol–water (15 : 15 : 5, v/v); the resulting aqueous phase (200 μl) is dried by vacuum centrifugation, redissolved in 50 μl of chloroform–methanol–12 N HCl (200 : 100 : 1, v/v), and spotted onto a TLC plate [either Whatman (Clifton,NJ) silica gel 150 A or Merck (Darmstadt, Germany) silica gel 60 plates which have been pretreated with 60 mM EDTA, 2% tartarate (w/v), and 50% ethanol (pH 8.0) and then dried at room temperature]. The plates are then developed in chloroform–methanol–water–15 N ammonium hydroxide (90 : 90 : 22 : 7, v/v). Phospholipids are visualized with iodine vapor or by spraying with 3% cupric acetate and 8% phosphoric acid (w/v) and baking for 40 min at 120°. Reaction products are visualized by autoradiography and identified by comparing to unlabeled standard phospholipids. Radioactivity in each phospholipid is determined by scraping the silica gel corresponding to the ^{32}P-labeled phospholipid and quantitating by scintillation counting.

Preparation of Human Erythrocyte Cytosol and Membranes

Fresh human erythrocytes are isolated from sedimentation (1 g) through phosphate-buffered saline (PBS) (4°) containing 0.75% dextran, 1 mM adenosine, and 10 mM glucose (pH 7.5). The erythrocytes are washed

---

[12] F. B. St. C. Palmer, *J. Lipid Res.* **22**, 1296 (1981).

3 more times in the same buffer by centrifugation at 600 g. The packed cells (800–1000 ml) are normally lysed in a volume of 6 liters of lysis buffer (5 mM sodium phosphate, 1 mM EDTA, 2 mM 2-mercaptoethanol, 0.1 mM DFP, 20 μg/ml PMSF, and 10 μg/ml of each pepstatin A and leupeptin), pH 8.0, at 4°. The membranes are washed free of cytosol using a Pellicon Cassette Filtration System (Millipore, Bedford, MA). After lysis, the membranes are washed with lysis buffer using a Pellicon Cassette Filtration System until they are white.

Purification of Cytosolic Phosphatidylinositol-4-Phosphate 5-Kinase

The lysate in 6–8 liters is precipitated with 43% ammonium sulfate (pH 6.8). The pellet is dissolved in 500 ml of lysis buffer (pH 7.5) containing 250 mM NaCl dialyzed overnight, centrifuged, and the supernatant applied to a phosphocellulose column (20–30 ml) equilibrated with the same buffer. The column is washed with lysis buffer containing 250 mM NaCl until the flow-through has an absorbance at 280 nm of 0.05 or less; the column is then washed with 100 ml of lysis buffer containing 500 mM NaCl and eluted with a 500–1400 mM NaCl gradient (Fig. 1). The fractions are assayed for activity and analyzed by SDS–PAGE (according to Laemmli,[13] with either 10% acrylamide or 7–15% linear gradient separating gels, both with 4% stacking gels). By comparing the activity profile and the silver-stained SDS–polyacrylamide gels, it is clear that the cytosolic PIP kinase corresponds to protein bands of 53 and 43 kDa. The 53- and 43-kDa bands in this pool of activity are always greater than 95% of the total protein, and in most preparations no other bands are observed by silver staining. Quantitation of the cytosolic PIP kinase purification is shown in Table I.

Preparation of Antibodies; Western Blot

Our first attempt to raise antibodies against the cytosolic PIP kinase was not successful; thus, to make the cytosolic PIP kinase more antigenic, it was coupled to keyhole limpet hemocyanin (KLH). KLH (2 mg) in 200 μl of water is combined with 3 mg of pure cytosolic PIP kinase in 500 μl of 0.1 M 4-morpholinoethanesulfonic acid (MES) and 0.9 M NaCl (pH 4.7); immediately, 50 μl of 10 mg/ml 1-ethyl-3-(3-dimethylaminopropyl) carbodiimide hydrochloride in water is added, and the solution is incubated for 3 h at room temperature. After coupling, the conjugated PIP kinase is dialyzed against PBS.[14] The coupling is analyzed by SDS–PAGE and

[13] U. K. Laemmli, *Nature (London)* **227**, 680 (1972).
[14] E. Harlow and D. Lane, "Antibodies: A Laboratory Manual," p. 56. Cold Spring Harbor Laboratory, Cold Spring Harbor, New York, 1988.

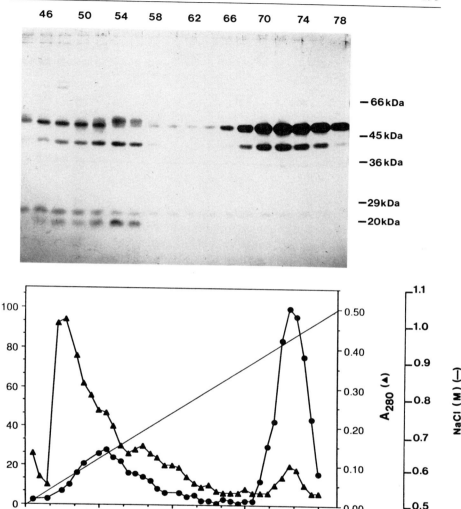

Fig. 1. Elution profile of cytosolic PIP kinase from the phosphocellulose cation-exchange column. The bottom graph shows the elution of PIP kinase activity, protein, and the NaCl gradient. The top photograph shows a silver-stained SDS–polyacrylamide gel of fractions 44 through 78 of the phosphocellulose elution. The peak of PIP kinase activity is at fraction 72.

TABLE I
PURIFICATION OF CYTOSOLIC PHOSPHATIDYLINOSITOL-4-PHOSPHATE 5-KINASE

Step	Protein (mg)	Activity[a] (nmol/min)	Specific activity (nmol/min/mg)	Yield (%)	Purification (-fold)
Lysate	2.2×10^5	840	0.0038	100	1
Ammonium sulfate precipitate	1.8×10^4	440	0.024	53	6.6
Phosphocellulose pool	5.8	75	13	9	3400

[a] Assayed with 80 μM PIP.

routinely shows only high molecular mass (>200 kDa) aggregates of proteins. Four rabbits are immunized and boosted with 200 μg of PIP kinase conjugated to KLH using the above method. After the initial boost, the antibody titers are further increased by boosting each week with 100 μg of pure cytosolic PIP kinase in PBS; this is done by intravenous injection. After the second intravenous boost, all four rabbits should show a response to the PIP kinase (titers > 1/4000).

Antibodies are affinity purified using a cytosolic PIP kinase-Sepharose affinity column. The affinity column is prepared by cyanogen bromide activation of Sepharose 4B.[15] Antisera (50 ml) are applied to a 5-ml column at a flow rate of 10 ml/hr, and the column is washed with 0.5 M NaCl and 10 mM sodium phosphate (pH 7.5) and then eluted with 1.0 M acetic acid. The pure antibody is dialyzed against isotonic KCl buffer and concentrated using a Centricon concentrator (Amicon, Danvers, MA).

Western blot analysis is done either with affinity-purified antibodies or with antiserum. For Western blotting, proteins are transferred from SDS–polyacrylamide slab gels to nitrocellulose paper.[16,17] After transfer, the nitrocellulose sheet [Schleicher & Schuell (Keene, NH) BASS, 0.45 μm] is washed in 150 mM NaCl, 50 mM Tris, 5 mM NaN$_3$ (pH 7.4) (Tris–saline), then blocked for 24 hr in 5% bovine serum albumin (BSA) in the same buffer. Affinity-purified antibodies (20 μg/ml) or antiserum (1/10 dilution in Tris–saline with 5% BSA) is used for blotting, and protein A labeled with ^{125}I (0.5 Ci/mg) at a concentration of 5 × 10^7 counts/min (cpm)/ml in 5% BSA is used to detect antibody. The sheets are washed with six to eight changes of 100 ml of Tris–saline over 6–12 hr, dried in

[15] S. C. March, I. Parikh, and P. Cuatrecasas, *Anal. Biochem.* **60**, 149 (1974).

[16] R. A. Anderson, I. C. Correas, C. E. Mazzucco, J. D. Castle, and V. T. Marchesi, *J. Cell. Biochem.* **37**, 269 (1988).

[17] H. Towbin, T. Staehlin, and J. Gordon, *Proc. Natl. Acad. Sci. U.S.A.* **76**, 4350 (1979).

between filter paper, and autoradiographed with Kodak (Rochester, NY) XAR-5 film.

Purification of Types I and II Membrane-Bound Phosphatidylinositol-4-Phosphate 5-Kinases

The membrane-bound PIP kinase activity is eluted from erythrocyte ghosts with 0.6 M NaCl, 5 mM sodium phosphate, 5 mM EDTA, 0.1 mM DFP, 0.1 mM PMSF, and 3 μg/ml of leupeptin and pepstatin A (pH 7.5) at 0° for 30 min. The membranes are pelleted by centrifugation at 250,000 g, and the supernatant, which contains greater than 80% of the membrane-bound PIP kinase activity, is dialyzed against lysis buffer containing 200 mM NaCl and 10% (v/v) glycerol (pH 7.5) and applied to a phosphocellulose column equilibrated with the same buffer (all at 4°). The membrane-bound PIP kinases are eluted with a 200–1400 mM NaCl gradient. The fractions are assayed for PIP kinase activity and analyzed by SDS–PAGE. As shown in Fig. 2, the PIP kinase activity elutes in two peaks, at 0.6 and 1.0 M NaCl. The first peak of PIP kinase activity (type I PIP kinase), which elutes at 0.6 M NaCl, is not pure. However, the second peak of PIP kinase activity, which elutes at 1.0 M NaCl, is quite pure and corresponds to 53- and 43-kDa bands. The preparation shown in Fig. 2 is a representative preparation which has a minor contamination by the α and the β chain of spectrin (230 and 220 kDa) and 100- and 105-kDa bands, which are adducin.[18] Quantitation of the purification of the membrane-bound PIP kinases is shown in Table II.

Properties

Cytosolic and Membrane-Bound Type II Phosphatidylinositol-4-Phosphate 5-Kinases Are Indistinguishable. The cytosolic and the membrane-bound type II PIP kinases have very similar properties. Both elute from the phosphocellulose column with 1.0 M NaCl, and on SDS–polyacrylamide gels both are 53 kDa with a 43-kDa minor band. To determine the extent to which these PIP kinases are related, they were compared by two-dimensional ^{125}I-labeled peptide mapping.[13,19] The kinases were compared using both chymotryptic and tryptic peptide maps. The peptide maps (Fig. 3) show that the cytosolic PIP kinase is indistinguishable from the membrane-bound type II PIP kinase. Both have a 43-kDa protein which

[18] D. W. Cleveland, S. G. Fisher, M. W. Kirschner, and U. K. Laemmli, *J. Biol. Chem.* **252,** 1102 (1977).

[19] D. W. Speicher, J. S. Morrow, W. J. Knowles, and V. T. Marchesi, *J. Biol. Chem.* **257,** 9093 (1982).

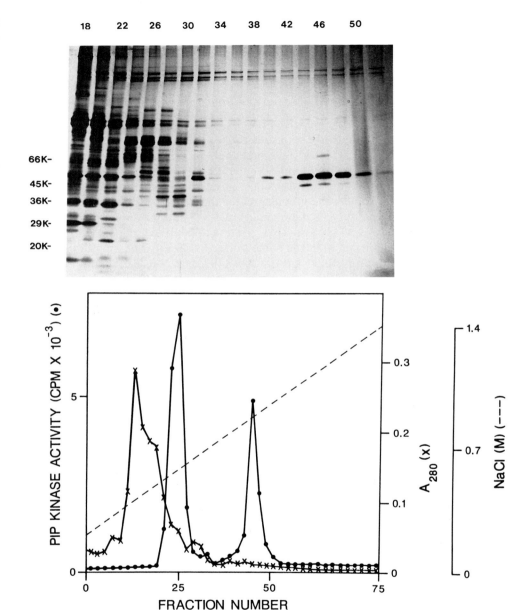

Fɪɢ. 2. Elution profile of membrane-bound PIP kinase activities from the phosphocellulose ion-exchange column. The bottom graph shows the elution of PIP kinase activity, protein, and the NaCl gradient. The top photograph shows a silver-stained SDS–polyacrylamide gel of fractions 18 through 52. Two peaks of PIP kinase activity are eluted from the phosphocellulose column; the first peak of activity (type I PIP kinase) has maximal activity in fraction 26; the second peak of PIP kinase activity (type II PIP kinase) has maximal activity in fraction 46.

TABLE II
MEMBRANE-BOUND PHOSPHATIDYLINOSITOL-4-PHOSPHATE 5-KINASE PURIFICATION

Step	Protein (mg)	Activity[a] (nmol/min)	Specific activity (nmol/min/mg)	Yield (%)	Purification (-fold)
0.6 M NaCl extract	440	14.0	0.03	100	1
Type I pool of activity	4.2	4.5	1.1	32	37
Type II pool of activity	0.5	7.4	15	53	520

[a] Assayed with 80 μM PIP.

copurifies with the 53-kDa major protein band. When the 43-kDa band was peptide mapped, it had a pattern of peptides which was a subset of the cytosolic and the membrane-bound type II PIP kinases but missing three tryptic peptides found in the map of the 53-kDa type II PIP kinases. This suggests that the 43-kDa band is a proteolytic fragment of the 53-kDa protein. Peptide mapping demonstrated that the cytosolic and the membrane-bound type II PIP kinases are indistinguishable and thus will be referred to as type II PIP kinases.

   Type I Phosphatidylinositol-4-Phosphate 5-Kinase Is Distinct from Type II Enzyme. Rabbit polyclonal antibodies were raised to the cytosolic type II PIP kinase, and the antibodies were affinity purified. This antibody was used to determine if the type I PIP kinase is related to the type II PIP kinases. Immunoreactivity with a 53-kDa band was shown in all the purified PIP kinases, the cytosol, and the membranes using the affinity-purified anticytosolic PIP kinase antibodies (data not shown), but this immunoreactivity was lower for the type I PIP kinase when the same amount of PIP kinase activity was compared. Moreover, when the elution of type II PIP kinase from phosphocellulose chromatography of the membrane extract was quantitated using the anticytosolic PIP kinase antibody, the type II PIP kinase eluted with a symmetrical peak which corresponds with type II PIP kinase activity but had only a minor overlap with the type I peak of PIP kinase activity (Fig. 4).

   To determine if the type I pool of PIP kinase activity could contain an activator of the type II PIP kinase, the type I pool was mixed with the type II pool of activity; the resulting activity was always the sum of the activity of the individual pools.

   Affinity-purified cytosolic PIP kinase antibodies are potent inhibitors of type II PIP kinase activity. As shown in Fig. 5, these antibodies inhibit greater than 98% of type II PIP kinase activity but do not inhibit kinase

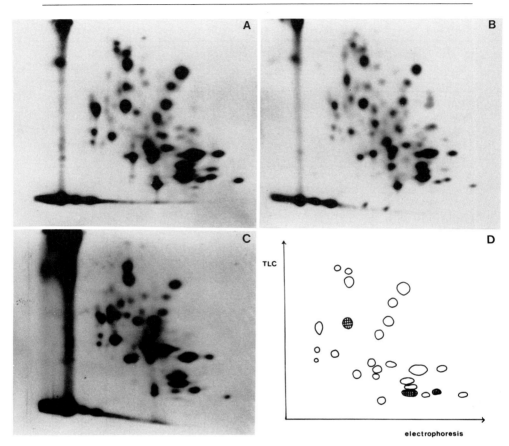

FIG. 3. Two-dimensional ^{125}I-labeled peptide maps of the cytosolic PIP kinase (A), the type II membrane-bound PIP kinase (B), and the 43-kDa protein (C) which copurifies with both PIP kinases. (D) Composite drawing of peptides in common (open symbols) and peptides which are missing in the 43-kDa protein (hatched).

activity in the phosphocellulose pool of type I PIP kinase. This is further evidence that type I PIP kinase is distinct from the type II PIP kinase.

Effect of PIP Kinase Modulators. Stock solutions of spermine hydrochloride and heparin were made up immediately before use and diluted to the final concentrations in the reaction vial. The effect of spermine on PIP kinase activities was assayed at 2 mM MgCl$_2$, and the effect of heparin was assayed at 5 mM MgCl$_2$; all other conditions are the same as above.

Previously, spermine has been shown to be a potent stimulator of PIP kinase activity in soluble preparations and on membranes.[3,10] Spermine

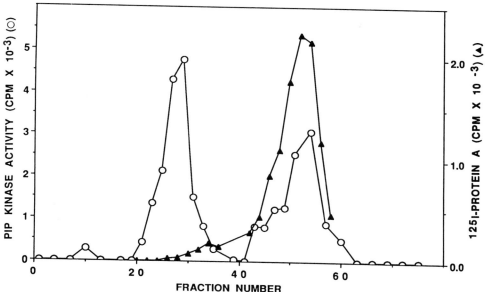

FIG. 4. The type I peak of PIP kinase activity does not correspond to the elution of type II PIP kinase from the phosphocellulose column. The top photograph shows an autoradiogram of a Western blot using anti-type II PIP kinase antibodies. For Western blotting, 40 μl of the fraction volume was applied to a SDS-7–15% polyacrylamide gel; after electrophoresis, the protein was transferred to nitrocellulose and blotted with ^{125}I-labeled protein A. After autoradiography, the 53-kDa band corresponding to the type II PIP kinase was excised and quantitated by gamma counting. The elution profile of PIP kinase activity from the phosphocellulose column and the amount of type II PIP kinase detected by antibody and ^{125}I-labeled protein A (▲) are shown.

also stimulates erythroid PIP kinase activity but is specific for type I PIP kinase. The type I PIP kinase increases by as much as 4-fold with 2 mM spermine, whereas spermine does not have any effect on the type II PIP kinases. In contrast, heparin has been shown to inhibit PIP kinase activity in some preparations.[7] The erythroid type I PIP kinase is also stimulated by low concentrations of heparin, whereas the type II PIP kinases are inhibited.

Kinetic Properties of PIP Kinases. The different forms of PIP kinase were compared functionally by determining their kinetic parameters to-

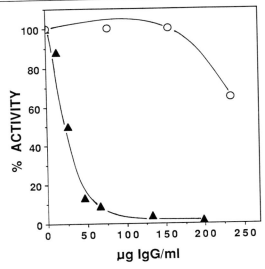

F<small>IG</small>. 5. Antitype II PIP kinase antibodies potently inhibit type II PIP kinases but do not affect type I PIP kinase. Effect of anticytosolic PIP kinase antibodies on type I PIP kinase activity (○) and type II PIP kinase activity (▲). For inhibition of 99% of the cytosolic and membrane-bound type II PIP kinase activity, 2 μg of affinity-purified IgG was required for each 1 μg of pure type II PIP kinase. At this ratio of antibody to type I PIP kinase, there was no detectable loss of activity.

ward both ATP and PIP. PIP was either used in a micellar form or, to model the phosphorylation of PIP in membranes more closely, PIP was incorporated into phospholipid liposomes. The following composition was the best substrate for both types of PIP kinase: 45% phosphatidyletha-nolamine (PE), 30% phosphatidylserine (PS), 12% sphingomyelin, and 1% PIP (in mol %).

To begin study of the assembly of the PIP kinases onto native membranes, the different forms of PIP kinase were reconstituted onto ghosts, inside-out vesicles (IOVs), or inside-out erythrocyte membranes which had been stripped of extrinsic proteins (sIOVs). Table III shows a summary of the results obtained. Type I PIP kinase has 10 times more affinity for PIP in a micellar form than does type II. When the activity of the PIP kinases toward PIP in a phospholipid membrane (liposomes) is compared, the type I PIP kinase has a 30-fold greater affinity toward PIP in liposomes than do the type II PIP kinases. Both types of PIP kinases have a higher affinity toward PIP in liposomes than toward PIP in micelles.

However, the most important observation was made when the different forms of PIP kinases were incubated with ghosts, IOVs, or sIOVs. In comparing the activities of types I and II PIP kinase toward intrinsic PIP, it became clear that the type I PIP kinase was the only form active toward

TABLE III
KINETIC PROPERTIES OF PHOSPHATIDYLINOSITOL-4-
PHOSPHATE 5-KINASES

Substrate	K_m (μM)	
	Type I	Type II
PIP micelles	6	60
PIP liposomes	1.4	40
PIP in membranes	1.2	No activity
ATP	25	5

membranes. The type II PIP kinases are not active under these conditions, even when 30-fold greater PIP kinase activity was added to membranes as compared with type I PIP kinase. Type I PIP kinase has the same K_m toward PIP in liposomes as toward PIP in sIOVs, but the V_{max} toward sIOVs is 15-fold less when compared with liposomes.

The type I PIP kinase has an apparent K_m of 25 μM ATP, whereas the type II PIP kinases have K_m values for ATP below 5 μM. Previously, it has been reported that some PIP kinases utilize GTP as a phosphate donor.[3] Both types I and II PIP kinases utilize GTP as a phosphate donor (data not shown). Indeed, with 50 μM GTP, both kinases more actively phosphorylate PIP compared with 50 μM ATP as the phosphate donor. With GTP, the type I PIP kinase activity increases 2-fold and the type II PIP kinase increases by 5-fold compared with ATP.

Conclusions

Three forms of PIP kinase have been purified from erythrocytes. Of these PIP kinases, the type II membrane-bound and the cytosolic forms have indistinguishable two-dimensional peptide maps, suggesting that they have the same primary sequence. Further, antibodies raised to the cytosolic PIP kinase cannot distinguish between the cytosolic and the membrane-bound type II PIP kinase. Functionally, the type II membrane-bound and the cytosolic forms appear identical: both have the same K_m for ATP and PIP, are inhibited by heparin, are not stimulated by spermine, have low activity toward PIP in liposomes, and have no activity toward intrinsic PIP in native membranes. In contrast, type I PIP kinase is structurally and functionally distinct from the type II PIP kinases: type I PIP kinase activity is not affected by the anticytosolic antibodies, is highly stimulated by both heparin and spermine, has a high affinity for PIP in micelles or incorporated into phospholipid liposomes, and is specifically the only form able to phosphorylate PIP in isolated membranes.

Although the type II PIP kinase does not phosphorylate PIP in isolated membranes, this does not demonstrate that it is inactive toward erythrocyte membranes *in vivo*. Instead, it is likely that the conditions for type II PIP kinase assembly on membranes are not known.

In all cells and tissues so far studied, the PIP kinase was found both as a cytosolic and as a membrane-bound enzyme.[3-10] Here we have shown that the erythrocyte cytosolic and the membrane-bound type II PIP kinases have identical or very similar structural properties. Functionally, this observation is important since, when activated, the cytosolic PIP kinase may function as a reserve pool which is assembled on membranes when increased PIP$_2$ production is required. This may occur in response to agonist stimulation of phosphoinositide turnover; such a mechanism would be analogous to the interaction of cytosolic forms of phospholipase C and protein kinase C with membranes.[1-3]

It appears that the PIP kinases are regulated by distinct pathways. Phospholipase C is regulated by GTP-binding proteins[20] or by tyrosine phosphorylation,[21,22] and analogous mechanisms may regulate PIP kinase activity toward membranes.

Acknowledgments

This work was supported by Grant GM 38906 from the National Institutes of Health and by American Heart Association Grant 880934.

[20] S. T. Taylor, J. A. Smith, and J. H. Exton, *J. Biol. Chem.* **265**, 17150 (1990).
[21] G. Todderud, M. I. Wahl, S. G. Rhee, and G. Carpenter, *Science* **249**, 296 (1990).
[22] T. Mustelin, K. M. Coggeshall, N. Isakov, and A. Altman, *Science* **247**, (1990).

[22] Phosphatidylinositol 4-Kinase and Phosphatidylinositol-4-Phosphate 5-Kinase from Bovine Brain Membranes

By ALBRECHT MORITZ, JAN WESTERMAN, PIERRE N. E. DE GRAAN, and KAREL W. A. WIRTZ

Introduction

The hormone-stimulated hydrolysis of phosphatidylinositol 4,5-bisphosphate (PIP$_2$) is a key event in the receptor-mediated signal transduction pathway.[1-4] Replenishment of the hormone-sensitive PIP$_2$ pool re-

[1] M. J. Berridge and R. F. Irvine, *Nature (London)* **312**, 315 (1984).
[2] Y. Nishizuka, *Nature (London)* **308**, 693 (1984).
[3] M. J. Berridge, *Annu. Rev. Biochem.* **56**, 159 (1987).

quires the sequential phosphorylation of phosphatidylinositol (PI) by phosphatidylinositol 4-kinase (PI kinase) and phosphatidylinositol-4-phosphate 5-kinase (PIP kinase). PI kinase activity in cells is almost exclusively membrane bound. The enzyme has been purified from bovine brain myelin,[5] bovine uteri,[6] rat brain,[7] porcine liver microsomes,[8] A431 cells,[9] and *Saccharomyces cerevisiae*.[10] The enzyme from rat brain[7] has a subunit molecular weight of 80,000, that from bovine uteri, porcine liver, and A431 cells[6,8,9] a M_r of 55,000, that from bovine brain myelin[5] a M_r of 45,000, and that from yeast[10] a M_r of 35,000. In addition to PI kinase that phosphorylates the *myo*-inositol moiety at the 4'-position, phosphatidylinositol 3-kinase activity has been detected in fibroblasts.[11] The latter PI kinase has been partially purified from the cytosol of bovine brain and has a molecular weight of 85,000.[12] In contrast to PI kinase, PIP kinase activity occurs in a soluble and membrane-bound form. From human erythrocytes both a cytosolic PIP kinase[13] and one of two membrane-bound forms[13,14] have been purified extensively (for details, see elsewhere in this volume[15]). A soluble form of PIP kinase with a molecular weight of 45,000 has also been extensively purified from rat brain.[16,17] Recently we reported the purification of a PIP kinase from bovine brain membranes.[18] This protein has an apparent molecular weight of 110,000.[18]

In this chapter, we report on PI and PIP kinases from bovine brain membranes. First, a simple and very sensitive assay is described to moni-

[4] J. R. Williamson and C. A. Hansen, *in* "Biochemical Action of Hormones" (G. Litwack, ed.), Vol. 14, p.29. Academic Press, Orlando, Florida, 1987.

[5] A. R. Saltiel, J. A. Fox, P. Sherline, N. Sahyouni, and P. Cuatrecasas, *Biochem. J.* **241**, 759 (1987).

[6] F. D. Porter,. Y.-S. Li, and T. F. Deuel, *J. Biol. Chem.* **263**, 8989 (1988).

[7] A. Yamakawa and T. Takenawa, *J. Biol. Chem.* **263**, 17555 (1988).

[8] W.-M. Hou, Z.-L. Zhang, and H.-H. Tai, *Biochim Biophys. Acta* **959**, 67 (1988).

[9] D. H. Walker, N. Dougherty, and L. J. Pike, *Biochemistry* **27**, 6504 (1988).

[10] C. J. Belunis, M. Bae-Lee, M. J. Kelley, and G. M. Carman, *J. Biol. Chem.* **263**, 18897 (1988).

[11] M. Whitman, C. P. Downes, M. Keeler, T. Keller, and L. Cantley, *Nature (London)* **332**, 644 (1988).

[12] S. J. Morgan, A. D. Smith, and P. J. Parker, *Eur. J. Biochem.* **191**, 761 (1990).

[13] L. E. Ling, J. T. Schulz, and L. C. Cantley, *J. Biol. Chem.* **264**, 5080 (1989).

[14] C. E. Bazenet, A. Ruiz Ruano, J. L. Brockman, and R. A. Anderson, *J. Biol. Chem.* **265**, 18012 (1990).

[15] C. E. Bazenet and R. A. Anderson, this volume [21].

[16] C. J. Van Dongen, H. Zwiers, and W. H. Gispen, *Biochem. J.* **223**, 197 (1984).

[17] C. J. Van Dongen, J. W. Kok, L. H. Schrama, A. B. Oestreicher, and W. H. Gispen, *Biochem. J.* **233**, 859 (1986).

[18] A. Moritz, P. N. E. de Graan, P. F. Ekhart, W. H. Gispen, and K. W. A. Wirtz, *J. Neurochem.* **54**, 351 (1990).

tor PI kinase activity during its purification. The assay is based on the detection of pyrene-labeled PIP (Pyr-PIP), which is produced from sn-2-pyrenylacyl-labeled PI (Pyr-PI) on incubation with PI kinase in the presence of ATP. In addition, we describe the partial purification of PI kinase from bovine brain membranes based on the method described earlier.[19] This purification yielded a very active fraction that was suitable for the preparative synthesis of Pyr-PIP from Pyr-PI.[19] In addition, an assay for PIP kinase activity is presented which is a modification of the method described earlier.[18] As mentioned above, we have purified a membrane-bound PIP kinase from bovine brain membranes.[18] Here we report the partial purification and properties of another PIP kinase from the same source. Some properties of the partially purified enzyme are presented and compared with properties of other PIP kinases.

Materials

Phosphocellulose (P-11) and diethylaminoethyl (DEAE)-cellulose (DE-52) were purchased from Whatman (Maidstone, UK). Phenyl-Sepharose, Ultrogel AcA 44, and ATP-agarose (AGATP type 3) were from Pharmacia LKB (Uppsala, Sweden). Thin-layer chromatography (TLC) gel plates were from Merck (Darmstadt, Germany). ATP was purchased from Boehringer (Mannheim, Germany), and bovine serum albumin from Calbiochem (San Diego, CA). [γ-^{32}P]ATP (specific activity 3000 Ci/mmol) was obtained from Amersham (UK). PIP and carrier lipid (Folch fraction I from bovine brain) were from Sigma (St. Louis, MO). Phosphatidylserine (PS) was purified from bovine brain.[19a] Pyr-PI was synthesized from yeast PI as described.[20,21]

Fluorescent Assay for Phosphatidylinositol 4-Kinase

Principle. The fluorescent substrate Pyr-PI is phosphorylated to Pyr-PIP on incubation with ATP and PI kinase. After separation by TLC, Pyr-PIP is visualized under UV light.

Procedure. The assay is carried out in Eppendorf tubes in an incubation medium containing final concentrations of 50 mM Tris-HCl (pH 7.4), 100 mM MgCl$_2$, 1 mM EGTA, 1 mM EDTA, 100 μM Pyr-PI (suspended by

[19] T. W. J. Gadella, A. Moritz, J. Westerman, and K. W. A. Wirtz, *Biochemistry* **29**, 3389 (1990).

[19a] H. S. Hendrickson and C. E. Ballou, *J. Biol. Chem.* **239**, 1369 (1964).

[20] P. J. Somerharju and K. W. A. Wirtz, *Chem. Phys. Lipids* **30**, 82 (1982).

[21] P. J. Somerharju, J. A. Virtanen, K. K. Eklund, P. Vaino, and P. K. J. Kikkunen, *Biochemistry* **24**, 2773 (1985).

vortexing), 0.25 M sucrose, 0.1% polyethylene glycol (PEG) 20,000, 0.3% 2-mercaptoethanol, 0.2% Triton X-100, and 1 mM ATP (final volume of 25 μl). The reaction is started by addition of the enzyme solution (5 μl), then continued for 90 min at 30°. The reaction is stopped by the addition of 0.5 M HCl (5 μl). Chloroform (50 μl) is added, and lipids are extracted from the incubation mixture by vortexing. After centrifugation in an Eppendorf 5414 centrifuge at maximum speed for 1 min, the upper phase is removed with a Hamilton syringe. Methanol (80 μl) is added to obtain a homogeneous solution. The solution is applied on an oxalate-impregnated TLC plate[21a] while N$_2$ is blown over the plate until the lipid-containing spots are dry (10 min). The TLC plate is developed in chloroform–acetone–methanol–acetic acid–water (40 : 15 : 15 : 12 : 7.5, by volume) for 60 min to separate the lipids. Pyr-PIP is detected under UV light (356 nm).

Comments. Sucrose, EGTA, EDTA, PEG 20,000, and 2-mercaptoethanol are added to the assay medium in order to stabilize the PI kinase activity.[19] The advantage of this assay is the rapid identification of column fractions that contain enzyme activity.

Other Assays. In general, the activity of PI kinase is monitored by measuring the incorporation of radioactive phosphate from [γ-^{32}P]ATP into PI (see Refs. 5–10). After the phosphorylation reaction, two-phase partitioning of the incubation mixture is performed to separate radioactive PIP from unreacted [γ-^{32}P]ATP. To remove the last traces of radioactive contaminants, [^{32}P]PIP is then separated by TLC or paper chromatography.[5–8,10] Separation of PIP on neomycin–glass beads has also been reported.[9] After chromatography, radioactive PIP is quantified by liquid scintillation counting. In some studies, [^{32}P]PIP was quantified directly after two-phase partitioning without additional chromatography.[7,8]

Purification of Phosphatidylinositol 4-Kinase

Recently the enzymatic synthesis of fluorescent Pyr-PIP from Pyr-PI was described.[19] This synthesis was based on the application of a partially purified, highly active preparation of PI kinase from bovine brain membranes. Here we present a modification of the purification method used to obtain this PI kinase preparation.[19]

Preparation of Membrane Fraction. All steps are carried out at 0–4°. Fifteen fresh bovine brains are collected on ice at a local slaughterhouse. The brains (5 kg wet weight) are freed of blood vessels and connective tissues. A 33% (w/v) homogenate is prepared in 50 mM Tris-HCl (pH 7.4),

---

[21a] J. Jolles, H. Zwiers, A. Dekker, K. W. A. Wirtz, and W. H. Gispen, *Biochem. J.* **194**, 283 (1981).

1 mM EDTA, 0.32 M sucrose, and 0.3% (v/v) 2-mercaptoethanol (buffer A) containing the protease inhibitors leupeptin (0.5 mg/liter), soybean and lima bean trypsin inhibitors (1 mg/liter each), and phenylmethylsulfonyl fluoride (PMSF; 1 mM). Homogenization is carried out with an Ystral homogenizer (type 40/34, Dottingen, Germany) for 1 min at moderate speed, with the clearance adjusted to its maximum. The homogenate is centrifuged at 14,000 g for 1 hr. The membrane pellet is washed once by rehomogenization (33% homogenate) in buffer A containing 1 mM PMSF, then sedimented by centrifugation at 14,000 g for 1 hr. To release peripheral proteins, the membrane pellet is subjected to a salt wash by rehomogenization (33% homogenate) in 50 mM Tris-HCl (pH 8.0), 1 mM EDTA, 200 mM NaCl, 1 mM PMSF, and 0.3% (v/v) 2-mercaptoethanol. After stirring for 30 min, the membrane suspension is centrifuged at 14,000 g for 1 hr. The membrane pellet (wet weight ~5000 g) containing PI kinase activity is stored at $-20°$ after freezing in CO_2 acetone. The ensuing supernatant is used for the purification of PIP kinase (see below).

Extraction of Enzyme from Membranes. An aliquot of 120 g of the washed membrane pellet is homogenized (25% homogenate) in 50 mM Tris-HCl (pH 7.4), 1 mM EDTA, 0.3% (v/v) 2-mercaptoethanol, 0.1% PEG 20,000, 75 mM KCl, 1% Triton X-100, and 0.1 mM PMSF. Homogenization is carried out with a Waring blender for 20 sec. After stirring for 30 min, the suspension is centrifuged for 90 min at 16,000 g, and the supernatant containing PI kinase activity is collected. Additional stabilizing components are added to the supernatant (volume 400 ml) resulting in a final buffer composition of 50 mM Tris-HCl (pH 7.4), 1 mM EDTA, 0.25 M sucrose, 0.3% (v/v) 2-mercaptoethanol, 1% Triton X-100, 0.1% PEG 20,000, 30 mM KCl, 50 μM ATP, and 0.1 mM PMSF (buffer B).

Column Chromatography. The membrane extract is applied to a DEAE-cellulose column (32 × 2.2 cm) equilibrated with 10 volumes of buffer B. After equilibration with 10 volumes of buffer B, a phosphocellulose column (6.3 × 1.1 cm) is coupled in tandem with the first column. With this procedure, the bulk of the PI kinase activity appears in the flow-through of the DEAE-cellulose column and is bound directly to the phosphocellulose. The flow rate is 14 ml/hr. After washing with 250 ml of buffer B the two columns are uncoupled, and washing of the phosphocellulose is continued with 10 volumes of buffer B. PI kinase is eluted from the phosphocellulose column with 1M NaCl in buffer B at a flow rate of 14 ml/hr, and the active fractions are pooled (5 ml).

Comments. By this procedure a highly active preparation of PI kinase is obtained in 3 days. Per unit volume the preparation has a 30-fold higher PI kinase activity than the membrane extract (starting material). The preparation is free of lipase activity. During the purification, sucrose,

EDTA, PEG 20,000, Triton X-100, ATP, and PMSF are added to the buffers in order to stabilize enzymatic activity. At 4° the enzyme remains stable for several days. Storage of the enzyme at −20° in 50% glycerol preserves the enzymatic activity for several months. In the above procedure phosphocellulose is used as an affinity resin. In purifications where PI kinase was purified to apparent homogeneity, several other types of affinity resins have been used, including PI-Sepharose,[5] heparin-agarose,[9] and reactive dye affinity resins for nucleotide-binding enzymes.[7–9]

Substrate Specificity. The enzymes from bovine uteri,[6] rat brain,[7] porcine liver microsomes,[8] and A431 cells[9] phosphorylate PI but not PIP. In contrast, PI kinase from bovine brain myelin phosphorylates both substrates.[5] PI kinase from bovine uteri phosphorylated lyso-PI at about one-tenth of the rate observed with PI.[6] The PI kinase from rat brain was also tested for diacylglycerol kinase activity; no phosphorylation of diacylglycerol was observed.[7] Analysis of the PIP produced demonstrated that PI kinase from A431 cells and from bovine uteri specifically phosphorylate PI at the 4 position.[6,9] Various nucleotides were tested as phosphoryl group donors for PI kinase from bovine uteri. 2-Deoxy-ATP was utilized by the enzyme as effectively as ATP. Other nucleotides, including GTP, were also utilized to some extent.[6] PI kinase from rat brain membranes did not utilize GTP as a phosphoryl group donor.[7]

Assay for Phosphatidylinositol-4-Phosphate 5-Kinase

Principle. The assay measures the incorporation of radioactive phosphate from [γ-^{32}P]ATP into PIP. Radioactive PIP$_2$ is quantified after separation by TLC. In general, all assays described for PIP kinase are based on this principle (for examples, see Refs. 13–18 and 22–24).

Procedure. The assay procedure is a slight modification of the method described previously.[18] A one-to-one mixture of PIP and PS is dried under N$_2$ from a chloroform solution and subsequently exposed to the vacuum of a lyophilizer for 2 hr. The substrate (0.16 μmol total phospholipid) is suspended in 0.6 ml buffer consisting of 50 mM Tris-HCl (pH 7.4), 0.333 M sucrose, 0.133% PEG 20,000, 150 mM NaCl, and bovine serum albumin (0.67 mg/ml) and sonicated with a Branson probe sonifier (output 50 W, Branson, Danbury, CT) under an N$_2$ atmosphere for 3 min (5 sec on, 10 sec off) on ice. Phosphorylation of PIP is carried out in glass tubes in an assay medium (final volume 50 μl) containing final concentrations of 50

[22] C. Cochet and E. M. Chambaz, *Biochem. J.* **237,** 25 (1986).
[23] T. Urumov and O. H. Wieland, *Biochim. Biophys. Acta* **1052,** 152 (1990).
[24] E. S. Husebye and T. Flatmark, *Biochim. Biophys. Acta* **1010,** 250 (1989).

mM Tris-HCl (pH 7.4), 100 mM NaCl, 15 mM MgCl$_2$, 1 mM EGTA, 80 μM PIP, 80 μM PS, bovine serum albumin (0.4 mg/ml), 0.25 M sucrose, 0.1% PEG 20,000, 0.04% Triton X-100, and 50 μM [γ-^{32}P]ATP (0.5–1.5 μCi/assay). After preincubation for 5 min at 30°, the reaction is started by addition of the enzyme solution (10 μl), and continued for 8 min at 30°.

The reaction is stopped by addition of 3 ml of chloroform–methanol–concentrated HCl (200 : 100 : 0.75, by volume). A mixture of carrier lipids consisting of PI, PIP, PIP$_2$, and PS (40–200 mmol each) is added, and two-phase separation is induced by addition of 0.6 M HCl (0.6 ml) followed by vortexing. After centrifugation, the upper phase is discarded, and the lower phase is washed twice with 1.5 ml of chloroform–methanol–0.6 M HCl (3 : 48 : 47, by volume). Then the lower phase is dried by a stream of N$_2$ at 50°, and the lipid residue is redissolved in 60 μl of ice-cold chloroform–methanol–water (75 : 25 : 2, by volume) by vortexing. An aliquot of 30 μl is applied on an oxalate-impregnated TLC plate,[21a] and the lipids are separated by developing the plate in chloroform–acetone–methanol–acetic acid–water (40 : 15 : 15 : 12 : 7.5, by volume) for 60 min. Lipids are visualized by iodine staining. After sublimation of the iodine, the area containing [^{32}P]PIP$_2$ is scraped from the plate and counted by liquid scintillation spectrometry after addition of 3 ml of xylofluor (Packard, Meridan, CT).

Comments. Sucrose and PEG 20,000 in the assay medium are required to stabilize the enzyme during the phosphorylation reaction. Before the lipid substrate is dissolved in buffer, it is extensively dried on the lyophilizer to remove the last traces of chloroform. This drying step is essential for obtaining reproducible activity measurements. PS is added to the assay to enhance PIP$_2$ formation.[22] Sonication of the lipids is performed with interruptions to prevent their degradation.[25] Under all circumstances, PIP$_2$ formation was linear with protein concentration and with time for at least 10 min.

Purification of Phosphatidylinositol-4-Phosphate 5-Kinase

Table I summarizes the various steps of the purification; all manipulations are performed at 4°.

Stabilization of Membrane Supernatant. The supernatant obtained after salt extraction of bovine brain membranes (see Purification of Phosphatidylinositol 4-Kinase) is used for the purification of PIP kinase. This supernatant is stabilized by addition of 0.2 volumes of buffer containing 50 mM Tris-HCl (pH 8.0), 1mM EDTA, 1.5 M sucrose, 0.3% (v/v) 2-mercaptoethanol, 6% Triton X-100, 0.6% PEG 20,000, 300 μM ATP, and 200 mM NaCl.

[25] M. Toner, G. Viano, A. McLaughlin, and S. McLaughlin, *Biochemistry* **27,** 7435 (1988).

TABLE I
PURIFICATION OF PHOSPHATIDYLINOSITOL-4-PHOSPHATE 5-KINASE FROM
BOVINE BRAIN MEMBRANES

Purification step	Protein (mg)	Total activity (nmol/min)	Specific activity[a] (nmol/min/mg)	Purification (-fold)	Yield (%)
Membrane supernatant	21,900	591	0.027	1	100
Phosphocellulose (stepwise)	1,302	482	0.37	13	81
Phosphocellulose (gradient)	406	313	0.77	28	54
Phenyl-Sepharose	187	318	1.7	63	54
Ultrogel AcA 44	4.9	215	43.9	1625	36
DEAE-cellulose, ATP-agarose	0.36	61.6	171	6333	10

[a] Activity is determined at 30°.

Phosphocellulose Chromatography: Stepwise Elution. Phosphocellulose (300 g wet weight) is equilibrated in 50 mM Tris-HCl (pH 8.0) and 200 mM NaCl, then added to the membrane supernatant (volume 11 liters). The slurry is stirred overnight; after settling for 90 min, the supernatant is decanted. The phosphocellulose is washed twice with 6 liters of 50 mM Tris-HCl (pH 8.0), 1 mM EDTA, 0.25 M sucrose, 0.3% (v/v) 2-mercapto-ethanol, 1% Triton X-100, 0.1% PEG 20,000, 50 μM ATP, and 0.1 mM PMSF (buffer C) containing 200 mM NaCl. The washed slurry is poured into a column (18 × 4.8 cm), and PIP kinase activity is eluted with buffer C containing 1 M NaCl, at a flow rate of 30 ml/hr.

Phosphocellulose Chromatography: Gradient Elution. The active fractions are pooled (volume 310 ml), diluted with buffer C until the ionic strength is 150 mM NaCl, and then applied to a phosphocellulose column (15 × 1.8 cm) equilibrated with 10 volumes of buffer C containing 150 mM NaCl. The enzyme is eluted with a linear gradient from 150 to 1500 mM NaCl in buffer C (volume 460 ml) at a flow rate of 16 ml/hr. The activity elutes at about 800 mM NaCl.

Phenyl-Sepharose Chromatography. The active fractions are pooled (volume 78 ml), adjusted to 1.5 M NaCl with solid NaCl, and diluted 5 times with 50 mM Tris-HCl (pH 8.0), 1 mM EDTA, 0.25 M sucrose, 0.3% (v/v) 2-mercaptoethanol, 0.1% PEG 20,000, 50 μM ATP, 0.1 mM PMSF, and 1.5 M NaCl (buffer D), to lower the Triton X-100 concentration to 0.2%. The enzyme solution is applied to a phenyl-Sepharose column (18 × 1.8 cm) which is equilibrated and eluted with buffer D containing 0.1% Triton X-100, at a flow rate of 16 ml/hr. PIP kinase activity appears in the flow-through of the column.

Gel Filtration on Ultrogel AcA 44. The active fraction from the phenyl-

Sepharose column is dialyzed overnight against 10 volumes of buffer C and loaded on a phosphocellulose column (14 × 1.4 cm) equilibrated with 10 volumes of buffer C containing 150 mM NaCl. A concentrated solution (52 ml) of PIP kinase activity is eluted from the column with 1 M NaCl in buffer C at a flow rate of 8 ml/hr. This solution is applied in two runs (2 × 26 ml) to an Ultrogel AcA 44 column (172 × 2.1 cm), which is equilibrated and eluted with buffer C containing 150 mM NaCl, at a flow rate of 22 ml/hr. Enzyme activity elutes from the column just ahead of the salt peak.

DEAE-Cellulose/ATP-Agarose Chromatography. The active fractions are pooled (77 ml), diluted with buffer C to 70 mM NaCl, and then applied to a DEAE-cellulose column (10.5 × 1.4 cm) which is equilibrated and eluted with buffer C containing 70 mM NaCl, at a flow rate of 8 ml/hr. The activity appearing in the flow-through of the column is used. The active fraction is applied to an ATP-agarose column (8 × 1.3 cm) equilibrated with 10 volumes of buffer C containing 70 mM NaCl. The enzyme is eluted from the column with a linear gradient from 70 to 1000 mM NaCl in buffer C (total volume of 160 ml), at a flow rate of 8 ml/hr. PIP kinase activity is eluted at about 500 mM NaCl. This final step results in a 6333-fold purified protein that has a specific activity of 171 nmol/min/mg protein at 30°.

Comments. In all buffers used EDTA, sucrose, NaCl, 2-mercaptoethanol, Triton X-100, PEG 20,000, PMSF, and ATP are included to prevent a rapid loss of enzyme activity during purification. Attempts to elute the PIP kinase activity from the phosphocellulose column (first step) with a salt gradient were unsuccessful because of clogging of the column. Similar problems were encountered when the Triton X-100 concentration was below 1%. In contrast to the PIP kinase purified earlier,[18] this form does not bind to DEAE-cellulose in the presence of 70 mM NaCl (last step). The final enzyme preparation still contains several protein bands, as revealed by silver staining after sodium dodecyl sulfate-polyacrylamide gel electrophoresis. Numerous attempts to obtain a homogeneous preparation have so far failed.

Properties of Phosphatidylinositol-4-Phosphate 5-Kinase

Storage and Stability. The purified enzyme is stored at −20° in buffer C containing 0.1% instead of 1% Triton X-100. Under these conditions, the enzyme retains its activity without significant losses for several months.

Substrate Specificity. The purified enzyme is highly specific for PIP; no phosphorylation of PI is observed. In the presence of deoxycholate, a condition optimal for measuring diacylglycerol kinase activity,[26] the activ-

---

[26] H. Kanoh, H. Kondoh, and T. Ono, *J. Biol. Chem.* **258,** 1767 (1983).

ity toward diacylglycerol is less than 1% of the activity observed with PIP as substrate. This specificity agrees with that reported for PIP kinase from rat brain cytosol.[22] Analysis of the product by the method described in Ref. 13 revealed that the purified enzyme phosphorylated PIP specifically at the 5-position.

Kinetic Parameters. Kinetic parameters were determined in the absence of PS and in the presence of 0.004% Triton X-100. Prior to the assay, the enzyme solution containing 0.1% Triton X-100 was diluted 5 times with buffer C without Triton X-100, and 10 μl of the ensuing solution was applied to the assay medium (final volume 50 μl). The final Triton X-100 concentration of 0.004% is far below the critical micelle concentration of 0.02%, which makes the assay conditions comparable to those used by others.[13,14,22,23] The K_m for PIP is 5 μM and for ATP 19 μM. For other PIP kinase preparations, K_m values for PIP ranged from 3 to 60 μM and for ATP from 2 to 130 μM.[13,14,22-24] In agreement with the reports by others,[13,14,27] PIP kinase activity is stimulated by PS ($\sim$2.5-fold) and inhibited by its product PIP$_2$. The enzyme activity shows a broad optimum around pH 7.0 and has optimal activity at 15 mM Mg^{2+}. The PIP kinases from the cytosol of rat brain and of bovine adrenal medulla, as well as the kinase from human erythrocytes, have an optimal activity around 10–30 mM Mg^{2+}.[13,16,22,28] This is far above the physiological Mg^{2+} concentration of about 2 mM. Yet, it has been reported that spermine greatly enhances the activity of PIP kinase from rat brain cytosol at physiological Mg^{2+} concentrations.[22,28]

[27] L. A. A. Van Rooijen, M. Rossowska, and N. G. Bazan, *Biochem. Biophys. Res. Commun.* **126**, 150 (1985).

[28] G. A. Lundberg, B. Jergil, and R. Sundler, *Eur. J. Biochem.* **161**, 257 (1986).

[23] Alkylglycerol Phosphotransferase

By Fred Snyder

Introduction

The phosphorylation of alkylglycerols derived from dietary sources or from catabolism of cellular ether-linked lipids leads to the formation of 1-alkyl-2-lyso-*sn*-glycero-3-phosphate (Fig. 1), the branch point intermediate in the *de novo* biosynthesis of plasmanylcholine (ethanolamine), and corresponding plasmalogenic analogs in membranes or platelet-activating factor (PAF). The enzyme responsible for this phosphorylation is ATP : 1-alkyl-

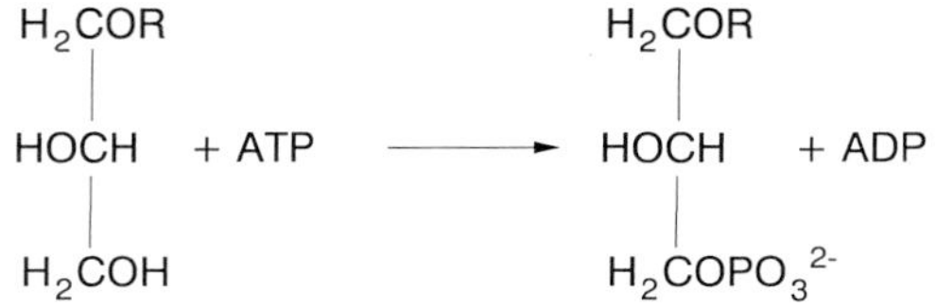

$$
\begin{array}{ccc}
\text{H}_2\text{COR} & & \text{H}_2\text{COR} \\
| & & | \\
\text{HOCH} & +\ \text{ATP} \longrightarrow \text{HOCH} & +\ \text{ADP} \\
| & & | \\
\text{H}_2\text{COH} & & \text{H}_2\text{COPO}_3{}^{2-}
\end{array}
$$

FIG. 1. Enzymatic reaction showing the phosphorylation of alkylglycerols by ATP: 1-alkyl-sn-glycerol phosphotransferase.

sn-glycerol phosphotransferase (EC 2.7.1.93, alkylglycerol kinase).[1,2] Alkylglycerol phosphotransferase makes it possible to bypass the reaction steps catalyzed by alkyldihydroxyacetone-phosphate (alkyl-DHAP) synthase (alkylglycerone-phosphate synthase) and NADPH: alkyl-DHAP oxidoreductase required for formation of the 1-alkyl-2-lyso-sn-glycero-3-phosphate intermediate in the *de novo* biosynthesis of ether-linked glycerolipids, including PAF.

Alkylglycerol phosphotransferase plays a very important role in nutritional experiments, both *in vivo*[3–9] and in cell culture,[10] where the intent is either to restore or to increase the cellular levels of ether-linked phospholipids. The alkylglycerols have often been fed as the diacetates, with the assumption that they are better tolerated than the esterified form normally encountered in the diet. After entering cells, the acetate groupings are hydrolyzed by lipases and the alkylglycerols produced can then serve as substrates for the phosphotransferase. The low level of plasmalogens in cells from patients with Zellweger syndrome (a peroxisomal deficiency disease) has been shown to be due to the absence of both DHAP acyltransferase and alkyl-DHAP synthase, two of the enzymes absolutely essential for the formation of the ether bond in glycerolipids. Thus, the feeding of alkylglycerols to Zellweger patients[8] makes it possible to increase the levels of ether-linked phospholipids in the defective cells since the alkylglycerol phosphotransferase and other enzymes of the ether lipid pathway

[1] K. Chae, C. Piantadosi, and F. Snyder, *Biochem. Biophys. Res. Commun.* **151,** 119 (1973).

[2] C. O. Rock and F. Snyder, *J. Biol. Chem.* **249,** 5382 (1974).

[3] Z. L. Bandi, H. K. Mangold, G. Holmer, and E. Aaes-Jorgensen, *FEBS Lett.* **12,** 217 (1971).

[4] F. Paltauf, *Biochim. Biophys. Acta* **239,** 38 (1971).

[5] A. K. Das and A. K. Hajra, *FEBS Lett.* **227,** 187 (1988).

[6] N. J. Weber, *Lipid Res.* **26,** 1412 (1985).

[7] I. Reichwald and H. K. Mangold, *Nutr. Metab.* **21**(Suppl. 1), 198 (1977).

[8] G. N. Wilson, R. G. Holmes, J. Custer, J. L. Lipkowitz, J. Stover, N. Datta, and A. K. Hajra, *Am. J. Med. Genet.* **24,** 69 (1986).

[9] M. L. Blank, E. A. Cress, Z. L. Smith, and F. Snyder, *Lipids* **26,** 166 (1991).

[10] M. C. Cabot and F. Snyder, *Biochim. Biophys. Acta* **617,** 410 (1980).

beyond alkyl-DHAP oxidoreductase, including the Δ^1-desaturase that forms ethanolamine plasmalogens, are present in the peroxisomal-defective cells. Such studies emphasize the quantitative contribution of alkylglycerol phosphotransferase versus alkyl-DHAP synthase in regulating the cellular levels of ether-linked glycerolipids. This chapter provides a concise description of the method for assaying the alkylglycerol phosphotransferase activity and a discussion of its properties.

Method for Measuring Enzyme Activity

Reagents and Substrates. Compounds required for the assay of the alkylglycerol phosphotransferase are ATP, coenzyme A (CoA), and 1-*O*-hexadecyl-*sn*-glycerol from Sigma (St. Louis, MO), and [γ-^{32}P]ATP from Du Pont–NEN (Boston, MA). If desired, the radiolabeled cosubstrate such as hexadecylglycerol can be prepared either by organic synthesis[11] or by Vitride reduction[12] of radiolabeled PAF (either [^{3}H]hexadecyl or [^{3}H]octadecyl groups at the *sn*-1 position) obtained from Du Pont–NEN.

Preparation of Enzyme Source. Tissues are homogenized in 0.125 M Tris buffer (pH 7.2) containing 0.25 M sucrose, 5 mM EDTA, and 5 mM dithiothreitol (DTT) using a Potter–Elvehjem apparatus.[2] Microsomal pellets are prepared by centrifugation of tissue homogenates at 105,000 g for 1 hr; the microsomal preparations are washed and recentrifuged twice in the homogenization medium. When stored at $-20°$, the microsomes maintain their enzyme activity for at least 2 months. Microsomal protein is determined by the method of Lowry *et al.*[13]

Enzyme Assay. The complete system for assaying alkylglycerol phosphotransferase activity consists of DTT (5 mM), ATP (6 mM), Mg^{2+} (5 mM), NaF (40 mM), ^{3}H-labeled hexadecyl- or octadecylglycerol (25 nmol added in 10 μl ethanol), Tris–maleate buffer (0.1 M, pH 7.1), and microsomes (45 μg protein). Incubations, carried out for 10 min at 37° in a Dubnoff metabolic shaker, are terminated by extracting the lipids with the Bligh and Dyer[14] procedure (Baxter, McGaw Park, IL), except that the methanol contains 2% acetic acid. Recovery of the radioactive product (alkylglycerophosphate) exceeds 95%. The assay system described can also be adapted to use [γ-^{32}P]ATP to monitor the formation of alkylglycerol [^{32}P]phosphate if unlabeled alkylglycerols are used as the cosubstrate.

---

[11] E. O. Oswald, C. Piantadosi, C. E. Anderson, and F. Snyder, *Lipids* **1**, 241 (1966).

[12] F. Snyder, M. L. Blank, and R. L. Wykle, *J. Biol. Chem.* **246**, 3639 (1971).

[13] O. H. Lowry, N. J. Rosebrough, A. L. Farr, and R. J. Randall, *J. Biol. Chem.* **193**, 265 (1951).

[14] E. G. Bligh and W. J. Dyer, *Can. J. Biochem. Physiol.* **37**, 911 (1959).

Also it should be noted that if CoA is included in the assay system, alkylacylglycerophosphate is formed as a product.

Identification of Products. Reaction products, based on the R_f of appropriate standards, can be resolved and isolated in pure form by thin-layer chromatography (TLC)[2] on silica gel HR (Sigma, St. Louis, MO) layers developed in chloroform–methanol–acetic acid (90 : 10 : 10, by volume) or in chloroform–methanol–NH$_4$OH (65 : 35 : 8, by volume). The alkylglyc-erophosphate migrates near the solvent front in the acidic system, but remains at the origin in the basic system. Adjustment of the methanol content of the TLC solvent mixtures can be used to alter the TLC migration of the phosphorylated product in either system. Treatment of the alkylglyc-erophosphate with either Vitride[12] or alkaline phosphatase[15] will generate the original substrate, alkylglycerols. Also, as mentioned above, the addition of CoA to the assay system can be used to demonstrate that the ether analog of phosphatidic acid (alkylacylglycerophosphate) can be formed from the alkylglycerophosphate, under the assay conditions described.[2]

Properties of Enzyme

Although the optimal pH of alkylglycerol phosphotransferase is 7.1, the pH profile is not very sharp. In fact 66% of the maximum enzyme activity is found within 1 pH unit of the pH optimum.[2]

The phosphotransferase activity is associated with the microsomal fraction, at least in rabbit Harderian glands, since both the greatest amount of total activity and the highest specific activity of the enzyme are found in the 105,000 g (60 min) fraction.[2] Moreover, the ratio of NADH–cyto-chrome-c reductase (a microsomal marker) and the alkylglycerol phospho-transferase is the same in both the microsomes and supernatant fraction. Since these two enzymes behave similarly during the subcellular fraction-ation procedures and because the bulk of the phosphotransferase is re-tained in the microsomal pellet, it has been concluded that the phospho-transferase activity found in the soluble fraction is due to microsomal contamination.[2] Most phosphotransferases are not generally associated with membranes, but other kinases that utilize lipophilic substrates such as diacylglycerols[16] and alkyldihydroxyacetone[17] have also been found to be membrane bound.

The alkylglycerol phosphotransferase is stereospecific for the sub-strates possessing an *sn*-1 alkyl chain, that is, 3-alkyl-*sn*-glycerols are not

[15] M. L. Blank and F. Snyder, *Biochemistry* **9,** 5034 (1970).
[16] L. E. Hokin and M. R. Hokin, *Biochim. Biophys. Acta* **67,** 470 (1963).
[17] K. Chae, C. Piantadosi, and F. Snyder, *J. Biol. Chem.* **248,** 6718 (1973).

a substrate. Also, alkylethylene glycols are not phosphorylated, and a thio-linked alkylglycerol produces only a trace amount of a more polar lipid in the assay system described.[2] In the substrate specificity studies, the interpretation of results obtained with monoacylglycerols was complicated because of the high lipase activity encountered in the Harderian gland preparations that were used as the source of the phosphotransferase.[2] However, it was noted that even extremely high concentrations (250 nmol) of rac-hexadecanoylglycerol and rac-1,2-dihexadecanoylglycerol had no effect on the phosphorylation of 1-alkyl-sn-glycerols under the conditions of the assay. Similarly, concentrations of rac-1-hexadecyl-2-hexadecanoylglycerol and 2-hexadecylglycerol 10-fold higher than alkylglycerols had no effect on the phosphorylation of alkylglycerols.[2]

Although firm conclusions regarding the specificity of the alkylglycerol phosphotransferase for the ester analogs are not possible from the Harderian gland experiments, it should be pointed out that monoacylglycerol phosphotransferases have been described in brain[18] and *Escherichia coli*.[19] However, the ability of these enzymes to use alkylglycerols has never been investigated. Nevertheless, at least for the ether-linked lipids, the phosphotransferase activity described in this chapter would appear to recognize only 1-alkyl-sn-glycerols as substrates.

Acknowledgments

This work was supported by the Office of Energy Research, U.S. Department of Energy (Contract No. DE-AC05-760R00033), the American Cancer Society (Grant BC-70V), the National Heart, Lung, and Blood Institute (Grant 27109-10), and the National Cancer Institute (CA-41642-05).

[18] R. A. Pieringer and L. E. Hokin, *J. Biol. Chem.* **237**, 653 (1962).
[19] R. A. Pieringer and R. S. Kunnes, *J. Biol. Chem.* **240**, 2833 (1965).

Section IV

Phosphatases

[24] Phosphatidate Phosphatase from Yeast Mitochondria

By GEORGE M. CARMAN and JENNIFER J. QUINLAN

Introduction

Phosphatidate phosphatase (3-*sn*-phosphatidate phosphohydrolase, EC 3.1.3.4) catalyzes the conversion of phosphatidate to diacylglycerol.[1]

$$\text{Phosphatidate} \rightarrow \text{diacylglycerol} + P_i$$

In the yeast *Saccharomyces cerevisiae* phosphatidate phosphatase is associated with the membrane and cytosolic fractions of the cell.[2,3] The enzyme plays an important role in the biosynthesis of phospholipids and triacylglycerols in *S. cerevisiae*.[4] A 91-kDa form of phosphatidate phosphatase has been purified from the total membrane fraction[5] and is described elsewhere in this series.[6] Immunoblot analysis of cell extracts using antibodies specific for the 91-kDa form of phosphatidate phosphatase revealed the existence of a 45-kDa form of the enzyme.[7] This immunoblot analysis also revealed that the 91-kDa enzyme is a proteolysis product of a 104-kDa enzyme.[7] The mitochondrial fraction contains the 45-kDa enzyme, whereas the microsomal fraction contains the 45- and 104-kDa enzymes.[7] The 45-kDa phosphatidate phosphatase is induced in yeast cells by inositol supplementation, whereas the 104-kDa enzyme is not affected by inositol.[7] Both forms of the enzyme are induced when cells enter the stationary phase of growth.[7]

The phosphatidate phosphatase 45-kDa enzyme has been purified from yeast mitochondria by a procedure similar to that used to purify the phosphatidate phosphatase 91-kDa enzyme from total membranes.[5,6] We describe here the purification and properties of the 45-kDa form of the enzyme.

---

[1] M. Kates, *Can. J. Biochem.* **35**, 575 (1955).

[2] K. Hosaka and S. Yamashita, *Biochim. Biophys. Acta* **796**, 102 (1984).

[3] K. R. Morlock, Y.-P. Lin, and G. M. Carman, *J. Bacteriol.* **170**, 3561 (1988).

[4] G. M. Carman and S. A. Henry, *Annu. Rev. Biochem.* **58**, 635 (1989).

[5] Y.-P. Lin and G. M. Carman, *J. Biol. Chem.* **264**, 8641 (1989).

[6] G. M. Carman and Y.-P. Lin, this series, Vol. 197, p. 548.

[7] K. R. Morlock, J. J. McLaughlin, Y.-P. Lin, and G. M. Carman, *J. Biol. Chem.* **266**, 3586 (1991).

Preparation of Substrates

[γ-^{32}P]Phosphatidate is synthesized enzymatically from [γ-^{32}P]ATP and diacylglycerol using *Escherichia coli* diacylglycerol kinase (Lipidex, Inc., Westfield, NJ) under the assay conditions described by Walsh and Bell.[8] The labeled substrate is purified by thin-layer chromatography.[3]

Assay Method

Phosphatidate phosphatase activity is routinely measured by following the release of water-soluble [^{32}P]P$_i$ from the chloroform-soluble 0.5 mM [γ-^{32}P]phosphatidate [1000–2000 counts/min (cpm)/nmol] in 50 mM Tris–maleate buffer (pH 7.0) containing 5 mM Triton X-100, 10 mM 2-mercaptoethanol, 2 mM MgCl$_2$, and enzyme protein in a total volume of 0.1 ml at 30°.[3,6] One unit of enzymatic activity is defined as the amount of enzyme that catalyzes the formation of 1 nmol of product per minute. Protein is determined by the method of Bradford.[9]

Growth of Yeast

Saccharomyces cerevisiae strain *ade5 MAT***a**[10] is used as a representative wild-type strain[11,12] for enzyme purification. Cells are grown in 1% yeast extract, 2% peptone, and 2% glucose (w/v) at 28° to late exponential phase, harvested by centrifugation, and stored at −80° as described.[13,14] We have not been able to purify phosphatidate phosphatase from the wild-type strain S288C, which has been used to purify other yeast phospholipid biosynthetic enzymes.[13–18]

Purification Procedure

All steps are performed at 5°.
Step 1: Preparation of Cell Extract. Cells (180 g) are disrupted with glass beads with a Bead-Beater (BioSpec Products, Bartlesville, OK) in

[8] J. P. Walsh and R. M. Bell, *J. Biol. Chem.* **261**, 6239 (1986).
[9] M. M. Bradford, *Anal. Biochem.* **72**, 248 (1976).
[10] M. R. Culbertson and S. A. Henry, *Genetics* **80**, 23 (1975).
[11] L. S. Klig, M. J. Homann, G. M. Carman, and S. A. Henry, *J. Bacteriol.* **162**, 1135 (1985).
[12] M. L. Greenberg, L. S. Klig, V. A. Letts, B. S. Loewy, and S. A. Henry, *J. Bacteriol.* **153**, 791 (1983).
[13] A. S. Fischl and G. M. Carman, *J. Bacteriol.* **154**, 304 (1983).
[14] G. M. Carman and A. S. Fischl, this volume [36].
[15] M. Bae-Lee and G. M. Carman, *J. Biol. Chem.* **259**, 10857 (1984).
[16] G. M. Carman and M. Bae-Lee, this volume [35].
[17] M. J. Kelley and G. M. Carman, *J. Biol. Chem.* **262**, 14563 (1987).
[18] G. M. Carman and M. J. Kelley, this volume [28].

50 mM Tris–maleate buffer (pH 7.0) containing 1 mM Na$_2$EDTA, 0.3 M sucrose, and 10 mM 2-mercaptoethanol as described previously.[13,14] Glass beads and unbroken cells are removed by centrifugation at 1500 g for 15 min to obtain the cell extract.

Step 2: Preparation of Mitochondrial Fraction. Crude mitochondria are collected from the cell extract by centrifugation at 32,000 g for 10 min. Mitochondrial pellets are washed with 50 mM Tris–maleate buffer (pH 7.0) containing 10 mM MgCl$_2$, 10 mM 2-mercaptoethanol, and 20% glycerol. Mitochondria are routinely frozen at −80° until the enzyme is purified.

Step 3: Preparation of Sodium Cholate Extract. Mitochondria are suspended in 50 mM Tris–maleate buffer (pH 7.0) containing 10 mM MgCl$_2$, 10 mM 2-mercaptoethanol, 20% glycerol, and 1% sodium cholate at a final protein concentration of 10 mg/ml. The suspension is incubated for 1 hr on a rotary shaker at 150 rpm. After the incubation, the suspension is centrifuged at 100,000 g for 90 min to obtain the sodium cholate extract.

Step 4: DE-53 Chromatography. A DE-53 (Whatman, Clifton, NJ) column (1.5 × 5.6 cm) is equilibrated with 5 column volumes of 50 mM Tris–maleate buffer (pH 7.0) containing 10 mM MgCl$_2$, 10 mM 2-mercapto-ethanol, and 20% glycerol followed by 1 column volume of the same buffer containing 1% sodium cholate (v/v). The enzyme is applied to the column followed by washing of the column with 4 column volumes of equilibration buffer containing 1% sodium cholate. The enzyme is eluted from the column with 10 column volumes of a linear NaCl gradient (0–0.3 M) in the same buffer. The peak of phosphatidate phosphatase activity elutes from the column at a NaCl concentration of about 0.1 M.

Step 5: Affi-Gel Blue Chromatography. An Affi-Gel Blue (Bio-Rad, Richmond, CA) column (1.0 × 6 cm) is equilibrated with 5 column volumes of 50 mM Tris–maleate buffer (pH 7.0) containing 10 mM MgCl$_2$, 10 mM 2-mercaptoethanol, 20% glycerol, and 0.1 M NaCl followed by equilibra-tion with 1 column volume of the same buffer containing 1% sodium cholate. The enzyme preparation from the previous step is applied to the column, which is then washed with 3.5 column volumes of equilibration buffer containing 0.3 M NaCl and 1% sodium cholate. Phosphatidate phosphatase is then eluted from the column with 9 column volumes of a linear NaCl gradient (0.3–1.0 M) in the same buffer at a flow rate of 30 ml/hr. The peak of phosphatidate phosphatase activity elutes from the column at a NaCl concentration of about 0.55 M. The enzyme preparation is desalted by dialysis against equilibration buffer.

Step 6: Hydroxylapatite Chromatography. A hydroxylapatite (BioGel HT, Bio-Rad) column (1.5 × 2.3 cm) is equilibrated with 10 mM potassium phosphate buffer (pH 7.0) containing 5 mM MgCl$_2$, 10 mM 2-mercapto-ethanol, 20% glycerol, and 1% sodium cholate. Dialyzed enzyme from the previous step is applied to the column. The column is washed with 2 column volumes of equilibration buffer followed by elution of phosphatidate phosphatase with 8 column volumes of a linear potassium phosphate gradient (10–150 mM) in equilibration buffer. The concentration of $MgCl_2$ is increased to 10 mM in the elution buffer. The peak of activity elutes from the column at a potassium phosphate concentration of about 95 mM. The most active fractions are pooled and used for the next step in the purification.

Step 7: Mono Q Chromatography. An anion-exchange Mono Q (Pharmacia LKB Biotechnology, Piscataway, NJ) column (0.5 × 5 cm) is equilibrated with 6 column volumes of 50 mM Tris–maleate buffer (pH 7.0) containing 10 mM $MgCl_2$, 10 mM 2-mercaptoethanol, 20% glycerol, 1% sodium cholate, and 0.1 M NaCl. The hydroxylapatite-purified enzyme is applied to the column. The column is washed with 8 column volumes of equilibration buffer followed by 2 column volumes of a linear NaCl gradient (0.1–0.17 M) in equilibration buffer. The enzyme is then eluted from the column with 10 column volumes of a linear NaCl gradient (0.17–0.4 M) in equilibration buffer. The peak of enzyme activity elutes at a NaCl concentration of about 0.2 M. Phosphatidate phosphatase activity is completely stable for at least 3 months of storage at $-80°$.

Enzyme Purity. A summary of the purification of the phosphatidate phosphatase 45-kDa enzyme is presented in Table I. The enzyme is purified 800-fold to a final specific activity of 2400 nmol/min/mg with an activity yield of 0.2%. The -fold purification and final specific activity of the 45-kDa form of phosphatidate phosphatase are considerably lower when

TABLE I

PURIFICATION OF PHOSPHATIDATE PHOSPHATASE FROM MITOCHONDRIAL FRACTION OF
Saccharomyces cerevisiae[a]

Purification step	Total units (nmol/min)	Protein (mg)	Specific activity (units/mg)	Purification (-fold)	Yield (%)
1. Cell extract	30,000	10,000	3	1	100
2. Mitochondria	6,000	909	6.6	2.2	20
3. Sodium cholate extract	4,200	350	12	4	14
4. DE-53	2,610	52	50	16.6	8.7
5. Affi-Gel Blue	1,200	4	300	100	4
6. Hydroxylapatite	480	1	480	160	1.6
7. Mono Q	60	0.025	2400	800	0.2

[a] Data from Morlock *et al.*[7]

compared to the 91-kDa form of the enzyme.[5,6] This is a reflection of the low activity yield of the 45-kDa enzyme during each step of the purification. It is not necessary to use the final Superose 12 chromatography step[5,6] to obtain a nearly homogeneous protein preparation. Phosphatidate phosphatase activity is associated with the 45-kDa protein on sodium dodecyl sulfate-polyacrylamide gels.[7]

Phosphatidate phosphatase 45-kDa enzyme is present during the purification of the phosphatidate phosphatase 91-kDa enzyme purified from total membranes. The 45-kDa enzyme is removed from the 91-kDa enzyme preparation during the final Superose 12 chromatography step.[5]

Properties of Phosphatidate Phosphatase

Maximum activity is observed between pH 6 and 7. The enzyme is dependent on magnesium ions, with maximum activity at 1 mM. The requirement for magnesium ions cannot be substituted by manganese, cobalt, or calcium ions. Maximal phosphatidate phosphatase activity is also dependent on the addition of 5 mM Triton X-100 (molar ratio of Triton X-100 to phosphatidate of 10 : 1). Phosphatidate phosphatase is unstable at temperatures above 30°, with total inactivation occurring after heating for 20 min at 50°. The enzyme is inhibited by p-chloromercuriphenylsulfonic acid (1 mM), N-ethylmaleimide (1 mM), phenylglyoxal (IC$_{50}$ 4.3 mM), and propranolol (IC$_{50}$ 0.95 mM).

The kinetic properties of the phosphatidate phosphatase 45-kDa enzyme have been examined with uniform mixed micelles containing Triton X-100 and phosphatidate.[7] Phosphatidate phosphatase displays saturation kinetics with respect to the bulk and surface concentrations of phosphatidate.[7] At a surface concentration of phosphatidate of 9 mol %, the K_m value for the bulk concentration of phosphatidate is 94 μM. At a bulk concentration of phosphatidate of 0.5 mM, the K_m value for the surface concentration of phosphatidate is 2.9 mol %. These results are consistent with the phosphatidate phosphatase 45-kDa enzyme following "surface dilution" kinetics.[19]

The properties of the purified phosphatidate phosphatase 45-kDa enzyme are similar to those of the 91-kDa enzyme[5,7,20] with respect to pH optimum, cofactor requirement, kinetic properties using Triton X-100–phosphatidate mixed micelles, temperature stability, and inhibition to various compounds. However, the 45- and 91-kDa forms of phosphatidate

[19] R. A. Deems, B. R. Eaton, and E. A. Dennis, *J. Biol. Chem.* **250**, 9013 (1975).
[20] Y.-P. Lin and G. M. Carman, *J. Biol. Chem.* **265**, 166 (1990).

phosphatase differ with respect to their isoelectric points and peptide fragments resulting from V8 proteolysis and cyanogen bromide cleavage.[7]

Acknowledgments

This work was supported by U.S. Public Health Service Grant GM-28140 from the National Institutes of Health, New Jersey State funds, and the Charles and Johanna Busch Memorial Fund.

[25] Phosphatidylglycerophosphate Phosphatase from *Escherichia coli*

By WILLIAM DOWHAN and CINDEE R. FUNK

Introduction

$$\text{Phosphatidylglycerophosphate} \rightarrow \text{phosphatidylglycerol} + P_i \qquad (1)$$
$$\text{Phosphatidic acid} \rightarrow \text{diacylglycerol} + P_i \qquad (2)$$
$$\text{Lysophosphatidic acid} \rightarrow 1\text{-acyl-}sn\text{-glycerol} + P_i \qquad (3)$$

Chang and Kennedy[1] reported a membrane-associated phosphatidylglycerophosphate phosphatase activity [reaction (1)] in *Escherichia coli* that was distinct from other phosphatases with specificities limited to water-soluble substrates. Icho and Raetz[2] determined that reaction (1) in crude extracts of *E. coli* was dependent on at least two gene products. The *pgpA* gene (mapping at 10 min) product (PGP-A) only catalyzes reaction (1), whereas the *pgpB* gene (mapping at 28 min) product (PGP-B) catalyzes all three reactions. Neither of these genes appears to encode the biosynthetic phosphatidylglycerophosphate phosphatase activity since strains with both genes inactivated by gene interruption[3] still synthesize near-normal *in vivo* levels of phosphatidylglycerol. Cell lysates made from double mutants still have about 50% of the normal levels of activity catalyzing reaction (1) when assayed at 30°; this residual activity (PGP-C) went unrecognized because, unlike the activity contributed by PGP-A and PGP-B, it is both temperature- and detergent-sensitive.[3] These extracts also contain significant levels of activity [reaction (3)] which is independent of the PGP-C activity, suggesting that a second lysophosphatidic acid

---

[1] Y.-Y. Chang and E. P. Kennedy, *J. Lipid Res.* **8**, 456 (1967).
[2] T. Icho and C. R. H. Raetz, *J. Bacteriol.* **153**, 722 (1983).
[3] C. R. Funk and W. Dowhan, *J. Bacteriol.* **174**, in press (1992).

phosphatase is present in *E. coli*. Both the *pgpA*[4] and *pgpB*[5] genes have been cloned, sequenced, and overexpressed on high copy number plasmids. Overexpression of these gene products results in the expected elevation in the above phosphatase activities.

Preparation of Substrates

Synthesis of Phosphatidylglycero[^{32}P]phosphate. The synthesis of phosphatidylglycero[^{32}P]phosphate substrate requires phosphatidylglycerophosphate synthase from *E. coli*.[6] It is not necessary to use homogeneous enzyme for this procedure, but the enzyme must be free of the activity catalyzing reaction (1). This activity can be minimized by first starting with strains of *E. coli* which overexpress the synthase[6] and by then extracting the membrane-associated synthase with Triton X-100 in the presence of magnesium ions and sulfhydryl reagents, which minimizes the level of activity from reaction (1). Alternatively, activity from reaction (1) can be eliminated by using membranes from a strain in which the *pgpA* and *pgpB* genes have been inactivated and carrying out the synthesis at 42°.[3]

sn-Glycero-3-[^{32}P]phosphate is first made[1] by incubating the following mixture (5 ml) for 6 hr or longer at 37°: 3 m*M* glycerol, 2 m*M* [γ-^{32}P]ATP (1 mCi/mmol), 50 m*M* Tris-HCl (pH 8.0), 1 mg/ml bovine serum albumin, 10 m*M* MgCl$_2$, and 10 units of glycerokinase from *Candida mycoderma* (commercially available). This reaction mixture is then diluted to 7.5 ml by adding 2.5 ml of the following solution: 50 m*M* Tris-HCl (pH 8.0), 3% Triton X-100, 6 m*M* CDPdiacylglycerol, 280 m*M* MgCl$_2$, and 50 milliunits (mU) of phosphatidylglycerophosphate synthase. After at least a 6-hr incubation at 37°, the reaction mixture is thoroughly mixed with 20 ml of methanol, 10 ml of chloroform, and 10.5 ml of 0.5 *M* NaCl which is 0.1 *N* in HCl. After mixing with an additional 10 ml of chloroform, the organic (lower) and aqueous phases are separated by centrifugation for 5 min in a clinical centrifuge. The aqueous phase is shaken with an additional 6 ml of chloroform, and all of the chloroform phases are pooled, dried under a stream of N$_2$ at room temperature, and dissolved in 1 ml of chloroform. The radiolabeled product is separated from Triton X-100 by precipitation with 11 ml of acetone ($-20°$).[7] After standing at $-20°$ overnight the product is collected by centrifugation at 12,000 *g* for 20 min, dissolved in 1 ml of chloroform, and stored at $-20°$.

[4] T. Icho, *J. Bacteriol.* **170**, 5110 (1988).
[5] T. Icho, *J. Bacteriol.* **170**, 5117 (1988).
[6] W. Dowhan, this volume [37].
[7] M. Kates, *in* "Laboratory Techniques in Biochemistry and Molecular Biology" (T. S. Work and E. Work, eds.), Vol. 3, Part 2, p. 393. North-Holland Publ., Amsterdam, 1972.

The crude lipid product (before acetone precipitation) can also be purified[8] by chromatography on a 1.8 × 30 cm DEAE-cellulose[2,9] column (DE-52) equilibrated with chloroform–methanol–water (2 : 3 : 1, v/v). The resin is first suspended and washed several times in water to remove fines, then washed once with 5% sodium hydroxide, and rinsed with water until the wash water is neutral. The resin is then washed with 10% acetic acid, rinsed with water until the wash water is again neutral, washed with methanol, and finally washed with chloroform–methanol–water (2 : 3 : 1, v/v). The resin slurry is poured into a column and packed to a constant height under gravity. The crude lipid product is dissolved in 3 ml of chloroform–methanol–water (2 : 3 : 1, v/v) and adjusted to pH 7.0 with 1 M Tris free base and loaded onto the DE-52 column. The column is washed with 1–2 column volumes of the equilibration solvent followed by 2.5 column volumes of the same solvent containing 30 mM ammonium acetate. The radiolabeled product is then eluted with 2.5 column volumes (5 ml fractions) of the same solvent containing 60 mM ammonium acetate. The peak of radiolabel is pooled, mixed with 0.33 volumes of 0.5 N HCl, and adjusted to a final chloroform–methanol–water ratio of 2 : 2 : 1.8. The phases are separated by centrifugation and the lower phase taken to dryness as above; the product is dissolved in 1 ml of chloroform and stored at $-20°$. After column purification, the reaction yields about 8 μmol of labeled phosphatidylglycerophosphate.

Synthesis of [^{32}P]Phosphatidic Acid. [^{32}P]Phosphatidic acid is synthesized by a modification of several protocols.[10,11] Chloroform solutions of commercially available 1,2-dioleoyl-*sn*-glycerol (20 μmol) and bovine heart cardiolipin (8 μmol) are mixed and taken to dryness. The residue is thoroughly suspended by vortexing with a mixture containing 10 ml of water, 1 ml of 1 M imidazole hydrochloride (pH 6.6), 0.56 ml of 25% Triton X-100, and 80 μl of 0.25 M EGTA. The suspended lipid is then incubated another 10 min after the addition of 125 μl of 4 M LiCl, 24 μl of 0.25 M diethylenetriaminepentaacetic acid, 40 μl of 1 M dithiothreitol, 0.4 ml of 5 M NaCl, and 1 unit of *E. coli* diacylglycerol kinase (commercially available). Finally, the mixture is supplemented with 6.6 ml of water, 0.8 ml of 50 mM [γ-^{32}P]ATP (0.5 mCi/mmol), and the reaction started by the addition of 0.25 ml of 1 M MgCl$_2$. The mixture (20 ml) is incubated overnight at 25°. The lipids and detergent are separated from the remaining components by partitioning into the chloroform phase after mixing with 40 ml of metha-

[8] H. S. Hendrickson and C. E. Ballou, *J. Biol. Chem.* **239**, 1369 (1963).
[9] R. J. Tyhach, R. Engel, and B. E. Tropp, *J. Biol. Chem.* **251**, 6717 (1976).
[10] J. P. Walsh and R. M. Bell, *J. Biol. Chem.* **261**, 11021 (1986).
[11] E. G. Schneider and E. P. Kennedy, *J. Biol. Chem.* **248**, 3739 (1973).

nol, 40 ml of chloroform, and 16 ml of 0.5 M NaCl (0.1 N in HCl). After mixing and separation of the phases by centrifugation, the lower phase is taken to dryness at room temperature.

The radiolabeled product is further purified by one-dimensional silica gel thin-layer chromatography using chloroform–methanol–water (65 : 25 : 4, v/v). The [^{32}P]phosphatidic acid (R_f 0.26) is detected by autoradiography and the appropriate area scraped from the silica gel plate. The silica gel is extracted twice with 22 ml of chloroform–methanol–water (1 : 1 : 0.2, v/v) and the pooled extracting solvent mixed with 14 ml of 0.5 M NaCl (0.1 N in HCl). After separation of the lower phase, the upper phase is washed with 10 ml of chloroform and the pooled lower phases taken to dryness. The product is dissolved in 1 ml of chloroform and stored at $-20°$. About 60% of the original diacylglycerol is converted to radiolabeled phosphatidic acid, yielding about 12 μmol of product.

Synthesis of Lyso[^{32}P]phosphatidic Acid. 1-Acyl-*sn*-glycero[3-^{32}P] phosphate is synthesized using a modification of a method with porcine pancreatic phospholipase A$_2$.[12] The incubation mixture consists of 10 mM Tris-HCl (pH 9.4), 20 mM CaCl$_2$, 1.8 mM (7.2 μmol) [^{32}P]phosphatidic acid (as synthesized above), 0.23% Triton X-100, and 700 units of phospholipase A$_2$ in a final volume of 4.0 ml. The reaction is terminated after overnight incubation at room temperature, and the product (R_f 0.1) is isolated as described above for [^{32}P]phosphatidic acid. The conversion of [^{32}P]phosphatidic acid to lyso[^{32}P]phosphatidic acid is about 90% efficient, with a yield of 6.5 μmol of product.

Enzyme Assays

Phosphatidylglycerophosphate Phosphatase. Activity [reaction (1)] is determined by a modification[3] of the method Chang and Kennedy.[1] Phosphatidylglycero[^{32}P]phosphate [10^3–10^4 disintegrations/min (dpm)/nmol] is dried out of chloroform and suspended with vortexing to a final concentration of 0.2 mM in a 2-fold concentrated assay mixture consisting of 100 mM Tris-HCl (pH 7.4), 0.24% Triton X-100, 2 mM dithiothreitol, and 2 mM EDTA. Enzyme and water up to 40 μl, 50 μl of the above mixture, and 10 μl of 40 mM MgCl$_2$ are mixed on ice, and the mixture is incubated at 30° for 20 min. The reaction is terminated by the addition of 0.5 ml of methanol (0.1 N in HCl), 1.5 ml of chloroform, and 1.5 ml of 1 M MgCl$_2$. After the phases are mixed and separated by centrifugation, 1 ml (out of

[12] G. H. deHaas, A. J. Slotboom, M. G. van Oort, F. van den Wiele, W. Atsma, M. van Linde, and B. Roelofsen, *in* "Enzymes of Lipid Metabolism II" (L. Freysz, H. Dreyfus, R. Massorelli, and S. Gaft, eds.), p. 107. Plenum, New York, 1986.

2.1 ml) of the upper phase is counted for the released $^{32}PO_4$ in 10 ml of a scintillation fluid compatible with aqueous samples. One unit of phosphatase activity is defined as the amount of enzyme which releases 1 μmol/ min of water-soluble radiolabel under the above conditions.

Phosphatidic Acid Phosphatase. Activity [reaction (2)] is determined[3] as described for the phosphatidylglycerophosphate phosphatase activity above, except the substrate is [^{32}P]phosphatidic acid. One unit of phosphatidic acid phosphatase activity is defined as the amount of enzyme which releases 1 μmol of water-soluble radiolabel under the above conditions.

Lysophosphatidic Acid Phosphatase. Activity [reaction (3)] is determined as described by van den Bosch and Vagelos.[13] Lyso[^{32}P]phosphatidic acid (10^3–10^4 dpm/nmol) is dried out of chloroform and suspended with gentle vortexing to a final concentration of 0.2 mM in a 2-fold concentrated assay mixture consisting of 500 mM Tris-HCl (pH 7.0) and 10 mM dithiothreitol. Enzyme and water up to 40 μl, 50 μl of the above mixture, and 10 μl of 2 mM MgCl$_2$ are mixed on ice, and the mixture is incubated at 30° for 20 min. The reaction is terminated by the addition of 0.1 ml of 10% (w/v) trichloroacetic acid, 0.5 ml of 0.5 M NaCl (0.1 N in HCl), 0.8 ml of chloroform, and 0.8 ml of methanol. After the phases are mixed and separated by centrifugation, the aqueous phase is washed twice with 1.5 ml of chloroform before one-half of the aqueous phase (0.75 ml) is counted for the released $^{32}PO_4$ in 10 ml of scintillation fluid. One unit of phosphatase activity is defined as the amount of enzyme which releases 1 μmol/min of water-soluble radiolabel under the above conditions.

Properties of Activities

Phosphatidylglycerophosphate Phosphatase Activity. The activity[1] responsible for reaction (1) is associated with the membrane fraction of *E. coli* and appears to be a composite of the activities of at least three gene products[2,3]; therefore, many of the properties of this activity are based on assays performed on this composite of enzymes. The composite activity is sensitive to sulfhydryl reagents (2 mM N-ethylmaleimide and 5 mM HgCl$_2$) and freeze–thawing.[3] The highest specific activity is obtained from membranes freshly isolated in the presence of dithiothreitol. The activity is resistant to solubilization with Triton X-100 unless 5 mM EDTA is included in the solubilization buffer. NaF (5 mM) inhibits 80% of the activity. The apparent K_m for phosphatidylglycerophosphate is 80 μM. A 10-fold partial purification for this activity has been reported.[1] The pH and

[13] H. van den Bosch and P. R. Vagelos, *Biochim. Biophys. Acta* **218**, 233 (1970).

Mg^{2+} (Ca^{2+}, Mn^{2+}, and Zn^{2+} do not substitute), EDTA, and Triton X-100 concentrations are optimized in the assay described above.

Transforming wild-type cells with a multicopy number plasmid carrying either the *pgpA* or *pgpB* gene results in at least a 10-fold amplification of activity [reaction (1)][4,5] (Table I). Interruption of both genes reduces the phosphatase activity by 50% when the assay is carried out at 30°.[3] In the double mutants no phosphatase activity can be detected after preincubation (10 min) under the assay conditions at 42° followed by assaying at 42°; incubation in the absence of Triton X-100 followed by assaying at 30° results in the detection of some activity.[3] This third temperature-sensitive phosphatase activity (PGP-C), that is, the activity remaining after inactivation of the above two genes, may be the phosphatidylglycerol biosynthetic activity since the double mutants still have normal levels of phosphatidylglycerol although phosphatidylglycerophosphate levels are elevated more in cells grown at 42° than 30°, consistent with a greater temperature lability of the PGP-C-dependent phosphatase activity. Although the elevation of phosphatidylglycerophosphate in these mutants is great (15- to 50-fold) relative to wild-type cells, the absolute level of this phospholipid is less than 5% of the total.[2,3]

Phosphatidic Acid and Lysophosphatidic Acid Phosphatases. As can be seen from Table I, overproduction of the *pgpB* gene product increases the level of all three phosphatase activities. Interruption of the *pgpA* gene alone does not affect the level of activities catalyzing reactions (2) or (3), but interruption of *pgpB* alone lowers the level of the activity responsible for reaction (2) 10- to 15-fold and that of reaction (3) by 6-fold.[3] This result suggests that there may be additional genes encoding these activities. The assay conditions outlined above are optimal for these activities.

Localization of Activities. The activities catalyzing reactions (1)–(3)

TABLE I
PHOSPHATASE ACTIVITIES IN VARIOUS MUTANTS[a]

Genotype	Reaction (1) (mU)		Reaction (2) (mU)		Reaction (3) (mU)	
	30°	42°	30°	42°	30°	42°
Wild type	14	6	0.4	0.03	2.1	2.1
pgpA pgpB	10	0	0.01	0.01	0.45	0.45
WT/*pgpA*[+b]	190	—	0.25	—	3.0	—
WT/*pgpB*[+b]	130	—	1.7	—	7.3	—

[a] From C. R. Funk and W. Dowhan, *J. Bacteriol.* submitted (1991).
[b] Wild-type strain carrying the indicated gene on a multicopy number plasmid.

are localized to the cell membrane fraction[1,3,5] but have different distributions among the two membranes of the cell envelope. The contribution of PGP-A[5] and PGP-C[3] to reaction (1) is associated with the inner membrane. In the case of the PGP-B-associated activities, that of reaction (1) is equally distributed between the inner and outer membranes, whereas that of reaction (2) is almost exclusively localized to the outer membrane.[5] The distribution of the activity responsible for reaction (3) among the two membranes has not been determined.

[26] 1-Alkyl-2-acetyl-sn-glycero-3-phosphate Phosphatase

By Fred Snyder and Ten-ching Lee

Introduction

1-Alkyl-2-acetyl-sn-glycero-3-phosphate (alkylacetylglycerophosphate) phosphatase represents an intermediary linkage in the *de novo* pathway of platelet-activating factor (PAF) biosynthesis.[1] In this reaction 1-alkyl-2-acetyl-sn-glycerols are formed via dephosphorylation of 1-alkyl-2-acetyl-sn-glycero-3-phosphate (Fig. 1). The product is the immediate precursor of PAF, but alkylacetylglycerols are also known to activate protein kinase C directly,[2] modulate responses stimulated by diglycerides,[3] and induce cellular differentiation in HL-60 cells.[4] Thus, the alkylacetylglycerols appear to have additional cellular functions other than just serving as the immediate precursor of PAF.

Although many phosphatases are often classified as catabolic enzymes, some play an important role in biosynthetic pathways. This is certainly true for 1-alkyl-2-acetyl-sn-glycero-3-phosphate phosphatase since it represents the essential bridging metabolic step between the acetyltransferase and cholinephosphotransferase in the *de novo* network of PAF biosynthesis. Moreover, the alkylacetylglycerophosphate phosphatase appears to have different properties from other phosphatases involved in lipid metabolism (e.g., phosphatidate phosphatase).[1]

[1] T.-C. Lee, B. Malone, and F. Snyder, *J. Biol. Chem.* **263,** 1755 (1988).
[2] L. L. Stoll, P. H. Figard, N. R. Yerram, M. A. Yorek, and A. A. Spector, *Cell Regul.* **1,** 13 (1989).
[3] D. A. Bass, L. C. McCall, and R. L. Wykle, *J. Biol. Chem.* **263,** 19610 (1988).
[4] M. J. C. McNamara, J. D. Schmitt, R. L. Wykle, and L. W. Daniel, *Biochem. Biophys. Res. Commun.* **122,** 824 (1984).

METHODS IN ENZYMOLOGY, VOL. 209

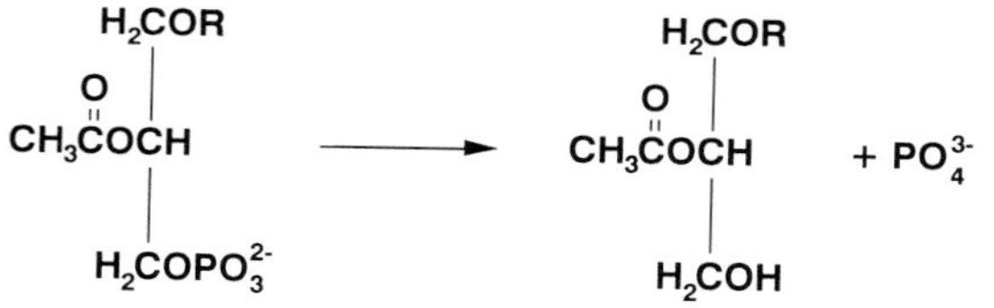

$$
\begin{array}{ccc}
\text{H}_2\text{COR} & & \text{H}_2\text{COR} \\
\quad\quad | & & \quad\quad | \\
\overset{\text{O}}{\underset{\|}{\text{CH}_3\text{COCH}}} & \longrightarrow & \overset{\text{O}}{\underset{\|}{\text{CH}_3\text{COCH}}} \quad + \text{PO}_4^{3-} \\
\quad\quad | & & \quad\quad | \\
\text{H}_2\text{COPO}_3^{2-} & & \text{H}_2\text{COH}
\end{array}
$$

FIG. 1. Reaction showing the dephosphorylation of 1-alkyl-2-acetyl-sn-glycero-3-phosphate by a specific phosphatase in the *de novo* pathway of PAF biosynthesis.

Method for Measuring Enzyme Activity

Reagents and Substrates. The following compounds are required for the preparation of substrates and analysis of products in the phosphatase assay: 1-hexadecyl-2-acetyl-GPC (sn-glycero-3-phosphocholine) and cacodylic acid (Sigma, St. Louis, MO); 1-octadecyl-2-acetyl-GPC (Bubendorf-Schweiz); and [1′,2′-³H]hexadecyl-2-acetyl-GPC (59.5 Ci/mmol), (Du Pont–NEN, Boston, MA). Phospholipase D is prepared from fresh cabbage according to the procedure of Hayashi.[5]

Both the labeled and unlabeled 1-hexadecyl-2-acetyl-sn-glycero-3-phosphate and 1-octadecyl-2-acetyl-sn-glycero-3-phosphate are made by treating the corresponding 1,2-diradyl-GPC analogs with a fresh batch of phospholipase D.[6] Thin-layer chromatography (TLC) can be used to purify the phospholipase D products as previously reported.[7] This involves the use of silica gel H (Sigma) layers prepared with a solution of 10 mM Na$_2$CO$_3$. The chromatograms are developed in chloroform–methanol–acetic acid–saline (50 : 25 : 8 : 6, by volume); alkylacetylglycerophosphate has an R_f of 0.59 in this system.

Preparation of Enzyme Source. Isolation of microsomes from rat spleens and other tissues can be done according to procedures described previously (see [23], this volume), except 0.25 M sucrose is used in both the homogenization and suspension media.[1] Protein is determined by the method of Lowry *et al.*[8] with bovine serum albumin as a standard.

Enzyme Assay. The standard incubation mixture for assaying the alkylacetylglycerophosphate phosphatase[1] consists of 0.1 M cacodylate buffer (pH 6.6–7.0), 200 μM [³H]hexadecylacetylglycerophosphate (0.1 μCi) dissolved in 5–10 μl of ethanol, and microsomal protein (enzyme) in a final volume of 0.25 ml. Incubations are carried out at 37°C. Rates of the

---

[5] O. Hayashi, this series, Vol. 1, p. 660.

[6] S. F. Yang, S. Freer, and A. A. Benson, *J. Biol. Chem.* **242**, 477 (1967).

[7] T.-C. Lee, B. Malone, and F. Snyder, *J. Biol. Chem.* **261**, 5373 (1986).

[8] O. H. Lowry, N. J. Rosebrough, A. L. Farr, and R. J. Randall, *J. Biol. Chem.* **193**, 265 (1951).

reaction with spleen microsomes are linear up to 30 μg of protein through 30 min. Reactions are terminated by extracting the lipids as described by Bligh and Dyer,[9] except 2% acetic acid is included in the methanol and 0.05 N HCl in the aqueous phase to maximize the extraction of both the unreacted substrate and products.

Identification of Products. The alkylacetylglycerols and the deacetylated products, alkylglycerols, are analyzed by TLC using silica gel G (Sigma) layers developed in chloroform–methanol–acetic acid (96 : 4 : 1, by volume). In this system,[1] [³H]hexadecylacetylglycerol has an R_f of 0.49 and [³H]hexadecylglycerol an R_f of 0.27. Radioassay of the TLC zones containing these lipids is done by liquid scintillation spectrometry.[10] Further identification of the alkylacetylglycerols[11,12] and alkylglycerols[13] can be achieved by preparing benzoate derivatives of both products for subsequent analysis by high-performance liquid chromatography (HPLC). However, before HPLC is performed the benzoylation products should be purified by TLC on silica gel G layers developed in hexane–diethyl ether (75 : 25, v/v).

Properties of Enzyme

Alkylacetylglycerophosphate phosphatase has an optimal pH of 6.6–7.0 as determined in both homogenates and microsomal fractions of rat spleens.[1] In contrast, dioleoylglycerophosphate (i.e., phosphatidate) phosphatase in microsomes was found to have an optimal pH of 6.2, compared to 7.4 for the alkylacetylglycerophosphate phosphatase measured under conditions where both substrates were dissolved in ethanol.[1] The difference in the optimal pH for the assay of alkylacetylglycerophosphate phosphatase in ethanol is unknown, but the ethanol was required to solubilize the dioleoylglycerophosphate substrate.

Subcellular distribution studies have shown the bulk of the phosphatase activity for alkylacetylglycerophosphate is located in the microsomal fraction (~53%). However, a significant portion of the activity is also found in the mitochondria (~28%). The specific activity of the phosphatase in the microsomal and mitochondrial factions was enriched only 2.6- and 2.2-fold, respectively, over the postnuclear fraction.[1]

[9] E. G. Bligh and W. J. Dyer, *Can. J. Biochem. Physiol.* **37**, 911 (1959).

[10] F. Snyder, in "The Current Status of Liquid Scintillation Counting" (E. D. Bransome, ed.), p. 248. Grune & Stratton, New York, 1970.

[11] M. L. Blank, E. A. Cress, and F. Snyder, *J. Chromatogr.* **392**, 421 (1987).

[12] M. L. Blank, E. A. Cress, V. Fitzgerald, and F. Snyder, *J. Chromatogr.* **508**, 382 (1990).

[13] M. L. Blank, E. A. Cress, T.-C. Lee, N. Stephens, C. Piantadosi, and F. Snyder, *Anal. Biochem.* **133**, 430 (1983).

TABLE I

SUBSTRATE SPECIFICITY OF ALKYLACETYLGLYCEROPHOSPHATE
PHOSPHATASE IN RAT SPLEEN MICROSOMES[a]

Substrate	Position of structural variation on glycerol moiety	% of control
Hexadecylacetyl-GP	sn-1	100
Octadecylacetyl-GP	sn-1	90
Palmitoylacetyl-GP	sn-1	92
Hexadecylpropionyl-GP	sn-2	173
Hexadecylbutyryl-GP	sn-2	110
Hexadecylhexanoyl-GP	sn-2	114
Hexadecyllyso-GP	sn-2	36
Dioleoyl-GP	sn-1/sn-2	20

[a] Average of two experiments with duplicate determinations. GP designates sn-glycero-3-phosphate. The data are a summary of results from a previously published paper.[1]

Table I summarizes the effect of variations in the structural features of alkylacetylglycerophosphate as a substrate for the *de novo* phosphatase. Alkylacetylglycerophosphate with either 16:0 or 18:0 carbon chains and the 16:0 acyl analog are essentially equal as substrates for the phosphatase. Also the phosphatase preferentially utilizes alkylpropionylglycerophosphate but otherwise exhibits very little difference in the degree of specificity for other short-chain moieties (2:0, 4:0, 6:0) esterified at the sn-2 position. Alkyllysoglycerophosphate also serves as a substrate for the phosphatase, but it is utilized at a considerably lower rate (64% less) than controls. Dioleoylglycerophosphate had 80% less activity than the control values obtained with alkylacetylglycerophosphate as the substrate, and it also exhibited different properties with respect to temperature and detergent sensitivities than when alkylacetylglycerophosphate was the substrate. The activity of alkylacetylglycerophosphate phosphatase rapidly declines at temperatures higher than 37°, whereas the activity of dioleoylglycerophosphate phosphatase increases up to 60°; however, both activities are completely abolished at 80°. Also deoxycholate (12 m*M*) causes a 60% decrease in the alkylacetylglycerophosphate phosphatase but actually stimulates the dioleoylglycerophosphate phosphatase activity. At a concentration of 60 m*M* deoxycholate, the alkylacetylglycerophosphate phosphatase activity is completely inhibited, whereas the dioleoylglycerophosphate phosphatase activity still possesses half of its original activity. These results strongly suggest that the alkylacetylglycerophosphate phosphatase activity is distinctly different from the well-known phosphatidate phosphatase activity[1] involved in the biosynthesis of choline- and ethanolamine-containing phospholipids.

An additional difference in the properties between the enzyme activities that utilize alkylacetylglycerophosphate and dioleoylglycerophosphate as substrates is their response to magnesium or calcium.[1] Alkylacetylglycerophosphate phosphatase activity is inhibited by both Ca^{2+} and Mg^{2+} (>0.5 mM). The activity of dioleoylglycerophosphate phosphatase is also inhibited by Ca^{2+} at all concentrations tested (up to 10 mM), but at low levels of magnesium (0.1 mM) the dioleoylglycerophosphate phosphatase activity is stimulated. Nevertheless, higher concentrations of Mg^{2+} (5 mM) also inhibit the dioleoylglycerophosphate phosphatase activity to 56% of the control value, but the inhibitory effect is not as much as for the alkylacetylglycerophosphate phosphatase (26% of the control value).

The activity of alkylacetylglycerophosphate phosphatase in the *de novo* pathway of PAF biosynthesis has been found in all rat tissues studied.[1] Microsomes from the kidney medulla appear to possess the highest activity, but brain, spleen, kidney cortex, and lung also exhibit relatively high activities. Microsomal preparations from heart and liver have the lowest phosphatase activities. The significance of the wide tissue distribution of this novel phosphatase activity is not understood at the present time. However, studies to date emphasize that all of the enzyme activities in the *de novo* route of PAF biosynthesis occur in most mammalian cells.

Acknowledgments

This work was supported by the Office of Energy Research, U.S. Department of Energy (Contract No. DE-AC05-760R00033), the American Cancer Society (Grant BC-70V), and the National Heart, Lung, and Blood Institute (Grant 35495-04A1).

Section V

Cytidylyltransferases

[27] Purification of CDPdiacylglycerol Synthase from *Escherichia coli*

By CARL P. SPARROW

Introduction

CDPdiacylglycerol synthase (CTP:phosphatidate cytidylyltransferase; EC 2.7.7.41, phosphatidate cytidylyltransferase) catalyzes the activation of phosphatidic acid by CTP to form CDPdiacylglycerol and inorganic pyrophosphate. This activity is essential for all phospholipid biosynthesis in *Escherichia coli*[1] and for the biosynthesis of acidic phospholipids in mammalian cells.[2]

The enzymes of phospholipid synthesis are usually membrane bound and present in relatively low levels, making their purification difficult. The purification of *E. coli* CDPdiacylglycerol synthase is greatly aided by enzyme overproduction driven by an expression plasmid.[3] Starting with cells that overproduce the enzyme 50-fold, essentially homogeneous enzyme can be obtained after chromatographic procedures that yield 160-fold purification.[4] The enzyme can be partially purified from wild-type cells using the same procedures.

Assay of CDPdiacylglycerol Synthase Activity

The assay follows the conversion of [α-^{32}P]dCTP to chloroform-soluble material, dependent on phosphatidic acid. The standard assay mixture contains 100 mM potassium phosphate, pH 7.4, 0.2% (w/v) Triton X-100, 1 mM phosphatidic acid, 1 mg/ml of bovine serum albumin, 10 mM MgCl$_2$, 5 mM [α-^{32}P]dCTP ($\sim$2 Ci/mol) and 0.25 mM dithiothreitol. Stock solutions of each component are stored separately and then mixed together on the day of assay. To prevent precipitation of magnesium phosphatidate it is essential to mix the Triton X-100 and phosphatidic acid prior to addition of the magnesium chloride. The reaction is performed in 13 × 100 mm Pyrex tubes with Teflon-faced screw caps (Corning, Corning, NY). The reaction is initiated by adding 10 μl of enzyme source to 40 μl of a mixture of the other assay components.

[1] C. R. H. Raetz, *Annu. Rev. Genet.* **20**, 253 (1986).
[2] J. D. Esko and C. R. H. Raetz, *in* "The Enzymes" (P. D. Boyer, ed.), 3rd Ed., Vol., 16, p. 208. Academic Press, New York, 1983.
[3] T. Icho, C. P. Sparrow, and C. R. H. Raetz, *J. Biol. Chem.* **260**, 12078 (1985).
[4] C. P. Sparrow and C. R. H. Raetz, *J. Biol. Chem.* **260**, 12084 (1985).

METHODS IN ENZYMOLOGY, VOL. 209

After 10 min at 30°, the reaction is stopped by adding 2 ml of chloroform–methanol (1 : 1, v/v). The tubes can be stored in this state indefinitely if they are tightly capped. The extraction is finished by adding 0.85 ml of 1 M NaCl, pH 2. This creates a two-phase system.[5] The samples are mixed well and spun briefly to separate the phases. The upper phase, the interface, and any interfacial precipitate is aspirated and discarded. One milliliter of the lower phase (which is 50% of the total lower phase) is transferred to a glass vial, the solvent is evaporated, and radioactivity is quantitated by liquid scintillation spectrometry.

One unit of activity is defined as the amount of enzyme needed to convert 1 nmol of dCTP to chloroform-soluble material per minute. Under these conditions, the reaction is linear with time and enzyme up to about 5–10 nmol of product per 10 min. The reaction is linear to larger conversions with partially purified enzyme, presumably because there are fewer activities competing for the dCTP substrate. The most appropriate blank for the assay is a tube containing enzyme but no phosphatidic acid. Similar or identical blanks are obtained from tubes containing phosphatidic acid but no enzyme.

The standard assay is sensitive to the molecular species of phosphatidic acid used. Dioleoylphosphatidic acid, palmitoyloleoylphosphatidic acid, or phosphatidic acid made from egg phosphatidylcholine give optimal results, whereas dipalmitoylphosphatidic acid gives rates less than 10% of optimum.[4] This fact is the cause of the disagreement in the literature concerning the level of CDPdiacylglycerol synthase activity in wild-type crude extracts.[4,6]

Purification of CDPdiacylglycerol Synthase

A summary of the purification of CDPdiacylglycerol synthase is given in Table I. The behavior of CDPdiacylglycerol synthase during column chromatography is influenced greatly by detergents and by the presence of EDTA. Although the physical basis of these effects is unknown, the detergents are useful tools for purifying the enzyme. Furthermore, CDPdiacylglycerol synthase is fairly stable in the presence of dithiothreitol, even at room temperature, and therefore the columns are run on the laboratory bench (20°–24°). After elution, column fractions are stored on ice.

Tris buffers are prepared with chloride as the counterion. Concentrations in units of percent are all (w/v). Protein is assayed after precipitation as described by Peterson,[7] except that samples containing high concentra-

[5] E. G. Bligh and J. J. Dyer, *Can. J. Biochem. Physiol.* **37**, 911 (1959).
[6] K. E. Langley and E. P. Kennedy, *J. Bacteriol.* **136**, 85 (1978).
[7] G. L. Peterson, *Anal. Biochem.* **83**, 346 (1977).

TABLE I
PURIFICATION OF CDPDIACYLGLYCEROL SYNTHASE FROM *Escherichia coli*[a]

Step	Total protein (mg)	Specific activity (units/mg)	Yield (%)	Requirements for step[b]
Crude extract	1656	192	(100)	(Plasmid)
Membranes	676	466	99	Mg^{2+}
Solubilized membranes	589	346	64	EDTA, OG[c]
First DEAE column	288	607	55	EDTA
Second DEAE column	10.3	9,102	30	Zwittergent 3-12
Hydroxylapatite column	2.2	30,786	21	P_i gradient, OG

[a] Adapted from C. P. Sparrow and C. R. H. Raetz, *J. Biol. Chem.* **260**, 12084 (1985).
[b] See text for detailed discussion of requirements.
[c] OG, *n*-Octyl-β-D-glucopyranoside.

tions of Zwittergent cannot be precipitated. For these samples appropriate blanks must be assayed to subtract the signal generated by dithiothreitol.

Growth of Cells and Preparation of Crude Extract. The strain DH1/pCD100 overproduces CDPdiacylglycerol synthase about 50-fold.[3] This strain is grown in LB medium[8] and harvested in log phase. The cell paste (22 g) is resuspended in 9 volumes of 0.1 M potassium phosphate, pH 7.4, containing 10 mM magnesium sulfate, and the cells are broken by passage through a Mantlin-Gaulin press (9000 psi). The material is centrifuged at 5000 g for 15 min at 4° to remove unbroken cells. The supernatant is the crude extract.

Membrane Preparation and Solubilization of Enzyme. The crude extract is ultracentrifuged at 180,000 g for 90 min at 4°, and the membrane pellet is resuspended in 30–50 ml of 10 mM Tris, pH 8.0, containing 0.25 mM dithiothreitol. Resuspension can be accomplished using either a Teflon–glass homogenizer or a needle and syringe. CDPdiacylglycerol synthase is solubilized by adjusting the membrane fraction to 6–7 membrane protein/ml in 10 mM Tris, pH 8.5, 10 mM EDTA, 0.25 mM dithiothreitol, 1.5% Triton X-100, 1.5% *n*-octyl-β-D-glucopyranoside (OG), and 10% glycerol. This mixture is stirred for 1 hr on ice and then spun at 180,000 g for 60 min. The supernatant is the solubilized membrane fraction.

DEAE-cellulose Column Chromatography. CDPdiacylglycerol synthase binds to DEAE-cellulose (DE-52, Whatman, Clifton, NJ) in the absence of EDTA but not in the presence of EDTA. This property is exploited in the following purification. The solubilized membrane fraction

[8] J. H. Miller, "Experiments in Molecular Genetics." Cold Spring Harbor Laboratory, Cold Spring Harbor, New York, 1972.

(~600 mg protein) is loaded onto a DE-52 column, 4 cm wide and 7 cm high, which is preequilibrated with 10 mM Tris, pH 8.0, 10 mM EDTA, 2.0% Triton X-100, 10% glycerol, and 0.25 mM dithiothreitol. The column is washed with the same buffer, at a flow rate of 8 ml/min. Because only the unretained material is desired, a few fractions of large volume are collected: typically, the first 50 ml can be discarded, and the next 120–150 ml contains the CDPdiacylglycerol synthase activity. This is the DEAE run-through fraction. This material can be stored frozen with no loss of activity.

The run-through fraction is dialyzed at 4° against at least 20 volumes of buffer T, which is 10 mM Tris, pH 8.0, 1% Triton X-100, and 0.25 mM dithiothreitol. This material is loaded onto a second DE-52 column, 2.6 cm wide and 4 cm high, preequilibrated in buffer T. This column is run at 5 ml/min. After loading, the column is washed sequentially with 10 ml of buffer T, then 10 ml of buffer LT, which is 10 mM Tris, pH 8.0, 0.2% Triton X-100, and 0.25 mM dithiothreitol. Significant amounts of protein, but little CDPdiacylglycerol synthase, is then removed from the column by washing with 60 ml of buffer LT containing 70 mM KCl followed by 20 ml of buffer LT. The CDPdiacylglycerol synthase is then step-eluted with 63 ml of 10 mM Tris, pH 8.0, 4% Zwittergent 3-12 (Calbiochem, San Diego, CA), 25 mM KCl, and 0.25 mM dithiothreitol. The first 10–15 ml of eluant contains no activity and is discarded; the next 50 ml is collected in one fraction, called the Zwittergent enzyme preparation.

Hydroxylapatite Column Chromatography. The Zwittergent enzyme preparation is diluted 1 : 1 with buffer T and loaded onto a hydroxylapatite column (Bio-Rad, Richmond, CA, BioSil A, 200–400 mesh). The column, 1.4 cm wide and 3.4 cm high, is preequilibrated with buffer T and run at 4 ml/min. After loading, the column is washed with 10 ml of buffer T, and then the detergent is changed by washing with 5 ml of 10 mM Tris, pH 8.0, 10 mM EDTA, 1.2% OG, and 0.25 mM dithiothreitol. The load and wash are collected as one fraction. A phosphate elution is begun with 38 ml of 0.2 M phosphate, pH 8.0, 10 mM EDTA, 1.2% OG, and 0.25 mM dithiothreitol, and collected as one fraction. (The stock phosphate buffer contains sodium and potassium cations in a ratio of 1 : 2.) Then a linear phosphate gradient of 110 ml, 0.2–0.9 M, elutes the CDPdiacylglycerol synthase. The gradient solutions also contain 10 mM EDTA, 1.2% OG, and 0.25 mM dithiothreitol. Fractions are collected and analyzed for CDPdiacylglycerol synthase and total protein content. A typical run of this column is shown in Fig. 1. The latter half of the peak of CDPdiacylglycerol synthase activity is usually pure. In some preparations the early fractions are contaminated with a protein that migrates slightly slower than CDPdi-

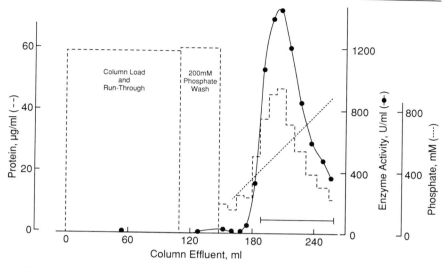

FIG. 1. Hydroxylapatite chromatography of CDPdiacylglycerol synthase. The Zwitter-gent-eluted enzyme from the second DEAE column is chromatographed on hydroxylapatite as described in the text. Fractions are assayed for protein and CDPdiacylglycerol synthase activity. The width of the plateaus for protein concentration indicate the volume of the fraction. The bar at the lower right indicates the fractions pooled. [Adapted from C. P. Sparrow and C. R. H. Raetz, *J. Biol. Chem.* **260**, 12084 (1985).]

acylglycerol synthase on sodium dodecyl sulfate (SDS)–polyacrylamide gels.

Handling and Storage of Purified CDPdiacylglycerol Synthase

The pooled peak fractions from the hydroxylapatite column can be concentrated by ultrafiltration on Amicon (Danvers, MA) PM10 membranes. This concentrates the OG as well as the enzyme. The pure enzyme can be dialyzed, but to maintain enzyme activity the dialyzing buffer must contain 1% OG, 0.25 mM dithiothreitol, and at least 200 mM salt (e.g., NaCl, potassium phosphate). Purified enzyme can be stored frozen for months if quick-frozen in liquid nitrogen. Activity will be lost if the enzyme is diluted into assay cocktail (phosphate buffer containing albumin) prior to freezing. The Zwittergent-eluted enzyme fraction from the second DEAE column loses all activity on freezing.

Characteristics of Purified CDPdiacylglycerol Synthase

Pure CDPdiacylglycerol synthase has an apparent minimum subunit molecular mass of about 27,000 daltons as judged by SDS–polyacrylamide gel electrophoresis.[4] This is close to the mass of 27,570 daltons predicted from the gene sequence.[3]

The reaction mechanism of CDPdiacylglycerol synthase is sequential.[4] The enzyme does not hydrolyze CDPdiacylglycerol.[4] The apparent kinetic constants are as follows: K_m for phosphatidic acid, 0.28 mM; K_m for dCTP, 0.58 mM; V_{max}, 55 μmol/min/mg.[4] These constants may be somewhat dependent on assay conditions, because CDPdiacylglycerol synthase displays surface dilution kinetics.[4]

CDPdiacylglycerol synthase will utilize either CTP or dCTP but no other nucleotide triphosphates.[4] The enzyme is also specific for long-chain phosphatidic acid; 1-acyl-*sn*-glycero-3-phosphate is not a substrate, and phosphatidic acids with acyl chains shorter than 16 carbons are poor substrates.[4] The enzyme has a broad pH optimum between pH 7 and 8. CDPdiacylglycerol synthase can be reconstituted into phospholipid vesicles that retain activity.[4]

The forward reaction of CDPdiacylglycerol synthase does not go to completion; under standard assay conditions, the equilibrium constant is about 0.2.[4] The reaction can be driven closer to completion *in vitro* by the addition of inorganic pyrophosphatase. For example, optimal conditions for synthesizing radioactive dCDPdiacylglycerol are 100 mM Tris, pH 7.5, 200 mM KCl, 1 mg/ml bovine serum albumin, 1 mM phosphatidic acid, 0.3% Triton X-100, 10 mM MgCl$_2$, 10 $\mu$$M$ [α-^{32}P]dCTP, 150 units/ml pure CDPdiacylglycerol synthase, and 30 units/ml of inorganic pyrophosphatase. After 3 hr at 30°, up to 80% of the [α-^{32}P]dCTP will be converted to [α-^{32}P]dCDPdiacylglycerol. The radioactive lipid product can be purified as previously described for unlabeled CDPdiacylglycerol.[4]

[28] CDPdiacylglycerol Synthase from Yeast

By GEORGE M. CARMAN and MICHAEL J. KELLEY

Introduction

CDPdiacylglycerol synthase (CTP : phosphatidate cytidylyltransferase; EC 2.7.7.41) catalyzes the conversion of phosphatidate to CDPdiacylglycerol.[1] The phospholipid product of the reaction, CDPdiacylglycerol,

$$CTP + phosphatidate \rightarrow PP_i + CDPdiacylglycerol$$

is the source of the phosphatidyl moiety in the primary (phosphatidyletha-nolamine methylation) pathway for the synthesis of the major phospholip-ids in *Saccharomyces cerevisiae*.[2] The main site of CDPdiacylglycerol synthase activity is the mitochondria.[3,4] About 20% of CDPdiacylglycerol synthase activity is associated with microsomes,[4] which is contributed mainly by plasma membrane-associated activity.[5] The expression of CDPdiacylglycerol synthase is regulated by inositol alone and in combina-tion with serine, ethanolamine, and choline.[6–8] Mitochondrial-associated CDPdiacylglycerol synthase has been purified to apparent homogeneity.[4] The major purification of the enzyme is achieved by CDPdiacylglycerol-Sepharose affinity chromatography. Binding of the enzyme to the affinity resin occurs because the CDPdiacylglycerol synthase reaction is favored in the reverse direction.[4,9] In this chapter we describe the purification and properties of the enzyme.

Preparation of Phosphatidate, CDPdiacylglycerol, and CDPdiacylglycerol-Sepharose

Phosphatidate is prepared from soybean lecithin using cabbage phos-pholipase D.[10,11] CDPdiacylglycerol is prepared from phosphatidate and CMPmorpholidate by the method of Agranoff and Suomi[12] with the modi-fications of Carman and Fischl.[13] The CDPdiacylglycerol-Sepharose resin[14] used for CDPdiacylglycerol synthase purification is the high-capac-

[1] J. R. Carter and E. P. Kennedy, *J. Lipid Res.* **7**, 678 (1966).
[2] G. M. Carman and S. A. Henry, *Annu. Rev. Biochem.* **58**, 635 (1989).
[3] K. Kuchler, G. Daum, and F. Paltauf, *J. Bacteriol.* **165**, 901 (1986).
[4] M. J. Kelley and G. M. Carman, *J. Biol. Chem.* **262**, 14563 (1987).
[5] A. J. Kinney and G. M. Carman, *J. Bacteriol.* **172**, 4115 (1990).
[6] M. J. Homann, S. A. Henry, and G. M. Carman, *J. Bacteriol.* **163**, 1265 (1985).
[7] M. J. Homann, A. M. Bailis, S. A. Henry, and G. M. Carman, *J. Bacteriol.* **169**, 3276 (1987).
[8] L. S. Klig, M. J. Homann, S. Kohlwein, M. J. Kelley, S. A. Henry, and G. M. Carman, *J. Bacteriol.* **170**, 1878 (1988).
[9] G. Belendiuk, D. Mangnall, B. Tung, J. Westley, and G. S. Getz, *J. Biol. Chem.* **253**, 4555 (1978).
[10] M. Kates and P. S. Sastry, this series, Vol. 14, p. 197.
[11] W. Dowhan and T. Larson, this series, Vol. 71, p. 561.
[12] B. W. Agranoff and W. D. Suomi, *Biochem. Prep.* **10**, 47 (1963).
[13] G. M. Carman and A. S. Fischl, *J. Food Biochem.* **4**, 53 (1980).
[14] T. J. Larson, T. Hirabayashi, and W. Dowhan, *Biochemistry* **15**, 974 (1976).

ity affinity resin prepared by Carman and Fischl.[15] The preparations of CDPdiacylglycerol and CDPdiacylglycerol-Sepharose are described elsewhere in this volume.[15]

Assay Method

CDPdiacylglycerol synthase activity is measured by following the incorporation of 1.0 mM [5-^{3}H]CTP [10,000 counts/min (cpm)/nmol] into CDPdiacylglycerol in 50 mM Tris–maleate buffer (pH 6.5) containing 20 mM MgCl$_2$, 15 mM Triton X-100, 0.5 mM phosphatidate, and enzyme protein in a total volume of 0.1 ml at 30°.[4,9] The reaction is terminated by the addition of 0.5 ml of 0.1 N HCl in methanol. Chloroform (1 ml) and 1 M MgCl$_2$ (1.5 ml) are added, the system is mixed, and the phases are separated by a 2-min centrifugation at 100 g. A 0.5-ml sample of the chloroform phase is removed and taken to dryness on an 80° water bath. Betafluor (4 ml) is added to the sample, and radioactivity is determined by scintillation counting. The chloroform-soluble phospholipid product is identified with standard CDPdiacylglycerol by thin-layer chromatography.[9]

One unit of enzymatic activity is defined as the amount of enzyme that catalyzes the formation of 1 nmol of product per minute under the assay conditions described above. Protein concentration is determined by the method of Bradford[16] using bovine serum albumin as the standard. Dilute protein samples are concentrated prior to the protein assay as described by Wessel and Flugge.[17] Buffers which were identical to those containing the protein samples are used as blanks.

Growth of Yeast

Wild-type $S.$ $cerevisiae$ strain S288C (α gal2) is used for enzyme purification. Cells are grown in 1% yeast extract, 2% peptone, and 2% glucose at 28° to late exponential phase, harvested by centrifugation, and stored at −80°.[15,18]

Purification Procedure

All steps are performed at 5°.

Step 1: Preparation of Cell Extract. Cells (50 g) are disrupted with glass beads with a Bead-Beater (BioSpec Products, Bartlesville, OK) in

---

[15] G. M. Carman and A. S. Fischl, this volume [36].
[16] M. M. Bradford, *Anal. Biochem.* **72,** 248 (1976).
[17] D. Wessel and U. I. Flugge, *Anal. Biochem.* **138,** 141 (1984).
[18] A. S. Fischl and G. M. Carman, *J. Bacteriol.* **154,** 304 (1983).

50 mM Tris-HCl buffer (pH 7.5) containing 1 mM Na_2EDTA, 0.3 M sucrose, and 10 mM 2-mercaptoethanol as previously described.[15,18] Glass beads and unbroken cells are removed by centrifugation at 1500 g for 5 min to obtain the cell extract (supernatant).

Step 2: Preparation of Mitochondria. CDPdiacylglycerol synthase activity is enriched 4- to 5-fold in the mitochondrial fraction over the cell extract and is therefore used as the source of enzyme. Crude mitochondria are collected from cell extracts by differential centrifugation.[9] Mitochondrial pellets are suspended in 50 mM Tris-HCl buffer (pH 7.5) containing 10 mM 2-mercaptoethanol and 20% glycerol (w/v) at a protein concentration of 20 mg/ml.

Step 3: Solubilization with Triton X-100. The mitochondrial fraction is suspended in 50 mM Tris–maleate buffer (pH 6.5) containing 0.1 M KCl, 10 mM $MgCl_2$, 10% glycerol, and 1% Triton X-100 at a final protein concentration of 10 mg/ml. The suspension is incubated for 2 hr on a rotary shaker at 150 rpm. After the incubation, the suspension is centrifuged at 27,000 g for 30 min to obtain the solubilized (supernatant) fraction. The Triton X-100 extract is stored at $-80°$. On thawing, a precipitate forms in the extract which is removed by centrifugation at 27,000 g for 10 min.

Step 4: CDPdiacylglycerol-Sepharose Chromatography. A CDPdiacylglycerol-Sepharose column (1.5 × 15 cm) is equilibrated with 500 ml of chromatography buffer [Tris–maleate (pH 6.5), 10 mM $MgCl_2$, 10% glycerol, and 0.5% Triton X-100]. The Triton X-100-solubilized enzyme (36 ml, one-fourth of the preparation) is applied to the column in 12-ml aliquots. To effect enzyme binding, each aliquot is incubated in the column for 20 min before addition of the next aliquot. Binding of CDPdiacylglycerol synthase to the affinity column is dependent on the presence of magnesium ions and Triton X-100 in the chromatography buffer. After the Triton X-100 extract is applied, the column is washed with 250 ml of chromatography buffer containing 0.5 M NaCl followed by 250 ml of chromatography buffer containing 1% Triton X-100. The column is then saturated with chromatography buffer containing 5 mM CTP and 2 M NaCl and incubated for 1 hr. CDPdiacylglycerol synthase is eluted from the column with this buffer at a flow rate of 20 ml/hr. Elution of the enzyme from the resin is dependent on both the substrate CTP and NaCl in the elution buffer. In the absence of either of these components, enzyme dissociation from the column is negligible. Fractions containing activity are pooled and desalted on a Sephadex G-25 column (3 × 12.5 cm) equilibrated with chromatography buffer.

Step 5: Hydroxylapatite Chromatography. A hydroxylapatite (BioGel HT, Bio-Rad, Richmond, CA) column (2.5 × 3.5 cm) is equilibrated with

TABLE I
PURIFICATION OF CDPDIACYLGLYCEROL SYNTHASE FROM Saccharomyces cerevisiae[a]

Purification	Total units (nmol/min)	Protein (mg)	Specific activity (units/mg)	Purification (-fold)	Yield (%)
1. Cell extract	2700	4427	0.61	1.0	100
2. Mitochondria	2106	744	2.83	4.64	78
3. Triton X-100 extract	1602	300	5.33	8.74	59
4. CDPdiacylglycerol-Sepharose	1362	1.57	868	1422	50
5. Hydroxylapatite	866	0.61	1409	2309	32

[a] Data are from Kelley and Carman.[4]

120 ml of chromatography buffer. Desalted enzyme from the previous step is applied to the column at a flow rate of 10 ml/hr. The column is washed with 120 ml of 0.1 M potassium phosphate buffer (pH 6.5) containing 5 mM $MgCl_2$, 10% glycerol, and 0.5% Triton X-100. CDPdiacylglycerol synthase is then eluted from the column with a 100-ml linear potassium phosphate gradient (0.2–0.5 M) containing 5 mM $MgCl_2$, 10% glycerol, and 0.5% Triton X-100 at a flow rate of 10 ml/hr. The peak of enzyme activity elutes from the column at a phosphate concentration of about 0.3 M. The hydroxylapatite-purified enzyme loses about 70% of the activity after 2 weeks of storage at $-80°$. To stabilize CDPdiacylglycerol synthase activity, CTP (1 mM) is added to the purified enzyme preparation. In the presence of CTP, the purified enzyme is 90–100% stable for at least 3 months of storage at $-80°$ and to at least two cycles of freezing and thawing.

Enzyme Purity. The five-step purification scheme summarized in Table I results in an apparently homogeneous preparation of CDPdiacylglycerol synthase. Enzyme purity is assessed by native[19] and sodium dodecyl sulfate-polyacrylamide gel electrophoresis.[20] Native gels (6% polyacrylamide) contain 0.5% Triton X-100 and 1 mM CTP. The enzyme preparation does not contain measurable phosphatidylserine synthase, phosphatidylinositol synthase, or phosphatidylglycerophosphate synthase activities, which are also associated with the mitochondrial fraction of *S. cerevisiae*.[3] The enzyme is purified 2309-fold over the cell extract, with an activity yield of 32%.

[19] B. Davis, *Ann. N.Y. Acad. Sci.* **121**, 404 (1964).
[20] U. K. Laemmli, *Nature (London)* **227**, 680 (1970).

Molecular Weight

The subunit molecular weight of CDPdiacylglycerol synthase is 56,000 as determined by sodium dodecyl sulfate-polyacrylamide gel electrophoresis.[4] Radiation inactivation analysis[21,22] of the native mitochondrial-bound enzyme and purified enzyme suggests that functional CDPdiacylglycerol synthase has a target size or native molecular weight of 114,000. Thus, it appears that the native enzyme is a dimer composed of identical subunits with molecular weights of 56,000.[4]

Properties of CDPdiacylglycerol Synthase

Maximum CDPdiacylglycerol synthase activity is dependent on $MgCl_2$ and Triton X-100 at pH 6.5. The energy of activation is 9 kcal/mol, the enzyme is labile above 30°, and thioreactive agents inhibit activity. CDPdiacylglycerol synthase follows typical saturation kinetics toward CTP and phosphatidate when activity is measured with a mixed micelle substrate of Triton X-100 and phosphatidate.[4] The true K_m values for CTP and phosphatidate are 1 and 0.5 mM, respectively, and the V_{max} is 4.7 μmol/min/mg. The results of kinetic experiments and the ability of the enzyme to catalyze a variety of isotopic exchange reactions suggest that CDPdiacylglycerol synthase catalyzes a sequential Bi–Bi reaction. CDPdiacylglycerol synthase binds to CTP prior to phosphatidate, and PP_i is released prior to CDPdiacylglycerol in the reaction sequence.[4] dCTP[4,9] and thiophosphatidate[23] are competitive inhibitors of the enzyme.

Synthetic and Analytical Uses

Pure CDPdiacylglycerol synthase can be used to synthesize radiolabeled CDPdiacylglycerol from phosphatidate and labeled CTP.[4] Radiolabeled CDPdiacylglycerol has been used in conjunction with pure phoshatidylinositol synthase[18] to determine quantitatively the intracellular inositol concentration of yeast.[24]

Acknowledgments

This work was supported by U.S. Public Health Service Grant GM-28140 from the National Institutes of Health, New Jersey State funds, and the Charles and Johanna Busch Memorial Fund.

[21] E. S. Kempner and W. Schlegel, *Anal. Biochem.* **92**, 2 (1979).

[22] G. R. Kepner and R. I. Macey, *Biochim. Biophys. Acta* **163**, 188 (1968).

[23] S. I. Bonnel, Y.-P. Lin, M. J. Kelley, G. M. Carman, and J. Eichberg, *Biochim. Biophys. Acta* **1005**, 289 (1989).

[24] M. J. Kelley, A. M. Bailis, S. A. Henry, and G. M. Carman, *J. Biol. Chem.* **263**, 18078 (1988).

[29] Choline-Phosphate Cytidylyltransferase

By PAUL A. WEINHOLD and DOUGLAS A. FELDMAN

Introduction

Choline-phosphate cytidylyltransferase (EC 2.7.7.15, CTP : choline-phosphate cytidylyltransferase) catalyzes a major rate-determining step in the biosynthesis of phosphatidylcholine in mammalian cells.[1,2] Both cytosolic and membrane fractions contain choline-phosphate cytidylyltransferase activity. This series previously presented discussions on the partial purification and properties of cytosolic cytidylyltransferase.[3,4] Purification of the enzyme from liver cytosol using techniques appropriate for soluble enzymes has not been successful. This was primarily due to the inability to recover enzyme activity after chromatographic procedures. The use of detergents in the purification process enabled us to purify the enzyme from liver cytosol.[5,6]

Assay Method

Enzyme activity is determined by measuring the formation of radioactive CDPcholine from phospho[*methyl*-^{14}C]choline. Two methods have been used to separate CDPcholine from phosphocholine: adsorption of CDPcholine by charcoal[3,7] or separation of CDPcholine from phosphocholine by thin-layer chromatography.[4] We developed a charcoal adsorption method that is rapid, sensitive, and reproducible. The ability to perform many assays (50–100) per day is an additional advantage.

Reagents

Assay buffer: 50 mM imidazole, 150 mM KCl, 2 mM EDTA, pH 7.0
Phospho[*methyl*-^{14}C]choline: 16 mM, 1000–1500 disintegrations/min (dpm)/nmol, prepared in assay buffer. Phospho[*methyl*-^{14}C]choline is obtained from either New England Nuclear (Boston, MA) or

[1] S. L. Pelech and D. E. Vance, *Biochim. Biophys. Acta* **779**, 217 (1984).
[2] L. B. M. Tijburg, M. J. H. Geelen, and L. M. G. van Golde, *Biochim. Biophys. Acta* **1004**, 1 (1989).
[3] G. B. Ansell and T. Chojnacki, this series, Vol. 14, p. 121.
[4] D. E. Vance, S. D. Pelech, and P. C. Choy, this series, Vol. 71, p. 576.
[5] P. A. Weinhold, M. E. Rounsifer, and D. A. Feldman, *J. Biol. Chem.* **261**, 5104 (1986).
[6] D. A. Feldman and P. A. Weinhold, *J. Biol. Chem.* **262**, 9075 (1987).
[7] R. Sleight and C. Kent, *J. Biol. Chem.* **258**, 831 (1983).

Amersham (Arlington Heights, IL). Alternatively, radioactive phosphocholine can be prepared as described by Vance *et al.*[4]

CTP–magnesium acetate: 15 mM CTP, 60 mM magnesium acetate prepared in assay buffer

Phosphatidylcholine–oleic acid: 2.5 mM phosphatidylcholine, 2.5 mM oleic acid; preparation of this mixture is described below

CDPcholine recovery standard: 1.0 mM [*methyl*-^{14}C]CDPcholine (1000 dpm/nmol) prepared in water

Trichloroacetic acid (TCA)–phosphocholine: 10% TCA (w/v), 150 mM phosphocholine

Charcoal (acid washed, Sigma, St. Louis, MO, C-5510) 6 g of charcoal is suspended in 100 ml of water. The suspension is stirred 30 min before use and prepared fresh each day.

Charcoal elution solvent: 1160 ml water–1880 ml ethanol–110 ml 28% NH$_4$OH

Preparation of Phosphatidylcholine–Oleic Acid Liposomes

Stock solutions of phosphatidylcholine (100 mg/ml chloroform, from egg yolk, Type X1-E, Sigma) and oleic acid (100 mM in hexane) are stored at $-20°$ in tightly capped containers. To prepare 1.0 ml of stock liposomes (1 : 1 molar ratio of phosphatidylcholine–oleic acid), 20 μl of stock phosphatidylcholine and 25 μl of stock oleic acid are added to a glass tube. The solvent is evaporated with a stream of N$_2$ at room temperature. Assay buffer (1.0 ml) is added. The mixture is sonicated while the tube is immersed in a beaker of room temperature water, using a Sonifier fitted with a microtip probe set to a 50 W output for 12 min. The liposome suspension should be slightly cloudy but translucent. The liposome suspension is prepared fresh daily and kept at room temperature.

Assay Procedure

The following reagents are added for each assay: 10 μl phospho[*methyl*-^{14}C]choline, 20 μl CTP–magnesium acetate, 20 μl phosphatidylcholine–oleic acid, and 50 μl of enzyme plus assay buffer. This results in 1.6 mM phosphocholine, 3.0 mM CTP, 12.0 mM Mg^{2+}, 2.0 mM EDTA, 150 mM KCl, 500 μM phosphatidylcholine, 500 μM oleic acid, and 50 mM imidazole, pH 7.0, in the final assay mixture. By using glass tubes 13 $\times$ 125 mm for the reaction the charcoal adsorption procedure for CDPcholine isolation can be conducted in the same tube. The reaction is started by the addition of enzyme. The reaction is incubated at 37°. The addition of 100 μl of TCA–phosphocholine stops the reaction. At this point, the reaction tubes can be stored at 4°. Duplicate blank reaction tubes contain all reactants but no enzyme. Duplicate tubes for CDPcholine recovery are prepared by adding 10 μl of [methyl-^{14}C]CDPcholine recovery standard to complete assay mixtures. Enzyme is added to the recovery tubes, and TCA–phosphocholine is added immediately.

CDPcholine is recovered from the reactions by adsorption to charcoal. The charcoal is collected, washed, and eluted using a multiple sample filtration manifold designed to hold scintillation vials (Hoefer Scientific, San Francisco, CA, Model FH225V). A 25-mm cellulose acetate filter with a pore size of 0.45 μm (Millipore, Bedford, MA, HAWP, HA) is used. The charcoal suspension is stirred in a beaker continuously while 1.0-ml aliquots are removed and added to each reaction tube. After 30 min at room temperature, 7.0 ml of water is added to each tube. The charcoal suspension is mixed vigorously. The charcoal is collected by centrifugation at room temperature for 5 min at 2000 rpm. The supernatant is discarded into an appropriate radioactive waste container. The charcoal pellet is suspended in 7.0 ml of water, transferred to wells, and collected under reduced pressure. The tube is rinsed with 7.0 ml of water which is then transferred to the well. The charcoal pad is washed 3 times with 7 ml of water. The vacuum is off when adding the water. The collected wash water is discarded, and 20-ml scintillation vials are placed in the manifold collection rack. Add 1.0 ml of elution solvent and allow 5 min for extraction; then apply vacuum and collect the extract into scintillation vials. The extraction is repeated three more times. Add 10 ml of scintillation fluid and measure the radioactivity. We use Safety Solve from Research Products International, Inc. (Mt. Prospect, IL). Most of the newer biodegradable scintillation fluids are unsatisfactory because they are unable to form a single phase with 4.0 ml of the extraction solvent.

Blanks (no enzyme or zero time) range from 80 to 120 dpm. The recovery of CDPcholine is measured with each set of assays. The recovery is very reproducible for a particular lot of charcoal (67.9 ± 0.5% for 10 samples run at the same time; 68.5 ± 1.5% for recovery experiments performed over 2 months). We use the recovery factor for each set of assays to correct the data to 100% recovery. Occasionally, the commercially obtained phospho[methyl-^{14}C]choline contains small amounts of radioactive contaminants that cause increased blanks. This can be removed by treating the stock phospho[*methyl*-^{14}C]choline with charcoal.

Enzyme Assay in Homogenate and Membrane Fractions

The rate of the cytidylyltransferase reaction is constant for at least 60 min when cytosolic preparation, partially purified enzyme, or purified enzyme is used. Similarly, the rate is proportional to the amount of soluble enzyme preparation over a reasonable range of protein. However, crude homogenate or membrane-containing fractions produce constant rates for only short incubation times. The rate of the reaction also is proportional to the amount of preparation only at low protein concentrations. These restrictions result in the production of low amounts of CDPcholine and a loss in sensitivity of the assay. This problem has been attributed to the presence of enzymes that degrade CTP and cytidine esters.[3] Under the present assay conditions, CDPcholine is not degraded. However, a high nucleotidase activity that rapidly hydrolyzes CTP exists in membrane-containing preparations. This can be effectively prevented by including ADP in the assay.[8] We add 6.0 mM ADP and an additional 3 mM magnesium acetate to all assays that contain membranes.

Lipid Requirements for Cytidylyltransferase Assay

A variety of lipids have been used by investigators for the assay of cytidylyltransferase. It has been common practice to include lipid in assays involving soluble forms of cytidylyltransferase. However, it has been considered unnecessary by some investigators to include lipid in assays for membrane cytidylyltransferase. Although membrane cytidylyltransferase activity is not increased appreciably by phospholipids, fatty acids stimulate the activity significantly (30–40%); however, fatty acids also cause inhibition at higher concentrations. We find that the phosphatidylcholine–oleic acid liposomes used in the assay of soluble enzyme also produce apparent maximal activities for membrane preparations. The activities are usually 10–20% higher than can be obtained with oleic acid alone.

Enzyme Purification

Reagents

Buffer A: 50 mM Tris-HCl, 150 mM NaCl, 1.0 mM EDTA, 2.0 mM dithiothreitol (DTT), 0.025% sodium azide, pH 7.4

Buffer B: 50 mM Tris-HCl, 1.0 mM EDTA, 2 mM DTT, 0.025% sodium azide

Phenylmethylsulfonyl fluoride (PMSF): 200 mM in 2-propanol

Phosphatidylcholine–oleic acid emulsion: 5 mM phosphatidylcholine, 10 mM oleic acid. Prepare the emulsion by mixing 4 ml of phosphatidylcholine (100 mg/ml egg phosphatidylcholine in chloroform) and 10 ml oleic acid (100 mM in hexane) in a 500-ml round-bottomed

---

[8] P. A. Weinhold, L. Charles, and D. A. Feldman, *Biochim. Biophys. Acta*, in press (1991).

flask. The organic solvents are evaporated under reduced pressure by rotary evaporation for 60 min at 37°–40°. Tip the flask up and down during the initial evaporation to produce a thin film of lipid on the inside of the flask. Add 100 ml of buffer A. The mixture is stirred using a magnetic stir bar in the presence of a few glass marbles to remove the lipid from the side of the flask. Then the lipid mixture is sonicated for 60 min with the flask submerged in a large reservoir of water at room temperature.

Procedure

Preparation of Liver Cytosol. The abdomen of an anesthetized rat (200–250 g, Crl:CD BR, VAF/Plus, Charles River Labs., Wilmington, MA) is opened, and the vena cava is ligated above the kidney. The chest is opened. An 18-gauge stomach tube needle is inserted into the vena cava just below the heart and secured with a suture. The hepatic vein is cut, and the liver is perfused thoroughly with 0.9% NaCl (w/v) using a pump at a flow rate of about 30 ml/min. All subsequent procedures are performed at 4°. Livers from 7–8 rats (90–100 g) are minced with a pair of scissors and homogenized in buffer A (5 ml/g liver). A volume of PMSF stock solution is added to the homogenate to give a final PMSF concentration of 1.0 mM.

Cytosol is prepared by sequential centrifugation of the homogenate at 10,000 g for 20 min and 100,000 g for 60 min. The protein concentration in cytosol is measured using the Coomassie protein reagent (Pierce Chemical Co., Rockford, IL) with bovine serum albumin as standard. We dilute the cytosol with buffer A to a final protein level of 6 mg/ml.

Acid Precipitation and Detergent Extraction. Phosphatidylcho-line–oleic emulsion is added slowly to the cytosol stirred at 4° (10 ml added to 90 ml cytosol). This gives a final concentration of 0.5 mM phosphatidyl-choline–1.0 mM oleic acid. Stir the mixture for 20 min at 4°.

The pH of the cytosol–lipid mixture is reduced to 5.0 by slow addition of 1.0 M acetic acid. The mixture is stirred for an additional 10 min at 4°. The precipitated protein is collected by centrifugation at 10,000 g for 20 min. Discard the supernatant and suspend the precipitate in 400 ml buffer A. The protein precipitate is again collected by centrifugation at 10,000 g for 20 min. Octylglucopyranoside (Sigma) is dissolved at a final concentration of 20 mM in buffer A that contains 1 mM PMSF. The protein precipitate is suspended in a volume of octylglucoside buffer that equals 40% of the volume of cytosol used for acid precipitation (including added lipid). We stir the suspension for 30 min at room temperature. Cytidylyltransfer-ase is recovered in the octylglucoside extract following centrifugation at

10,000 g for 20 min. Approximately 60–70% of the cytidylyltransferase activity in the cytosol is recovered in the octylglucoside extract.

DEAE-Sepharose Column Chromatography. A 1.6 × 15 cm column of DEAE-Sepharose (Pharmacia, Piscataway, NJ) is packed and equilibrated with 30 ml buffer A at a flow rate of 50 ml/hr. All subsequent procedures are performed using a flow rate of 30 ml/hr.

The octylglucoside extract is applied to the column followed by 120 ml buffer A. The column is washed with a 300-ml linear gradient of 150 to 300 mM NaCl in buffer B, followed by 240 ml of 300 mM NaCl in buffer B. Cytidylyltransferase activity is eluted with 75 ml of buffer B that contains 50 mM octylglucoside and 400 mM NaCl. Usually, 70–80% of the activity applied to the column is recovered in the octylglucoside peak. A peak of cytidylyltransferase activity also is eluted in the 150–300 mM NaCl gradient. The amount of activity in this peak seems to be related to the amount of lipid in the extract applied to the column. For example, if the activity recovered in the peak from the gradient is mixed with phosphatidylcholine–oleic acid to give the original 0.5 mM phosphatidylcholine–1.0 mM oleic acid, cytidylyltransferase activity now elutes with octylglucoside when chromatographed on a new DEAE-Sepharose column. However, if the peak is rerun on a DEAE-Sepharose column without the addition of lipid the enzyme activity is eluted again in the NaCl gradient.

Hydroxylapatite Column Chromatography. Suspend hydroxylapatite (Bio-Rad, Richmond, CA) in buffer A and allow the suspension to settle for 2 min. The supernatant is aspirated, and the process is repeated several times to remove fine particles. A 1.6 × 7.5 cm column is packed by gravity, with the column outlet closed. The column is equilibrated with 150 ml buffer A at a flow rate of 30 ml/hr.

Add buffer A to the pooled enzyme from the DEAE column (2 volumes of buffer A to 1 volume of enzyme). The diluted enzyme pool is applied to the hydroxylapatite column followed by 150 ml buffer A containing 0.15 M potassium phosphate, pH 7.5. Wash the column with 100 ml of buffer A–0.20 M potassium phosphate, pH 7.5, followed by sequential elution with a 150-ml linear gradient of 0 to 50 mM octylglucoside in buffer A–0.20 M potassium phosphate, pH 7.5; 75 ml of 50 mM octylglucoside in buffer A–0.20 M potassium phosphate, pH 7.5; and 60 ml of 100 mM octylglucoside in buffer A–0.20 M potassium phosphate, pH 7.5. Cytidylyltransferase activity is eluted with 100 ml buffer A–0.20 M potassium phosphate–0.03% Triton X-100, pH 7.5.

Approximately 35–45% of the cytidylyltransferase activity applied to the column is recovered in the Triton-eluted peak. The specific activity of the enzyme is 12–15 μmol/min/mg protein. This is about a 2000-fold purification with a yield of 8–15%. The enzyme at this step of purification gives a single protein band on nondenaturing polyacrylamide electrophoresis run in the presence of 0.1% Nonidet P-40 (NP-40). Enzyme activity is associated with the protein band from the electrophoretic gel. Sodium dodecyl sulfate (SDS)-polyacrylamide electrophoresis and Coomassie blue staining reveal two protein bands: one with an M_r of about 45,000, the other with an M_r of 38,000. Chromatography on a second hydroxylapatite column separates these proteins. The catalytic activity is associated with the protein containing the 45,000 M_r subunit.

The second column of hydroxylapatite (0.8 × 5 cm) is packed by gravity and equilibrated with 25 ml of buffer A. The fractions from the first hydroxylapatite column that contain cytidylyltransferase activity are pooled and diluted with 5 volumes of buffer A containing 0.03% Triton X-100. The diluted enzyme pool is applied to the hydroxylapatite column at a flow rate of 20 ml/hr. The column is washed with 20 ml buffer A containing 50 mM potassium phosphate and 0.03% Triton. Cytidylyltransferase activity is eluted with 25 ml buffer A containing 0.20 M potassium phosphate, 0.03% Triton X-100. A summary of the purification is shown in Table I.

Hydroxylapatite column chromatography is also useful to exchange detergents. For example, after the diluted enzyme is applied to the column, the column is washed with 50 mM phosphate in buffer A without Triton X-100. Enzyme activity is now eluted with buffer A containing 0.20 M phosphate, 100 mM octylglucoside. Since octylglucoside is readily removed by dialysis, this method may be useful in reconstitution experiments.

Purity. Cytidylyltransferase from the second hydroxylapatite column contains a single protein when examined by SDS-polyacrylamide electrophoresis. The specific activity is 47.5 μmol/min/mg protein. The protein content is determined by measuring the intensity of Coomassie-stained proteins after SDS-polyacrylamide electrophoresis.[6]

Sanghera and Vance[9] reported purification of cytidylyltransferase using the present method, except they used a Mono Q column in place of the last hydroxylapatite column. They reported a final specific activity of 34.2 μmol/min/mg protein. Two additional laboratories have purified cytidylyltransferase using our method with minor modification.[10,11] In all cases the preparations gave essentially a single protein band on SDS-polyacrylamide electrophoresis.

[9] J. S. Sanghera and D. E. Vance, *J. Biol. Chem.* **264,** 1215 (1989).
[10] R. Cornell, *J. Biol. Chem.* **264,** 9077 (1989).
[11] J. D. Watkins and C. Kent, *J. Biol. Chem.* **265,** 2190 (1990).

TABLE I

CYTIDYLYLTRANSFERASE PURIFICATION[a]

Step	Protein (mg)	Activity (nmol/min)	Specific activity (nmol/min/mg)	Purification (-fold)	Yield (%)
Cytosol	3129 ± 168	17,510 ± 1677	5.7 ± 0.07	1	100
Octylglucoside Extract	87 ± 6	11,920 ± 822	139 ± 14	25 ± 2	69 ± 2.5
DEAE-Sepharose	11 ± 1.3	7,080 ± 560	639 ± 53	115 ± 9	42 ± 5
Hydroxylapatite I	0.09 ± 0.03	1,043 ± 260	12,250 ± 1390	2180 ± 160	6.4 ± 2
Hydroxylapatite II	0.011	542	47,500	8333	3.3

[a] The results through the hydroxylapatite I step are averages from six preparations. The results for the hydroxylapatite II step are from a single preparation.

Properties

Molecular Weight. Purified cytidylyltransferase contains a single subunit when analyzed by SDS-polyacrylamide electrophoresis. The apparent molecular weight of this subunit is 44,500 when determined from a calibration curve of standard proteins separated on a 12.5% gel. Analysis on a 4–30% gradient gel gives an M_r of 44,700. Others have reported an M_r of 42,000[9,10] from electrophoretic analysis. A cytidylyltransferase cDNA has been cloned from rat liver using an oligonucleotide corresponding to a peptide sequence from purified cytidylyltransferase.[12] A molecular weight of 41,720 was predicted from the cDNA sequence. Direct estimation of the native molecular weight of purified cytidylyltransferase by gel-filtration chromatography is not possible because purified cytidylyltransferase binds tightly to the chromatographic matrix. Gel-filtration chromatography can only be performed in the presence of detergents.

We calculated a molecular weight of 97,000 ± 10,000 for cytidylyltransferase in liver cytosol.[13] This calculation used Stokes radii obtained from gel filtration and sedimentation coefficients from glycerol density centrifugation. Purified cytidylyltransferase and cytidylyltransferase from liver cytosol move to the same position in the glycerol gradient.[14] These results suggest that the native M_r of the purified cytidylyltransferase is approximately 97,000. This is consistent with a dimeric structure containing two 45,000 subunits. Cornell[10] reached a similar conclusion from cross-linking studies with purified enzyme.

Requirements for Lipids. Purified cytidylyltransferase requires the addition of exogenous lipid for maximal activity. Approximately 10–15% of maximal activity is obtained in the absence of added lipid. Maximal activity is obtained with phosphatidylcholine–oleic acid liposomes (1 : 1 molar ratio) at a final concentration of 100 μM total lipid (50 μM oleic acid). Phosphatidylcholine or oleic acid added separately do not stimulate activity appreciably. Phosphatidylcholine–oleic acid liposomes with a greater than 60 mol % of oleic acid strongly inhibit the reaction. Phosphatidylcholine–oleic acid liposomes containing 9 mol % oleic acid produce 80–90% of maximal activity at concentrations of 500 μM total lipid. Similar stimulation of activity is obtained with phosphatidylcholine liposomes containing 9 mol % phosphatidylglycerol or 9 mol % phosphatidylinositol if the total lipid concentration is 500 μM (45 μM phosphatidylglycerol or phosphati-

[12] G. B. Kalmar, R. J. Kay, A. Lechance, R. Abersold, and R. B. Cornell, *Proc. Natl. Acad. Sci. U.S.A.* **87**, 6029 (1990).

[13] P. A. Weinhold, M. E. Rounsifer, L. Charles, and D. A. Feldman, *Biochim. Biophys. Acta* **1006**, 299 (1989).

[14] D. A. Feldman and P. A. Weinhold, unpublished results, 1989.

dylinositol). Phosphatidylglycerol alone provides about 40% maximal activity at 45 μM. This activity remains unchanged at higher concentrations of phosphatidylglycerol. Phosphatidylglycerol stimulates enzyme activity 15–20% at concentrations as low as 0.5 μM. The activation by lipids is associated with the formation of enzyme–lipid aggregates that migrate on glycerol density gradients as discrete complexes.

Purified cytidylyltransferase has been reported to be inhibited by sphingolipids.[15] This inhibition is reversible by activating phospholipids.

Kinetic Properties. Purified cytidylyltransferase exhibits maximal activity at pH 7.0 in either Tris-HCl or imidazole-HCl buffers. The activity at pH 6.5 or 7.5 is nearly the same as at pH 7.0. The K_m values are 0.24 mM for phosphocholine and 0.22 mM for CTP. These values are obtained from secondary plots of apparent V_{max} (obtained from primary double-reciprocal plots) versus the corresponding concentrations of phosphocholine or CTP.

Inhibitors. The enzyme is inhibited by 5,5′-dithiobis-2-nitrobenzoic acid (DTNB), N-ethylmaleimide (NEM), and p-chloromercuribenzoate (pCMB).[5] Iodoacetamide does not cause inhibition at concentrations up to 1.5 mM. pCMB is the most potent, producing 80% inhibition at 0.6 μM. NEM, on the other hand, gives about 80% inhibition at 1.5 mM. The inhibition is not reversed by DTT (20 mM) or by mixture of DTT and 2-mercaptoethanol. Both NEM and pCMB appear to interact with the same type of sulfhydryl at short incubations. Both CTP and phosphocholine provide protection from the inhibition by NEM. This result suggests that the type of sulfhydryl reacting at short incubations may be at or near the active site of the enzyme.

Cytidylyltransferase is inhibited by phosphate.[5] Maximal inhibition of 60% occurs at 120 mM phosphate. One-half maximal inhibition is obtained at 20 mM phosphate. The inhibition is noncompetitive toward either substrate but is reversible since inhibition can be released by dilution. Because the enzyme is eluted from hydroxylapatite with buffers containing 200 mM phosphate, less than 5 μl of enzyme is used to measure enzyme activity in order to avoid significant phosphate inhibition.

Stability and Storage Conditions. Purified cytidylyltransferase is stable for several months at $-70°$ in buffer A containing 0.03% Triton X-100 and 200 mM phosphate. The enzyme is less stable at 4°, with a loss of about 20% per day. Removal of detergents by DEAE-Sepharose chromatography[16] leads to a rapid loss of activity and self-aggregation. The enzyme,

[15] P. S. Sohal and R. M. Cornell, *J. Biol. Chem.* **265**, 11746 (1990).
[16] R. Cornell, *J. Biol. Chem.* **264**, 9077 (1989).

even in the presence of detergents, has a high tendency to bind to glass and plastic surfaces. This can result in large losses when the enzyme is diluted and transferred to other containers. The use of siliconized glass tubes significantly reduces the surface binding. We use Surfasil (Pierce), diluted 1 : 10 in hexane, to coat the inside of glass tubes.

[30] Ethanolamine-Phosphate Cytidylyltransferase

By LILIAN B. M. TIJBURG, PIETER S. VERMEULEN, and LAMBERT M. G. VAN GOLDE

Introduction

$$CTP + phosphoethanolamine \rightleftharpoons CDPethanolamine + PP_i$$

CTP : phosphoethanolamine cytidylyltransferase[1] (EC 2.7.7.14, ethanolamine-phosphate cytidylyltransferase) is generally considered the key regulatory enzyme of the biosynthesis of phosphatidylethanolamine via the CDPethanolamine pathway.[2,3] The assay and partial purification of this cytosolic enzyme have previously been described by Sundler.[4] Recently, we have modified the original purification procedure to achieve a 1200-fold purification of the enzyme.

Assay

Principle. The assay is based on the conversion of phospho[1,2-^{14}C]ethanolamine to CDP[1,2-^{14}C]ethanolamine. Radioactive substrate and reaction product are separated by thin-layer chromatography, which is followed by determination of the radioactivity in CDPethanolamine.

Reagents

Tris buffer, 40 mM, adjusted to pH 7.8 with 1 M HCl, containing 20 mM MgCl$_2$

Dithiothreitol, 100 mM, freshly prepared

[1] E. P. Kennedy and S. B. Weiss, *J. Biol. Chem.* **222**, 193 (1956).
[2] R. Sundler and B. Åkesson, *J. Biol. Chem.* **250**, 3359 (1975).
[3] L. B. M. Tijburg, M. J. H. Geelen, and L. M. G. van Golde, *Biochim. Biophys. Acta* **1004**, 1 (1989).
[4] R. Sundler, *J. Biol. Chem.* **250**, 8585 (1975).

CTP, 40 mM, adjusted to pH 7 with 1 M NaOH

Phospho[1,2-^{14}C]ethanolamine, 10 mM (0.5 mCi/mmol)

Because labeled phosphoethanolamine is not commercially available, we prepare the ^{14}C derivative from [1,2-^{14}C]ethanolamine (4 mCi/mmol). Ethanolamine kinase that is required for this procedure is partially purified from rat liver, as described by Porter and Kent.[5] Rat livers (30 g) are homogenized in 4 volumes of 0.15 M KCl, 20 mM Tris-HCl (pH 7.5), 2 mM mercaptoethanol, 0.5 mM phenylmethylsulfonyl fluoride, and 1 mM EDTA. Following acidification of the cytosol to pH 5.2 and subsequent centrifugation, the enzyme is isolated by $(NH_4)_2SO_4$ precipitation.[5] The $(NH_4)_2SO_4$-precipitable fraction (30–45%) is redissolved in 20 ml of 10 mM Tris-HCl (pH 7.5), 2 mM mercaptoethanol, and 1 mM EDTA at a protein concentration of approximately 25 mg/ml and, subsequently, dialyzed against the same buffer. Partially purified ethanolamine kinase from 30 g rat liver is sufficient for the preparation of 100 μCi phosphoethanolamine.

Five microcuries of [1,2-^{14}C]ethanolamine is added to a 40-ml tube, and the solvent is evaporated under a stream of nitrogen. The reaction mixture contains the following in a total volume of 5 ml: 10 mM Tris-HCl (pH 8.5), 10 mM MgCl$_2$, 4 mM ATP, 0.25 mM (4 mCi/mmol) [1,2-^{14}C]ethanolamine, and 1 ml partially purified ethanolamine kinase. The reaction is carried out for 2 hr at 37° and terminated by the addition of 18.75 ml methanol–chloroform (2 : 1, v/v).[6] After 60 min, 6.25 ml chloroform and 6.25 ml water are added, and the mixture is stirred for 5 min. After the phases have been separated by centrifugation, the upper phase, containing labeled phosphoethanolamine and unreacted ethanolamine, is collected. The lower chloroform phase is washed twice with 10 ml methanol–water (5 : 4, v/v). The aqueous extracts of 20 incubations are combined and evaporated to dryness under reduced pressure using a rotary evaporator.

The residue is dissolved in 25 ml distilled water, and the pH of the solution is adjusted to pH 8 with 1 M NaOH. The mixture is applied to a Dowex 1-X4 (200–400 mesh, formate form) column (1.5 × 25 cm). The column is flushed with 100 ml water to remove labeled ethanolamine; subsequently, phosphoethanolamine is eluted with 200 ml of 20 mM formic acid. The fractions containing labeled phosphoethanolamine are combined and evaporated to dryness under a stream of nitrogen at 40°. The residue, containing pure phosphoethanolamine, is dissolved in water at a concentration of 5 μCi/ml. Unlabeled phosphoethanolamine is added to give the desired specific activity. The yield is 70–80%. The purity of [^{14}C]phospho-

---

[5] T. J. Porter and C. Kent, this volume [15].

[6] L. B. M. Tijburg, M. Houweling, M. J. H. Geelen, and L. M. G. van Golde, *Biochim. Biophys. Acta* **959**, 1 (1988).

ethanolamine is more than 99%, as assessed by thin-layer chromatography on silica gel G plates (Merck, Darmstadt, Germany) eluted with methanol–0.5% NaCl–ammonium hydroxide (50 : 50 : 5, v/v).

Procedure. The assay mixture contains the following in a final volume of 100 μl: Tris-HCl (2 μmol, 50 μl), MgCl$_2$ (1 μmol), dithiothreitol (0.5 μmol, 5 μl), CTP (0.2 μmol, 5 μl), phospho[1,2-^{14}C]ethanolamine (0.1 μmol, 10 μl), bovine serum albumin (25 μg), and distilled water (0–30 μl). The mixture is prewarmed at 37°, and the assay is started by the addition of enzyme (0–30 μl). The reaction is carried out for 15 min and stopped by immersing the tube for 2 min into a boiling water bath. The tubes are centrifuged for 5 min at 5000 g. Fifty microliters of the supernatant is spotted on a silica gel G thin-layer chromatography plate, and 0.3 μmol CDPethanolamine and 1 μmol phosphoethanolamine are applied as carrier. The plates are eluted with methanol–0.5% NaCl–ammonium hydroxide (50 : 50 : 5, v/v). After the solvent has evaporated from the plate, ethanolamine-containing compounds are visualized by spraying with a solution of 0.2% (w/v) ninhydrin in 96% ethanol, followed by heating for 5 min at 100°. CDPethanolamine is scraped from the plate into a scintillation vial, and the radioactivity is determined by liquid scintillation counting. The reaction is linear with protein concentration between 0 and 150 μg cytosolic protein and linear with time for at least 30 min. One unit of enzyme activity is defined as 1 μmol of CDPethanolamine formed per minute.

Purification Procedure

Ammonium Sulfate Precipitation. Livers from 6 rats (body weight 200–250 g), previously perfused free of blood with saline, are homogenized with 5 strokes of a motor-driven Potter–Elvehjem homogenizer in 4 volumes of 150 mM NaCl, 20 mM Tris-HCl (pH 7.8), 1 mM EDTA, 1 mM phenylmethylsulfonyl fluoride, 10 mg/liter benzamidine, 1 mg/liter leupeptin, and 0.7 mg/liter pepstatin A. All procedures are performed at 4°. The homogenate is centrifuged at 10,000 g for 20 min and the pellet discarded. After centrifugation at 105,000 g for 70 min, the resulting supernatant is collected and the pH readjusted to pH 7.2 with 0.1 M NaOH. The solution is brought to 25% of saturation with solid (NH$_4$)$_2$SO$_4$ over a period of 10 min with rapid stirring. After 15 min, the precipitate is removed by centrifugation at 20,000 g for 20 min, and further (NH$_4$)$_2$SO$_4$ is added to bring the supernatant to 40% saturation. After stirring for 10 min, the mixture is subjected to centrifugation as described above. The pellet is dissolved in approximately 80 ml of 20 mM Tris-HCl (pH 7.6), 1 mM EDTA, and 1 mM dithiothreitol (buffer A) and desalted by dialyzing against the same buffer.

DE-52 Chromatography. After the solution has been clarified by centrifugation at 20,000 g for 5 min, the sample is loaded onto a DE-52 column (2.65 × 50 cm), equilibrated with buffer A, at a flow rate of 55 ml/hr. The column is flushed with 400 ml buffer A. The enzyme is eluted from the column with a 500-ml linear salt gradient of 0 to 0.3 M NaCl in buffer A, followed by 100 ml of 0.3 M NaCl in the same buffer. Phosphoethanolamine cytidylyltransferase activity elutes at a salt concentration of around 0.15 M NaCl. The activity peak is pooled and dialyzed overnight against 20 mM Tris-HCl (pH 7.6), 1 mM EDTA, 1 mM dithiothreitol, and 10% (v/v) ethylene glycol (buffer B).

Matrex Gel Red A Chromatography. The pooled enzyme is applied to a column of Matrex Gel Red A (1.3 × 39 cm), equilibrated with buffer B. After a 150-ml wash with buffer B, the enzyme is eluted with 450 ml of a linear gradient of 0 to 1.0 M KCl in the same buffer, at a flow rate of 20 ml/hr. This procedure yields an activity peak eluting between 0.3 and 0.5 M KCl (Fig. 1).

Octyl-Sepharose CL-4B Chromatography. Fractions from the Matrex Gel Red A column containing cytidylyltransferase activity are pooled and directly pumped onto an Octyl-Sepharose CL-4B column (2.6 × 20 cm). The column is flushed with 75 ml 0.5 M NaCl, 25 mM Tris-HCl (pH 7.5),

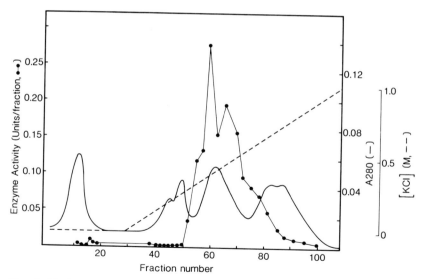

FIG. 1. Elution of CTP : phosphoethanolamine cytidylyltransferase from a Matrex Gel Red A column. Fractions of 6 ml were collected, and the enzyme activity was determined. For experimental details, see text.

TABLE I
PURIFICATION OF ETHANOLAMINE-PHOSPHATE CYTIDYLYLTRANSFERASE
FROM RAT LIVER

Fraction	Protein (mg)	Total activity (units)	Specific activity (units/mg $\times 10^3$)	Purification (-fold)	Recovery (%)
Cytosol	4880	18.4	3.77	1	100
25–40% (NH$_4$)$_2$SO$_4$	1190	16.4	13.8	3.7	89
DE-52	209	14.4	70.1	19	78
Matrex Gel Red A	58	13.1	228	60	71
Octyl-Sepharose	3.04	4.73	1560	414	25
Matrex Gel Blue A	0.37	1.62	4380	1162	8.8

0.5 mM EDTA, 1 mM dithiothreitol and 10% (v/v) ethyleneglycol at a flow rate of 40 ml/hr. Subsequently, the column is batchwise eluted with 75 ml of 15 mM Tris-HCl (pH 7.5), 0.5 mM EDTA, and 1 mM dithiothreitol (buffer C) containing 10% ethyleneglycol, followed by 75 ml buffer C with 45% and 75 ml of buffer C with 70% ethyleneglycol at flow rates of 40, 30, and 20 ml/hr, respectively.

Matrex Gel Blue A Pseudoaffinity Chromatography. Enzyme activity recovered from the Octyl-Sepharose column is pooled and dialyzed against 25 mM Tris-HCl (pH 8.2), 0.5 mM EDTA, 1 mM dithiothreitol and 5% (v/v) ethyleneglycol (buffer D). The enzyme is applied to a Matrex Gel Blue A column (2.6 $\times$ 20 cm) at a flow rate of 35 ml/hr. After loading the sample the pump is stopped for 15 min to allow proteins to bind to the column. The column is then flushed with 100 ml of buffer D. Since cytidylyltransferase does not bind to the column, the enzyme activity is recovered in the flushing step. This procedure results in one activity peak of pure phosphoethanolamine cytidylyltransferase.

Fractions containing cytidylyltransferase activity are collected and concentrated approximately 10 times by dialysis against 20 volumes of 20 mM Tris-HCl (pH 7.8), 0.5 mM EDTA, 2 mM dithioerythritol, 5% (v/v) ethyleneglycol, and 30% (g/v) polyethyleneglycol-6000. Concentrated enzyme is dialyzed against the same buffer omitting polyethyleneglycol-6000 and supplemented with 10% (v/v) glycerol. The enzyme preparation is stored at $-70°$.

The results of a typical purification procedure of CTP: phosphoethanolamine cytidylyltransferase are summarized in Table I. Polyacrylamide gel electrophoresis (PAGE) under nondenaturing conditions, followed by sodium dodecyl sulfate–PAGE, reveals one single band at 49 kDa containing the cytidylyltransferase activity. The specific activity of the 1160-fold purified fraction is a factor of 3.8 higher than that reported previously.[4]

Properties

pH Optimum. Phosphoethanolamine cytidylyltransferase has a sharp optimum at pH 7.8 and a broader one, with lower maximal activity, around pH 6.[4]

Stability. The enzyme has a limited stability below pH 7. The activity of purified enzyme is dependent on the presence of dithiothreitol in the assay mixture.[4] In the 105,000 g supernatant the enzyme is quite stable toward freezing and thawing. Purified enzyme can be stabilized by the addition of 10% (v/v) glycerol or 2% bovine serum albumin and stored at $-20°$ for at least 4 weeks without significant loss of activity.

Molecular Weight. The enzyme has a molecular weight of 100,000 to 120,000 as estimated by Superose 12 gel filtration. This value is in line with that reported by Sundler using Sephadex G-200 chromatography.[4]

Kinetic Properties. In contrast to CTP : phosphocholine cytidylyltransferase, which is an ambiquitous enzyme,[3] phosphoethanolamine cytidylyltransferase is localized predominantly in the cytosol. Although a striking feature of phosphocholine cytidylyltransferase is its requirement of lipid for significant activity,[7] addition of lipids does not stimulate the activity of cytosolic phosphoethanolamine cytidylyltransferase. The Michaelis constants for CTP and phosphoethanolamine are 53 and 65 μM, respectively. An ordered sequential reaction mechanism has been proposed, with CTP being the first substrate to bind to the enzyme and CDPethanolamine being the last product to be released.[4] In addition to CTP, phosphoethanolamine cytidylyltransferase can also use deoxyCTP as a substrate and form deoxyCDPethanolamine.[8]

---

[7] P. A. Weinhold and D. A. Feldman, this volume [29].
[8] E. P. Kennedy, L. F. Borkenhagen, and S. W. Smith, *J. Biol. Chem.* **234,** 1998 (1959).

Section VI

Phosphocholine/Phosphoethanolamine Phosphotransferases

[31] Cholinephosphotransferase from Mammalian Sources

By ROSEMARY B. CORNELL

Introduction

Cholinephosphotransferase (CDPcholine : 1,2-diacylglycerol choline-phosphotransferase, EC 2.7.8.2) catalyzes the final reaction in the synthesis of phosphatidylcholine. It is an integral membrane enzyme found

$$\text{CDPcholine} + sn\text{-1,2-diacylglycerol} \rightarrow \text{CMP} + \text{phosphatidylcholine}$$

predominantly in the endoplasmic reticulum but may also be associated with Golgi, mitochondria, or nuclear membranes in some tissues.[1-4] Cholinephosphotransferase (CPT) activity is distinct from ethanolaminephosphotransferase (EPT) activity in several respects, including sensitivity to detergents, heat, CMP, cation selectivity,[5] and trypsin.[6] CPT and EPT activities have been separated chromatographically[5] and genetically.[7]

Assay Method

Principle. A method is described utilizing microsomes and a diacylglycerol suspension produced by cosonicating diacylglycerol, Tween 20, and phospholipids. Preparation of the substrate in this manner yields the highest reported specific activity.[8] CPT can also be assayed in microsomes using endogenously generated membrane-bound diacylglycerol by stimulation of *de novo* synthesis of diacylglycerol[9] or by treatment with phospholipase C[10] or CMP.[11] An assay for CPT using permeabilized HeLa cells has been described.[12] In the method described below [*methyl*-14C]CDPcholine incorporation into phosphatidylcholine is monitored. The water-soluble

[1] C. L. Jelsema and D. J. Morre, *J. Biol. Chem.* **253**, 7960 (1978).
[2] M. D. Sikpi and S. K. Das, *Biochim. Biophys. Acta* **899**, 35 (1987).
[3] J. E. Vance and D. E. Vance, *J. Biol. Chem.* **263**, 5898 (1988).
[4] R. R. Baker and H. Y. Chang, *Can. J. Biochem.* **60**, 724 (1982).
[5] K.-M. O and P. C. Choy, *Biochem. Cell Biol.* **67**, 680 (1989).
[6] R. Coleman and R. M. Bell, *J. Biol. Chem.* **252**, 3050 (1977).
[7] M. Polokoff, D. C. Wing, and C. R. H. Raetz, *J. Biol. Chem.* **256**, 7687 (1981).
[8] J. C. Miller and P. A. Weinhold, *J. Biol. Chem.* **256**, 12662 (1981).
[9] H. Ide and P. A. Weinhold, *J. Biol. Chem.* **257**, 14926 (1982).
[10] M. G. Sarzala and L. M. G. van Golde, *Biochim. Biophys. Acta* **441**, 423 (1976).
[11] H. Kanoh and K. Ohno, *Biochim. Biophys. Acta* **326**, 17 (1973).
[12] P. Lim, R. B. Cornell, and D. E. Vance, *Biochem. Cell Biol.* **64**, 692 (1986).

substrate is separated from the insoluble product by a Bligh–Dyer extraction.[13]

Reagents

	μl (μg)/assay (50 μl)	Final concentration
1 *M* Tris-HCl, pH 8.5	2.5 μl	50 m*M*
0.2 *M* MgCl$_2$	2.5 μl	10 m*M*
10 m*M* EGTA	2.5 μl	0.5 m*M*
8 m*M* [*methyl*-^{14}C]CDPcholine, specific radioactivity 0.5	2.5 μl	0.4 m*M*
16 m*M* *sn*-1,2-Diolein emulsion	7.5 μl	2.4 m*M*
Microsomes	30 μg protein	
Water	To 50 μl	

Procedure

1. [*methyl*-^{14}C]CDPcholine is purchased from New England Nuclear (Boston, MA, 40–60 mCi/mmol) and diluted with unlabeled CDPcholine (Sigma, St. Louis, MO) to the above concentration and specific radioactivity.
2. An emulsion of 1,2-diolein (Sigma), asolectin (soy phospholipids, Associated Concentrates, Woodside, NY), and Tween 20 (Sigma) is prepared as follows:
   a. Five milligrams (~8 μmol) diolein from a chloroform stock is dried down under nitrogen in a small round-bottomed glass tube or flask. Then 0.25 mg Tween 20 is added from a 25 mg/ml aqueous stock, and the mixture is vortexed vigorously.
   b. Asolectin (50 mg/ml water) is vortexed vigorously for 30 min into a homogeneous suspension. One-half milliliter asolectin suspension is added to the diolein–Tween mixture.
   c. The complete mixture is sonicated under nitrogen for about 5 min using a microtip probe sonicator (e.g., Heat Systems, Plainview, NY, 375 W), with the sample immersed in an ice bath. A suspension appearing homogeneous to the eye results. The emulsion can be used for a few days without a noticeable effect on enzyme activity when stored under nitrogen at 4°, but prolonged storage is accompanied by generation of the 1,3-isomer and inhomogeneity of the suspension.
3. The reaction is set up in disposable 13 × 100 mm glass tubes. Tris, MgCl$_2$, and EGTA are added from a cocktail followed by water,

[13] E. G. Bligh and W. J. Dyer, *Can. J. Biochem. Physiol.* **37,** 911 (1959).

enzyme, and diolein. Tubes are vortexed and placed in a 37° shaking water bath. The reaction is initiated with CDPcholine. Diolein and zero-time "blanks" are included. The reaction is terminated after 15 min by addition of 1.5 ml methanol–chloroform (2 : 1, v/v).

4. ^{14}C-Labeled lipids are extracted by addition of 0.3 ml water, 0.5 ml chloroform, and 0.5 ml water in sequence with thorough vortexing after each addition. After centrifugation at 2500 rpm for 5 min, the upper layer is removed and the lower chloroform layer is washed with 2 ml theoretical upper phase (methanol–water–chloroform, 48 : 47 : 3, v/v). The washed chloroform layer is transferred to a scintillation vial, the solvent is evaporated under nitrogen, scintillation fluid is added, and the sample is counted.

The zero-time blank has typically less than 50 disintegrations/min (dpm). Samples lacking exogenous diolein have significant radioactivity depending on the microsome source.[14] More than 95% of the radioactivity in the chloroform phase is associated with phosphatidylcholine (PC). The specific activity of CPT in rat liver microsomes when assayed as described is approximately 30 nmol PC formed/min/mg protein. The K_m for CDPcholine is around 200 μM using rat liver microsomes and diacylglycerol prepared as above. The K_m for diacylglycerol is approximately 150 μM.[8] Reactions are linear in the range of 5–30 min and 0 to 50 μg rat liver microsomal protein.

Solubilization and Partial Purification

Kanoh and Ohno reported the first solubilization procedure for rat liver microsomal CPT.[15] CPT was released from the membrane by sonication in a medium containing 5 mM deoxycholate [approximately the critical micelle concentration (CMC)], and 20% glycerol at pH 8.5. Higher concentrations of deoxycholate led to inactivation. The high pH was required for solubilization. The solubilized preparation was reportedly stable to storage at −20°; however, the lipid content of the final preparation was very high (lipid/protein weight ratio of 1.3). Further treatment with Triton X-100 led to inactivation. O and Choy partially solubilized CPT activity from hamster liver microsomes using Triton QS-15.[16] The detergent inactivated CPT, but some activity was regained after dialysis.

Attempts to purify the solubilized transferase have met with very limited success. Kanoh and Ohno reported a 4-fold purification, to a

---

[14] H. Kanoh and K. Ohno, this series, Vol. 71, p. 536.
[15] H. Kanoh and K. Ohno, *Eur. J. Biochem.* **66**, 201 (1976).
[16] K.-M. O and P. C. Choy, *Lipids* **25**, 122 (1990).

specific activity of 21 nmol/min/mg, by a two-step sonication and centrifugation protocol using 4–5 mM deoxycholate.[15] This is described in detail elsewhere in this series.[14] O and Choy reported a 7-fold purification to a specific activity of 3.7 nmol/min/mg after DEAE-Sepharose and Sepharose 6B chromatography.[16] Both partially purified preparations appeared to be large lipid–protein aggregates or mixed detergent micelles, judging from the gel filtration elution positions. Further dissolution attempts with higher detergent concentrations led to irreversible inactivation.[15,16]

The lack of progress in the purification of mammalian CPT can be attributed to the failure to find conditions suitable for stabilizing the enzyme in the absence of a lipid environment.

Reconstitution of Cholinephosphotransferase Activity after Detergent Inactivation

Cornell and MacLennan compared the detergent concentration curves for inactivation of CPT and solubilization of sarcoplasmic reticulum membranes containing CPT.[17] Detergent concentrations required to convert membrane vesicles to detergent-mixed micelles as assessed by loss of turbidity inevitably led to greater than 90% inactivation of CPT. Cholate, deoxycholate, sodium dodecyl sulfate (SDS), Tween 20, octylglucoside, and Triton X-100 were tested.[17] Rat liver CPT is also severely inhibited by 3-[(3-cholamidopropyl)dimethylammonio]-1propanesulfonate (CHAPS) and Zwittergent solubilization of microsomes.

At relatively low detergent concentrations (detergent/protein weight ratio less than 3), even though the solubilized enzyme is inactive, the activity can be reconstituted by addition of a 5-fold excess of crude lipid, such as asolectin, and removal of detergent by dialysis, gel filtration, Bio-Beads (Bio-Rad, Richmond, CA), or other means effective for the particular detergent. Reconstitution from Triton, octylglucoside, deoxycholate, or cholate solutions has been successful. Solubilization at high detergent/protein ratios (>10, w/w) results in irreversible inactivation unless diacylglycerol is present during both the solubilization and reconstitution steps. Glycerol and diacylglycerol also stabilize rat brain EPT activity. Inclusion of these compounds in column chromatography buffers was a key factor in the 37-fold purification of EPT from rat brain. The stabilization by diacylglycerol may be an example of substrate stabilization of the active conformation of the enzyme. An example of a reconstitution protocol using cholate to solubilize microsomes is outlined below.

Solubilization. One milliliter (4 mg) microsomal protein in 10 mM Tris-

---

[17] R. B. Cornell and D. H. MacLennan, *Biochim. Biophys. Acta* **821,** 97 (1985).

HCl, pH 8, 0.1 mM phenylmethylsulfonyl fluoride (PMSF), 20% glycerol (T/G buffer) is added to 1 ml of a suspension containing 40 mg/ml ($\sim$100 mM) cholate, 40% glycerol, 30 mg/ml asolectin, and 10 mg/ml diacylglycerol. The mixture is vortexed and incubated 30 min at 0°–4° followed by centrifugation at 4° for 30 min at 200,000 g. Pellets are minute or undetectable.

Reconstitution. All steps are performed at 4°. Twenty milligrams of additional asolectin is added to the supernatant. The detergent is removed in two steps. (1) The sample is applied to a 20 ml Sephadex G-25 column equilibrated and eluted with T/G buffer. The turbid void volume is collected. (2) The turbid fractions from the Sephadex column are dialyzed against 100 volumes of T/G buffer for at least 24 hr. Bio-Beads SM-2 (Bio-Rad) can be added to the dialysis buffer to absorb detergent and accelerate its removal from the sample. Centrifugation of the samples as above results in sedimentation of the CPT activity. The pellets can be resuspended in a small volume of T/G buffer using a glass–Teflon type tissue grinder and stored at −70°.

Properties

Detergent Sensitivity. According to several reports low, subsolubilizing concentrations of detergents such as Tween 20,[18] Triton,[17,18] lysolecithin,[19] and deoxycholate[15] stimulate CPT activity. The stimulation is probably related to their ability to disperse diacylglycerol, and it would depend on the manner of presentation of the diacylglycerol. Membrane-solubilizing concentrations of detergent (i.e., above the CMC or above a detergent/lipid weight ratio of 2) lead to nearly complete inactivation.[5,17] Several types of anionic, zwitterionic, and nonionic class A and B detergents have been tried. The effects of cationic detergents, however, have not been reported. CPT may be one of the most detergent-sensitive enzymes to confront lipid enzymologists. Its acute detergent sensitivity suggests that sites critical to its activity are specifically dependent on a phospholipid environment.

Metal Ion Requirement. The CPT reaction is selective for Mg^{2+} in most animal tissues, although Mn^{2+} can substitute at least partly in the reaction catalyzed by partially purified CPT from hamster liver[5] and platelet microsomal CPT.[20] Mg^{2+} or Mn^{2+} in the range of 5–10 mM is optimal.

---

[18] G. Arthur, S. Tam, and P. C. Choy, *Can. J. Biochem. Cell Biol.* **62**, 1059 (1984).
[19] S. Parthasarathy and W. J. Baumann, *Biochem. Biophys. Res. Commun.* **91**, 637 (1979).
[20] S. Taniguchi, S. Morikawa, H. Hayashi, K. Fujii, H. Mori, M. Fujiwara, and M. Fugiwara, *J. Biochem.* (*Tokyo*) **100**, 485 (1986).

Ca^{2+} is inhibitory in the standard assay in the 10–100 μM range, and the inhibition is competitive with Mg^{2+} or Mn^{2+}.[20]

Topography on Microsomal Membrane. The resistance to extraction from membranes with high salt, EDTA, or low detergent concentrations and the inactivation that coincides with detergent-mediated solubilization of membrane bilayers suggest that CPT is an integral membrane protein.[17] Since CDPcholine is impermeable to intact microsomal membranes, one would anticipate that the active site of CPT faces the cytosol. Evidence supporting this notion includes inactivation by membrane-impermeant proteases or mercury–dextran.[21-23]

Reversibility. Incubation of microsomes with CMP inhibits the incorporation of [^{14}C]CDPcholine into PC by stimulation of the back-reaction of CPT.[24] CMP at 0.5 mM inhibits rat liver microsomal CPT 50% when assayed by the method described above. The back-reaction can be used to generate endogenous diacylglycerol. The K_m for CMP in the back-reaction ranges from 0.18 to 0.35 mM depending on the source.[11,25] Pitfalls associated with this procedure have been described in an earlier volume of this series.[14]

[21] R. Coleman and R. M. Bell, *J. Cell Biol.* **76**, 245 (1978).
[22] D. E. Vance, P. C. Choy, S. B. Farren, P. Lim, and W. J. Schneider, *Nature (London)* **270**, 268 (1977).
[23] L. M. Ballas and R. M. Bell, *Biochim. Biophys. Acta* **602**, 578 (1980).
[24] G. Goracci, P. Gresele, G. Arienti, P. Porrovecchio, G. Nenci, and G. Porcellati, *Lipids* **18**, 179 (1983).
[25] G. Goracci, E. Francescangeli, L. Horrocks, and G. Porcellati, *Biochim. Biophys. Acta* **664**, 373 (1981).

[32] Choline- and Ethanolaminephosphotransferases from *Saccharomyces cerevisiae*

By Russell H. Hjelmstad and Robert M. Bell

Introduction

Phosphatidylcholine (PC) and phosphatidylethanolamine (PE), the principal phospholipids of eukaryotic membranes, are synthesized from the common precursor *sn*-1,2-diacylglycerol and CDPcholine or CDPethanolamine, respectively, in reactions catalyzed by membrane-bound amino alcohol phosphotransferases.[1-3] Comparative enzymological studies in

[1] E. P. Kennedy and S. B. Weiss, *J. Biol. Chem.* **222**, 193 (1956).

Saccharomyces cerevisiae[4] and higher eukaryotic cells[3] have suggested the presence of distinct microsomal cholinephosphotransferase (EC 2.7.8.2) and ethanolaminephosphotransferase (EC 2.7.8.1) enzymes. In addition, a distinct cholinephosphotransferase may function in 1-alkyl-2-acetyl-glycerophosphocholine (platelet-activating factor, PAF)[5] synthesis.

The yeast *S. cerevisiae* provides an excellent genetic system in which to pursue in-depth studies on the structure, function, and regulation of choline- and ethanolaminephosphotransferases. Mutants defective in cholinephosphotransferase (*cpt* mutants) and ethanolaminephosphotransferase (*ept* mutants) activities were isolated,[6,7] and the corresponding structural genes for two amino alcohol phosphotransferases have been cloned, sequenced, and used to generate chromosomal null mutations.[6-9] The *CPT1* gene product is primarily a cholinephosphotransferase, whereas the *EPT1* gene product catalyzes both choline- and ethanolaminephosphotransferase reactions. The inferred gene products (*CPT1*: 407 amino acids, 46,000 Da; *EPT1*: 391 amino acids, 44,525 Da) bear 54% amino acid identity, are highly hydrophobic, and have similar predicted secondary structures including seven putative membrane-spanning segments. Together, the *CPT1* and *EPT1* gene products make up the complete set of amino alcohol phosphotransferases active in yeast,[10] permitting the study of the individual gene products in membranes prepared from strains bearing a null mutation in the cognate gene. The mixed micellar assays described here were developed to perform detailed enzymological studies of the individual *CPT1* and *EPT1* gene products.

Assay Method

Principle. Yeast choline- and ethanolaminephosphotransferase activities are determined as the amount of ^{32}P incorporated from [^{32}P]CDPcholine or [^{32}P]CDPethanolamine into PC or PE, respectively. The activities are measured in Triton X-100 mixed micelles containing *sn*-1,2-dioleoyl-

[2] C. R. H. Raetz and J. D. Esko, *in* "The Enzymes" (P. D. Boyer, ed.), 3rd Ed., vol. 16, p. 207. Academic Press, New York, 1983.

[3] R. M. Bell, *Annu. Rev. Biochem.* **49**, 459 (1980).

[4] A. K. Percy, M. A. Carson, J. F. Moore, and C. J. Waechter, *Arch. Biochem. Biophys.* **230**, 69 (1984).

[5] D. S. Woodard, T. Lee, and F. Snyder, *J. Biol. Chem.* **262**, 2520 (1987).

[6] R. H. Hjelmstad and R. M. Bell, *J. Biol. Chem.* **262**, 3909 (1987).

[7] R. H. Hjelmstad and R. M. Bell, *J. Biol. Chem.* **263**, 19748 (1988).

[8] R. H. Hjelmstad and R. M. Bell, *J. Biol. Chem.* **265**, 1755 (1990).

[9] R. H. Hjelmstad and R. M. Bell, *J. Biol. Chem.* **266**, 5094 (1991).

[10] R. H. Hjelmstad and R. M. Bell, *J. Biol. Chem.* **266**, 4357 (1991).

glycerol and dioleoyl-PC. The product [^{32}P]PC or [^{32}P]PE is extracted into acidified organic solvent and quantitated by scintillation counting. Delivery of the water-insoluble diacylglycerol substrate in the homogeneous micellar phase overcomes problems encountered in previous assays[4,6] arising from the physical properties of this substrate and effectively dilutes endogenous diacylglycerols present in yeast membrane preparations. Endogenous phospholipids are similarly subject to surface dilution, and complete dependence on a phospholipid activator is observed. The concentration dependencies of lipid substrates and cofactors as well as water-soluble components can thus be systematically studied using the mixed micellar assays.

Reagents. 4-Morpholinepropanesulfonic acid (MOPS), CDPcholine, CDPethanolamine, and MgCl$_2$ are from Sigma Chemical Company (St. Louis, MO). Triton X-100 (10% aqueous detergent in sealed ampules) is purchased from Pierce Chemical Company (Rockford, IL). L-α-Dioleoylphosphatidylcholine is obtained from Avanti Polar Lipids (Birmingham, AL). *sn*-1,2-Dioleoylglycerol is prepared from L-α-dioleoyl-PC by phospholipase C digestion[11] and quantitated as detailed elsewhere in this volume.[12] Other reagents are obtained or prepared as previously described.[6,7,10] Both dioleoylglycerol (~25–30 mM) and dioleoyl-PC (20 mg/ml, 25.4 mM) are stored as chloroform solutions at $-20°$ under argon.

Radiolabeled Substrates. [*methyl*-^{14}C]CDPcholine and [*ethanolamine*-1,2-^{14}C]CDPethanolamine are commercially available but are expensive, and the available specific activity of the ^{14}C compounds is limiting for some experiments. We have developed inexpensive and reliable procedures for the synthesis of [^{32}P]CDPcholine and [^{32}P]CDPethanolamine from [^{32}P]P$_i$. Complete experimental protocols for these radiochemical syntheses have been published.[6,10] Briefly, choline [^{32}P]phosphate or ethanolamine [^{32}P]phosphate is prepared from [^{32}P]P$_i$ by including choline or ethanolamine and yeast choline kinase in a standard enzymatic [γ-^{32}P]ATP-generating system. The product amino alcohol [^{32}P]phosphates are purified by anion-exchange chromatography. The ethanolamine [^{32}P]phosphate is converted to the benzyl carbamate derivative, and the protected amine is purified by cation-exchange chromatography. The choline [^{32}P]phosphate and ethanolamine [^{32}P]phosphate benzyl carbamates are then used in nucleophilic displacement reactions with CMPmorpholidate to give [^{32}P]CDPcholine and [^{32}P]CDPethanolamine benzyl carbamates. The latter product is deprotected, and both [^{32}P]CDPamino alcohols are purified by anion-exchange chromatography.[6,10]

---

[11] R. D. Mavis, R. M. Bell, and P. R. Vagelos, *J. Biol. Chem.* **247**, 2835 (1972).

[12] J. P. Walsh and R. M. Bell, this volume [17].

The concentrations of CDPcholine and CDPethanolamine are determined by the phosphate assay of Ames and Dubin[13] and ultraviolet absorbance measurements at 280 nm using an extinction coefficient of 12.8 mM^{-1} cm^{-1} for the CMP moieties. Both substrates may be prepared at initial radiochemical specific activities such that they remain useful for periods of 3 months. The more stable CDP[*methyl*-³H]choline can also be prepared using a modification of this method.[6] [*methyl*-³H]Choline may be purchased from New England Nuclear (Boston, MA).

The enzymatic and chemical synthesis of radiolabeled substrates may be generalized to prepare [³²P]CDPdimethylethanolamine and [³²P]CDPmonomethylethanolamine.[10] The yeast choline kinase utilizes the dimethylethanolamine and monomethylethanolamine substrates poorly but will efficiently convert them to their corresponding [³²P]phosphates in the presence of regenerating ATP.

Enzyme Preparations. Membranes are prepared from yeast cultures and the protein concentration determined as previously described.[7] Aliquots stored at −70° (1–10 mg/ml protein) are thawed immediately prior to use and diluted using isolation buffer (20%, w/v, glycerol, 50 mM MOPS, pH 7.5, and 1 mM EDTA) to the appropriate concentration such that the desired amount may be delivered in a volume of 10 μl. Membranes containing solely the *CPT1* gene product activity are prepared from strain HJ051 which bears a chromosomal insertional/deletional mutation in the *EPT1* locus generated by gene disruption.[7] Similarly, membranes exclusively containing the *EPT1* gene product activity are derived from HJ001 which contains an insertional mutation in the *CPT1* locus.[8] The wild-type parental strain DBY746[6] contains a mixture of both enzymes. Both of these strains have been shown to exhibit true null phenotypes with respect to the disrupted genes.[8,9]

Assay Procedure. The cholinephosphotransferase assay contains the following in a total volume of 200 μl: 50 mM MOPS–NaOH (pH 7.5), 20 mM MgCl₂, 0.45% (w/v, 6.5 mM) Triton X-100, 10 mol % *sn*-1,2-dioleoylglycerol, 10 mol % dioleoyl-PC, and 0.5 mM [³²P]CDPcholine. The ethanolaminephosphotransferase contains precisely the same constituents except that 0.1 mM [³²P]CDPethanolamine is used rather than the [³²P]CDPcholine. The mole fractions of lipid components in the micellar phase are estimated as the mole lipid/total mole detergent ratio, neglecting the monomer concentration of Triton X-100 (0.3 mM), which is less than 5% of the total Triton X-100 concentration.[14]

A 4× assay mix is prepared which contains 1.8% (w/v) Triton X-100

[13] B. N. Ames and D. T. Dubin, *J. Biol. Chem.* **235,** 769 (1960).
[14] A. Helenius, D. R. McCaslin, E. Fries, and C. Tanford, this series, Vol. 56, p. 734.

(27.7 mM), 200 mM MOPS–NaOH (pH 7.5), 80 mM MgCl$_2$, and 10 mol % each of dioleoylglycerol and dioleoyl-PC; 50 μl is prepared for each assay to be performed. Based on the amount of Triton X-100 to be used in the 4× solution [μmol Triton X-100 = 27.7 mM × volume 4× solution (ml)], volumes of dioleoylglycerol and dioleoyl-PC stock solutions in chloroform are pipetted into a glass screw-capped test tube using a Hamilton syringe such that 10 mol % of each are present (μmol lipid = 0.1 × μmol Triton X-100). The chloroform is dried under a stream of dry nitrogen. The remaining constituents of the 4× stock solution are then added to the dried lipids using stock solutions (10%, w/v, Triton X-100, 1 M MOPS–NaOH, pH 7.5, and 1 M MgCl$_2$). Solubilization of the lipids into detergent is achieved by alternating vigorous vortexing, bath sonication, and shaking in a 37° water bath until the solution is clear and no visible lipid film remains on the bottom of the tube.

[^{32}P]CDPcholine and [^{32}P]CDPethanolamine are prepared as 4× substrate solutions of 2.0 and 0.4 mM, respectively. The radiolabeled compounds are stored as aqueous solutions at −20° at concentrations of approximately 5 mM. Their radioactive specific activity is initially 500–1000 mCi/mmol. The 4× stock solutions are prepared using the ^{32}P-labeled compounds and unlabeled CDPamino alcohols (also stored as ~5 mM solutions) such that the final specific activity is 40 mCi/mmol. For some experiments, higher specific activities are required and may be easily prepared by varying the relative amounts of labeled and unlabeled compounds used to prepare the 4× substrate solution.

To perform the assay, 50 μl of the 4× assay mix is dispensed into each individual assay tube. Water is then added to bring the volume of each tube to 140 μl. Membranes (10 μl, 5–100 μg protein) are then added to each tube followed by brief vortexing. The mixture is allowed to stand at room temperature for 5 min, after which the reaction is initiated by the addition of 50 μl of 4× [^{32}P]CDPcholine or [^{32}P]CDPethanolamine. After 10 min at 25°, the reaction is terminated by the addition of 0.6 ml of 1% perchloric acid and 3 ml of methanol–chloroform (2 : 1, v/v). Phases are broken by addition of 1 ml each of chloroform and 1% perchloric acid followed by brief centrifugation at high speed in a clinical centrifuge. The upper phase is removed, and the lower phase is washed with two successive 2-ml portions of 1% perchloric acid, each wash being followed by repeated centrifugation and removal of the upper phase. One milliliter of the final lower chloroform phase is pipetted into a scintillation vial, the chloroform is removed by heating in a heating block, and the radioactivity is determined by scintillation counting in 4 ml of Aquasol-2 (New England Nuclear). Greater than 99% of the chloroform-soluble radioactivity produced in the choline- and ethanolaminephosphotransferase assays comigrates with authentic PC and PE, respectively, on analysis by thin-layer chromatography.[10] Both assays are linear with respect to time and protein in the ranges of 0–30 min and 0–100 μg of membrane protein.

Variation of water-soluble components can be accommodated in the assay by utilizing the volume ordinarily added as water to the assay. When variation of lipid components is desired, the lipid to be varied is omitted from the 4× assay mix and is instead delivered to each assay tube in chloroform. After removal of chloroform by drying, the variable lipid is solubilized by the 4× assay mix containing the nonvariable lipid, detergent, and water-soluble components by vortexing and shaking as described above. We have observed that solubilization of dioleoylglycerol into Triton X-100 mixed micelles is more easily achieved in the presence of phospholipid and that when this is not possible, greater solubilization efforts are required.

Properties of the Yeast CPT1 and EPT1 Gene Products

Diacylglycerol Dependencies. Both the CPT1 and EPT1 gene products exhibit Michaelis–Menten kinetics with respect to the mole fraction of dioleoylglycerol present in the micellar phase. It is essential to express the concentration of surface active components in this way since mixed micellar assays exhibit "surface dilution kinetics."[15] Because purified preparations of these enzymes are not yet available, the crude membrane preparations used to measure their activities contain appreciable endogenous diacylglycerol which is also subject to surface dilution. As the amount of membrane protein added to the assay is decreased, the mole fraction of endogenous diacylglycerol in the assay becomes negligible, and complete dependence on exogenous diacylglycerol is achieved. Less than 20 μg of membrane protein should be used in experiments in which dilution of endogenous diacylglycerol is important.

Phospholipid Activator Requirement. Nonlinearity with protein concentration in both assays suggested the presence of a dissociable enzyme activator substance present in the membranes; this requirement could be satisfied by exogenous PC.[10] Like endogenous diacylglycerol, endogenous phospholipid is effectively diluted into the micellar phase, and its mole fraction becomes negligible at low membrane protein concentrations; absolute requirements for a phospholipid activator can be demonstrated for the activities of both the CPT1 and EPT1 gene products. Less than 10 μg of membrane protein must be used to achieve complete dependence on

[15] E. A. Dennis, in "The Enzymes" (P. D. Boyer, ed.), 3rd Ed., Vol. 16, p. 307. Academic Press, New York, 1983.

TABLE I
KINETIC PARAMETERS OF AMINO ALCOHOL PHOSPHOTRANSFERASE ACTIVITIES OF *CPT1*
AND *EPT1* GENE PRODUCTS[a]

Gene product(s) (strain)	K_m (μM)				V_{max} (nmol/min/mg protein)			
	ChoPT	DMEPT	MMEPT	EthPT	ChoPT	DMEPT	MMEPT	EthPT
Wild-type mixture (DBY746)	109	27	26	20	0.78	0.44	0.30	1.26
EPT1 (HJ001)	120	29	29	22	0.62	0.42	0.27	1.35
CPT1 (HJ051)	100	137	—[b]	—	0.20	0.07	—	—

[a] Kinetic parameters were obtained from double-reciprocal plots. ChoPT, Cholinephosphotransferase; DMEPT, dimethylethanolaminephosphotransferase; MMEPT, monomethylethanolaminephosphotransferase; EthPT, ethanolaminephosphotransferase. From Hjelmstad and Bell.[10]

[b] Could not be evaluated.

exogenous phospholipid. Several naturally occurring phospholipids satisfy the phospholipid activator requirement of these enzymes to varying extents.[10]

CDPamino Alcohol Specificities. The activities of the *CPT1* and *EPT1* gene products have been evaluated using a series of CDPamino alcohol substrates including CDPcholine, CDPdimethylethanolamine, CDPmonomethylethanolamine, and CDPethanolamine.[10] The *CPT1* gene product

TABLE II
DIFFERENTIAL PROPERTIES OF *CPT1* AND *EPT1* GENE PRODUCTS

Property (units)	*EPT1* Gene product		*CPT1* Gene product, cholinephosphotransferase activity
	Ethanolaminephosphotransferase activity	Cholinephosphotransferase activity	
K_m, dioleoylglycerol (mol %)	3.3	2.9	8.0
K_m, CDPcholine (μM)	—	120	110
K_m, CDPethanolamine (μM)	22	—	—
K_a, Mg^{2+} (mM)	2.6	2.6	0.7
Phospholipid activation by PE (%)[a]	13	65	65
Inhibition by CMP (% remaining activity)[b]	18	16	71

[a] Activity obtained using 10 mol % PE is expressed as a percentage of that measured when 10 mol % PC is used.

[b] Value reported represents the percent remaining activity in the presence of 1 mM CMP relative to no addition.

primarily utilizes CDPcholine and CDPethanolamine, whereas the *EPT1* gene product utilizes all four derivatives to appreciable extents (Table I). The *CPT1* gene product does, however, have measurable activity when CDPethanolamine or CDPmonomethylethanolamine are employed. Monomethylethanolamine- and dimethylethanolaminephosphotransferase activities solely represent alternative substrate specificities of the *CPT1* and *EPT1* gene products since *cpt1 ept1* double-mutant strains completely lack these activities.[10]

Mg^{2+} *Dependencies.* The activities of both the *CPT1* and *EPT1* gene products exhibit absolute requirements for a divalent metal ion.

Differential Properties of the CPT1 and EPT1 Gene Products. The properties of the *CPT1* and *EPT1* gene product-dependent cholinephosphotransferase activities and the *EPT1* gene product-dependent ethanolaminephosphotransferase activity are summarized in Table II. Intrinsic properties of the two enzymes which are independent of the CDPamino alcohol employed include the apparent K_m for dioleoylglycerol, the apparent K_a for Mg^{2+}, and inhibition by CMP. The latter property, which occurs by an undetermined mechanism, is useful in assessing the relative contributions of the *CPT1* and *EPT1* gene products to total cholinephosphotransferase activity in membranes containing a mixture of the two enzymes.[10] The *CPT1* and *EPT1* gene products also exhibit differing properties with respect to phospholipid activation; however, this difference is dependent on the CDPamino alcohol employed, and the cholinephosphotransferase activities of both enzymes are indistinguishable with respect to this property.

[33] 1-Alkyl-2-acetyl-sn-glycerol Cholinephosphotransferase

By Ten-Ching Lee and Fred Snyder

Introduction

A dithiothreitol (DTT)-insensitive cholinephosphotransferase (EC 2.7.8.16, 1-alkyl-2-acetyl-*sn*-glycerol cholinephosphotransferase) catalyzes the transfer of phosphocholine from CDPcholine to 1-alkyl-2-acetyl-*sn*-glycerols (Fig. 1) in the final step of the biosynthesis of platelet-activating factor (PAF) in the *de novo* pathway[1,2]; in fact the term DTT-insensitive

[1] W. Renooij and F. Snyder, *Biochim. Biophys. Acta* **663**, 545 (1981).
[2] D. S. Woodard, T.-C. Lee, and F. Snyder, *J. Biol. Chem.* **262**, 2520 (1987).

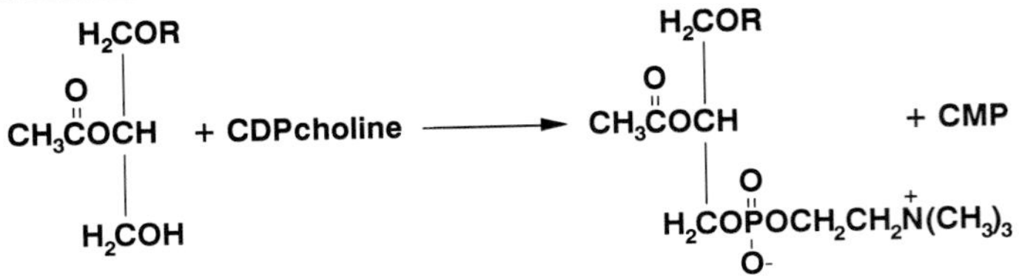

FIG. 1. Final reaction step in the *de novo* synthesis of PAF catalyzed by CDPcholine : 1-alkyl-2-acetyl-*sn*-glycerol cholinephosphotransferase.

is not totally descriptive since DTT actually causes a slight but significant stimulation of the activity. In contrast, DTT strongly inhibits the cholinephosphotransferase activity that catalyzes the formation of phosphatidylcholine from diacylglycerols and CDPcholine. Therefore, the terms DTT-insensitive and DTT-sensitive are useful in distinguishing these two types of cholinephosphotransferases. Because the two types of cholinephosphotransferase activity differ in a number of other properties it would appear that different proteins might be responsible for synthesizing PAF and phosphatidylcholine.[1,2] However, since the DTT-insensitive cholinephosphotransferase has not yet been solubilized or purified from its membrane environment, the question of whether a single or two separate catalytic proteins are involved in the expression of these cholinephosphotransferase activities remains to be resolved.

The importance of the DTT-insensitive cholinephosphotransferase in maintaining PAF at physiological levels cannot be underestimated since the cosubstrate (CDPcholine) formed by cytidylyltransferase can be rate-limiting in the synthesis of PAF by the *de novo* route.[3,4] Yet, an additional function of the DTT-insensitive cholinephosphotransferase could be its regulation of the cellular levels of alkylacetylglycerols, since the latter can also induce biological responses by influencing protein kinase C activity[5,6] and by inducing cell differentiation (e.g., converting HL-60 cells to macrophage-like cells[7]). Obviously, the DTT-insensitive cholinephosphotransferase occupies a central position in the PAF *de novo* pathway since it can utilize one type of mediator (alkylacetylglycerols) as a substrate to produce another (PAF) with completely different activities.

[3] M. L. Blank, Y. J. Lee, E. A. Cress, and F. Snyder, *J. Biol. Chem.* **263**, 5656 (1988).
[4] T.-C. Lee, B. Malone, M. L. Blank, V. Fitzgerald, and F. Snyder, *J. Biol. Chem.* **265**, 9181 (1990).
[5] L. L. Stoll, P. H. Figard, N. R. Yerram, M. A. Yorek, and A. A. Spector, *Cell Regul.* **1**, 13 (1989).
[6] D. A. Bass, L. C. McCall, and R. L. Wykle, *J. Biol. Chem.* **263**, 19610 (1988).
[7] M. J. C. McNamara, J. D. Schmitt, R. L. Wykle, and L. W. Daniel, *Biochem. Biophys. Res. Commun.* **122**, 824 (1984).

Method for Measuring Enzyme Activity

Reagents and Substrates. The following reagents are required to measure the DTT-insensitive cholinephosphotransferase: Tris-HCl, EGTA, MgCl$_2$, and DTT (Sigma Chemical Co., St. Louis, MO); CDP[*methyl*-^{14}C]choline (Du Pont/New England Nuclear Products, Boston, MA); and 1-hexadecyl-2-acetyl-*sn*-glycerol (Novabiochem AG, Läufelfingen, Switzerland).

Preparation of Enzyme Source. Routinely, microsomes from different tissues of adult male CDF rats (8–12 weeks old) are isolated by homogenizing the tissues at 4° in 3–4 volumes of 0.25 *M* sucrose–10 m*M* Tris-HCl–1 m*M* EDTA (pH 7.4) with four strokes of a Potter–Elvehjem homogenizer.[1] The homogenate is first centrifuged at 20,000 *g* for 10 min. Microsomes are next isolated by centrifugation of the 20,000 *g* supernatant at 100,000 *g* for 60 min. Microsomal preparations are washed once with 0.25 *M* Tris-HCl (pH 8.0) and once more with 0.25 *M* sucrose–10 m*M* Tris-HCl (pH 7.4) to remove any adsorbed proteins. The final pellet, suspended in an appropriate volume of 0.25 *M* sucrose–20 m*M* Tris-HCl (pH 7.4), is stored at −20° and assayed for enzyme activity within 1 week.

Enzyme Assay. Incubations are carried out at 37° for 5–10 min in an incubation mixture containing 1-hexadecyl-2-acetyl-*sn*-glycerol (200 nmol in 25 μl ethanol), 0.2 ml Tris-HCl (0.5 *M*, pH 8.0), 0.1 ml each of EGTA (5 m*M*), MgCl$_2$ (100 m*M*), bovine serum albumin (BSA) (10 mg/ml), DTT (100 m*M*), CDP[*methyl*-^{14}C]choline (1 m*M*, 1 mCi/mmol), and an aliquot of microsomal proteins (up to 100 μg); the final volume of the assay system is 1 ml. Reactions are stopped by lipid extraction according to a modified method (including 2% acetic acid in methanol) of Bligh and Dyer.[8]

Identification of Products. Labeled lipid products are separated by thin-layer chromatography (TLC) on silica gel HR (Sigma) layers in a solvent system of chloroform–methanol–concentrated NH$_4$OH–water (60:35:8:3.3, v/v). Products are identified by cochromatography with known standards; the R_f values of 1-alkyl-2-lyso-*sn*-glycerophosphocholine, PAF, and phosphatidylcholine are 0.27, 0.35, and 0.47, respectively. Each lipid fraction on the chromatoplate is visualized with I$_2$ vapor, and the desired zones are then scraped into counting vials for radioassay by liquid scintillation spectrometry.[9]

---

[8] E. G. Bligh and W. J. Dyer, *Can. J. Biochem. Physiol.* **37**, 911 (1959).
[9] F. Snyder, in "The Current Status of Liquid Scintillation Counting" (E. D. Bransome, ed.), p. 248. Grune & Stratton, New York, 1970.

Properties of Enzyme

The DTT-insensitive cholinephosphotransferase is present in a variety of rat tissues,[1,10,11] human neutrophils,[12] rabbit platelets,[13] and chick retina.[14] The enzyme is located in the microsomal fraction of liver, spleen, and brain[1,10] with an optimal activity observed at pH 8.0.[1,2,10] The enzyme activity in the inner medulla of rats is significantly increased when the neutral lipid substrate, 1-hexadecyl-2-acetyl-sn-glycerol, is solubilized in 2.5% ethanol.[2] However, the presence of 2.5% ethanol in the incubation system inhibits the DTT-insensitive cholinephosphotransferase from rat brain, and, therefore, the enzyme activity in this system must be determined by suspending the lipid substrate in 0.02% Tween 20 with sonication.[10]

Studies on the substrate specificity of the DTT-insensitive cholinephosphotransferase toward different alkyl chains at the sn-1 position reveal that the enzyme produces higher activities with 16 : 0 and 18 : 1 substrates than with an 18 : 0 substrate.[1,2] In contrast, oleoylacetylglycerol gives reaction rates comparable to that of the alkyl analog, hexadecylacetylglycerol.[2] In addition, the DTT-insensitive cholinephosphotranferase prefers substrates with acetyl or propionyl groups at the sn-2 position; the 2-butyryl analog is utilized at only half the rate, and 2-acetamide and 2-methoxy analogs are poor substrates.[2]

The DTT-insensitive cholinephosphotransferase requires Mg^{2+} (10–20 mM) for activity, but Mn^{2+} can partially substitute for Mg^{2+}. On the other hand, Ca^{2+} concentrations greater than 10 μM inhibit the activity.[3] Of all the sulfhydryl reagents (i.e., DTT, mercaptoethanol, reduced glutathione, and cysteine) tested,[2] only DTT exerts a differential effect on the cholinephosphotransferase activities responsible for the synthesis of PAF and phosphatidylcholine. In microsomal preparations from the rat inner medulla, PAF synthesis is slightly but significantly increased by DTT at concentrations above 5 mM, whereas phosphatidylcholine synthesis is completely inhibited by DTT concentrations in the range of 5–10 mM.[2] The cholinephosphotranferase that synthesizes long-chain choline phos-

[10] E. Francescangeli and G. Goracci, *Biochem. Biophys. Res. Commun.* **161,** 107 (1989).

[11] S. Fernandez-Gallardo, M. A. Gijon, M. C. Garcia, E. Cano, and M. Sanchez-Crespo, *Biochem. J.* **254,** 707 (1988).

[12] F. Alonso, M. Garcia Gil, M. Sanchez-Crespo, and J. M. Mato, *J. Biol. Chem.* **257,** 3376 (1982).

[13] T.-C. Lee, B. Malone, and F. Snyder, *in* "Platelet-Activating Factor in Immune Responses and Renal Diseases" (P. Braquet, K. H. Hsieh, E. Pirotzky, and J. M. Mencia-Huerta, eds.), p. 1. Excerpta Medica Asia Ltd., Hong Kong, 1989.

[14] F. Bussolino, F. Gremo, C. Tetta, G. P. Pescarmona, and G. Camussi, *J. Biol. Chem.* **261,** 16502 (1986).

phoglycerides in rat brain is less sensitive to DTT inhibition, and only 50% of the enzyme activity is inhibited by 80 mM DTT.[10] Based on results showing that the cholinephosphotransferase activities which synthesize PAF and phosphatidylcholine have different sensitivities toward DTT, detergents, ethanol, and temperature, it appears that two separate cholinephosphotransferase activities exist.[2]

The DTT-insensitive cholinephosphotransferase[12,13] along with the two other enzymes[13] (acetyl-CoA : 1-alkyl-2-lyso-sn-glycero-3-phosphate acetyltransferase and alkylacetylglycerophosphate phosphohydrolase) involved in the *de novo* pathway of PAF biosynthesis are not affected by stimuli that increase the activity of acetyl-CoA : 1-alkyl-2-lyso-sn-glycero-3-phosphocholine acetyltransferase in the remodeling pathway of PAF biosynthesis. However, Bussolino et al.[14,15] have reported that the DTT-insensitive cholinephosphotransferase activity in the retina and embryonic retina from chickens is activated by acetylcholine and dopamine. Noteworthy is the finding that these neurotransmitters have no effect on the acetyltransferase of the remodeling pathway. Under conditions where the injection of 2-bromoethylamine hydrobromide into rats induces a specific papillary necrosis associated with the inner medulla, PAF levels in blood are reduced and, interestingly, only the DTT-insensitive cholinephosphotransferase, but not the acetyltransferase and phosphohydrolase of the *de novo* pathway, is significantly decreased in the inner medulla.[16] These results indicate the DTT-insensitive cholinephosphotransferase, in addition to the regulatory roles of the acetyltransferase and cytidylyltransferase, can also play an important role in the control of PAF levels in the *de novo* pathway.

Acknowledgments

This work was supported by the Office of Energy Research, U.S. Department of Energy (Contract No. DE-AC05-760R00033), the American Cancer Society (Grant BC-70V), and the National Heart, Lung, and Blood Institute (Grant 35495-04A1).

[15] F. Bussolino, G. Pescarmona, G. Camussi, and F. Gremo, *J. Neurochem.* **51,** 1755 (1988).
[16] T.-C. Lee, B. Malone, D. Woodard, and F. Snyder, *Biochem. Biophys. Res. Commun.* **163,** 1002 (1989).

Section VII

Synthases

[34] Phosphatidylserine Synthase from *Escherichia coli*

By WILLIAM DOWHAN

Introduction

(d)CDPdiacylglycerol + L-serine → phosphatidylserine + (d)CMP

Phosphatidylserine synthases (EC 2.7.8.8, CDPdiacylglycerol–L-serine *O*-phosphatidyltransferase) from several gram-negative bacteria have been partially characterized and seem to be very similar in their physical and catalytic properties[1,2]; the enzyme from *Escherichia coli* has been extensively studied and characterized. Phospholipid biosynthetic enzymes in bacteria are generally found associated with membranes, but the phosphatidylserine synthases from gram-negative bacteria are found associated with the ribosomal fraction of cell lysates.[1,2] The same reaction is catalyzed by the phosphatidylserine synthases from *Saccharomyces cerevisiae*[3] and the gram-positive *bacilli*,[4,5] but the enzymes from these organisms, unlike the enzymes from gram-negative bacteria, are dependent on added divalent metal ions for activity and are membrane associated. A similar CDPdiacylglycerol-dependent phosphatidylserine synthase activity has never been found in higher eukaryotic cells.

Assay Method

The enzyme is routinely assayed by following the incorporation of radiolabeled L-serine into chloroform-soluble material in the presence of a nonionic detergent such as Triton X-100; the ribose and deoxyribose forms of the liponucleotide are equivalent substrates.[6] In crude extracts either [³H]serine or [3-¹⁴C]serine should be used since the product is rapidly converted to phosphatidylethanolamine by the presence of the phosphatidylserine decarboxylase[7]; alternatively [U-¹⁴C]serine can be used if the decarboxylase is inhibited by inclusion of 10 m*M* hydroxylamine in the assay. Highly purified preparations of the enzyme can be assayed

[1] A. Dutt and W. Dowhan, *J. Bacteriol.* **132**, 159 (1977).
[2] C. R. H. Raetz and E. P. Kennedy, *J. Biol. Chem.* **247**, 2008 (1972).
[3] G. M. Carman and M. Bae-Lee, this volume [35].
[4] A. Dutt and W. Dowhan, *J. Bacteriol.* **147**, 535 (1981).
[5] A. Dutt and W. Dowhan, *Biochemistry* **24**, 1073 (1985).
[6] T. J. Larson and W. Dowhan, *Biochemistry* **15**, 5212 (1976).
[7] W. Dowhan, W. T. Wickner, and E. P. Kennedy, *J. Biol. Chem.* **249**, 3079 (1974).

spectrophotometrically by coupling the continuous release of CMP to the oxidation of NADH in the presence of ATP and phosphoenolpyruvate via the sequential actions of CMP kinase, pyruvate kinase, and lactate dehydrogenase.[8,9]

Procedure.[6] If necessary the enzyme is diluted with 0.1 M potassium phosphate buffer (pH 7.4) containing 0.1% (1.6 mM) Triton X-100 and 1 mg/ml of bovine serum albumin (BSA). Dilution and assay are done in polypropylene tubes since the dilute, purified enzyme tends to adhere to glass; serum albumin and high protein concentrations minimize this problem. The final concentrations of the components of the assay mixture are 0.67 mM CDPdiacylglycerol (synthesized from egg yolk phosphatidylcholine as described elsewhere in this volume[10]), 0.5 mM L-serine (radiolabeled as desired to a specific activity of 2–4 μCi/μmol), 0.1% Triton X-100, 0.1 M potassium phosphate buffer (pH 7.4), and 1 mg/ml BSA; an ionic strength of 0.3 or higher is optimal.[9] The reaction is initiated by adding 10 μl of an enzyme solution to 50 μl of a 1.2-fold concentrated stock solution of the above assay mixture. The reaction mixture is incubated in a 12-ml polypropylene tube at 30° for 10 min. The reaction is stopped by adding, in the following order, 0.5 ml of methanol (0.1 N in HCl), 1.5 ml chloroform, and 3.0 ml of 1 M MgCl$_2$. After thorough mixing of the two phases and separation of the phases by centrifugation at 1000 g_{av} in a clinical centrifuge at room temperature, 1 ml of the chloroform phase (lower phase) is removed, evaporated to dryness at 65°, and counted for radioactivity using any commercially available scintillation fluid. One unit of activity is defined as the amount of enzyme which converts 1 μmol of L-serine into chloroform-soluble material in 1 min under the above conditions.

The assay is linear for the pure enzyme up to about 50% conversion of the limiting substrate. When CDPdiacylglycerol is the limiting substrate, total conversion approaches only 95%, probably because of an inherent hydrolase activity which is 1% of the synthase rate. In crude preparations the linearity may fall off before reaching 50% conversion depending on the degree of contamination by the endogenous, membrane-associated CDP-diacylglycerol hydrolase activity[11]; this activity can be inhibited by 1 mM ATP.

[8] G. M. Carman and W. Dowhan, *J. Lipid Res.* **19,** 519 (1978).
[9] G. M. Carman and W. Dowhan, *J. Biol. Chem.* **254,** 8391 (1979).
[10] G. M. Carman and A. S. Fischl, this volume [36].
[11] C. R. H. Raetz, C. B. Hirschberg, W. Dowhan, W. T. Wickner, and E. P. Kennedy, *J. Biol. Chem.* **247,** 2245 (1972).

TABLE I
PURIFICATION OF PHOSPHATIDYLSERINE SYNTHASE[a]

Step	Total protein	Specific activity (units/mg)	Yield (%)
1. Cell-free extract	3.1 g	0.80	100
2. Cell supernatant	3.0 g	0.87	104
3. Phosphocellulose	ND[b]	ND	70
4. DEAE-Sephadex	34 mg	39	54

[a] Starting with 20 g of cell paste from E. coli strain JA200/pPS3155-λ as described by A. Ohta, K. Waggoner, K. Louie, and W. Dowhan, J. Biol. Chem. **256**, 219 (1981).
[b] Not determined.

Purification Procedure

It is possible to obtain near gram quantities of highly purified enzyme by utilizing a purification scheme which relies on specific elution of the enzyme from phosphocellulose using the liponucleotide substrate[6] and starting with a strain of E. coli containing a plasmid-borne copy of the structural gene (pss) which directs the overproduction of the enzyme 100- to 200-fold.[12] The structural gene for the phosphatidylserine synthase has been introduced into a derivative of plasmid pBR322 (pPS3155-λ) which carries the NOP region of λ phage[13]; the NOP region contains the λ origin of replication, a temperature-sensitive λ repressor, and the essential λ promoters. At 30° this plasmid replicates from the normal pBR322 origin and is carried at about 11 copies per cell. Induction at 42° results in inactivation of the λ repressor and an increase in the copy number of the plasmid to greater than 100 per cell. Because the production of the pss gene product is gene dose-dependent, cells carrying this plasmid grown at 42° have greatly elevated levels of phosphatidylserine synthase. Except where indicated the following steps (taken from Ohta et al.[12] and summarized in Table I) are carried out at 4°.

Growth of Cells. Cells can be grown in batches of 1 liter or less in shaking flasks or in 100-liter amounts in a fermentor using SB broth [12 g/liter of Bacto-tryptone (Difco, Detroit, MI), 25 g/liter of yeast extract, 0.1 M potassium phosphate (pH 7.0), and 0.5% glucose] supplemented with 50 μg/ml ampicillin (the plasmid drug marker). Strain JA200/pPS3155-λ is grown to mid-log phase (OD_{550} 1.5) at 30°, rapidly shifted to 42° for 30 min, and then incubated at 37° for 4 hr before harvesting the cells; high aeration and rapid shaking are maintained throughout the growth period. The yield

[12] A. Ohta, K. Waggoner, K. Louie, and W. Dowhan, J. Biol. Chem. **256**, 219 (1981).
[13] R. N. Rao and S. G. Rogers, Gene **3**, 247 (1978).

of cells is about 12 g/liter of wet weight cells, with a specific activity for the phosphatidylserine synthase ranging between 100- and 200-fold over that from strain JA200 lacking the plasmid. The cell paste can be stored at $-80°$ for years.

Step 1: Cell-Free Extract. Cell paste (20 g) of strain JA200/pPS3155-λ is suspended in 70 ml of buffer containing 0.1 M potassium phosphate (pH 7.4) and broken by sonication or passed through a French pressure cell at 1000 psi; the volume is adjusted to 140 ml with the same buffer.

Step 2: Cell Supernatant. The cell supernatant is obtained by centrifugation at 13,500 g_{av} for 2 hr. This treatment maximizes the recovery of ribosomal-bound enzyme with removal of membranes and cell debris. The supernatant is adjusted to 0.2% in Triton X-100 and 10% in glycerol.

Step 3: Phosphocellulose Column. The following chromatography is carried out at room temperature. A 5 × 6 cm Whatman (Clifton, NJ) P-11 phosphocellulose column is activated by washing with acid and base (see Kurland *et al.*[14]), equilibrated with 50 mM potassium phosphate (pH 7.4) containing 0.1% Triton X-100, 10% glycerol, and 1 mg/ml BSA, and finally washed with several column volumes of the same buffer lacking albumin; the albumin increases the yield of enzyme by blocking tight-binding sites on the column. The cell supernatant is applied to the column followed by washes of 300 ml of buffer A [0.1 M potassium phosphate (pH 7.4), 1% Triton X-100, 10% glycerol, and 0.5 mM dithiothreitol] containing 0.65 M NaCl and 200 ml of buffer B (buffer A at 0.1% Triton X-100 and containing 0.5 M NaCl). The enzyme is eluted from the column at 200 ml/hr using 180 ml of buffer B containing 0.4 mM CDPdiacylglycerol followed by 100 ml of buffer B. At this point the enzyme, in a volume of about 140 ml, is essentially homogeneous as judged by sodium dodecyl sulfate-polyacrylamide gel electrophoresis.

Step 4: DEAE-Sephadex Column. The pooled fractions from the P-11 column are dialyzed overnight against 800 ml of buffer C [20 mM potassium phosphate (pH 7.0), 0.2% Triton X-100, 10% glycerol, and 0.5 mM dithiothreitol]. The dialyzed enzyme is applied at 250 ml/hr to a 2.5 × 21 cm DEAE-Sephadex column equilibrated with buffer C. The column is washed with 200 ml of buffer C containing 0.1 M NaCl followed by elution of the enzyme at 60 ml/hr with buffer C containing 1.2 M NaCl. The pooled peak of activity (10–15 ml) is dialyzed against buffer A for storage at $-80°$.

Purity. As judged by sodium dodecyl sulfate-polyacrylamide gel electrophoresis, the preparation after Step 4 is over 95% homogeneous and has the same mobility (M_r 54,000) and specific activity as the enzyme prepared from cells lacking the overproducing plasmid.[6,12] The preparation

---

[14] C. G. Kurland, S. J. S. Hardy, and G. Mora, this series, Vol. 20, p. 381.

still contains significant amounts of CDPdiacylglycerol which binds to the DEAE-Sephadex column and elutes with the enzyme; with time, at 4° or above, this lipid is converted by the enzyme to phosphatidic acid and phosphatidylglycerol (see below). The bulk of the lipid can be removed by sedimentation of the enzyme through a 5 to 20% (w/v) glycerol gradient at 200,000 g_{av} for 12 hr in the presence of 0.1 M potassium phosphate (pH 7.4) as described by Carman and Dowhan[9]; Triton X-100 (0.2%) or octylglucoside (30 mM) should be added back to the pooled peak of enzyme for storage.

Properties

Stability. The enzyme is stable for several years at $-80°$ and several months at 4° when stored in buffer A. At ionic strengths below 0.1 the enzyme tends to precipitate, with irreversible loss of enzymatic activity. Although the enzyme is not isolated as a membrane protein, it does have a low affinity for nonionic detergent micelles (see below), and the presence of nonionic detergents (Triton X-100 or octylglucoside) above their critical micelle concentrations increases the stability of the enzyme. In the absence of detergent and in the presence of 0.1 M potassium phosphate buffer (pH 7.4), the enzyme is stable for several days at 4°. The enzyme tends to adhere to negatively charged surfaces, and thus dilute solutions stored in glass containers lose activity rapidly; plastic containers are used when working with dilute solutions of the enzyme.[6] There are no known specific inhibitors of the enzyme.

Activity and Detergent Dependence. The enzyme is dependent on nonionic detergent for activity and is sensitive to the ratio of detergent (above its critical micelle concentration) to lipid substrate.[9] At 0.1 mM CDPdiacylglycerol the optimum activity occurs at a Triton to substrate molar ratio of 8 : 1. Activity is lower at either higher or lower ratios; the same is true at other fixed concentrations of substrate. At an 8 : 1 molar ratio the lipid substrate is completely dispersed in mixed micelles with the detergent, whereas below this ratio the substrate may exist partially in structures which are not catalytically competent or the higher surface concentration of substrate in the micelles may be inhibitory. The apparent reduction in V_{max} at higher ratios appears to be due to dilution of the substrate within the mixed micelle surface, which reduces the affinity of the enzyme for the micelle (see below).

As shown in Fig. 1A, the enzyme follows normal Michaelis–Menten kinetics if the activity is plotted as a function of the total concentration of detergent plus lipid substrate at fixed molar ratios of detergent to substrate. In such experiments the apparent V_{max} approaches the true V_{max} as the ratio

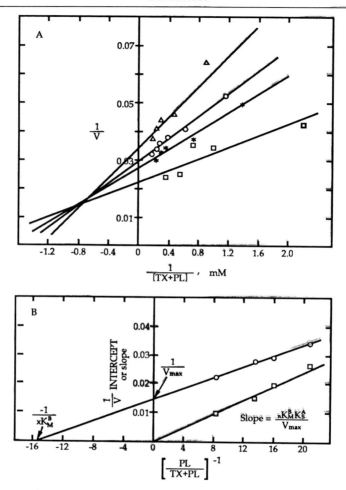

FIG. 1. Activity of phosphatidylserine synthase toward CDPdiacylglycerol (PL) in mixed micelles with Triton X-100 (TX). (A) The reciprocal of the velocity of the reaction (μmol/min/mg) is plotted as a function of the reciprocal of the total concentration of PL and TX at varying molar ratios of TX : PL: $\square$, 8 : 1; *, 12 : 1; $\bigcirc$, 16 : 1; $\triangle$, 20 : 1. (B) Replot of the $1/V$ intercepts ($\bigcirc$) and the slopes ($\square$) from (A) as a function of the mole fraction of phospholipid in the mixed micelles can be used to determine the functions shown in the figure and summarized in Table I. [Data from G. M. Carman and W. Dowhan, *J. Biol. Chem.* **254**, 8391 (1979).]

TABLE II

KINETIC CONSTANTS FOR PHOSPHATIDYLSERINE SYNTHASE[a]

Substrate	V_{max} (μmol/min/mg)	nK_s^A/x (mM)	xK_m^B (mole fraction)
CDP-1,2-dipalmitoyl-L-glycerol	71	1.4	0.065
CDPdiacyl-L-glycerol	79	1.2	0.064
CDP-1,2-dipalmitoyl-DL-glycerol[b]	50	3.4	0.10
CDP-1,2-dipalmitoyl-DL-glycerol[c]	50	3.2	0.058
CDP-1,2-dicaproyl-DL-glycerol[b]	90	3.3	0.06
CDP-1,2-dicaproyl-DL-glycerol[c]	90	3.3	0.03

[a] From G. M. Carman and W. Dowhan, *J. Biol. Chem.* **254**, 8391 (1979).
[b] Concentration of both stereoisomers considered in the phospholipid term.
[c] Only the L isomer considered in the phospholipid term and the D isomer considered in the inert Triton X-100 surface.

is increased. Replotting of the data (Fig. 1B) leads to the determination of three kinetic constants (see Warner and Dennis[15] for a discussion) at saturating concentrations of L-serine (1 mM): the true V_{max} of 80 μmol/min/mg; K_s^A (total concentration of detergent plus substrate), which is a measure of the affinity of the enzyme for the mixed micelle surface; and the binding constant, K_m^B, for the substrate on the micelle surface. The values[15] for x and n were assumed to remain constant and represent the area of a substrate or a Triton molecule on the micelle surface and the area of the enzyme binding site on the micelle surface, respectively. Using such an analysis it is possible to assess the specificity for a number of lipid substrates (Table II). The V_{max}, K_s^A, and K_m^B values are independent of fatty acid composition, and the enzyme is specific for the naturally occurring isomer of the phosphatidyl moiety. The D isomer is neither a substrate nor a competitive inhibitor but appears to be recognized as part of the inert surface of the micelle (no effect on K_m^B). Monoacylated lipids are neither substrates nor inhibitors of the enzyme.

Reaction Mechanism. Analysis of the steric course of replacement of CMP by serine has established that the enzyme catalyzes synthesis of phosphatidylserine with retention of configuration at the phosphorus of the phosphatidyl moiety,[16] consistent with a two-step Ping-Pong reaction mechanism involving a covalently bound enzyme–phosphatidyl intermediate. The enzyme catalyzes the following reactions,[6,17] which is also consis-

[15] T. G. Warner and E. A. Dennis, *J. Biol. Chem.* **250**, 8004 (1975).
[16] C. R. H. Raetz, G. M. Carman, W. Dowhan, R.-T. Jiang, W. Waszkuc, W. Loffredo, and M.-D. Tsai, *Biochemistry* **26**, 4022 (1987).
[17] C. R. H. Raetz and E. P. Kennedy, *J. Biol. Chem.* **249**, 5038 (1974).

tent with this conclusion: the exchange between serine and phosphatidylserine and between CMP and CDPdiacylglycerol without the presence of other substrates or products; the hydrolysis of either phosphatidylserine or CDPdiacylglycerol to form phosphatidic acid; and the ability of glycerol (1 M) or sn-glycero-3-phosphate (0.8 mM) to act as a phosphatidyl acceptor in the presence of a large excess of enzyme.

Physical Properties. The enzyme is unique among phospholipid biosynthetic enzymes of *E. coli* in that it is not membrane associated in cell-free extracts but is tightly bound to ribosomes.[2,18] The association is primarily ionic since it can be disrupted by buffers with an ionic strength of 1.0 or greater. The polyamine spermidine in the physiological range (2–10 mM) also dissociates the enzyme from ribosomes. Dissociation is not brought about by nonionic detergents or the normal ionic strength of the assay conditions, but addition of either lipid substrate or lipid product under the assay conditions effects complete dissociation. These results suggest that the enzyme is associated with ribosomes primarily through ionic components of its active site and explains its high affinity for negatively charged surfaces.

The effect of nonionic detergent, ionic strength, and CDPdiacylglycerol on the physical state of the enzyme has been investigated using changes in the sedimentation properties of the enzyme.[9] At the ionic strength of the normal assay mixture (0.6) and in the absence of detergent and lipid substrate, the enzyme sediments as a large monodisperse molecule with an M_r of 500,000. Addition of Triton X-100 above its critical micelle concentration and in large excess over enzyme results in a polydisperse sedimentation pattern (M_r from 100,000 to 500,000) for the enzyme, consistent with weak interaction of the enzyme with the detergent micelles. Addition of increasing concentrations of CDPdiacylglycerol in the presence of a fixed Triton X-100 concentration (1.6 mM) results in a progressive increase in a monodisperse species with an M_r of 100,000. Optimal formation of the latter species occurs at a molar ratio of detergent to substrate of 8 : 1 (optimal for assay). Increasing the substrate concentration further but at a molar ratio less than 8 : 1 results in a polydisperse pattern for the enzyme, consistent with the substrate dilution within the micelle surface. A similar effect of ionic strength on sedimentation pattern of the enzyme, in the presence of detergent and substrate, is also seen which parallels the effect on enzymatic activity; that is, the species with an M_r of 100,000 is favored as the optimal ionic strength for assay is approached. Comparison of the sedimentation properties of the various forms of the enzyme in glycerol gradients made in either D_2O or H_2O indicates that the species

---

[18] K. Louie and W. Dowhan, *J. Biol. Chem.* **255,** 1124 (1980).

with an M_r of 100,000 has a density less than that of protein, consistent with it being a complex of protein and detergent–lipid mixed micelle, whereas the species with an M_r of 500,000 has the density of protein.[19] Although the multimeric structure of the enzyme cannot be determined from the above results, the species with an M_r of 100,000 must be some multiple of the 54,000 subunit of the enzyme associated with a detergent–lipid substrate mixed micelle.

These physical properties of the enzyme coupled with the kinetic analysis of the enzyme support a model in which catalysis occurs at the surface of a hydrophobic–hydrophilic interface. The affinity of the enzyme for such a surface is dependent on the presence and concentration of the negatively charged lipid substrate on that surface. Therefore, the two binding constants kinetically determined for the enzyme[9] appear to have some validity in physical terms since the enzyme does show low affinity for a micelle surface (K_s^A), but the presence of the lipid substrate in that surface (K_m^B) results in very tight binding to the micelle; dilution of the lipid substrate within the surface decreases both surface affinity and enzymatic activity.

The affinity of the enzyme for detergent micelles supplemented with the lipid substrate also extends to membranes of *E. coli* supplemented with various phospholipids. As noted above, the enzyme preferentially associates with the ribosomal fraction rather than the membrane fraction of cell lysates. However, both enzyme synthesized *in vitro* or the enzyme naturally present in cell lysates can be induced to associate preferentially with *E. coli* membranes which have been enriched in either CDPdiacylglycerol or phosphatidylserine[20]; further enrichment with phosphatidylethanolamine does not induce membrane association, but the acidic phospholipids phosphatidylglycerol and cardiolipin do induce membrane association to some extent. The membrane-associated enzyme is kinetically competent, as evidenced by the formation of phosphatidylserine when serine is added to CDPdiacylglycerol-supplemented membranes. These results further support the above model for the action of this enzyme both *in vivo* and *in vitro*.

Subunit Structure. Based on the complete DNA sequence of the *pss* gene and verification by partial sequencing of the isolated protein, the enzyme is composed of a single monomer of 452 amino acids (Fig. 2) and has a molecular mass of 52,817 Da,[21] which is consistent with the minimum size of the gene product expressed *in vitro*[20] and the mobility of the protein

---

[19] W. Dowhan, unpublished observation (1982).
[20] K. Louie, Y.-C. Chen, and W. Dowhan, *J. Bacteriol.* **165**, 805 (1986).
[21] A. DeChavigny, P. N. Heacock, and W. Dowhan, *J. Biol. Chem.* **266**, 5323 (1991).

```
MLSKFKRNKH  QQHLAQLPKI  SQSVDDVDFF  YRPADFRETL   40
LEKIASAKQR  ICIVALYLEQ  DDGGKGILNA  LYEAKRQDDP   80
ELDVRVLVDW  HRAQRGRIGA  AASNTNADWY  CRMAQENPGV  120
DVPVYGVPIN  TREALGVLHF  KGFIIDDSVL  YSGASLNDVY  160
LHQHDNIAYD  RYHLIRNRKM  SDIMFEWVTQ  NIMNGRGVNR  200
LDDVNRPKSP  EIKNDIRLFR  QELRDAAYHF  QGDADNDQLS  240
VTPLVGLGKS  SLLNKTIFHL  MPCAEQKLTI  CTPYFNLPAI  280
LVRNIIQFVR  EGKKVEIIVG  DKTANDFYIS  EDEPFKIIGA  320
LPYLYEINLR  RFLSRLQYYV  NTDQLVVRLW  KDDDNTYHLK  360
GMWVDDKWML  ITGNNLNPRA  WRLDLENAIL  IHDPQLELAP  400
QREKELELIR  EHTTIVKHYR  DLQSIADYPV  KVRKLIRRLR  440
RIRIDRLISR  IL                                  452
```

FIG. 2. Amino acid sequence for the phosphatidylserine synthase from *E. coli* as predicted from the DNA sequence of the *pss* gene [summarized from A. DeChavigny, P. N. Heacock, and W. Dowhan, *J. Biol. Chem.* **266**, 5323 (1991)].

on sodium dodecyl sulfate-polyacrylamide gel electrophoresis[6]; as noted above the native enzyme is some unknown multimer of this subunit. The amino-terminal methionine is not blocked. The predicted amino acid composition (Table III) is not unusual, and a hydrophobicity analysis of the linear sequence reveals two hydrophobic domains (residues 120–155 and 240–285) which may be involved in either substrate interaction or membrane association. The majority of the amino acid sequence is hydrophilic in nature and is characterized by an enrichment in basic amino acids at both the amino (5 of 9 amino acids) and carboxyl (10 of 22 amino acids) termini. These highly basic domains may be involved in association of the enzyme with negatively charged surfaces and the negatively charged lipid substrates.

Properties of Mutants. Several mutants in the *pss* gene (mapping at 56 min of the *E. coli* chromosome) have been isolated which result in a conditional lethal, temperature-sensitive growth phenotype.[22–24] These mutant strains stop growing after 4 to 5 generations at the restrictive temperature (42°) at which point the level of phosphatidylethanolamine declines from its normal level of 75% of the total phospholipid to 35%; phosphatidylserine does not accumulate owing to the activity of phosphatidylserine decarboxylase. The enzyme isolated from such mutants is also temperature-labile, consistent with the growth phenotype. Supplementation of the growth medium with 20 m*M* MgCl$_2$ suppresses the temperature-

[22] A. Ohta and I. Shibuya, *J. Bacteriol.* **132**, 434 (1977).
[23] C. R. H. Raetz, G. A. Kantor, M. Nishijima, and K. F. Newman, *J. Bacteriol.* **139**, 544 (1979).
[24] I. Shibuya, C. Miyazaki, and A. Ohta, *J. Bacteriol.* **161**, 1086 (1985).

TABLE III
PREDICTED AMINO ACID COMPOSITION OF
PHOSPHATIDYLSERINE SYNTHASE[a]

Amino acid	Residues per 52,800 Da
Asp	39
Asn	26
Thr	14
Ser	17
Glu	21
Gln	21
Pro	17
Gly	20
Ala	27
Cys	4
Val	30
Met	8
Ile	37
Leu	51
Tyr	19
Phe	15
His	13
Lys	26
Trp	7
Arg	40

[a] Summarized from A. DeChavigny, P. N. Heacock, and W. Dowhan, *J. Biol. Chem.* **266,** 5323 (1991)

sensitive growth phenotype of the mutants without returning the level of phosphatidylethanolamine to wild-type levels.

Complete inactivation of the *pss* gene by insertion of a drug marker (*pss* :: *kan*) has shown that the residual phosphatidylethanolamine present in the above mutants results from the residual activity of the temperature-sensitive *pss* gene product.[21] Strains carrying the *pss* :: *kan* allele also stop growing when phosphatidylethanolamine levels reach 35%, but in the presence of millimolar concentrations of $Ca^{2+} > Mg^{2+} > Sr^{2+}$ the strains grow near normally (in the order of effectiveness indicated) in rich medium. In contrast to bacteria carrying the temperature-sensitive point mutations grown in the presence of Mg^{2+}, the level of phosphatidylethanolamine is less than 0.01% of the total phospholipid of the cell. The ratio of phospholipid (which is now almost exclusively phosphatidylglycerol and cardiolipin) to membrane protein and the phospholipid fatty acid composition are nearly identical to those of wild-type cells. Therefore, although phosphatidylethanolamine is an essential phospholipid under normal laboratory growth conditions, the nonspecific nature of the substitution by several divalent metal ions for this major membrane phospholipid suggests a primarily structural role for phosphatidylethanolamine. One possibility may be its role as an inert membrane matrix since this phospholipid carries no net charge; divalent metal ions could be acting to neutralize the high negative charge density of the remaining phospholipids. A second possibility may be the need for a phospholipid which can form nonbilayer structures possibly necessary in such processes as membrane fusion and translocation of macromolecules across membranes. This property of phosphatidylethanolamine[25] could be substituted for by cardiolipin, which is known to display nonbilayer structures[26] in the presence of divalent metal ions with the same order of effectiveness as the three divalent metal ions which suppress the growth phenotype of the *pss* :: *kan* allele.

[25] P. R. Cullis and B. de Kruijff, *Biochim. Biophys. Acta* **513**, 31 (1978).
[26] I. Vasilenko, B. de Kruijff, and A. J. Verkleij, *Biochim. Biophys. Acta* **684**, 282 (1982).

[35] Phosphatidylserine Synthase from Yeast

By GEORGE M. CARMAN and MYONGSUK BAE-LEE

Introduction

Phosphatidylserine synthase (CDPdiacylglycerol–L-serine O-phosphatidyltransferase, EC 2.7.8.8) catalyzes the incorporation of serine into

$$\text{CDPdiacylglycerol} + \text{serine} \rightarrow \text{phosphatidylserine} + \text{CMP}$$

phosphatidylserine.[1] The enzyme plays an important role in the regulation of phospholipid biosynthesis in the yeast *Saccharomyces cerevisiae*.[2] Phosphatidylserine synthase activity is associated with the mitochondrial and microsomal fractions of *S. cerevisiae*.[3,4] Phosphatidylserine synthase expression is regulated by inositol alone and in concert with serine, etha-

[1] J. N. Kanfer and E. P. Kennedy, *J. Biol. Chem.* **239**, 1720 (1964).
[2] G. M. Carman and S. A. Henry, *Annu. Rev. Biochem.* **58**, 635 (1989).
[3] G. S. Cobon, P. D. Crowfoot, and A. W. Linnane, *Biochem. J.* **144**, 265 (1974).
[4] K. Kuchler, G. Daum, and F. Paltauf, *J. Bacteriol.* **165**, 901 (1986).

nolamine, and choline.[5–8] Microsomal-associated phosphatidylserine synthase has been purified to near homogeneity using CDPdiacylglycerol-
Sepharose affinity chromatography.[9] We describe here the purification,
reconstitution, and properties of the enzyme.

Preparation of CDPdiacylglycerol and CDPdiacylglycerol-Sepharose

CDPdiacylglycerol is prepared from phosphatidic acid and CMPmorpholidate by the method of Agranoff and Suomi[10] with the modifications
of Carman and Fischl.[11] The CDPdiacylglycerol-Sepharose affinity resin
of Dowhan and co-workers[12,13] is synthesized as described by Fischl and
Carman[14] with the following modifications. The incubation times for coupling the adipic acid spacer arm to activated Sepharose 4B, NaIO$_4$ oxidation of CDPdiacylglycerol, and coupling the oxidized derivative of CDPdiacylglycerol to the adipic acid-Sepharose resin are 1 day, 1 day, and 1
day, respectively. These incubation times result in a low-capacity affinity
resin.[14] The preparations of CDPdiacylglycerol and CDPdiacylglycerol-
Sepharose are described elsewhere in this volume.[15]

Assay Method

Phosphatidylserine synthase activity is routinely measured by following the incorporation of 0.5 mM L-[3-^{3}H]serine [10,000 counts/min (cpm)/
nmol] into phosphatidylserine in 50 mM Tris-HCl buffer (pH 8.0) containing 0.6 mM MnCl$_2$, 0.2 mM CDPdiacylglycerol, 4 mM Triton X-100,
and enzyme protein in a total volume of 0.1 ml at 30°.[9,16] Purified phosphatidylserine synthase reconstituted in phospholipid vesicles is measured in
the absence of Triton X-100 and added CDPdiacylglycerol. The reaction
is terminated by the addition of 0.5 ml of 0.1 N HCl in methanol. Chloroform (1 ml) and 1 M MgCl$_2$ (1.5 ml) are added, the system is mixed, and

[5] L. S. Klig, M. J. Homann, G. M. Carman, and S. A. Henry, *J. Bacteriol.* **162,** 1135 (1985).
[6] M. J. Homann, A. M. Bailis, S. A. Henry, and G. M. Carman, *J. Bacteriol.* **169,** 3276 (1987).
[7] M. A. Poole, M. J. Homann, M. Bae-Lee, and G. M. Carman, *J. Bacteriol.* **168,** 668 (1986).
[8] A. M. Bailis, M. A. Poole, G. M. Carman, and S. A. Henry, *Mol. Cell. Biol.* **7,** 167 (1987).
[9] M. Bae-Lee and G. M. Carman, *J. Biol. Chem.* **259,** 10857 (1984).
[10] B. W. Agranoff and W. D. Suomi, *Biochem. Prep.* **10,** 47 (1963).
[11] G. M. Carman and A. S. Fischl, *J. Food Biochem.* **4,** 53 (1980).
[12] T. J. Larson, T. Hirabayashi, and W. Dowhan, *Biochemistry* **15,** 974 (1976).
[13] W. Dowhan and T. Hirabayashi, this series, Vol. 71, p. 555.
[14] A. S. Fischl and G. M. Carman, *J. Bacteriol.* **154,** 304 (1983).
[15] G. M. Carman and A. S. Fischl, this volume [36].
[16] G. M. Carman and J. Matas, *Can. J. Microbiol.* **27,** 1140 (1981).

the phases are separated by a 2-min centrifugation at 100 g. A 0.5-ml sample of the chloroform phase is removed and taken to dryness on an 80° water bath. Betafluor (4 ml) is added to the sample, and radioactivity is determined by scintillation counting. The chloroform-soluble phospholipid product of the reaction phosphatidylserine is analyzed by thin-layer chromatography with standard phosphatidylserine.[9,16] Alternatively, activity is measured by following the release of radiolabeled CMP from radiolabeled CDPdiacylglycerol.[16]

One unit of enzymatic activity is defined as the amount of enzyme that catalyzes the formation of 1 nmol of product per minute. Protein is determined by the method of Bradford.[17] Buffers which are identical to those containing the protein samples are used as blanks. The presence of Triton X-100 does not interfere with the protein determination, provided the blank contains a final concentration of detergent identical to that of the sample.[9]

Electroblotting of Phosphatidylserine Synthase Activity[18]

The procedure for the electroblotting of phosphatidylserine synthase is similar to that used for the electroblotting of phosphatidylinositol synthase.[15] Briefly, the enzyme is dialyzed for 3 hr at 8° against sodium dodecyl sulfate-polyacrylamide gel electrophoresis treatment buffer. The enzyme is subjected to sodium dodecyl sulfate-polyacrylamide gel electrophoresis and transferred to nitrocellulose paper. Enzyme on the nitrocellulose paper is renatured in 50 mM Tris-HCl (pH 8.0), 0.6 mM MnCl$_2$, 30 mM MgCl$_2$, 1 mM CDPdiacylglycerol, 0.5% Triton X-100, 10 mM 2-mercaptoethanol, 20% glycerol, 3% bovine serum albumin for 1 hr at 8°. The presence of CDPdiacylglycerol and cofactors in the renaturation buffer stabilizes phosphatidylserine synthase activity. The nitrocellulose paper is used for enzyme assays. The recovery of phosphatidylserine synthase activity is about 10%. This procedure allows the confirmation of the molecular weight of phosphatidylserine synthase.

Growth of Yeast

Wild-type $S.$ $cerevisiae$ strain S288C (α $gal2$) is used for enzyme purification. Cells are grown in 1% yeast extract, 2% peptone, and 2% glucose at 28° to late exponential phase, harvested by centrifugation, and stored at $-80°$.[14,15]

---

[17] M. M. Bradford, $Anal.$ $Biochem.$ **72**, 248 (1976).
[18] M. A. Poole, A. S. Fischl, and G. M. Carman, $J.$ $Bacteriol.$ **161**, 772 (1985).

Purification Procedure

All steps are performed at 5°.

Step 1: Preparation of Microsomes. Cells (75 g, wet weight) are disrupted with glass beads in 50 mM Tris-HCl buffer (pH 7.5) containing 1 mM Na_2EDTA, 0.3 M sucrose, and 10 mM 2-mercaptoethanol.[14,15] Glass beads and unbroken cells are removed by centrifugation at 1500 g for 10 min to obtain the cell extract. Microsomes are collected from the cell extract by differential centrifugation.[14,15] Microsomes are washed with and resuspended in 50 mM Tris-HCl buffer (pH 7.5) containing 10 mM 2-mercaptoethanol and 20% glycerol (w/v).

Step 2: Preparation of Triton X-100 Extract. The microsome fraction is suspended in 50 mM Tris-HCl buffer (pH 8.0) containing 2 mM $MnCl_2$, 30 mM $MgCl_2$, 10 mM 2-mercaptoethanol, 0.5 M KCl, 20% glycerol (w/v), and 1% Triton X-100 at a final protein concentration of 10 mg/ml. After incubation for 1 hr at 5°, the suspension is centrifuged at 100,000 g for 2 hr to obtain the solubilized (supernatant) fraction. Solubilization of the enzyme from microsomes with 1% Triton X-100 is dependent on the presence of $MnCl_2$ in the solubilization buffer.[14,16]

Step 3: CDPdiacylglycerol-Sepharose Chromatography. A CDPdiacylglycerol-Sepharose column (0.9 × 4 cm) is equilibrated with 50 ml of 50 mM Tris-HCl buffer (pH 8.0) containing 0.6 mM $MnCl_2$, 30 mM $MgCl_2$, 20% glycerol, 10 mM 2-mercaptoethanol, and 0.5% Triton X-100. The Triton X-100 extract is applied to the affinity column in 1.5-ml samples. To effect enzyme binding, each sample is incubated in the column for 10 min before the addition of the next sample. After the entire Triton X-100 extract is applied, the column is washed with 30 ml of equilibration buffer containing 1 M NaCl. The column is then saturated with equilibration buffer containing 1 mM CDPdiacylglycerol and 1 M NaCl and incubated for 1 hr. Phosphatidylserine synthase is eluted from the column with this buffer at a flow rate of 1 ml/min. Fractions containing activity are pooled and desalted by dialysis or Sephadex G-25 chromatography using 20 mM Tris-HCl buffer (pH 8.0) containing 0.3 mM $MnCl_2$, 10 mM $MgCl_2$, 20% glycerol, 10 mM 2-mercaptoethanol, and 0.5% Triton X-100.

Step 4: DE-53 Chromatography. A DE-53 (Whatman, Clifton, NJ) column (0.6 × 4 cm) is equilibrated with 20 ml of 20 mM Tris-HCl buffer (pH 8.0) containing 0.3 mM $MnCl_2$, 10 mM $MgCl_2$, 20% glycerol, 10 mM 2-mercaptoethanol, and 0.5% Triton X-100. Desalted enzyme from the previous step is applied to the column at a flow rate of 1 ml/min. The column is washed with 10 ml of equilibration buffer followed by elution of the enzyme with equilibration buffer containing 1 M NaCl. Fractions

TABLE I

PURIFICATION OF PHOSPHATIDYLSERINE SYNTHASE FROM *Saccharomyces cerevisiae*[a]

Purification step	Total units (nmol/min)	Protein (mg)	Specific activity (units/mg)	Purification (-fold)	Yield (%)
1. Microsomes	591	493	1.2	1	100
2. Triton X-100 extract	579	241	2.4	2	98
3. CDPdiacylglycerol-Sepharose	457	0.34	1340	1120	77
4. DE-53 chromatography	421	0.18	2300	1920	71

[a] Data are from Bae-Lee and Carman.[9]

containing activity are pooled and desalted as described above. The purified enzyme is 100% stable for at least 6 months when stored at −80°.

Enzyme Purity. The four-step purification scheme summarized in Table I results in a nearly homogeneous preparation of phosphatidylserine synthase as evidenced by native[19] and sodium dodecyl sulfate-polyacrylamide gel electrophoresis.[9] Native polyacrylamide gels contain 0.5% Triton X-100. The enzyme preparation does not contain the CDPdiacylglycerol-dependent enzymes phosphatidylinositol synthase, CDPdiacylglycerol synthase, nor phosphatidylglycerophosphate synthase. The subunit molecular weight of the purified enzyme is 23,000.[9,18] When the enzyme is partially purified in the presence of the protease inhibitor phenylmethylsulfonyl fluoride (PMSF), a protein with a molecular weight of 30,000 as well as the 23,000 subunit is isolated.[20] The 23,000 subunit of phosphatidylserine synthase is a proteolytic cleavage product of the 30,000 subunit of phosphatidylserine synthase.[20] The enzyme is purified about 2000-fold over the microsomal fraction and about 5000-fold relative to the activity in cell extracts. Phosphatidylserine synthase can also be purified from strain VAL2C(YEp*CHO1*),[9] which bears a hybrid plasmid that directs overproduction of the enzyme.[21] The phosphatidylserine synthase-overproducing strain facilitates the acquisition of larger amounts of pure enzyme.[9]

[19] M. A. Poole, Ph.D. Thesis, Rutgers University, New Brunswick, New Jersey (1986).
[20] K. Kiyono, K. Miura, Y. Kushima, T. Hikiji, M. Fukushima, I. Shibuya, and A. Ohta, *J. Biochem. (Tokyo)* **102**, 1089 (1987).
[21] V. A. Letts, L. S. Klig, M. Bae-Lee, G. M. Carman, and S. A. Henry, *Proc. Natl. Acad. Sci. U.S.A.* **80**, 7279 (1983).

Reconstitution of Phosphatidylserine Synthase

Pure phosphatidylserine synthase has been reconstituted into unilamellar phospholipid vesicles containing its substrate CDPdiacylglycerol.[22] The procedure is similar to that used for the reconstitution of phosphatidylinositol synthase[23] with some modifications. The reconstitution procedure is described elsewhere in this volume.[15] Pure phosphatidylserine synthase is reconstituted into vesicles directly from mixed micelles containing pure enzyme, phospholipid, Triton X-100, and octylglucoside. The final concentrations of Triton X-100, octylglucoside, and phospholipids in the reconstitution mixture are 2.0, 150, and 10.6 mM, respectively. The final molar ratios of octylglucoside to Triton X-100, octylglucoside to phospholipids, and phospholipids to CDPdiacylglycerol are 75 : 1, 14 : 1, and 16 : 1, respectively. Octylglucoside inhibits phosphatidylserine synthase, and activity cannot be measured in the reconstitution mixture.

The reconstitution mixture is immediately passed through a Sephadex G-50 superfine column (1.5 × 32 cm) equilibrated with 50 mM Tris-HCl buffer (pH 7.5) containing 1 mM MnCl$_2$, 10 mM 2-mercaptoethanol, 250 mM NaCl, and 10% glycerol at a flow rate of 20 ml/hr at 5°. The column is pretreated with a 1 mg/ml sonicated suspension of phospholipids (phosphatidylcholine – phosphatidylethanolamine – phosphatidylinositol – phosphatidylserine, 3 : 2 : 2 : 1, v/v), which increases the yields of reconstituted activity. Reconstituted phosphatidylserine synthase elutes from the column in the void volume.

The average size of vesicles containing phosphatidylserine synthase is 90 nm in diameter.[22] Each vesicle contains about 1 molecule of enzyme, which is reconstituted asymmetrically with 80–86% of its active site facing outward.[22] Maximum reconstituted phosphatidylserine synthase activity is obtained with vesicles containing phosphatidylcholine, phosphatidylethanolamine, phosphatidylinositol, and phosphatidylserine,[22] the four major phospholipids in wild-type *S. cerevisiae*.[5,24]

Properties of Phosphatidylserine Synthase

Optimum phosphatidylserine activity requires either MnCl$_2$ (0.6 mM) or MgCl$_2$ (20 mM) at the pH optimum of 8.0 using a mixed micelle assay of Triton X-100 and CDPdiacylglycerol.[9] The manganese requirement of

[22] J. M. Hromy and G. M. Carman, *J. Biol. Chem.* **261**, 15572 (1986).
[23] A. S. Fischl, M. J. Homann, M. A. Poole, and G. M. Carman, *J. Biol. Chem.* **261**, 3178 (1986).
[24] M. L. Greenberg, L. S. Klig, V. A. Letts, B. S. Loewy, and S. A. Henry, *J. Bacteriol.* **153**, 791 (1983).

the enzyme reconstituted into phospholipid vesicles is 5 mM.[22] The K_m value for serine is 0.83 mM as determined with a mixed micelle substrate of Triton X-100 and CDPdiacylglycerol.[25] Phosphatidylserine synthase activity is dependent on the bulk (83 μM)[25] and surface (2.9 mol %)[26] concentrations of CDPdiacylglycerol, characteristic of surface dilution kinetics.[27] Based on the results of kinetic experiments,[25] the ability of the enzyme to catalyze isotopic exchange reactions between substrates and products,[23] and a stereochemical analysis of the reaction using ^{31}P nuclear magnetic resonance spectroscopy,[28] the enzyme catalyzes a Bi–Bi sequential reaction mechanism. Phosphatidylserine synthase binds to CDPdiacylglycerol before serine, and phosphatidylserine is released prior to CMP in the reaction sequence.[9] Phosphatidylserine synthase uses dCDPdiacylglycerol as well as CDPdiacylglycerol as substrate.[9] Phosphatidylserine synthase is labile above 40° and is inactivated by thioreactive agents.[9]

Inositol (K_i 65 μM) is a noncompetitive inhibitor of phosphatidylserine synthase,[25] whereas cardiolipin (K_i 0.7 mol %) and diacylglycerol (7 mol %) are competitive inhibitors of the enzyme.[26] Phosphatidate (K_a 0.033 mol %), phosphatidylcholine (K_a 3.4 mol %), and phosphatidylinositol (K_a 3.2 mol %) are activators of phosphatidylserine synthase activity.[26] Phosphatidylserine synthase (23,000 subunit[29] and 30,000 subunit) is phosphorylated by cAMP-dependent protein kinase.[29] The enzyme has one phosphorylation site, a serine residue, which results in a 60–70% reduction in enzyme activity.[29] The phosphorylation of phosphatidylserine synthase results in a decrease in the rate phosphatidylserine synthesis *in vivo*.[30]

Synthetic and Analytical Uses

Pure phosphatidylserine synthase can be used to synthesize radiolabeled phosphatidylserine from CDPdiacylglycerol and labeled serine.[9] The enzyme can be used to synthesize various fatty acyl derivatives of phosphatidylserine using the appropriate fatty acyl derivatives of CDPdiacylglycerol as substrate. Pure phosphatidylserine synthase can also be used

[25] M. J. Kelley, A. M. Bailis, S. A. Henry, and G. M. Carman, *J. Biol. Chem.* **263**, 18078 (1988).

[26] M. Bae-Lee and G. M. Carman, *J. Biol. Chem.* **265**, 7221 (1990).

[27] R. A. Deems, B. R. Eaton, and E. A. Dennis, *J. Biol. Chem.* **250**, 9013 (1975).

[28] C. R. H. Raetz, G. M. Carman, W. Dowhan, R.-T. Jiang, W. Waszkuc, W. Loffredo, and M.-D. Tsai, *Biochemistry* **26**, 4022 (1987).

[29] A. J. Kinney and G. M. Carman, *Proc. Natl. Acad. Sci. U.S.A.* **85**, 7962 (1988).

[30] A. J. Kinney, M. Bae-Lee, S. Singh Panghaal, M. J. Kelley, P. M. Gaynor, and G. M. Carman, *J. Bacteriol.* **172**, 1133 (1990).

to determine quantitatively the serine concentration by following the formation of radiolabeled CMP from radiolabeled CDPdiacylglycerol.

Acknowledgments

This work was supported by U.S. Public Health Service Grant GM-28140 from the National Institutes of Health, New Jersey State funds, and the Charles and Johanna Busch Memorial Fund.

[36] Phosphatidylinositol Synthase from Yeast

By George M. Carman and Anthony S. Fischl

Introduction

Phosphatidylinositol synthase (CDPdiacylglycerol–*myo*-inositol 3-phosphatidyltransferase, EC 2.7.8.11) catalyzes the incorporation of

$$\text{CDP-diacylglycerol} + \text{inositol} \rightarrow \text{phosphatidylinositol} + \text{CMP}$$

inositol into phosphatidylinositol.[1] The enzyme[2] and its product phosphatidylinositol[3,4] are essential to the growth of the yeast *Saccharomyces cerevisiae*. Phosphatidylinositol synthase activity is associated with the mitochondrial,[5,6] microsomal,[5,6] and plasma membrane[7] fractions of *S. cerevisiae*. Microsome-associated phosphatidylinositol synthase has been purified to near homogeneity, primarily by CDPdiacylglycerol-Sepharose affinity chromatography.[8] We describe here the purification, reconstitution, and properties of the enzyme.

Preparation of CDPdiacylglycerol

CDPdiacylglycerol is prepared from phosphatidic acid and CMPmorpholidate by the method of Agranoff and Suomi[9] with the modifications of

[1] H. Paulus and E. P. Kennedy, *J. Biol. Chem.* **235**, 1303 (1960).
[2] J. Nikawa, T. Kodaki, and S. Yamashita, *J. Biol. Chem.* **262**, 4876 (1987).
[3] S. A. Henry, K. D. Atkinson, A. J. Kolat, and M. R. Culbertson, *J. Bacteriol.* **130**, 472 (1977).
[4] G. W. Becker and R. L. Lester, *J. Biol. Chem.* **252**, 8684 (1977).
[5] G. S. Cobon, P. D. Crowfoot, and A. W. Linnane, *Biochem. J.* **144**, 265 (1974).
[6] K. Kuchler, G. Daum, and F. Paltauf, *J. Bacteriol.* **165**, 901 (1986).
[7] A. J. Kinney and G. M. Carman, *J. Bacteriol.* **172**, 4115 (1990).
[8] A. S. Fischl and G. M. Carman, *J. Bacteriol.* **154**, 304 (1983).
[9] B. W. Agranoff and W. D. Suomi, *Biochem. Prep.* **10**, 47 (1963).

Carman and Fischl.[10] Phosphatidic acid is prepared from soybean lecithin by reaction with cabbage phospholipase D as previously described.[11,12] One gram of phosphatidic acid (free acid)[12] is dissolved in 10 ml of anhydrous chloroform. The catalyst 4-dimethylaminopyridine (0.7 g) and CMPmorpholidate (0.8 g) are added to the mixture without mixing. CMPmorpholidate, which is stored at $-20°$, is allowed to equilibrate to room temperature in a desiccator prior to its introduction into the reaction mixture. The reaction mixture is incubated for 2 days at $37°$.

Following incubation, 20 ml of methanol–water (1 : 1, v/v) is added, the upper phase is adjusted to pH 2 with 1 N HCl, and the phases are separated by centrifugation at 100 g for 2 min at room temperature. Unreacted CMPmorpholidate is removed in the aqueous phase during this phase partition. The chloroform phase is evaporated to dryness under reduced pressure using a rotary evaporator, and the residue is dissolved in 5 ml of chloroform–pyridine–formic acid (50 : 30 : 7, v/v). This mixture is applied to a silica gel column (2 × 20 cm) equilibrated with the same solvent. The column is washed first with 100 ml of chloroform–pyridine–formic acid (50 : 30 : 7, v/v) followed by 200 ml of chloroform–methanol (9 : 1, v/v). The 4-dimethylaminopyridine and unreacted phosphatidic acid are removed from the column during these washing steps. CDPdiacylglycerol is then eluted from the column with 150 ml of chloroform–methanol–glacial acetic acid–water (50 : 30 : 4 : 8, v/v). The solvent is evaporated *in vacuo*.

The residue is dissolved in 10 ml of chloroform and mixed with 20 ml methanol–water (1 : 1, v/v). The upper phase is adjusted to pH 2 with 1 N HCl. The mixture is centrifuged at 100 g for 2 min to separate the phases. The lower chloroform phase is washed twice with 20 ml methanol–water (1 : 1, v/v). The chloroform phase is then adjusted to pH 7.0 with 1 N NH_4OH in methanol to make the diammonium salt of CDPdiacylglycerol. The solvent is evaporated *in vacuo,* and the CDPdiacylglycerol is dispersed in water by sonication to concentrations between 10 and 30 mM. The concentration of CDPdiacylglycerol is determined from its molar extinction coefficient (12.8 × 10³).[10,12] The yield of CDPdiacylglycerol from phosphatidic acid is typically around 40%. The compound is stable at $-20°$ for several months.

Preparation of CDPdiacylglycerol-Sepharose

The $NaIO_4$-oxidized derivative of CDPdiacylglycerol is covalently attached to Sepharose 4B via an adipic acid dihydrazide spacer arm as

[10] G. M. Carman and A. S. Fischl, *J. Food Biochem.* **4,** 53 (1980).
[11] M. Kates and P. S. Sastry, this series, Vol. **14,** p. 197.
[12] W. Dowhan and T. Larson, this series, Vol. **71,** p. 561.

described by Dowhan and co-workers[13,14] with the modifications of Fischl and Carman.[8] Sepharose 4B is activated with cyanogen bromide in acetonitrile.[15] The incubation times for coupling the adipic acid spacer arm to activated Sepharose 4B, NaIO$_4$ oxidation of CDPdiacylglycerol, and coupling the oxidized derivative of CDPdiacylglycerol to the adipic acid-Sepharose resin are 2 days, 2 days, and 3 days, respectively. These modifications increase the concentration of the bound, oxidized CDPdiacylglycerol derivative from 2 to 30 μmol of ligand per milliliter of resin.[8]

Assay Method

Phosphatidylinositol synthase activity is routinely measured by following the incorporation of 0.5 mM myo-[2-^{3}H]inositol [10,000 counts/min (cpm)/nmol] into phosphatidylinositol in 50 mM Tris-HCl buffer (pH 8.0) containing 2 mM MnCl$_2$, 0.2 mM CDPdiacylglycerol, 2.4 mM Triton X-100, and enzyme protein in a total volume of 0.1 ml at 30°.[8,16] Purified phosphatidylinositol synthase reconstituted in phospholipid vesicles is measured in the absence of Triton X-100 and added CDPdiacylglycerol. The reaction is terminated by the addition of 0.5 ml of 0.1 N HCl in methanol. Chloroform (1 ml) and 1 M MgCl$_2$ (1.5 ml) are added, the system is mixed, and the phases are separated by a 2-min centrifugation at 100 g. A 0.5-ml sample of the chloroform phase is removed and taken to dryness on an 80° water bath. Betafluor (4 ml) is added to the sample, and radioactivity is determined by scintillation counting. The chloroform-soluble phospholipid product of the reaction, phosphatidylinositol, is analyzed by thin-layer chromatography with standard phosphatidylinositol.[8,16] Alternatively, activity is measured by following the release of radiolabeled CMP from radiolabeled CDPdiacylglycerol.[16]

One unit of enzymatic activity is defined as the amount of enzyme that catalyzes the formation of 1 nmol of product per minute. Protein is determined by the method of Bradford.[17] Buffers which are identical to those containing the protein samples are used as blanks. The presence of Triton X-100 does not interfere with the protein determination, provided the blank contains a final concentration of detergent identical to that of the sample.[8]

[13] T. J. Larson, T. Hirabayashi, and W. Dowhan, *Biochemistry* **15**, 974 (1976).
[14] W. Dowhan and T. Hirabayashi, this series, Vol. **71**, p. 555.
[15] S. C. March, I. Parikh, and P. Cuatrecasas, *Anal. Biochem.* **60**, 149 (1974).
[16] G. M. Carman and J. Matas, *Can. J. Microbiol.* **27**, 1140 (1981).
[17] M. M. Bradford, *Anal. Biochem.* **72**, 248 (1976).

Electroblotting of Phosphatidylinositol Synthase Activity[18]

Samples for electrophoresis are dialyzed for 3 hr at 8° against 62.5 mM Tris-HCl buffer (pH 6.8) containing 2% sodium dodecyl sulfate, 50% glycerol, and 5% 2-mercaptoethanol. Two microliters of 1% methyl green is added to samples as a transfer indicator, and sodium dodecyl sulfate-polyacrylamide gel electrophoresis[19] is performed with 10% gels. After electrophoresis, proteins are transferred electrophoretically to nitrocellulose.[20–22] After transfer, the nitrocellulose sheet is cut in half. One half is stained with Coomassie blue, and the other half is soaked in renaturation buffer [50 mM Tris-HCl (pH 8.0), 30 mM $MgCl_2$, 1 mM CDPdiacylglycerol, 0.5% Triton X-100, 10 mM 2-mercaptoethanol, 20% glycerol, 3% bovine serum albumin] for 1 hr at 8°. The presence of CDPdiacylglycerol and cofactors in the renaturation buffer stabilizes phosphatidylinositol synthase activity. Each lane is cut into 0.5-cm strips and used for enzyme assays. The recovery of phosphatidylinositol synthase activity is low (~0.7%). This is due to the treatments of the enzyme due to electrophoresis and transfer. This procedure allows the confirmation of the molecular weight of phosphatidylinositol synthase.

Growth of Yeast

Wild-type *S. cerevisiae* strain S288C (*α gal2*) is used for enzyme purification. Cells are grown at 28° in 30 liters of medium containing 1% yeast extract, 2% peptone, and 2% glucose in a 50-liter fermentor. Air at a rate of 15 liters/min is passed through the culture, which is stirred at 320 rpm. The culture is grown to late exponential phase to an equivalent density of 40 mg (wet weight) per milliliter. Cells are harvested by centrifugation in a Sharples AS-16 Supercentrifuge at 13,000 *g* at a flow rate of 4 liters/min at 5°. The cell paste is frozen and stored at −80°.

Purification Procedure

All steps are performed at 5°.

Step 1: Preparation of Microsomes. Cells (100 g, wet weight) are washed in 50 mM Tris-HCl buffer (pH 7.5) containing 1 mM Na_2EDTA, 0.3 M sucrose, and 10 mM 2-mercaptoethanol and are resuspended in 100

[18] M. A. Poole, A. S. Fischl, and G. M. Carman, *J. Bacteriol.* **161**, 772 (1985).

[19] U. K. Laemmli, *Nature (London)* **227**, 680 (1970).

[20] W. Burnette, *Anal. Biochem.* **112**, 195 (1981).

[21] A. Haid and M. Suissa, this series, Vol. **96**, p. 192.

[22] H. Towbin, T. Staehlin, and J. Gordon, *Proc. Natl. Acad. Sci. U.S.A.* **76**, 4350 (1979).

ml of the same buffer. The cell suspension is mixed with 300 g of prechilled glass beads (0.5 mm in diameter) and disrupted by homogenization in a Bead-Beater (BioSpec Products, Bartlesville, OK) for five 1-min bursts, with a 4-min pause between bursts. The homogenate is brought to a final volume of 300 ml and then centrifuged at 1500 g for 10 min. The pellet is discarded, and the supernatant is centrifuged at 27,000 g for 10 min. The supernatant is saved, and the pellet is resuspended in 200 ml of the same buffer and recentrifuged at 27,000 g for 10 min. The pellet is discarded and the supernatant combined with the first 27,000 g supernatant. This material is then centrifuged at 100,000 g for 2 hr to obtain the microsome fraction. Microsomes are washed with 100 ml of 50 mM Tris-HCl buffer (pH 7.5) containing 10 mM 2-mercaptoethanol and 20% glycerol (w/v) and resuspended in the same buffer.

Step 2: Preparation of Triton X-100 Extract. The microsome fraction is suspended in 50 mM Tris-HCl buffer (pH 8.0) containing 30 mM MgCl$_2$, 10 mM 2-mercaptoethanol, 20% glycerol (w/v), and 1% Triton X-100 at a final protein concentration of 10 mg/ml. After incubation for 1 hr at 5°, the suspension is centrifuged at 100,000 g for 2 hr to obtain the solubilized (supernatant) fraction. Under these solubilization conditions, microsome-associated phosphatidylserine synthase is not solubilized.[8] If 2 mM MnCl$_2$ is substituted for MgCl$_2$ in the solubilization buffer, phosphatidylinositol synthase and phosphatidylserine synthase activities are released from microsomes.[16]

Step 3: CDPdiacylglycerol-Sepharose Chromatography. CDPdiacylglycerol-Sepharose (25 ml) is equilibrated with 12 volumes of 50 mM Tris-HCl buffer (pH 8.0) containing 30 mM MgCl$_2$, 10 mM 2-mercaptoethanol, 20% glycerol, and 1% Triton X-100. The Triton X-100 extract is mixed in batches with the affinity resin and allowed to react for 1 hr with shaking. This mixture is then poured into a column (1.5 × 14 cm), and the run-through material is recirculated through the column. The column is washed with 20 column volumes of 50 mM Tris-HCl buffer (pH 8.0) containing 30 mM MgCl$_2$, 10 mM 2-mercaptoethanol, 20% glycerol, 0.5% Triton X-100, and 0.5 M NH$_4$Cl. The column is then saturated with 50 mM Tris-HCl buffer (pH 8.0) containing 30 mM MgCl$_2$, 10 mM 2-mercaptoethanol, 20% glycerol, 0.5% Triton X-100, 0.65 mM CDPdiacylglycerol, and 0.8 M hydroxylamine hydrochloride and allowed to react for 1 hr. Phosphatidylinositol synthase is eluted from the resin with 2 column volumes of this buffer at a flow rate of 20 ml/hr. Fractions containing activity are pooled and desalted by either Sephadex G-25 gel filtration or dialysis using 20 mM Tris-HCl buffer (pH 7.4) containing 1 mM dithiothreitol, 20% glycerol, and 0.5% Triton X-100.

Step 4: Chromatofocusing. A column (0.9 × 24 cm) of chromatofocus-

TABLE I

PURIFICATION OF PHOSPHATIDYLINOSITOL SYNTHASE FROM *Saccharomyces cerevisiae*[a]

Purification step	Total units (nmol/min)	Protein (mg)	Specific activity (units/mg)	Purification (-fold)	Yield (%)
1. Microsomes	2000	2500	0.8	1	100
2. Triton X-100 extract	1500	1250	1.2	1.5	75
3. CDPdiacylglycerol-Sepharose	1200	2.5	480	600	60
4. Chromatofocusing run-through	1200	1.5	800	1000	60

[a] Data from Fischl and Carman.[8]

ing resin (PBE 94 from Pharmacia LKB Biotechnology, Inc., Piscataway, NJ) is equilibrated with 20 mM Tris-HCl buffer (pH 7.4) containing 1 mM dithiothreitol, 20% glycerol, and 0.5% Triton X-100. The desalted enzyme from the previous step is applied to the chromatofocusing column at a flow rate of 20 ml/hr. The column is then washed with 2 column volumes of equilibration buffer. The enzyme does not bind to the resin under these conditions and emerges in the run-through and wash fractions. The purified enzyme is concentration by ultrafiltration. The enzyme is 100% stable for at least 6 months and about 50% stable for 4 years when stored at $-80°$.

Enzyme Purity. The four-step purification scheme summarized in Table I results in a phosphatidylinositol synthase preparation that is nearly homogeneous as evidenced by native[23] and sodium dodecyl sulfate-polyacrylamide gel electrophoresis.[8] Native gels contain 0.5% Triton X-100. The phosphatidylinositol synthase preparation is devoid of other CDPdiacylglycerol-dependent enzymes including phosphatidylserine synthase, CDPdiacylglycerol synthase, and phosphatidylglycerophosphate synthase. The enzyme is purified 1000-fold over the microsomal fraction and 3300-fold relative to the activity in cell extracts. The subunit molecular weight of phosphatidylinositol synthase is 34,000 as determined by sodium dodecyl sulfate-polyacrylamide gel electrophoresis[8] and electroblotting of enzyme activity.[18]

Reconstitution of Phosphatidylinositol Synthase

Pure phosphatidylinositol synthase has been reconstituted into unilamellar phospholipid vesicles containing its substrate CDPdiacylglycerol.[24] Because phosphatidylinositol synthase is purified in the presence

[23] A. S. Fischl, Ph.D. Thesis, Rutgers University, New Brunswick, New Jersey (1986).

[24] A. S. Fischl, M. J. Homann, M. A. Poole, and G. M. Carman, *J. Biol. Chem.* **261**, 3178 (1986).

of Triton X-100, it is necessary to reconstitute the enzyme into vesicles directly from mixed micelles containing pure enzyme, phospholipid, Triton X-100, and octylglucoside. The reconstitution procedure is modeled after those described by Mimms et al.[25] and Green and Bell.[26]

Highly purified phospholipids (25 mg) in chloroform–methanol (9 : 1, v/v) are transferred to a test tube, and solvent is evaporated under a stream of nitrogen. Residual solvent is removed in vacuo for 2 hr. Dried phospholipids are suspended in 1 ml of 750 mM octylglucoside. A 0.10-ml sample of pure phosphatidylinositol synthase in 50 mM Tris-HCl buffer (pH 8.0) containing 7.8 mM Triton X-100, 10 mM 2-mercaptoethanol, and 10% glycerol is mixed with 0.1 ml of octylglucoside-solubilized phospholipids and 0.25 μmol of CDPdiacylglycerol. The mixture is then diluted to a final volume of 0.5 ml with 50 mM Tris-HCl (pH 7.5). The final concentrations of Triton X-100, octylglucoside, and phospholipids in this reconstitution mixture are 2.0, 150, and 5.5 mM, respectively. The final molar ratios of octylglucoside to Triton X-100, octylglucoside to phospholipids, and phospholipids to CDPdiacylglycerol are 75 : 1, 27 : 1, and 10 : 1, respectively.

The mixture is passed through a Sephadex G-50 superfine column (1.5 × 32 cm) equilibrated with 50 mM Tris-HCl buffer (pH 7.5) containing 1 mM MnCl$_2$, 10 mM 2-mercaptoethanol, 250 mM NaCl, and 10% glycerol at a flow rate of 20 ml/hr at 5°. Vesicles and phosphatidylinositol synthase emerge from the column in the void volume and are well separated from the detergents.[24] The average size of vesicles containing phosphatidylinositol synthase is 40 nm in diameter.[24] Each vesicle contains about 1 molecule of enzyme, which is reconstituted asymmetrically with 90% of its active site facing outward.[24] Maximum reconstituted phosphatidylinositol synthase activity is obtained with vesicles containing phosphatidylcholine, phosphatidylethanolamine, phosphatidylinositol, and phosphatidylserine,[24] the four major phospholipids in wild-type S. cerevisiae.[27,28]

Properties of Phosphatidylinositol Synthase

Maximum phosphatidylinositol synthase activity is dependent on either MnCl$_2$ (2 mM) or MgCl$_2$ (20 mM) at the pH optimum of 8.0 using a mixed micelle assay of Triton X-100 and CDPdiacylglycerol[8] or enzyme

[25] L. T. Mimms, G. Zamphighi, Y. Nozaki, C. Tanford, and J. A. Reynolds, *Biochemistry* **20**, 833 (1981).
[26] P. R. Green and R. M. Bell, *J. Biol. Chem.* **259**, 14688 (1984).
[27] L. S. Klig, M. J. Homann, G. M. Carman, and S. A. Henry, *J. Bacteriol.* **162**, 1135 (1985).
[28] M. L. Greenberg, L. S. Klig, V. A. Letts, B. S. Loewy, and S. A. Henry, *J. Bacteriol.* **153**, 791 (1983).

reconstituted into phospholipid vesicles.[24] The K_m value for inositol is 0.21 mM as determined with a mixed micelle substrate of Triton X-100 and CDPdiacylglycerol.[29] Phosphatidylinositol synthase activity is dependent on the bulk (66 μM)[29] and surface (3.1 mol %)[30] concentrations of CDPdiacylglycerol, characteristic of surface dilution kinetics.[31] Based on the results of kinetic experiments[29] and the ability of the enzyme to catalyze isotopic exchange reactions between substrates and products,[24] the enzyme catalyzes a Bi–Bi sequential reaction mechanism. Phosphatidylinositol synthase binds to CDPdiacylglycerol before inositol, and phosphatidylinositol is released prior to CMP in the reaction sequence.[24] Phosphatidylinositol synthase is labile above 60° and is inactivated by thioreactive agents.[8] The activation energy for the reaction is 35 kcal/mol.[8]

Synthetic and Analytical Uses

Pure phosphatidylinositol synthase can be used to synthesize radiolabeled phosphatidylinositol from CDPdiacylglycerol and labeled inositol. The enzyme can also be used to synthesize various fatty acyl derivatives of phosphatidylinositol using the appropriate fatty acyl derivatives of CDPdiacylglycerol as substrate. Pure phosphatidylinositol synthase has been used to determine quantitatively the intracellular inositol concentration of yeast by following the formation of radiolabeled CMP from radiolabeled CDPdiacylglycerol.[29]

Acknowledgments

This work was supported by U.S. Public Health Service Grant GM-28140 from the National Institute of Health, New Jersey State funds, and the Charles and Johanna Busch Memorial Fund.

[29] M. J. Kelley, A. M. Bailis, S. A. Henry, and G. M. Carman, *J. Biol. Chem.* **263,** 18078 (1988).
[30] M. Bae-Lee and G. M. Carman, *J. Biol. Chem.* **265,** 7221 (1990).
[31] R. A. Deems, B. R. Eaton, and E. A. Dennis, *J. Biol. Chem.* **250,** 9013 (1975).

[37] Phosphatidylglycerophosphate Synthase from *Escherichia coli*

By WILLIAM DOWHAN

Introduction

(d)CDPdiacylglycerol + *sn*-glycero-3-phosphate → phosphatidylglycerophosphate + CMP

The phosphatidylglycerophosphate synthase (EC 2.7.8.5, CDPdiacylglycerol–*sn*-glycerol-3-phosphate 3-phosphatidyltransferase) of *Escherichia coli* is tightly associated with the cytoplasmic membrane,[1] as shown for the enzyme from *Bacillus licheniformis*.[2] A similar activity has been reported to be associated with mitochondrial membranes from *Saccharomyces cerevisiae*[3] and higher eukaryotic cells.[4] Only the enzyme from *E. coli* has been purified to homogeneity and extensively studied.[5,6]

Assay Method

The enzyme catalyzes the displacement of (d)CMP from (d)CDPdiacylglycerol by *sn*-glycero-3-phosphate in the presence of a nonionic detergent and magnesium ion. Activity is usually determined by following the incorporation of radiolabeled glycerophosphate (^{3}H, ^{14}C, or ^{32}P) into chloroform-soluble material.[5] CDPdiacylglycerol is commercially available or can be chemically synthesized from egg yoke phosphatidylcholine.[7] In crude preparations the use of *sn*-glycero-3-[^{32}P]phosphate is not advisable because of the presence of several phosphatidylglycerophosphate phosphatase activities.[8–10] A spectrophotometric assay which couples the rate of release of CMP to NADH oxidation in the presence of ATP and phosphoenolpyruvate via CMP kinase, pyruvate kinase, and lactate dehy-

[1] Y.-Y. Chang and E. P. Kennedy, *J. Lipid Res.* **8**, 447 (1967).
[2] T. J. Larson, T. Hirabayashi, and W. Dowhan, *Biochemistry* **15**, 974 (1976).
[3] G. M. Carman and C. J. Belunis, *Can. J. Microbiol.* **29**, 1452 (1983).
[4] W. R. Bishop and R. M. Bell, *Annu. Rev. Cell Biol.* **4**, 579 (1988).
[5] T. Hirabayashi, T. J. Larson, and W. Dowhan, *Biochemistry* **15**, 5205 (1976).
[6] A. S. Gopalakrishnan, Y.-C. Chen, M. Temkin, and W. Dowhan, *J. Biol. Chem.* **261**, 1329 (1986).
[7] G. M. Carman and A. S. Fischl, this volume [36].
[8] T. Icho and C. R. H. Raetz, *J. Bacteriol.* **153**, 722 (1983).
[9] C. R. Funk and W. Dowhan, *J. Bacteriol.* **174**, in press (1992).
[10] W. Dowhan and C. R. Funk, this volume [25].

drogenase has been employed[11] with purified preparations of the phosphatidylserine synthase, CDPdiacylglycerol hydrolase, and phosphatidylglycerophosphate synthase of *E. coli*.

Procedure.[5] If necessary, enzyme for assay is diluted into 50 mM Tris-HCl buffer (pH 8.0) containing 0.1% Triton X-100. A 2-fold concentrated stock solution of the assay mixture containing 0.2 M Tris-HCl (pH 8.0), 2% Triton X-100, 0.4 mM CDPdiacylglycerol, and 1 mM *sn*-[U-^{14}C]glycero-3-phosphate (2–4 μCi/μmol) is prepared. The assay is carried out in 12-ml polypropylene tubes containing 50 μl of the stock assay mixture, enzyme, and water in a volume of 90 μl. The reaction is initiated at 37° by the addition of 10 μl of 1.0 M MgCl$_2$. After 10 min the reaction is stopped by the addition of 0.5 ml methanol (0.1 N in HCl), 1.5 ml chloroform, and 3.0 ml of 1 M MgCl$_2$. After thorough mixing of the two phases and separation of the phases by centrifugation at 1000 g_{av} in a clinical centrifuge, 1 ml of the chloroform phase is removed, evaporated to dryness at 65°, and counted for radioactivity using any commercially available scintillation fluid. One unit of activity is defined as the amount of enzyme which converts 1 μmol of *sn*-glycero-3-phosphate into chloroform-soluble product in 1 min under the above conditions.

Purification Procedure

The original purification scheme reported for this enzyme[5] has not given routinely reproducible results, primarily owing to the variability in the properties of one of the affinity resins employed (reduced affinity resin). A modification[6] of the original scheme has been reported which yields highly purified enzyme in low yield when starting with a strain of *E. coli* carrying a copy of a variant of the cloned *pgsA* locus (see below) on the multicopy number plasmid pPGL2134 (*pgsA–L*$^+$ *amp*^{R} ColE1^{ori}). The purification of the enzyme requires the synthesis of two chromatography resins, CDPdiacylglycerol-Sepharose 4B and Blue dextran 2000 attached to Sepharose 4B. The following buffers are used in the preparation of the enzyme: buffer A is composed of 50 mM Tris-HCl (pH 7.0), 10 mM MgCl$_2$, 2 mM dithiothreitol; buffer B is composed of buffer A containing 4% Triton X-100; buffer C is composed of buffer A plus 0.1% Triton X-100. Table I summarizes the results of the purification of the enzyme. The following procedures are carried out at 4° unless otherwise noted.

Synthesis of Affinity Resins. CDPdiacylglycerol-Sepharose is synthesized[2] by first activating[12] 100 ml of Sepharose 4B in 300 ml of water by

[11] G. M. Carman and W. Dowhan, *J. Lipid Res.* **19**, 519 (1978).
[12] S. C. March, I. Prikh, and P. Cuatrecasas, *Anal. Biochem.* **60**, 149 (1974).

TABLE I

PURIFICATION OF

PHOSPHATIDYLGLYCEROPHOSPHATE SYNTHASE[a]

Step	Total protein	Specific activity (units/mg)	Yield (%)
1. Cell-free extract	41 g	0.33	100
2. Membrane extract	6.3 g	1.2	57
3. Affinity column	ND[b]	ND	15
4. Chromatofocusing	ND	ND	5
5. Blue Dextran	2 mg	22	3

[a] Starting with 200 g of cell paste from *E. coli* strain JA200/pPGL2134 as described by A. S. Gopalakrishnan, Y.-C. Chen, M. Temkin, and W. Dowhan, *J. Biol. Chem.* **261**, 1329 (1986).

[b] Not determined.

adding 5 g of CNBr in 10 ml of acetonitrile; the mixture is stirred for 1 hr, maintaining the pH between 10.5 and 11.5 by dropwise addition of 8 N NaOH. The activated resin is washed with 2 liters of cold water and 500 ml of 0.1 M Na$_2$CO$_3$ (pH 9.5) and finally suspended and gently stirred in 100 ml of 0.1 M Na$_2$CO$_3$ (pH 9.5) with 9 g of adipic acid dihydrazide for 17 hr. Excess reagent is removed by washing with 1 M NaCl and water at room temperature. The vicinal hydroxyls of the ribose ring of CDPdiacylglycerol (0.36 mmol) are activated for attachment to the resin by oxidation to aldehydes by incubation with 0.47 mmol of NaIO$_4$ in 70 ml of sodium acetate (pH 4.5). Incubation is carried out in the dark for 1 hr at room temperature and then overnight at 0°; excess periodate is destroyed by adding 0.1 ml of glycerol. Oxidation to 90% can be verified by silica gel thin-layer chromatography using chloroform–methanol–glacial acetic acid–water (50 : 28 : 4 : 8, v/v) as the solvent; the product has an increased mobility relative to the starting material. The oxidized product is attached to the activated Sepharose 4B by gently mixing the two products overnight in 180 ml of 0.1 M sodium acetate (pH 5.0) containing 0.5% Triton X-100. The resin is extensively washed at room temperature, first with the incubation buffer containing 0.5 M KCl and then with water. The resin should contain 1–2 μmol of bound CDPdiacylglycerol per milliliter of Sepharose 4B as determined by phosphate analysis.[13]

    Blue dextran-Sepharose is prepared[14] by first activating 100 ml of Seph-

[13] G. R. Bartlett, *J. Biol. Chem.* **234**, 466 (1959).

[14] L. D. Ryan and C. S. Vestling, *Arch. Biochem. Biophys.* **160**, 279 (1974).

arose 4B as described above. The activated resin is stirred with 2 g of Blue dextran 2000 in 200 ml of 0.1 M Na$_2$CO$_3$ (pH 9.5) for 18 hr. Excess reagents are removed by extensive washing with cold water.

Growth of Cells. Strain JA200/pPGL2134 is grown in LB medium (10 g/liter of Bacto-tryptone (Difco, Detroit, MI), 5 g/liter of Bacto-yeast extract, 5 g/liter of NaCl) containing 100 μg/ml of ampicillin at 37° with aeration to late log phase of growth. The starting specific activity of this strain is 10-fold higher than for strains lacking the plasmid.

Step 1: Cell-free Extract. Cell paste (200 g) of strain JA200/pPGL2134 is suspended in 500 ml of buffer A and passed through a French pressure cell at 1000 psi to break the cells; cells can also be broken using sonication.

Step 2: Membrane Extract. The cell lysate is centrifuged at 105,000 g_{av} for 2 hr. The membrane pellet is suspended in 500 ml of buffer B and stirred for 2 hr before removing insoluble material by centrifugation for 2 hr at 105,000 g_{av}.

Step 3: Affinity Column. The Triton extract is gently stirred overnight with CDPdiacylglycerol-Sepharose at a ratio of 4 units of enzyme per milliliter of settled resin volume. The resin is then poured into a chromatography column with dimensions convenient for batch washing and elution. The resin is first washed with 10 column volumes of buffer C containing 3.5 M NaCl and then 4 column volumes of buffer C containing 0.8 N hydroxylamine hydrochloride (readjusted to pH 7.0 with NaOH); hydroxylamine is believed to release the covalently bound lipid ligand. After standing in the latter buffer overnight, the column is eluted with the latter buffer; the enzyme elutes in the void volume of the column. The pooled peak of activity is dialyzed against several changes of buffer C.

Step 4: Chromatofocusing. The dialyzed sample is adjusted to pH 9.4 with 0.25 M ethanolamine and applied to a PBE 94 (Pharmacia, Piscataway, NJ) chromatofocusing column (7–10 units of enzyme/ml packed resin volume) equilibrated with 0.25 M ethanolamine–acetate (pH 9.4) containing 1% Triton X-100 and 2 mM dithiothreitol. The column is washed with 2 column volumes of the equilibration buffer and then eluted with a 1 : 10 dilution of Polybuffer 96 (Pharmacia) adjusted to pH 6.0 with acetic acid and containing 1% Triton X-100 and 2 mM dithiothreitol. The enzymatic activity elutes in fractions with a pH of 9.1.

Step 5: Blue Dextran-Sepharose Chromatography. The pooled fractions from Step 4 are applied to a Blue dextran-Sepharose column (3 units of enzyme/ml of resin volume) in buffer C. The column is washed sequentially with 2 column volumes each of buffer C, buffer C containing 1 M NaCl, and buffer C without MgCl$_2$ but containing 0.6 mM CDPdiacylglycerol. The column is incubated overnight in the latter buffer and then eluted with the lipid-containing buffer until the enzymatic activity emerges.

Purity. Over 95% of the protein after Step 5 is a single band with an M_r of 19,000 as determined by sodium dodecyl sulfate-polyacrylamide gel electrophoresis using the method of Laemmli.[15] The same purification scheme for the enzyme from wild-type strains lacking the plasmid copy of the gene yields a preparation of 80% purity for the major 19,000-Da protein. The protein band from either preparation when eluted from a sodium dodecyl sulfate-polyacrylamide gel displays phosphatidylglycerophosphate synthase activity, thus establishing this species as the enzyme. The M_r of the protein depends on the denaturation conditions and the gel electrophoresis conditions[5,6] and can be as high as 24,000, which is close to the subunit molecular weight of the enzyme.

Properties

Stability.[5] Enzyme stored in buffer C is stable for at least 3 years at $-80°$ and several months at $4°$. The enzyme is unusually stable to heat, chaotropic agents, and denaturing detergents. The Triton X-100 extract retains full activity after treatment at $55°$ for 5 min, loses one-half its activity after 5 min at $60°$, and loses all activity after 5 min at $65°$. The enzyme retains all activity for up to 2 hr at $30°$ in the presence of buffer C containing 8 M urea and loses only one-half its activity in buffer C containing 1% sodium dodecyl sulfate; the combination of 1% sodium dodecyl sulfate and 4 M urea inactivates the enzyme. There are no known specific inhibitors of the enzyme.

Activity and Detergent Dependence. The detergent dependencies of the enzyme have been previously described in detail.[5] The activity is dependent on the presence of a nonionic detergent such as Triton X-100 (0.2–6% being optimal) and a divalent metal ion (100 mM Mg^{2+} being optimal); Mn^{2+} is much less effective, and Ca^{2+} is inhibitory. The dependence of activity on substrate levels follows normal Michaelis–Menten kinetics in the presence of detergent but, unlike the phosphatidylserine synthase[16] and phosphatidylserine decarboxylase,[17] is not sensitive to the molar ratio of detergent to lipid substrate. Kinetic data on the enzyme including isotope exchange data support an ordered sequential Bi–Bi reaction mechanism with the liponucleotide possibly binding first; the K_m

[15] U. K. Laemmli, *Nature* (*London*) **227**, 680 (1970).
[16] G. M. Carman and W. Dowhan, *J. Biol. Chem.* **254**, 8391 (1979).
[17] T. G. Warner and E. A. Dennis, *J. Biol. Chem.* **250**, 8004 (1975).

MQFNIPTLLT	LFRVILIPFF	VLVFYLPVTW	SPFAAALIFC	40
VAAVTDWFDG	FLARRWNQST	RFGAFLDPVA	DKVLVAIAMV	80
LVTEHYHSWW	VTLPAATMIA	REIIISALRE	WMAELGKRSS	120
VAVSWIGKVK	TTAQMVALAW	LLWRPNIWVE	YAGIALFFVA	160
AVLTLWSMLQ	YLSAARQICL	ISDRFGVIFS	KRSKVVKNIV	200
DSSRQVSRMQ	RIERRH			216
DSSGDAANLL	*I*			*211*

FIG. 1. Amino acid sequence of phosphatidylglycerophosphate synthase from *E. coli*. The residues in normal type are for the wild-type gene (*pgsA–N*) product, and the residues in italics (201–211) are for the product of the *pgsA–L* variant which differs from the wild-type gene product only at the carboxyl-terminal end. [From A. S. Gopalakrishnan, Y.-C. Chen, M. Temkin, and W. Dowhan, *J. Biol. Chem.* **261**, 1329 (1986).]

values for (d)CDPdiacylglycerol and *sn*-glycero-3-phosphate are 40 and 320 μM, respectively. The enzyme does not catalyze hydrolysis of the phosphate esters of either the lipid substrate or lipid product and does not facilitate isotope exchange between substrates and products unless three of the four reactants are present. These properties are consistent with a concerted reaction mechanism not requiring an enzyme-bound phosphatidyl intermediate.

Molecular Properties. In cloning the *pgsA* gene by complementation[18] of a structural gene mutation,[19] the wild-type gene (*pgsA–N*) as well as several variants were isolated. For some yet undetermined reason the wild-type gene is unstable when carried in multiple copies, resulting in disruptions and rearrangements of the cloned gene; one of the functional variants (*pgsA–L*) has an IS*1* transposon inserted into the 3′ end of the gene, which may have also disrupted a potential promoter for the neighboring *glyW* gene. DNA sequence analysis of both the *pgsA–N* and *pgsA–L* genes and partial protein sequence determination of the *pgsA–L* gene product[6] are consistent with the protein sequences shown in Fig. 1; the difference in sequence for the two gene products lies at the carboxyl-terminal end and appears to have little effect on the function of the enzyme or its catalytic properties.

The *pgsA–N* gene product has a minimum molecular mass of 24,800 Da and is made up of 216 amino acids (Fig. 1 and Table II); the *pgsA–L* gene product is 5 amino acids shorter and differs in sequence beginning with residue 204 of the wild-type gene product. The amino terminus of the native protein is blocked and, based on analysis of CNBr-generated peptides, may be lacking the initiation methionine, suggesting that the

[18] A. Ohta, K. Waggoner, A. Radominska-Pyrek, and W. Dowhan, *J. Bacteriol.* **147**, 552 (1981).

[19] M. Nishijima and C. R. H. Raetz, *J. Biol. Chem.* **254**, 7837 (1979).

TABLE II

PREDICTED AMINO ACID COMPOSITION OF
PHOSPHATIDYLGLYCEROPHOSPHATE SYNTHASES[a]

Amino acid	Number of residues/subunit	
	PGSN[b]	PGSL[c]
Asp	6	7
Asn	4	5
Thr	11	11
Ser	15	14
Glu	6	5
Gln	7	5
Pro	7	7
Gly	6	7
Ala	25	27
Cys	2	2
Val	24	23
Met	7	6
Ile	17	17
Leu	23	25
Tyr	4	4
Phe	15	15
His	3	2
Lys	7	7
Trp	11	11
Arg	16	11
Total	216	211

[a] From A. S. Gopalakrishnan, Y.-C. Chen, M. Temkin, and W. Dowhan, *J. Biol. Chem.* **261,** 1329 (1986).
[b] Product of the *pgsA–N* gene.
[c] Product of the *pgsA–L* gene.

amino terminus is blocked by the formation of pyroglutamic acid. The calculated isoelectric point of the enzyme is 8.9, consistent with its elution position from chromatofocusing columns.[6] Computer analysis of the protein sequence suggests at least four hydrophobic stretches long enough to span the membrane bilayer. The amino terminus of the protein is very hydrophobic, containing only one charged side chain out of 44 amino acids. In contrast the carboxyl terminus is hydrophilic and basic, with 10 basic residues and 3 acidic residues out of 41 amino acids; the carboxyl terminus of the *pgsA–L* gene product is hydrophobic.

Synthetic and Analytical Uses. Because the enzyme does not catalyze the hydrolysis of (d)CDPdiacylglycerol and the reaction goes to near

100% completion,[5] the enzyme can be used for the synthesis of various radiolabeled derivatives of phosphatidylglycerophosphate without further purification of the chloroform-soluble products; separation from Triton X-100 can be accomplished through precipitation of the lipid product out of chloroform at 4° by the addition of acetone to 90%. A detergent extract of membranes (as in Step 2 above) from a strain carrying plasmid pPGL2134 can be used as a source of the enzyme for synthetic purposes; the activity of the phosphatidylglycerophosphate phosphatases are low in extracts made under these conditions but can be further reduced by inactivation using sulfhydryl reagents[20] or by mutation.[8,9] Alternatively, the presence of phosphatidylglycerophosphate phosphatase activity (amplified by multiple copies of the appropriate gene[21,22]) in the crude extract made under the conditions of Step 2, but replacing $MgCl_2$ with 5 mM EDTA,[20] would lead to the synthesis of radiolabeled phosphatidylglycerol. Such crude preparations of the enzyme can be used to quantify the levels of cytidine liponucleotides in various preparations by carrying out assays using an excess of radiolabeled glycerophosphate.

Genetic Manipulation of *pgsA* Locus

Genetic studies have established that the phosphatidylglycerophosphate synthase is the only enzyme catalyzing the committed step to phosphatidylglycerol synthesis and that this gene product is essential to cell growth.[23] In a strain (HD38, *pgsA30::kan*) lacking a functional *pgsA* locus (mapping at 42 min of the *E. coli* chromosome), there is a direct relationship between the level of enzyme, phosphatidylglycerol content, and the growth rate of cells. Therefore, it appears that phosphatidylglycerol levels are growth limiting for some essential cell function(s); membrane integrity appears not to be compromised in such cells. By placing a functional copy of the *pgsA* gene under regulation of the *lac* operon in a genetic background lacking a functional *pgsA* gene (strain HDL1001), it is possible to regulate the cell content of phosphatidylglycerol as a function of the level in the growth medium of the inducer of the *lac* operon.[24] Using such a system it has been possible to prepare natural membranes of *E. coli* with defined changes in the ratio of phosphatidylethanolamine to phosphatidylglycerol.

Strain HDL1001 has proved useful in defining the role of acidic phospholipids in the process of protein translocation across the cytoplasmic

[20] Y.-Y. Chang and E. P. Kennedy, *J. Lipid Res.* **8**, 456 (1967).
[21] T. Icho, *J. Bacteriol.* **170**, 5110 (1988).
[22] T. Icho, *J. Bacteriol.* **170**, 5117 (1988).
[23] P. N. Heacock and W. Dowhan, *J. Biol. Chem.* **262**, 13044 (1987).
[24] P. N. Heacock and W. Dowhan, *J. Biol. Chem.* **264**, 14972 (1989).

membrane.[25,26] The use of this system can be extended to the study of the reconstitution *in vitro* of any other membrane-related process which might be dependent on the phospholipid composition of the membrane. It is also possible to study the effect of membrane phospholipid composition on processes *in vivo* by utilizing cells growing in different levels of the *lac* inducer. Although phosphatidylglycerol is essential to *E. coli*, cells not capable of synthesizing the major outer membrane lipoprotein (*lpp* gene product) of this organism can grow normally at 5–10% the level of this phospholipid in wild-type cells.[27] Because strain HDL1001 still expresses low levels of phosphatidylglycerophosphate in the absence of the *lac* inducer, introduction of a mutation in the *lpp* locus into strain HDL1001 has allowed the study of the effect of phospholipid composition on protein translocation[28] across the inner membrane of *E. coli* under conditions where growth rate is unaffected by phospholipid composition.

[25] T. de Vrije, R. L. de Swart, W. Dowhan, J. Tommassen, and B. de Kruijff, *Nature (London)* **334**, 173 (1988).
[26] R. Lill, W. Dowhan, and W. Wickner, *Cell (Cambridge, Mass.)* **60**, 271 (1990).
[27] Y. Asai, Y. Katayose, C. Hikita, A. Ohta, and I. Shibuya, *J. Bacteriol.* **171**, 6867 (1989).
[28] R. Kusters, W. Dowhan, and B. de Kruijff, *J. Biol. Chem.* **266**, 8659 (1991).

[38] Cardiolipin Synthase from *Escherichia coli*

By Isao Shibuya and Shuichi Hiraoka

Introduction

In *Escherichia coli,* cardiolipin is mainly synthesized by cardiolipin synthase, encoded by the *cls* gene,[1–3] via condensation of two molecules of phosphatidylglycerol, also yielding one molecule of glycerol.[4,5] This contrasts with the eukaryotic enzymes that condense CDPdiacylglycerol and phosphatidylglycerol to form cardiolipin and CMP.[6,7] The biological implication of the metabolically uneconomical pathway in *E. coli* is un-

[1] G. Pluschke, Y. Hirota, and P. Overath, *J. Biol. Chem.* **253**, 5048 (1978).
[2] A. Ohta, T. Obara, Y. Asami, and I. Shibuya, *J. Bacteriol.* **163**, 506 (1985).
[3] S. Nishijima, Y. Asami, N. Uetake, S. Yamagoe, A. Ohta, and I. Shibuya, *J. Bacteriol.* **170**, 775 (1988).
[4] C. B. Hirschberg and E. P. Kennedy, *Proc. Natl. Acad. Sci. U.S.A.* **69**, 648 (1972).
[5] E. T. Tunaitis and J. E. Cronan, Jr., *Arch. Biochem. Biophys.* **155**, 420 (1973).
[6] M. Schlame and K. Y. Hostetler, this volume [39].
[7] K. T. Tamai and M. L. Greenberg, *Biochim. Biophys. Acta* **1046**, 212 (1990).

METHODS IN ENZYMOLOGY, VOL. 209

known. The *E. coli* cardiolipin synthase is present exclusively in the envelope (crude membrane) fraction of the cell.[5,8] The *E. coli cls* null mutants that are completely devoid of cardiolipin synthase activity grow well and form small amounts of cardiolipin,[3] suggesting that *E. coli* possesses an alternative minor pathway to form cardiolipin. The most likely candidate responsible for this second pathway is phosphatidylserine synthase.[3,9] It has been suggested that cardiolipin synthase may catalyze practically reversible reactions, and its relaxed substrate specificity may allow the formation of lyso- and dilysocardiolipins from 2-acyllysophosphatidylglycerol[10] and a series of analogs of phosphatidylglycerol and cardiolipin if high concentrations of the appropriate straight-chain sugar alcohols are present in the culture medium.[11] The *cls* gene was cloned,[2] and cardiolipin synthase was amplified up to a 760-fold wild-type level. This enabled extensive purification and characterization of the synthase,[12] which had previously been difficult because of the low cellular content and very hydrophobic nature of the enzyme.

Assay for Cardiolipin Synthase

For the routine assay of *E. coli* cardiolipin synthase, we measure the radioactivity of glycerol, one of the reaction products formed from phosphatidyl[2-³H]glycerol or phosphatidyl[¹⁴C]glycerol, the substrate labeled in the nonacylated glycerol moiety. This is less time-consuming than measuring the radioactivity of cardiolipin formed on separation by thin-layer chromatography (TLC)[2,5,8] and is highly sensitive and specific for cardiolipin synthesis under the reaction conditions described below.[12]

Substrate Preparation

Phosphatidyl[2-³H]glycerol is most conveniently synthesized enzymatically using the envelope fraction from *E. coli* cells amplified with phosphatidylglycerophosphate (PGP) synthase.[2] Phosphatidyl[¹⁴C]glycerol, though more expensive, is also useful and can be prepared from *sn*-[¹⁴C]glycerol 3-phosphate in the same manner.

Procedure. To a 50-ml glass-stoppered centrifuge tube add 2 ml of reaction mixture containing 80 mM Tris-HCl (pH 7.5), 10 mM MgCl$_2$,

[8] R. Cole and P. Proulx, *Can. J. Biochem.* **55,** 1228 (1977).

[9] I. Shibuya, C. Miyazaki, and A. Ohta, *J. Bacteriol.* **161,** 1086 (1985).

[10] H. Homma and S. Nojima. *J. Biochem.* (*Tokyo*) **91,** 1103 (1982).

[11] I. Shibuya, S. Yamagoe, C. Miyazaki, H. Matsuzaki, and A. Ohta, *J. Bacteriol.* **161,** 473 (1985).

[12] S. Hiraoka, K. Nukui, N. Uetake, A. Ohta, and I. Shibuya, *J. Biochem.* (*Tokyo*) **110,** 443 (1991).

5 mM 2-mercaptoethanol, 1 mM EDTA, 0.2% Triton X-100 (w/v), 0.5 mM CDPdiacylglycerol ammonium salt, 0.2 mM *sn*-[2-^{3}H]glycerol 3-phosphate ammonium salt (300 μCi, 750 mCi/mmol), and a crude membrane preparation, corresponding to approximately 140 μg of protein, from *E. coli* cells harboring plasmid pMA1. The reaction proceeds for 1 hr at 30°.

Lipids are extracted from the reaction mixture by the standard Ames method,[13] separated on a silica gel plate (No. 5721, E. Merck AG, Darmstadt, Germany) as a band about 5 cm long, and developed one-dimensionally with chloroform–methanol–acetic acid (100 : 41.7 : 16.7, v/v). An autoradiogram is prepared by exposing the plate to X-ray film (Kodak, Rochester, NY, RP X-Omat) for 2 days, and the band corresponding to phosphatidylglycerol is scraped off the plate into a 50-ml glass-stoppered centrifuge tube. Lipids are extracted with 7.8 ml of chloroform–methanol–100 mM Tris-HCl containing 50 mM EDTA (pH 7.0) (10 : 20 : 8, v/v) at room temperature overnight, centrifuged briefly, and the silica gel residue is extracted again using the same procedure. The extracts are combined (17.5 ml), and 4.5 ml each of chloroform and 100 mM Tris-HCl containing 50 mM EDTA (pH 7.0) are added, mixed, and centrifuged, and the chloroform layer is saved.

Phosphorus content is determined by the method of Morrison,[14] and radioactivity is measured for a known portion of the final sample. Usually, about 100 μCi (33% yield of ^{3}H) of phosphatidyl[2-^{3}H]glycerol with a specific radioactivity of about 1.2 $\times$ 10^6 disintegrations/min (dpm) per nanomole is obtained. The phosphatidyl[2-^{3}H]glycerol is diluted with nonlabeled phosphatidylglycerol (prepared from egg lecithin, Serdary Research Laboratories, London, ON, Canada) and suspended in water with sonication to give 2–3 $\times$ 10^4 dpm/nmol for the cardiolipin synthase assay.

Materials. CDPdiacylglycerol is purchased (Serdary) or prepared from egg yolk lecithin as described by Larson *et al.*[15] Although the phosphatidylglycerol preparations described here have fatty acid residues from the egg yolk, they are efficiently utilized as substrate by *E. coli* cardiolipin synthase. For studies such as fatty acid specificity, however, CDPdiacylglycerols with the desired fatty acid residues[16] should be used.

Plasmid pMA1[2] is a pBR322 derivative harboring the *Hin*dIII–*Eco*RI fragment of pPG1-L[17] that bears a mutant *pgsA* allele (*pgsAL*) amplifiable in high copy number plasmids.[17,18] The wild-type *pgsA* gene is highly toxic on amplification. For preparation of the envelope fraction amplified with

[13] G. F. Ames, *J. Bacteriol.* **95**, 833 (1968).
[14] W. R. Morrison, *Anal. Biochem.* **7**, 218 (1964).
[15] T. J. Larson, T. Hirabayashi, and W. Dowhan, *Biochemistry* **15**, 974 (1976).
[16] W. C. McMurray and E. C. Javis, *Can. J. Biochem.* **58**, 771 (1980).
[17] A. Ohta, K. Waggoner, A. Radominska-Pyrek, and W. Dowhan, *J. Bacteriol.* **147**, 552 (1981).

this mutant PGP synthase, cells harboring pMA1 are grown to the early stationary phase in 50 ml of LB broth[19] supplemented with 50 μg/ml of ampicillin, collected by centrifugation, and washed twice with 50 mM Tris-HCl (pH 7.5). They are suspended in 2 ml of 50 mM Tris-HCl (pH 7.5) containing 10 mM 2-mercaptoethanol, and disrupted by sonication (typically three 1-min runs with 1-min intervals with a Branson Sonifier, Branson Sonic Power Co., Danbury, CT). After the removal of unbroken cells by centrifugation at 3000 g for 10 min, the crude membrane fraction is collected by centrifugation at 145,000 g for 1 hr, suspended in 0.3 ml of 50 mM Tris-HCl containing 10 mM 2-mercaptoethanol, and, if appropriate, stored frozen at $-70°$. The specific activity of PGP synthase is usually amplified about 50-fold in wild-type cells. Introduction of a defect in the *cls* gene into the cells or of AMP in the reaction mixture to inhibit CDPdiacylglycerol hydrolase does not affect the yield of the labeled substrate.

Notes. To study the various features of cardiolipin synthase, such as substrate specificity and reaction mechanism, phospholipids labeled in other positions are useful. [^{32}P]Phosphatidylglycerol is most conveniently prepared by growing an *E. coli pssA1 cls-1* double mutant (such as strain SD9[9]; for construction, see below) at 42° in a low phosphate synthetic medium, M56LP,[20] supplemented with [^{32}P]P$_i$, 10 mM MgCl$_2$ (total concentration 20 mM), and 400 mM sucrose, followed by extraction and separation as described above. Usually, more than 70% of the total-lipid phosphorus is in phosphatidylglycerol.[9] [*acyl*-^{14}C]Phospholipids are prepared by culturing a *pssA1* mutant (such as strain SD10[9]) in Pennassay broth (Difco, Detroit, MI) supplemented with 10 mM MgCl$_2$ and sodium [1-^{14}C]acetate at 42°. Approximately equal amounts of three major phospholipids are obtained in the exponential phase,[9] and cardiolipin content is increased up to 90% in the stationary phase.[11] In these cases, ^{14}C radioactivity is localized exclusively and evenly in two acyl chains of phospholipids, as assessed by chemical deacylation and phospholipase A digestion.

Assay Procedure

The following solutions are mixed in 10-ml stoppered polyethylene centrifuge tube: 100 μl of 800 mM potassium phosphate (pH 7.1; final, 400 mM, pH 7.0), 10 μl of 200 mM 2-mercaptoethanol (final, 10 mM), 6 μl of 0.5% Triton X-100 (final 0.015%), 10 μl of 20 mg/ml of bovine serum

[18] A. S. Gopalakrishnan, Y.-C. Chen, M. Temkin, and W. Dowhan, *J. Biol. Chem.* **261,** 1329 (1986).

[19] J. H. Miller, "Experiments in Molecular Genetics." Cold Spring Harbor Laboratory, Cold Spring Harbor, New York, 1972.

[20] T. M. McIntyre, B. K. Chamberlain, R. E. Webster, and R. M. Bell, *J. Biol. Chem.* **252,** 4487 (1977).

albumin (final, 1 mg/ml), an enzyme preparation (see below), and water to a volume of 180 μl. The tubes are preincubated at 37° for 2 min, and the reaction is started by adding 20 μl of 200 μM phosphatidyl[2-^{3}H]glycerol (26,000 dpm/nmol, final, 20 μM). For maximum assay sensitivity, the Triton concentration is critical and should be 0.015% both for crude envelope fractions and solubilized preparations. To measure the activity within the range of linear relationships between the reaction time and product formation, enzyme preparations are added to give 10–60 μg and 5–20 ng of protein for the wild-type envelope fractions and purified fractions, respectively.

The reaction proceeds for 10 min at 37° and is terminated by adding 1.6 ml of methanol, followed by the addition of 3.2 ml of chloroform and 1 ml of 0.88% KCl containing 0.1% glycerol. For zero-time controls, methanol is added prior to the addition of the substrate. Afer vigorous mixing to solubilize the precipitated phosphate, the two layers are separated by a brief centrifugation. One milliliter of the aqueous layer is withdrawn into a vial to which 10 ml of an aqueous counting scintillant (ACSII, Amersham, Little Chalfont, Buckinghamshire, UK) is added, and radioactivity is determined in a liquid scintillation spectrometer. One unit of enzyme activity is defined as 1 nmol of radioactive glycerol formed in 1 min and is calculated by the following equation: $(R_{10} - R_0) \times 0.24/S$, where S is the specific radioactivity of the substrate (dpm/nmol) and R_{10} and R_0 are the radioactivities (dpm) after 10 min of reaction and at zero time, respectively.

Notes. The assay method described here[12] is simpler and more sensitive than those described previously[2,4,5,8]; it employes more suitable reaction conditions (i.e., remarkable activation by phosphate and a critical activating concentration of Triton X-100) and eliminates a laborious TLC step. For separation of the two layers after the reaction, the present method employs the classic procedure of Folch *et al.*,[21] which usually gives better separation of glycerol from acidic phospholipids than does the generally used method of Ames.[13] Contamination of the radioactive lipids in the aqueous layer is usually negligible. The addition of carrier glycerol is essential for the quantitative recovery of the product glycerol; radioactive glycerol of more than 96% is usually present in the aqueous layer.

The reaction conditions described here are highly specific for cardiolipin synthase; possible side reactions such as the formation of acylphosphatidylglycerol,[22] formerly considered bisphosphatidic acid,[5,23] and phos-

[21] J. Folch, M. Lees, and G. H. Slone Stanley, *J. Biol. Chem.* **226,** 497 (1957).
[22] M. Nishijima, T. Sa-eki, Y. Tamori, O. Doi, and S. Nojima, *Biochim. Biophys. Acta* **528,** 107 (1978).
[23] G. Benns and P. Proulx, *Biochem. Biophys. Res. Commun.* **44,** 382 (1971).

pholipase reactions[24] are negligible.[12] The reverse reaction of cardiolipin synthesis, as suggested from *in vivo* observations,[11] is detectable when [*acyl*-^{14}C]cardiolipin is included in the reaction mixture but does not interfere with the usual assay.[12] However, the method is indirect, and care must be taken to confirm the stoichiometry in nontypical reactions such as those with cytoplasmic components or mutant membranes. In this regard, use of phosphatidylglycerols labeled with ^{14}C or ^{32}P in the phosphatidyl moiety and phosphatidyl[^{14}C]glycerol is useful.

It is important to note that the values obtained with the present, as well as with the previous, assay methods do not necessarily represent the absolute activity of cardiolipin synthase, because we do not know the behavior of endogenous nonlabeled phosphatidylglycerol. The endogenous and exogenous phosphatidylglycerol fractions do not seem to mix,[5] and further studies are needed to clarify the relationships between the two fractions of phosphatidylglycerol.

Amplification and Purification of Cardiolipin Synthase

Amplification System

For high-level overproduction of cardiolipin synthase, we use strain JM109 [*recA1 endA1 gyrA96 thi hsdR17 relA1 supE44* Δ(*lac-proAB*)/F′ (*traD36 proAB*+ *lacI*q *lacZ*ΔM15)][25] harboring a plasmid pNT6.[12] This 5.9-kilobase (kb) pBR322 derivative bears the *cls* gene, derived from pCL11[2] through pPD324,[12] downstream of the *tac* promoter,[26] so that the addition of isopropyl-β-D-thiogalactoside (IPTG) into the culture medium causes an overproduction of the *cls* gene product, cardiolipin synthase. The λP$_R$ promoter placed in front of the *cls* gene in combination with a temperature-sensitive *cI857* allele, instead of the *tac* promoter and IPTG, has also been used successfully in our laboratory. These levels of amplification cause lethal accumulation of synthase molecules in the cytoplasmic membrane; induction-resistant mutants, most of which have acquired down-mutations in the *cls* expression, frequently appear.[12]

Cell Culture and Preparation of Envelope Fraction

For the maximum amplification of cardiolipin synthase, strain JM109 harboring pNT6 is grown with vigorous shaking in a 300-ml Erlenmeyer flask equipped with a side arm for Klett measurement. Fifty milliliters of

---

[24] R. Cole and P. Proulx, *Can. J. Biochem.* **55**, 1228 (1977).

[25] C. Yanisch-Perron, J. Vieira, and J. Messing, *Gene* **33**, 103 (1985).

[26] H. A. de Boer, L. J. Comstock, and M. Vasser, *Proc. Natl. Acad. Sci. U.S.A.* **80**, 21 (1983).

LB medium is supplemented with 50 μg/ml of ampicillin. At Klett 50, IPTG is added to give the final concentration of 300 μM and cultured for 2 more hr. Under these conditions, the turbidity increase stops and cells lose viability. Cells are collected by centrifugation, washed with 50 mM Tris-HCl (pH 7.5), and resuspended in 2 ml of 50 mM Tris-HCl (pH 7.5) containing 10 mM 2-mercaptoethanol. They are disrupted by sonication as described for the PGP synthase preparation. Unbroken cells are removed by centrifugation at 3000 g for 10 min, and the envelope (crude membrane) fraction is collected by centrifugation at 39,000 g for 1 hr and suspended in 3 ml of the same Tris buffer. They can be stored frozen at −70° without an appreciable loss of enzyme activity.

For purification of cardiolipin synthase, cells are grown on a larger scale in a jar fermentor. In this case, strain JM109 harboring pNT6 is grown to Klett 50 in 4 liters of LB medium supplemented with 0.2% glucose and ampicillin (w/v), made 300 μM with IPTG, and then cultured for 4 hr. Cells are collected, washed, and suspended in 40 ml of buffer that contains 20 mM MgCl$_2$ and a few micrograms of DNase I. They are disrupted twice in a French pressure cell operated at 20,000 psi.

Notes. If assayed for cardiolipin synthase by the method described above, the envelope fraction of wild-type cells (JM109/pBR322) usually gives a specific activity (units per milligram of protein) of approximately 1, whereas those of JM109/pPD324, uninduced JM109/pNT6, and induced JM109/pNT6 are amplified 20-, 30-, and 750-fold, respectively.[12] In our hands, the extent of amplification is usually lower in a jar (~250-fold) than in Erlenmeyer flasks probably because of its nonoptimal aeration. Sodium dodecyl sulfate (SDS)–polyacrylamide gel electrophoresis[27] of the envelope fraction from the induced JM109/pNT6 cells reveals a massive accumulation of the 45-kDa *cls* gene product.[12]

Purification

Solubilization. Cardiolipin synthase is solubilized using a Teflon homogenizer by suspending the envelope fraction in buffer A, which contains 10 mM Tris-HCl (pH 7.5), 10 mM 2-mercaptoethanol, 1% Triton X-100 (w/v), and 20% sucrose (w/v), to give 10 mg of protein/ml and 1 mg Triton/mg of protein. The suspension is stirred on ice for 1 hr, and unsolubilized materials are removed by centrifugation at 145,000 g for 1 hr at 4°. With this extraction, the protein content decreases to about 20% and the specific activity of cardiolipin synthase increases about 2.5-fold. This extract can be stored frozen by virtue of the addition of sucrose.

[27] U. K. Laemmli, *Nature (London)* **227**, 680 (1970).

TABLE I
PURIFICATION OF *E. coli* CARDIOLIPIN SYNTHASE FROM OVERPRODUCING CELLS[a]

Fraction	Total protein (mg)	Total enzyme (units)	Specific activity (units/mg protein)
Envelope fraction	262	65,800	251
Triton extract	54.6	36,300	665
Phosphocellulose	5.46	26,200	4800

[a] Data from S. Hiraoka, K. Nukui, U. Uetake, A. Ohta, and I. Shibuya, *J. Biochem. (Tokyo)* **110,** 443 (1991). Strain JM109 harboring plasmid pNT6 was cultured in a jar fermentor and induced with IPTG. The envelope fraction was prepared, extracted with Triton X-100, and fractionated by phosphocellulose column chromatography as described in the text.

Column Chromatography. Phosphocellulose (Whatman, Clifton, NJ, P11) is pretreated according to the recommendation of the manufacturer to remove fines and convert the ammonium form to the free acid form. Typically, phosphocellulose is packed in a column (26×135 mm, 72 ml), equilibrated with buffer A, and charged with 25 ml of the solubilized supernatant. The column is first washed with 1.5 liters of 100 mM potassium phosphate (pH 7.5) containing 10 mM 2-mercaptoethanol and 0.2% Triton X-100 and then eluted with 600 mM potassium phosphate (pH 7.5) containing 500 mM KCl, 10 mM 2-mercaptoethanol, and 0.2% Triton X-100. The eluate is collected in 5-ml fractions and subjected to assays for protein and enzymatic activity. Active fractions (e.g., fractions 9 through 22) are combined and mixed with the same volume of 40% sucrose for storage at $-70°$.

From amplified cells grown in a jar, a preparation with a specific activity of about 5000 with a yield of 40% of the total activity in the envelope fraction is typically obtained (see Table I). This activity is about 10,000 times higher than that of unamplified whole cells. SDS–polyacrylamide gel electrophoresis of the preparation followed by scanning in a densitometer shows that the major protein band of 45 kDa contains about 80% of the total protein.[12] Our attempts to purify further the enzymatically active synthase with various chromatographic systems have not been successful.

Other Aspects of Cardiolipin Synthase

Possible Secondary Pathway to Form Cardiolipin

The observation that small but significant amounts of cardiolipin are present in null *cls* mutants in which cardiolipin synthase activity is not detectable suggests the presence of a minor pathway for cardiolipin formation other than cardiolipin synthase.[3] Phosphatidylserine synthase has been suggested to be responsible for this secondary pathway, since the residual cardiolipin content in *cls* mutants further decreases in the copresence of the *pssA1* allele and increases in the presence of a multicopy plasmid bearing the *pss* gene.[3,9] To prove this hypothesis and examine the essential nature of cardiolipin in *E. coli,* it should be useful to analyze a *pssA1 cls-1* double mutant. It is difficult, however, to construct a *cls*-null *pssA1* double mutant,[3] and *cls pss* double null mutants seem to be inviable.[28]

Construction of pssA1 cls-1 Double Mutant.[9] To construct mutants by P1 phage transduction,[19] it is convenient to use strains NK5151 (*trpB83*::Tn*10*) and NK6024 (*pheA18*::Tn*10*), N. Kleckner's strains obtained through the *E. coli* Genetic Stock Center (Yale University, New Haven, CT). A strain of desired genetic background is first transduced with the *trpB83*::Tn*10* allele by selecting tetracycline resistance, and then the *cls-1* allele is introduced by selecting tryptophan prototrophy followed by examination of cardiolipin content or cardiolipin synthase activity. To this *cls-1* mutant the *pssA1* allele is introduced after the introduction of the *pheA18*::Tn*10* allele. Cotransduction frequencies of *trp-cls* and *pheA-pss* are approximately 30 and 40%, respectively. Phosphatidylserine synthase is assayed for total cell lysates by the CDPdiacylglycerol-dependent incorporation of radiolabeled L-serine into chloroform-soluble material at 30° as described.[29,30]

Characterization of pssA1 cls-1 Double Mutants.[3,9] The phospholipid compositions of various strains with different functional levels of the *pss* gene product are determined by radioactivity measurements of the spots on two-dimensional thin-layer chromatograms for cells labeled uniformly with ^{32}P. Typically, 2 and 8 μCi/ml of [^{32}P]]phosphate are added to M56LP medium and NBY medium, respectively. The cardiolipin content is about 0.2 mol % of the total phospholipids for *cls-1* mutants and is 10-fold less in the double mutants, whereas that of a double mutant harboring a multicopy plasmid carrying the *pss* gene[30] is 0.3–0.5%.

Notes. A *cls*-null *pssA1* double mutant can be constructed by first covering the chromosomal *pssA1* allele with an F′ (F142) that carries the intact *pss* gene and by curing the F′ with acridine orange on introduction of the *cls::kan* allele.[3] This strain is, however, extremely weak.

[28] A. DeChavigny, P. N. Heacock, and W. Dowhan, *J. Biol. Chem.* **266,** 5323 (1991).
[29] A. Ohta and I. Shibuya, *J. Bacteriol.* **132,** 434 (1977).
[30] A. Ohta, K. Waggoner, K. Louie, and W. Dowhan, *J. Biol. Chem.* **256,** 2219 (1981).

Formation of Acid Phospholipid Analogs

When high concentrations of certain sugar alcohols are present together with Mg^{2+} in a broth medium, *E. coli pssA1* (*ts*) mutants grow well at 42° and accumulate analogs of phosphatidylglycerol and cardiolipin in which nonacylated glyceryl moieties are replaced with these sugar alcohols.[11] A genetic analysis has strongly suggested that these reactions are catalyzed by cardiolipin synthase.[11] They are useful for studies to understand the structure–function relationships of acidic phospholipids.

Procedure. Strain SD10 (*pssA1*)[9] is grown in broth medium NBY[11] supplemented with 600 m*M* D-mannitol and 20 m*M* $MgCl_2$ at 42° to late stationary phase. Lipids are extracted and separated as usual.[11] Typically, molar percentages of lipid phosphorus for phosphatidylmannitol, diphosphatidylmannitol, and phosphatidylethanolamine are 34, 60, and 6, respectively. Normal phosphatidylglycerol and cardiolipin are practically absent.

Notes. D-Arabinose, xylitol, erythritol, and L-threitol also give two types of analog lipids when added in the medium in place of D-mannitol, whereas sorbitol, galactitol, and ribitol give only the analogs of phosphatidylglycerol.[11] The assay conditions optimized for cardiolipin synthesis (see above) allow purified cardiolipin synthase preparations to form only small amounts of phosphatidylglycerol and phosphatidylglycerol analogs. Further study is necessary to find the optimum conditions for *in vitro* analog formation.

[39] Mammalian Cardiolipin Biosynthesis

By Michael Schlame and Karl Y. Hostetler

Introduction

Cardiolipin (diphosphatidylglycerol) is a specific component of the mitochondrial inner membrane, and it is the only major mitochondrial phospholipid which is not imported from extramitochondrial sites.[1] It interacts in specific ways with cytochrome oxidase,[2,3] the ADP/ATP carrier protein,[4] and serves as the membrane receptor for mitochondrial

[1] G. Daum, *Biochim. Biophys. Acta* **822,** 1 (1985).
[2] M. Fry and D. E. Green, *Biochem. Biophys. Res. Commun.* **93,** 1238 (1980).
[3] S. B. Vik, G. Georgevich, and R. A. Capaldi, *Proc. Natl. Acad. Sci. U.S.A.* **78,** 1456 (1981).
[4] K. Beyer and M. Klingenberg, *Biochemistry* **24,** 3821 (1985).

creatine phosphokinase.[5,6] Mammalian cardiolipin biosynthesis was shown to occur in rat liver mitochondria by phosphatidyl transfer from CDPdiglyceride to phosphatidylglycerol with the formation of CMP in a reaction which requires a divalent cation such as Mg^{2+}, Mn^{2+}, or Co^{2+}.[7-10] The mammalian reaction mechanism differs from bacterial cardiolipin synthesis where two phosphatidylglycerols are converted to cardiolipin with the liberation of glycerol,[9,11] whereas in yeast mitochondria the synthesis seems to follow the mammalian scheme.[12]

Mammalian cardiolipin synthase has not been purified to homogeneity, although McMurray and Jarvis[13] have reported solubilization of the enzyme. Cardiolipin synthase is a membrane protein, and its solubilization requires detergent.[13] This may be problematic since many detergents inhibit the enzyme activity.[8] In this chapter, we describe methodology for measuring the activity of cardiolipin synthase in mammalian mitochondria including preparation of radiolabeled substrates, detail assay methods, and provide some information regarding solubilization of the enzyme.

Synthesis of Radiolabeled Substrates

CDPdimyristoyl[U-14C]glycerol

Acylation of [U-14C]glycerol 3-Phosphate to [14C]Phosphatidic Acid. Three μmol (0.5 mCi) of disodium [U-14C]glycerol 3-phosphate (New England Nuclear, Boston, MA) is dissolved in pyridine–water (1 : 1, v/v) and converted to the pyridinium salt by elution through a Dowex 50W (pyridinium+) cation-exchange column (3 ml bed volume). The eluate is taken to dryness, lyophilized overnight, and finally dissolved in anhydrous pyridine. As much as 50% of the radioactive material may remain insoluble in pyridine and should be subjected to another Dowex (pyridinium+) treatment. Acylation of [U-14C]glycerol 3-phosphate is performed in 0.7 ml anhydrous pyridine in the presence of 50 μmol of myristic anhydride

[5] M. Müller, R. Moser, D. Cheneval, and E. Carafoli, *J. Biol. Chem.* **260**, 3839 (1985).
[6] M. Schlame and W. Augustin, *Biomed. Biochim. Acta* **44**, 1083 (1985).
[7] K. Y. Hostetler, H. van den Bosch, and L. L. M. van Deenen, *Biochim. Biophys. Acta* **239**, 113 (1971).
[8] K. Y. Hostetler and H. van den Bosch, *Biochim. Biophys. Acta* **260**, 380 (1972).
[9] K. H. Hostetler, H. van den Bosch, and L. L. M. van Deenen, *Biochim. Biophys. Acta* **260**, 507 (1972).
[10] K. Y. Hostetler, J. M. Galesloot, P. Boer, and H. van den Bosch, *Biochim. Biophys. Acta* **380**, 382 (1975).
[11] C. B. Hirschberg and E. P. Kennedy, *Proc. Natl. Acad. Sci. U.S.A.* **69**, 648 (1972).
[12] K. T. Tamai and M. L. Greenberg, *Biochim. Biophys. Acta* **1046**, 214 (1990).
[13] W. C. McMurray and E. C. Jarvis, *Can. J. Biochem.* **58**, 771 (1980).

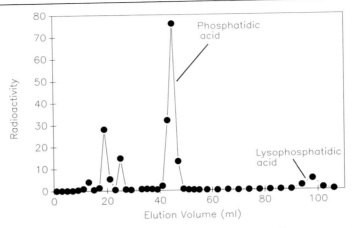

FIG. 1. HPLC purification of [^{14}C]PA after its synthesis from [U-^{14}C]glycerol 3-phosphate. The chloroform extract of the reaction mixture was separated on a Porasil column (10 μm, 7.8 × 300 mm, Waters, Milford, MA) developed with *n*-hexane–2-propanol–58% NH$_4$OH–water (430 : 570 : 15 : 85, by volume) at a flow rate of 2 ml/min. Fractions were collected, and 2-μl aliquots were withdrawn for scintillation counting. Radioactivity is given as disintegrations per minute (dpm) × 10^{-3}.

and 50 μmol of 4-(dimethylamino)pyridine. The reaction mixture is stirred for 2 days in a tightly stoppered tube containing argon atmosphere. Then water is added, and labeled phosphatidic acid ([^{14}C]PA) is extracted into chloroform and purified by high-performance liquid chromatography (HPLC) as shown in Fig. 1 (extent of [U-^{14}C]glycerol 3-phosphate acylation is 70%).

Synthesis of [^{14}C]Phosphatidic Acid Morpholidate. Phosphatidic acid is converted to its morpholidate as described by van Wijk *et al.*[14] [^{14}C]PA is supplemented with 5 μmol of unlabeled dimyristoyl-PA (disodium salt) and dissolved in 0.5 ml chloroform–*tert*-butanol (1 : 1, v/v). Two microliters of water and 5 μl of morpholine are added, and the mixture is stirred and gently refluxed for 2 hr using a Wheaton Microkit for chemical microsynthesis (Wheaton Scientific, Millville, NJ). During refluxing 20 μl of 0.16 M 1,3-dicyclohexylcarbodiimide in *tert*-butanol is added every 10 min. The reaction mixture is dried, dissolved in water, and extracted with chloroform. An aliquot of the extract is checked by high-performance thin-

---

[14] G. M. T. van Wijk, K. Y. Hostetler, and H. van den Bosch, submitted for publication (1991).

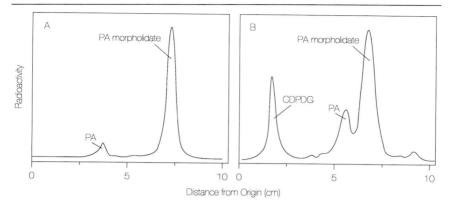

FIG. 2. HPTLC of the reaction products (A) after synthesis of [^{14}C]PA morpholidate and (B) after its further conversion to CDP[^{14}C]DG. The chromatograms were developed on precoated silica 60 HPTLC plates (E. M. Science, Gibbstown, NJ) and scanned by a Radio TLC Scanner (Model RS, Radiomatic Instruments, Tampa, FL). Solvents were chloroform–methanol–58% NH$_4$OH–water (70 : 38 : 4 : 6, by volume) in (A) and chloroform–acetone–methanol–acetic acid–water (10 : 4 : 2 : 2 : 1, by volume) in (B).

layer chromatography (HPTLC) (Fig. 2A, conversion rate 90%). Then the extract is evaporated and lyophilized overnight.

Condensation of CMP and [^{14}C]Phosphatidic Acid Morpholidate. CMP (25 mg, acid form) is poured into a 25-ml round-bottomed flask and suspended in pyridine–methanol–benzene (2 : 1 : 1, v/v), and the solvent is removed under reduced pressure (rotary evaporator) followed by lyophilization overnight. The lyophilized chloroform extract containing the [^{14}C]PA morpholidate is dissolved in 6 ml of dry pyridine and transferred to the round-bottomed flask containing the CMP. Methanol and benzene (0.5 ml each) are added to improve the solubility of CMP, and the mixture is subjected to repeated evaporation (rotary evaporator, 38°) and resolubilization in pyridine until an optimal solubility of all reactants is achieved. Finally, 1.5 ml anhydrous pyridine is added, and the reaction mixture is stirred at 38° for 3 days under an argon atmosphere. After 2 days the solvent is replaced by fresh anhydrous pyridine. The product, CDPdimyristoyl[^{14}C]glycerol, is purified by HPTLC (Fig. 2B, conversion rate 20–25%) and extracted from the silica gel with chloroform–methanol–10 mM HCl (1 : 2 : 0.8, v/v). The condensation of CMP and phosphatidic acid morpholidate was adapted from the procedure originally described by van Wijk *et al.*[14]

Phosphatidyl[U-^{14}C]glycerol

Phosphatidyl[U-^{14}C]glycerol is formed when rat liver[8] or rat lung mitochondria are incubated with CDPDG (CDPdiacylglycerol) and [U-^{14}C]glycerol 3-phosphate. Lung mitochondria have the advantage of higher enzymatic activity for PG (phosphatidylglycerol) formation than liver mitochondria, but the endogenous PG content of lung mitochondria is also higher, thus diluting the specific radioactivity of the product.

Four hundred nanomoles of CDPDG derived from egg yolk phosphatidylcholine (Serdary Laboratories, London, ON, Canada), 100 μCi of [U-^{14}C]glycerol 3-phosphate (specific activity 152 Ci/mol), 30 μmol Tris (pH 7.4), 10 μmol 2-mercaptoethanol, and 17 mg rat lung mitochondrial protein, prepared as previously described,[15] are incubated in a total volume of 1 ml for 2.5 hr at 37°. The lipids are extracted and separated by silicic acid column chromatography developed by a stepwise methanol gradient in chloroform.[16] [^{14}C]PG is eluted at 10% methanol. After this purification step [^{14}C]PG is free of [^{14}C]cardiolipin, the only radiolabeled lipid impurity. To further remove impurities of nonlabeled membrane phospholipids [^{14}C]PG should be chromatographed on precoated HPTLC plates (silica 60, E. M. Science, Gibbstown, NJ) developed with chloroform–methanol–water (65 : 25 : 4, v/v). The yield is routinely about 5 μCi phosphatidyl[U-^{14}C]glycerol. Alternatively ^{32}P-labeled substrates may be prepared using mutant *Escherichia coli* membranes according to the method of Tamai and Greenberg.[12]

Isolation of Mitochondria

Mitochondria should be isolated in a cold (4°) buffer containing 0.25 M sucrose, 5 mM Tris (pH 7.4), and 2 mM EDTA. The tissue is gently homogenized using a power-driven Potter–Elvehjem glass–Teflon homogenizer, and mitochondria are purified from the homogenate by differential centrifugation. Details of the isolation of mitochondria from different tissues are described elsewhere in this series.[17] Mitochondria at a concentration of 20–30 mg protein/ml may be kept frozen at $-70°$ for several months prior to assay of cardiolipin synthase.

[15] M. Schlame, B. Rustow, D. Kunze, H. Rabe, and G. Reichmann, *Biochem. J.* **240,** 247 (1986).
[16] C. C. Sweeley, this series, Vol. 14, p. 255.
[17] J. Nedergaard and B. Cannon, this series, Vol. 55, p. 3.

Assay Method

Solutions

> Buffer: 75 mM Tris, pH 9.0, 0.5 mM EGTA
> CDP[^{14}C]DG stock solution: 30 μM CDPdimyristoyl[U-^{14}C]glycerol (~50 dpm/pmol)
> CDPDG stock solution: 0.5 mM CDPDG (derived from egg yolk phosphatidylcholine)
> [^{14}C]PG stock solution: 80 μM phosphatidyl[U-^{14}C]glycerol (~30 dpm/pmol)
> PG stock solution: 10 mM PG (derived from egg yolk phosphatidylcholine)
> Rat liver mitochondria (15–25 mg protein/ml)
> 50 mM CoCl$_2$

Lipid stock solutions are made by sonication of the dried substance in water (Model W-225R sonicator, Ultrasonics, Inc., Farmingdale, NY, equipped with a microtip, operated at 20–30% of full power, with the tube being placed in an ice bath).

Procedure. The enzyme may be assayed using either radiolabeled PG or CDPDG. To prepare the incubation mixture the following additions are made: (1) 60 μl buffer, (2) 10 μl of CDP[^{14}C]DG (or CDPDG) stock solution, (3) 10 μl of PG (or [^{14}C]PG) stock solution, and (4) 10 μl mitochondria. The reaction is started by addition of 10 μl 50 mM CoCl$_2$, and incubation is performed for 2 hr at 37°. To stop the incubation, 2 ml methanol and 1 ml chloroform are added, and the lipids are extracted according to Bligh and Dyer[18] using 0.1 M HCl instead of water. The extracts are evaporated, redissolved in 0.35 ml chloroform–methanol (2:1, v/v), and 50 μg of unlabeled lipid standards (cardiolipin and either CDPDG or PG) are added in order to improve chromatographic separation and detection. The extracts are applied to activated silica gel 60 H (E. M. Science)/Florisil (49:1, w/w, J. T. Baker, Phillipsburg, NJ) plates (self-prepared) (0.5 mm thickness, 20 × 20 cm) which are marked into lanes and developed with chloroform–methanol–water (65:25:4, v/v). The lipids are stained with iodine vapor, and the cardiolipin spots are scraped into vials for scintillation counting. This TLC method results in a convenient and reproducible separation of cardiolipin (R_f 0.85) from PG (R_f 0.37) or CDPDG (R_f 0.24). The formation of [^{14}C]cardiolipin is linear with respect to incubation time up to 2 hr and with respect to the amount of mitochondrial protein up to 0.25 mg.

The assay of cardiolipin synthase with [^{14}C]CDPDG has several advan-

---

[18] E. G. Bligh and W. J. Dyer, *Can. J. Biochem.* **37**, 911 (1959).

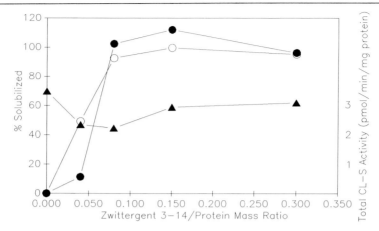

FIG. 3. Solubilization of cardiolipin synthase (CL-S) from rat liver mitochondria using the detergent Zwittergent 3-14. Mitochondria (final protein concentration ~5 mg/ml) were stirred for 1 hr at 4° in a solution containing 60 mM Tris (pH 7.5), 20% (v/v) glycerol, 0.4 mM EGTA, 0.4 M KCl, and varying amounts of Zwittergent 3-14. Solubilization of mitochondrial protein (open circles) and CL-S (filled circles) was quantitated by recovery in the supernatant after ultracentrifugation (1 hr, 140,000 g). Triangles show total CL-S activity (assayed with CDP[^{14}C]DG).

tages: much larger amounts of radioactive substrate can be obtained; the fatty acid composition of the substrate CDPDG can be rigidly controlled in contrast to the isolation of natural PG from mitochondria; and the analysis of CDPDG conversion to cardiolipin by TLC is much easier than TLC separation of PG and cardiolipin, which are close together in most systems.

Properties of Mitochondrial Cardiolipin Synthase

Although CDPDG is the preferred substrate of mitochondrial cardiolipin synthase, the enzyme does not possess an absolute liponucleotide specificity. It shows some activity with deoxyCDPDG, ADPDG, UDPDG,[9,19] as well as with liponucleotide analogs like 2',3'-dideoxycytidine diphosphodiglyceride or 2',3'-dideoxythymidine diphosphodiglyceride,[20] but it is virtually inactive in the presence of GDPDG.[19]

Although cardiolipin synthase is sensitive to detergents,[8] the active enzyme may be solubilized by Miranol H2M[13] or Zwittergent 3-14 (Fig.

[19] B. J. H. M. Poorthuis and K. Y. Hostetler, *Biochim. Biophys. Acta* **431**, 408 (1976).

[20] G. M. T. van Wijk, K. Y. Hostetler, M. Schlame, and H. van den Bosch, *Biochim. Biophys. Acta,* in press (1991).

3). The Zwittergent-solubilized enzyme has an alkaline pH optimum (pH 8–9) and its K_m values are 45 μM and 1.6 μM for phosphatidylglycerol and CDP-diacylglycerol, respectively.[21] The enzyme/Zwittergent micelle has an apparent molecular mass of approximately 60 kDa. Cardiolipin synthase is a strongly acidic protein since it requires high salt concentration to be eluted from Mono-Q anion exchanger.[21] The activity of both membrane-bound and solubilized cardiolipin synthase is dependent on the presence of divalent cations in the relative order $Co^{2+} > Mn^{2+} > Mg^{2+}$, whereas Ba^{2+}, Ca^{2+}, Cu^{2+}, Cd^{2+}, Hg^{2+}, and Zn^{2+} are inhibitory.[10,13] Mammalian cardiolipin synthase is different from the respective yeast enzyme by its cation specificity and its response to Triton X-100.[12]

[21] M. Schlame and K. Y. Hostetler, *J. Biol. Chem.*, in press (1991).

Section VIII

Phospholipid Transformations

[40] Serine–Ethanolamine Base-Exchange Enzyme from Rat Brain

By Julian N. Kanfer

Introduction

The base-exchange enzyme activities are widely distributed in mammalian cell membranes, and their characteristics have been reviewed.[1] These activities do not result in the net synthesis of a phospholipid, rather as the name implies they merely catalyze an exchange of the polar amino alcohol or L-serine of preexisting phospholipids according to the following reaction.

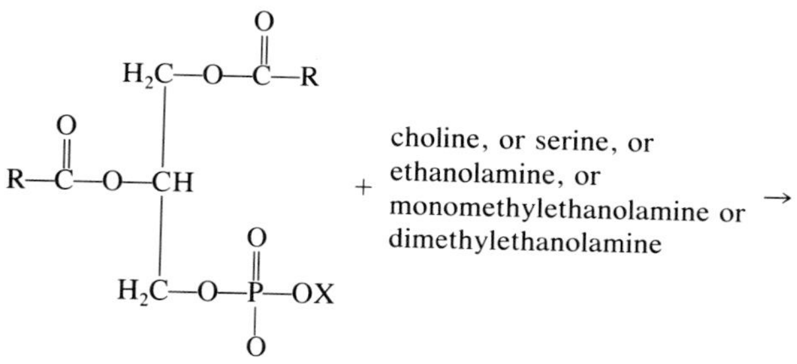

Acceptor phospholipid

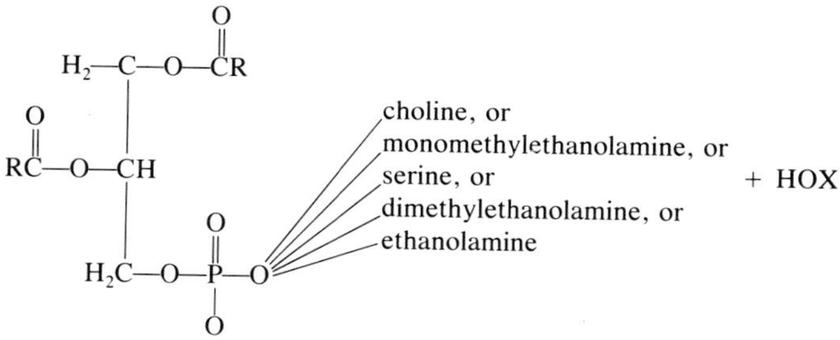

choline, or serine, or ethanolamine, or monomethylethanolamine or dimethylethanolamine

choline, or monomethylethanolamine, or serine, or dimethylethanolamine, or ethanolamine

+ HOX

The *de novo* pathway of phospholipid biosynthesis involving phosphorylated and cytidine nucleotide derivatives is the principal route for formation of the phospholipids possessing choline or ethanolamine or mono-

[1] J. N. Kanfer, *in* "Phosphatidylcholine Metabolism" (D. E. Vance, ed.), p. 65. CRC Press, Boca Raton, Florida, 1989.

methylethanolamine or dimethylethanolamine as the characteristic polar head groups. However, the serine base-exchange enzyme is the sole mechanism available for phosphatidylserine formation by mammalian tissues, and the partial purification of the enzyme was described in an earlier volume of this series.[2]

The assays for the base-exchange activities depend on solubility differences of the precursor amino alcohol employed as substrate and the product phospholipid corresponding to the given amino alcohol. The most frequently used approach based on this physical difference is a partitioning of an aqueous and an organic solution. Utilizing a radioactive water-soluble precursor which is quantitatively retained in the aqueous phase, the amount of radioactivity appearing in the organic phase is an accurate reflection of the base-exchange enzyme activity of an biological sample. Confirmation of the reliability of this approach is obtained from thin-layer chromatographic techniques demonstrating cochromatography of the expected product with authentic standards. Equally reliable results can be obtained by estimating the quantity of radioactivity present in a trichloroacetic acid (TCA) precipitate of an incubation mixture. Under these conditions the radioactive phospholipid produced by the base-exchange enzyme is coprecipitated with the proteins, and the water-soluble substrate is retained in an aqueous (TCA) phase. It is unnecessary to include an exogenous phospholipid acceptor in incubation mixtures containing membranes as the enzyme source because these contain adequate endogenous acceptor phospholipid substrates.

The notable features of the base-exchange enzyme activities are a stimulation by Ca^{2+}, a somewhat alkaline pH optimum, and a lack of requirement for an energy source.

Solubilization and Separation of Ethanolamine–Serine Base-Exchange Enzyme

The starting material is a rat brain particulate fraction enriched in microsomes.

Reagents

Solution A: 0.3% sodium cholate (w/v), 5 mM HEPES, pH 7.5, 1 mM 2-mercaptoethanol, 0.5% Miranol H2M (w/v), 20% glycerol
Solution B: 5 mM HEPES, pH 7.23, 20% glycerol, 1 mM EDTA, 1 mM 2-mercaptoethanol

[2] T. Miura, T. Taki, and J. N. Kanfer, this series, Vol. 71, p. 588.

Solution C: 5 mM HEPES, pH 7.23, 20% glycerol, 1 mM EDTA, 1 mM 2-mercaptoethanol, 1 M NaCl

Solution D: 5 mM HEPES, pH 7.23, 20% glycerol, 1 mM EDTA, 1 mM 2-mercaptoethanol, 0.1% Triton X-100

Solution E: 5 mM HEPES, pH 7.23, 20% glycerol, 1 mM EDTA, 1 mM 2-mercaptoethanol, 0.4% Triton X-100

Solution F: 5 mM HEPES, pH 7.23, 1 mM mercaptoethanol

Solution G: 1.5% polyacrylamide, 5% agarose containing 0.35 M Tris, pH 8.8, 1 mM EDTA, 1 mM mercaptoethanol, 20% glycerol, 0.01% N,N,N',N'-tetramethylenediamine, 0.014% ammonium persulfate

Solution H: 0.38 M glycine, 50 mM Tris, pH 8.3, 1 mM EDTA, 20% glycerol

[³H]Ethanolamine hydrochloride

L-[¹⁴C]Serine

Asolectin (soya phospholipids)

Nitrocellulose filter membranes HA, 0.45 μm, 2.5 cm

Trichloroacetic acid, 1 and 5%

Bovine serum albumin

CaCl$_2$

HEPES, pH 7.23

2-Mercaptoethanol

Sepharose CL-4B

Phenyl-Sepharose 4B

Glycerol

Procedure

The isolation of rat brain microsomes and solubilization of the base-exchange enzyme in solution A are carried out as previously described.[2]

Gel Filtration with Sepharose CL-4B. The solubilized microsomal extract, prepared from 25 one-month-old animals, is loaded onto a Sepharose CL-4B column (20 × 100 cm) preequilibrated with solution B. The column is eluted with solution B, and 10-ml fractions are collected. A typical profile is shown in Fig. 1. Tubes 61 to 71 are pooled, and sufficient solid NaCl is added to a final concentration of 1 M.

Phenyl-Sepharose 4B Column Chromatography. The sample from the Sepharose CL-4B column is applied to a phenyl-Sepharose column (25 ml bed volume) preequilibrated with solution C. The column is washed successively with 3 bed volumes of solution B, 2.5 bed volumes of solution D, and 3 bed volumes of solution E. The base-exchange enzyme activity is present in the fractions obtained with solutions D and E (Fig. 2). The

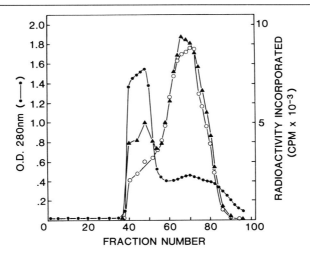

FIG. 1. Gel filtration of a solubilized microsomal extract containing the base-exchange enzymes on a Sepharose CL-4B column. Aliquots of 40 μl of each fraction were used for detecting serine (▲) and ethanolamine (○) base-exchange activities. Protein was measured by absorbance at 280 nm (●). Activity is expressed as counts incorporated per 18 min/40-μl sample into phospholipid from the radioactive substrates.

fractions containing the bulk of the activity, tubes 56 to 67, are pooled and concentrated with an Amicon (Danvers, MA) PM30 membrane.

Glycerol Gradient Sedimentation. Glycerol gradient sedimentation is performed by a slight modification of a published procedure.[3] A 2-ml sample from the phenyl-Sepharose column is layered onto a gradient composed of 2 ml of 36% glycerol in solution F, 3.3 ml of 34% glycerol in solution F, 2.2 ml of 30% glycerol in solution F, 1.4 ml of 26% glycerol in solution F, 1.3 ml of 21% glycerol in solution F, and 0.8 ml of 15% glycerol in solution F. Centrifugation is carried out at 30,000 rpm for 72 hr at 3° in a Beckman SW 40.1 Ti rotor, and 0.5-ml aliquots are withdrawn from the bottom of the centrifuge tubes. Results are shown in Fig. 3. The fractions with greatest activities, tubes 16–19, are pooled and concentrated with an Amicon PM30 membrane.

Nondenaturing Gel Electrophoresis. The concentrated enzyme solution, 60 μg protein in a 200 μl volume, is applied to a nondenaturing gel of solution G that had been prerun in solution H for 1 hr at 2 mA. The sample is subjected to 2 mA/tube for 2 hr and the gels sliced into 5-mm sections. The identical sections from each gel are pooled and dissociated

[3] M. K. Brakke and N. V. Pelt, *Anal. Biochem.* **38,** 56 (1970).

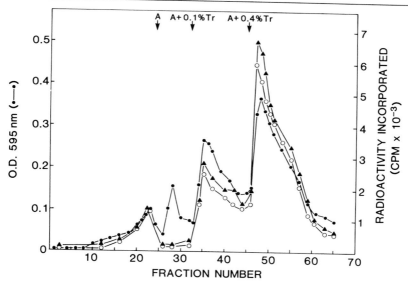

Fig. 2. Phenyl-Sepharose 4B column chromatography. The pooled Sepharose CL-4B fraction was adjusted to 1 M NaCl and then placed on the phenyl-Sepharose column, and elution was carried out with solution B, solution D, and solution E. The arrows indicate the change of solutions. Protein was measured by absorbance at 595 nm using a Coomassie blue staining assay (●), and 40-μl aliquots of each fraction were assayed for serine (▲) and ethanolamine (○) exchange enzyme activities.

by forcing through a 20-gauge needle, the suspension is dialyzed in the cold against solution B for 36 hr, and the eluted enzyme is recovered in a 100,000 g supernatant. The activities for the ethanolamine base-exchange enzyme and the serine base-exchange enzyme are given in Table I. There was on a single band detectable by Coomassie Brillant blue staining of polyacrylamide gels of the most highly purified enzyme preparation.

Assay Procedure

The incubation contains an asolectin microdispersion having 25 μg phosphate, either 20.2 nmol L-[^{14}C]serine (specific activity 30 nmol/μCi) or 7.2 nmol [^{3}H]ethanolamine (specific activity 40 nmol/μCi), 2 μmol CaCl$_2$, 10 μmol HEPES, pH 7.23, and enzyme preparation in a total volume of 240 μl. The incubation is carried out at 37° for 18 min and terminated with 1 ml of 1% TCA. The contents of each tube are transferred to a nitrocellulose membrane, and the membrane is washed with 25 ml

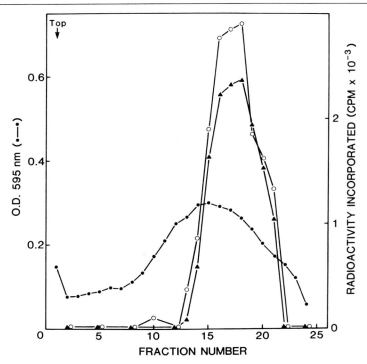

Fig. 3. Glycerol gradient sedimentation. Two-milliliter aliquots of the concentrated phe-nyl-Sepharose 4B fraction were layered onto the glycerol gradient and centrifuged at 30,000 rpm for 72 hr at 3°. Protein was measured by absorbance at 595 nm using a Coomassie blue staining assay (●). The incorporations of serine (▲) and ethanolamine (○) into phospholipids were measured.

chilled 5% TCA. Radioactivity retained on the dried membrane represent-ing product formation is determined by scintillation counting.

Properties of Purified Enzyme

The activity for the incorporation of both L-serine and ethanolamine was linear with time up to 90 min of incubation and with increasing protein up to 2 μg. The pH optimum with both substrates was 7.0, and Ca^{2+} was optimum at 8 to 10 mM. The K_m for L-serine was 0.11 mM, and for ethanolamine the K_m was 20 μM (Fig. 4). These were also found to be reciprocal competitive inhibitors of each other's incorporations, with K_i values nearly identical to their respective K_m values. The V_{max} value for L-serine was 330 nmol/mg/hr, and for ethanolamine the V_{max} was 40 nmol/

TABLE I
PURIFICATION OF A SERINE–ETHANOLAMINE BASE-EXCHANGE ENZYME

Step	Specific activity (nmol/mg protein/hr)		Recovery	
	Serine	Ethanolamine	Serine	Ethanolamine
Microsomes	10.9	15.4	100	100
Solubilized enzyme	5.6 (0.5)[a]	5.3 (0.34)	38	23
Sepharose 4B	150.2 (13.8)	121.5 (7.9)	139	80
Phenyl-Sepharose	131 (13.8)	107 (6.95)	34	20
Glycerol gradient	170 (15.6)	239 (15.5)	8.6	8.9
Agarose gel electrophoresis	240 (22.0)	406 (26.4)	1.9	2.3

[a] Values in parentheses are purification (-fold) compared to the microsomes.

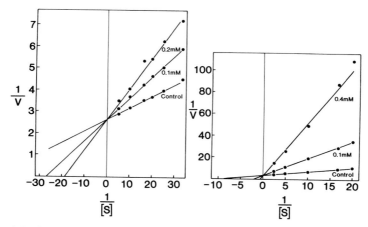

FIG. 4. (*Left*) Lineweaver–Burk plots of the effect of nonradioactive serine on the incorporation of ethanolamine by the purified enzyme preparation. The incubations were carried out with 0.4 μg of purified enzyme, varying concentrations of radioactive ethanolamine, and either 0.1 or 0.2 mM nonradioactive serine for 60 min. [S] is millimolar ethanolamine, and V is nanomoles incorporated per mg/hr. (*Right*) Lineweaver–Burk plots of the effect of nonradioactive ethanolamine on the incorporation of radioactive serine by the purified enzyme preparation. The incubations were carried out with 0.4 μg of purified protein, varying concentrations of radioactive serine, and either 0.1 or 0.4 mM nonradioactive ethanolamine for 60 min. [S] is millimolar serine, and V is nanomoles incorporated per mg/hr.

mg/hr. Pure phosphatidylethanolamine and asolectin, a commercial heterogeneous mixture of soybean phospholipids, were the best acceptors for both the L-serine and ethanolamine incorporations. Phosphatidylinositol, phosphatidylcholine, phosphatidylserine, phosphatidylglycerol, and lysophosphatidylcholine were inactive. There was no detectable choline base-exchange activity or phospholipase D activity in the purified enzyme. Details can be obtained in the original description of this activity.[4]

[4] T. T. Suzuki and J. N. Kanfer, *J. Biol. Chem.* **260**, 1394 (1985).

[41] Phosphatidylserine Decarboxylase from *Escherichia coli*

By WILLIAN DOWHAN and QIAO-XIN LI

Introduction

Phosphatidylserine → phosphatidylethanolamine + CO_2

The phosphatidylserine decarboxylase (EC 4.1.1.65, phosphatidyl-L-serine carboxy-lyase) of *Escherichia coli* is associated with the cytoplasmic membrane and requires nonionic detergent for solubilization and detection of enzymatic activity.[1] The enzyme catalyzes decarboxylation through formation of a Schiff base between the amine of the substrate and the carbonyl residue of its pyruvate prosthetic group.[2] The prosthetic group is covalently bound in an amide linkage to the amino terminus of the smaller (α subunit) of the two subunits of the enzyme.[3]

Assay Methods

The most widely used assay[1] (described below) for the phosphatidylserine decarboxylase measures the release of $^{14}CO_2$ from phosphatidyl-L-[1-^{14}C]serine. The radiolabeled substrate, which is not commercially available, can be synthesized from CDPdiacylglycerol and DL-[1-^{14}C]serine using the *E. coli* phosphatidylserine synthase.[4] The release of CO_2 from unlabeled substrate can also be followed by gas chromatography.[5] Finally, the conversion of radiolabeled phosphatidylserine, which is available com-

[1] W. Dowhan, W. T. Wickner, and E. P. Kennedy, *J. Biol. Chem.* **249**, 3079 (1974).
[2] M. Satre and E. Kennedy, *J. Biol. Chem.* **253**, 479 (1978).
[3] Q.-X. Li and W. Dowhan, *J. Biol. Chem.* **263**, 11516 (1988).
[4] W. Dowhan, this volume [34].
[5] T. G. Warner and E. A. Dennis, *J. Biol. Chem.* **250**, 8004 (1975).

mercially, to radiolabeled phosphatidylethanolamine has been used to follow the activity of the enzyme from yeast.[6]

Synthesis of Phosphatidyl-L-[1-^{14}C]serine. The synthesis is carried out in 1 ml of 0.1 M potassium phosphate (pH 7.4) containing 0.1% Triton X-100, 1 mg/ml bovine serum albumin (BSA), 0.45 mM CDPdiacylglycerol, 1 mM DL-[1-^{14}C]serine (only the L isomer is recognized by the synthase), and 2 units of *E. coli* phosphatidylserine synthase[4]; the radiolabeled serine is usually not diluted with unlabeled carrier. The mixture is incubated at 30° for 30 min and the reaction stopped by the addition of 4 ml of chloroform–methanol (1 : 2, v/v) followed by 2 ml of 0.5 M NaCl (0.1 N in HCl) and 2 ml of chloroform. After mixing, the phases are separated by centrifugation, and the chloroform phase is taken to dryness. The radiolabeled product is stored in chloroform at −20°. The phosphatidyl-L-[1-^{14}C]serine is diluted to the desired specific activity using any naturally occurring unlabeled phosphatidylserine. The lipids should be dispersed in water (adjusted to pH 7–8 with a weak base) by brief sonication.

Procedure. If necessary, enzyme for assay is diluted into 0.1 M potassium phosphate buffer (pH 7.0) containing 0.2% Triton X-100 and 1 mg/ml BSA. The assay is carried out in 25-ml Erlenmeyer flasks with center wells containing Whatman (Clifton, NJ) No. 1 filter papers (2 × 3 cm) and 50 μl of 2 N KOH. The assay mixture (final concentration) consists of 0.2% Triton X-100, 0.2 M potassium phosphate (pH 7.0), 0.15 mM phosphatidyl-L-[1-^{14}C]serine (60–100 μCi/mmol), 1 mg/ml BSA, and enzyme in a final volume of 0.5 ml. The flasks are stoppered with a rubber serum cap after addition of the enzyme and 0.1 ml each of 1 M potassium phosphate (pH 7.0), BSA (5 mg/ml), and 5% Triton X-100 to the area surrounding the center well; water is added to bring the volume to 0.4 ml. The reaction is initiated at 37° by the addition of 0.1 ml of 0.75 mM phosphatidyl-L-[1-^{14}C]serine through the rubber stopper using a syringe and is terminated after 10–30 min by the addition of 1 ml of 1 N HCl using a syringe. After an additional 20 min the KOH-impregnated filter papers are removed and counted for $^{14}CO_2$ in 10 ml of water-miscible scintillation solution containing 1 ml of water. Maximum efficiency of detection of radiolabel is achieved after 6 hr of incubation in the scintillation solution, but rough estimates of relative levels within a given set of assays can be obtained about 30 min after transfer of the filter papers. Nonspecific decarboxylation is accounted for by employing a nonenzyme control. The release of $^{14}CO_2$ is linear up to about 40% of total conversion to product.

Units. One unit is defined as the amount of enzyme which catalyzes the release of 1 μmol of $^{14}CO_2$ in 1 min under the above assay conditions.

[6] M. A. Carson, M. Emala, P. Hogsten, and C. J. Waechter, *J. Biol. Chem.* **259**, 6267 (1984).

Purification Procedure

Cell lysates of *E. coli* strain JA200, carrying the structural gene (*psd*) for the phosphatidylserine decarboxylase on the multicopy number plasmid pPSD2b (*psd*[+] *tet*[R] ColE1[ori]), has a specific activity for the enzyme which is 25-fold higher than that reported for wild-type strains of *E. coli*. Li and Dowhan[3] used this overproducing strain and a modification of a previously reported purification scheme[1] to obtain purified enzyme. The latter strain is grown at 37° with aeration to late log phase in LB medium (10 g/liter Bacto-tryptone, 5 g/liter Bacto-yeast extract, 5 g/liter NaCl, and 25 μg/ml tetracycline) and harvested by centrifugation. The cell paste can be stored at −80° or used immediately. In the following purification (see Table I) Steps 1–3 are carried out at 4°. The remainder of the steps are carried out at room temperature, but the enzyme is placed on ice as soon as possible. Mixing of large volumes and suspension of pellets is carried out with the aid of a Waring blender controlled by an external rheostat.

Step 1: Cell-Free Extract. Cell paste (250 g) of strain JA200/pPSD2b is suspended in 400 ml of buffer containing 0.1 M potassium phosphate (pH 6.8), 5 mM MgSO$_4$, and 10 mM 2-mercapoethanol. The suspension is passed through a French pressure cell at 1000 psi to break the cells; cells can also be broken using sonication.

Step 2: Membrane Extract. The cell lysate is centrifuged at 100,000 g_{av} for 5 hr. The membrane pellet is suspended in 250 ml of 0.1 M potassium phosphate (pH 7.2), 5% Triton X-100, and 10 mM 2-mercaptoethanol and incubated on ice for 12 hr. Insoluble material is removed from the membrane extract by centrifugation for 90 min at 100,000 g_{av}.

Step 3: Acetone Precipitation. Glycerol is added to the membrane

TABLE I
PURIFICATION OF PHOSPHATIDYLSERINE DECARBOXYLASE[a]

Step	Total protein	Specific activity (units/mg)	Yield (%)
1. Cell-free extract	24 g	0.69	100
2. Membrane extract	7.2 g	1.2	57
3. Acetone precipitation	7.2 g	1.1	52
4. DEAE-cellulose	1.9 g	5.2	58
5. Sephadex G-150	200 mg	35	43
6. QAE-Sephadex	70 mg	55	23

[a] Starting with 250 g of cell paste from *E. coli* strain JA200/pPSD2b [Q.-X. Li and W. Dowhan, unpublished data (1987)].

extract to a final concentration of 20% (w/v), and the pH is adjusted to 5.2 with 0.5 N acetic acid. Acetone ($-20°$) is rapidly added with stirring to the supernatant to a final concentration of 70% (v/v). The solution is immediately centrifuged at 14,000 g_{av} for 10 min. The pellet is suspended by stirring for 15 min in 750 ml of buffer containing 10 mM sodium acetate (pH 5.0), 1% Triton X-100, 10% (w/v) glycerol, and 10 mM 2-mercaptoethanol. Insoluble material is removed by centrifugation at 14,000 g_{av} for 20 min, and the supernatant is adjusted to pH 7.2 with 1.5 M Tris free base.

Step 4: DEAE-cellulose Chromatography. The supernatant from Step 3 is applied to a 5 × 25 cm Whatman DE-52 column equilibrated in buffer A [10 mM potassium phosphate (pH 7.2), 1% Triton X-100, and 10 mM 2-mercaptoethanol] at 250 ml/hr. The column is washed with 150 ml of buffer A and developed at 250 ml/hr with a 3-liter linear gradient from 0 to 0.6 M NaCl in buffer A. The enzyme elutes in a broad peak between 0.3 and 0.4 M NaCl.

The enzyme is concentrated using step elution from a second DE-52 column. First the pooled peak of activity from the first column is diluted with 2.5 volumes of water and passed at 600 ml/hr through a 2.5 × 4 cm DE-52 column equilibrated in buffer A. The enzyme is then slowly eluted from the second column in less than 5% of the volume of the pooled peak using buffer B (buffer A at 0.1% Triton X-100) containing 0.6 M NaCl.

Step 5: Sephadex G-150 Gel Filtration. The concentrated enzyme solution is applied to and eluted at 25 ml/hr from a 2.5 × 50 cm Sephadex G-150 column in buffer B containing 50 mM NaCl. The enzymatic activity emerges at the void volume of this column.

Step 6: QAE-Sephadex Chromatography. The pooled peak of enzymatic activity is adjusted to pH 6.0 with 0.5 N acetic acid and applied at 50 ml/hr to a 2.5 × 40 cm QAE-Sephadex (A-25) column equilibrated with buffer C [20 mM potassium phosphate (pH 6.0), 0.5% (w/v) Triton X-100, 10% glycerol, and 10 mM 2-mercaptoethanol] containing 50 mM NaCl. The column is developed at the same flow rate with a linear gradient of 50 mM to 0.5 M NaCl in buffer C. The enzyme emerges in a peak centered around 0.3 M NaCl. The pooled peak of enzymatic activity is concentrated as described for Step 4 above except the column size is reduced to 1.4 × 2.5 cm.

Step 7: High-Performance Liquid Chromatography. Utilization of the additional purification steps (sucrose density gradient centrifugation and 4% agarose gel filtration) previously reported for purification of the enzyme from nonoverproducing strains[1] does not significantly increase the degree of purification (85%) or specific activity (55 units/mg) over that previously reported. The preparation at this point is sufficiently pure for many enzymological studies. Material of higher purity for protein chemistry work or immunological studies can be obtained by subjecting the material after Step 6 to high-performance liquid chromatography (HPLC) under nondenaturing conditions.[3] The concentrated enzyme (~2 mg at 1 mg/ml) is adjusted to 2% in Triton X-100 and equilibrated by dialysis with buffer D [20 mM Tris-HCl (pH 7.4), 2% Triton X-100, 10% glycerol, and 10 mM 2-mercaptoethanol]; a large excess of Triton X-100 over protein is necessary to achieve further purification. The sample is applied (1 ml/min) to a Waters (Milford, MA) PROTEIN PAK DEAE-5PW column (7.5 × 75 mm) equilibrated with buffer D. The column is developed at the same flow rate with buffer D for 5 min, a linear increase of NaCl from 0 to 0.6 M in buffer D over 50 min, and then 5 min of buffer D containing 0.6 M NaCl. The enzyme elutes near 0.3 M NaCl in about 60% yield. Sodium dodecyl sulfate-polyacrylamide gel electrophoresis is used to determine the most appropriate samples to pool.

Purity. As judged by electrophoresis in 12.5% polyacrylamine gels in the presence of 6 M urea and 0.2% sodium dodecyl sulfate, the preparation after the HPLC step is over 95% homogeneous[3]; no specific activity data have been reported on this material. The enzyme is composed of two subunits. Under the above electrophoresis conditions the α and β subunits have M_r values of 15,000 and 28,000, respectively. The mobility of both the α subunit (M_r between 7500 and 16,000) and the β subunit (M_r between 28,000 and 35,000) is dependent on the electrophoresis conditions. The α subunit is poorly detected with both silver and Coomassie blue R-250 staining.

Properties

Stability. The phosphatidylserine decarboxylase is most stable in the pH range from 6.5 to 7.5 (either 0.1 M potassium phosphate or Tris-HCl buffers are routinely used) with 10% glycerol and a minimum concentration (above the critical micelle concentration) of a nonionic detergent (0.1% Triton X-100 or 25 mM octylglucoside) for dilute protein solutions (less than 100 μg/ml) and a proportionately higher amount of detergent for more concentrated protein solutions.[1,3] Under these conditions the enzyme is stable for several years at $-20°$ to $-80°$, several months at $0°$, and at least a week at room temperature. The membrane-associated enzyme loses only half its activity after incubation at $65°$ for 10 min but is much more labile after solubilization with Triton X-100, losing all activity in 2 min at $65°$.

Inhibitors. Because catalysis is mediated by the carbonyl moiety of the pyruvate prosthetic group,[2,3] the enzyme is sensitive to a variety of carbonyl reagents such as $NaBH_4$, $NaCNBH_3$ (in the presence of an amine), hydroxylamine, and hydrazines; the presence of Triton X-100 is necessary to assure that inactivation is a direct result of modification of the prosthetic group. At pH 8.0 the enzyme is rapidly inactivated by 10 mM NaBH$_4$ but is stable in the presence of NaCNBH$_3$. In the presence of the latter reducing agent and either 0.5 M ammonium phosphate (pH 7.6) or 0.1 mM phosphatidylserine, the enzyme is inactivated with either conversion of the prosthetic group to alanine or reduction of the Schiff base between the keto moiety of pyruvate and the α-amine group of the substrate, respectively.[3]

Subunit Structure. The two subunits of the enzyme can be separated by HPLC under denaturing conditions on a Vydac C$_{18}$ column (2.5 × 250 mm) at room temperature.[3] An enzyme preparation (2 mg) after Step 6 is precipitated by adjusting the solution to 10% (w/v) in trichloroacetic acid, washed with acetone to remove Triton X-100, dissolved in 70% formic acid, and applied to the column. The column is developed (1 ml/min) with a two-stage linear gradient of acetonitrile in 1% aqueous trifluoroacetic acid. In the first stage acetonitrile is increased from 0 to 45% in 10 min followed by an increase to 90% in 30 min. Protein is monitored by absorbance at 216 nm. The α subunit elutes at 55.5% acetonitrile followed by the β subunit at 66% acetonitrile.

Based on protein and DNA sequence information[3] and studies on the synthesis of phosphatidylserine decarboxylase both *in vivo* and *in vitro*,[7] the enzyme is first made (Fig. 1) as a 35,893-Da proenzyme (π subunit) which is processed to a 7332-Da α subunit and a 28,579-Da β subunit in a posttranslation event which results in conversion of Ser-254 to the covalently bound pyruvate prosthetic group; the amino-terminal methionine of the β subunit is blocked presumably by a formyl moiety. The subunits in the purified enzyme are in a 1 : 1 molar ratio. The quaternary structure of the enzyme is not known, but based on gel filtration and sucrose gradient sedimentation data in the presence and absence of detergent (see below), the native enzyme is composed of a multimer of the heterodimer.[1]

The predicted amino acid composition[3] (Table II), which is in good agreement with the determined composition, revealed no unusual properties for the β subunit, but it showed a lack of cysteine, methionine, tyrosine, tryptophan, and arginine in the α subunit. A hydrophobicity analysis of the linear sequence reveals one major hydrophobic domain (amino acid residues 180–205) within the β subunit which may be the point of membrane association for the enzyme; a minor hydrophobic domain is centered around residue 120, and the amino-terminal end is neither strongly hydrophobic nor hydrophilic. The domain surrounding the processing site of the proenzyme (Ser-254) displays hydrophobic character, suggesting that

[7] Q.-X. Li and W. Dowhan, *J. Biol. Chem.* **265**, 4111 (1990).

```
fMLNSFKLSLQYILPKLWLTRLAGWGASKRA      30
GWLTKLVIDLFVKYYKVDMKEAQKPDTASY       60
RTFNEFFVRPLRDEVRPIDTDPNVLVMPAD       90
GVISQLGKIEEDKILQAKGHNYSLEALLAG      120
NYLMADLFRNGTFVTTYLSPRDYHRVHMPC      150
NGILREMIYVPGDLFSVNHLTAQNVPNLFA      180
RNERVICLFDTEFGPMAQILVGATIVGSIE      210
TVWAGTITPPREGIIKRWTWPAGENDGSVA      240
LLKGQEMGRFKLGSTVINLFAPGKVNLVEQ      270
LESLSVTKIGQPLAVSTETFVTPDAEPAPL      300
PAEEIEAEHDASPLVDDKKDQV              322
```

π Subunit

```
fMLNSFKLSLQYILPKLWLTRLAGWGASKRA      30        Prv-TVINLFAPGKVNLVEQLESLSVTKIGQPL      30
GWLTKLVIDLFVKYYKVDMKEAQKPDTASY       60            AVSTETFVTPDAEPAPLPAEEIEAEHDASP      60
RTFNEFFVRPLRDEVRPIDTDPNVLVMPAD       90                 LVDDKKDQV                      69
GVISQLGKIEEDKILQAKGHNYSLEALLAG      120
NYLMADLFRNGTFVTTYLSPRDYHRVHMPC      150                α Subunit
NGILREMIYVPGDLFSVNHLTAQNVPNLFA      180
RNERVICLFDTEFGPMAQILVGATIVGSIE      210
TVWAGTITPPREGIIKRWTWPAGENDGSVA      240
LLKGQEMGRFKLG                       253
```

β Subunit

Fig. 1. Amino acid sequence of prophosphatidylserine decarboxylase (π) and the mature subunits (α and β) formed by posttranslational cleavage between Gly-253 and Ser-254 resulting in conversion of Ser-254 (bold type and underlined) of the π subunit to the amino-terminal pyruvate (Prv) of the α subunit. The π and β subunits are blocked at their amino termini by formylmethionine (fM). [Summarized from Q.-X. Li and W. Dowhan, *J. Biol. Chem.* **263**, 11516 (1988).]

it is buried either in the membrane or in the interior of the protein. The α subunit has a strong hydrophilic character.

Physical Properties. The physical properties of the enzyme are dependent on the aggregation state of the enzyme and the presence of detergents.[1] Most studies have been carried out using Triton X-100 (critical micelle concentration[5] 0.02% or 0.32 mM) although octylglucoside can substitute provided it is used above its critical micelle concentration[8] (25

[8] R. Rosevear, T. VanAken, J. Baxter, and S. Ferguson-Miller, *Biochemistry* **19**, 4108 (1980).

TABLE II
PREDICTED AMINO ACID COMPOSITION OF
PHOSPHATIDYLSERINE DECARBOXYLASE
SUBUNITS[a]

Amino acid	Number of residues/subunit		
	α	β	π
Asp	5	13	18
Asn	2	12	14
Thr	5	15	20
Ser	4[b]	10	15[b]
Glu	8	13	21
Gln	3	7	10
Pro	7	14	21
Gly	2	20	22
Ala	7	17	24
Cys	0	2	2
Val	8	18	26
Met	0	8	8
Ile	3	15	18
Leu	7	29	36
Tyr	0	9	9
Phe	2	12	14
His	1	4	5
Lys	4	14	18
Trp	0	6	6
Arg	0	15	15
Pyr	1[b]	0	0
Total	69	253	322

[a] Summarized from Q.-X. Li and W. Dowhan, *J. Biol. Chem.* **263**, 11516 (1988).
[b] α Subunit corrected for conversion of Ser-254 of the π subunit to pyruvate.

mM). The enzyme sediments in sucrose density gradients as a distinct molecular species (M_r 170,000) in the absence of Triton X-100 with full recovery of activity, suggesting a multimer of at least four heterodimers for the enzyme in the absence of detergent; gel filtration analysis in the absence of detergent to support this conclusion has not been reported. Sedimentation in the presence of increasing amounts of Triton X-100 (in large molar excess over the amount of enzyme) shifts the sedimentation properties to that of a molecule of M_r 75,000; gel filtration under similar conditions suggests an M_r over 200,000. These changes in the physical properties of the enzyme are similar to that seen for other membrane proteins which associate with detergent micelles.[9,10] Therefore, the phosphatidylserine decarboxylase heterodimers appear to form a defined multimeric structure in the absence of detergent and associate with detergent micelles to form a large detergent micelle–protein complex which is still made up of some multiple of the heterodimer. Therefore, in working with the enzyme, especially during purification, it is important to maintain detergent concentrations above the critical micelle concentration and in sufficient excess to accommodate all of the protein at one molecule or less per micelle.[1,3]

Activity and Detergent Dependence. The enzyme is highly specific for the naturally occurring isomers of phosphatidylserine although there appears to be little preference for the nature of the acyl chain length or the degree of saturation.[1] The major requirements for catalysis are association of the substrate with a hydrophobic–hydrophilic interface (a membrane *in vivo* or a nonionic detergent micelle *in vitro*), a fully acylated *sn*-glycero-3-phosphate backbone, and most likely the L-serine isomer of phosphatidylserine. Precaution must be taken in interpreting results from experiments in which the enzyme is used as a reagent for decarboxylating phosphatidylserine associated with natural membranes because the bulk of the evidence suggests that under *in vitro* conditions exogenously added enzyme is inactive against membrane-associated substrate and only decarboxylates substrate which is in association with detergent micelles.[5] Lysophosphatidylserine and the *sn*-glycerol 1-phosphate isomer of the backbone are neither substrates nor inhibitors.[1] Because replacement of the leaving CO_2 moiety with a proton occurs with retention of configuration around the α carbon of serine,[11] it is unlikely that the D-serine isomer of phosphatidylserine would be a substrate.

The absolute dependence on nonionic detergent (ionic detergents even in the presence of nonionic detergents are inhibitory) for enzymatic activity *in vitro* is a function of the critical micelle concentration of the detergent and the mole fraction of the substrate in the detergent micelles.[5] Although a linear relationship between enzyme and rate of decarboxylation is found under the standard assay conditions, optimal activity is observed with 6 mM phosphatidylserine at a molar ratio of Triton X-100 to phosphatidylserine of 6 : 1. Deviating either above or below this ratio at other fixed phosphatidylserine concentrations results in a reduction in the apparent catalytic efficiency of the enzyme. Below this ratio the substrate is no longer homogeneously dispersed in mixed micelles with Triton X-100 but exists

[9] A. Helenius and K. Simons, *J. Biol. Chem.* **247,** 3656 (1972).
[10] S. Clarke, *J. Biol. Chem.* **250,** 5459 (1975).
[11] Z. No, C. R. Sanders II, W. Dowhan, and M.-D. Tsai, *Bioorg. Chem.* **16,** 184 (1988).

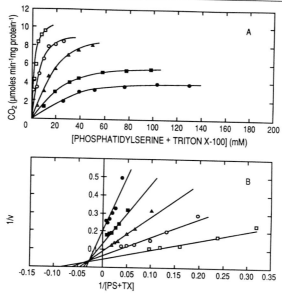

FIG. 2. (A) Dependence of phosphatidylserine decarboxylase on the sum of the phosphatidylserine (PS) and Triton X-100 (TX) concentrations at varying molar ratios of Triton to phosphatidylserine: (□) 6 : 1, (○) 16 : 1, (▲) 32 : 1, (■) 64 : 1, and (●) 128 : 1. The phosphatidylserine concentrations employed were similar at all molar ratios except at 128 : 1 where the highest concentration was 1.0 mM rather than 1.5 mM. (B) Lineweaver–Burk representation of each plot. [From T. G. Warner and E. A. Dennis, *J. Biol. Chem.* **250**, 8004 (1975).]

in an additional form of increased size (presumably liposomes containing detergent) which either is not kinetically competent or is less kinetically competent. Although increasing this ratio is also inhibitory, the apparent reduction in V_{max} appears to be related to a reduction in the mole fraction of the phosphatidylserine on the surface of the mixed micelles, which is the kinetically competent substrate. As shown in Fig. 2, the enzyme displays Michaelis–Menten saturation kinetics if the total concentration of substrate plus detergent at a fixed mole fraction of substrate (with respect to detergent) is varied. In such experiments the apparent V_{max} approaches the true V_{max} as the mole fraction of substrate is increased. Two additional kinetic parameters can be derived from such an analysis (see Ref. 5): the surface-binding constant, K_s^A (40 mM, total concentration of detergent plus substrate), which is a measure of the affinity of the enzyme for the mixed micelle surface; and the binding constant, K_m^B (0.03, mole fraction units), for the substrate on the micelle surface. Although there has been some controversy concerning the interpretation of this

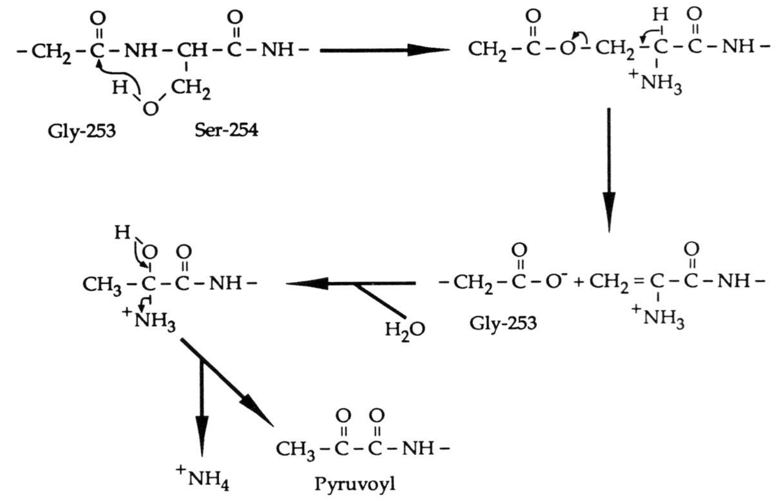

FIG. 3. Mechanism of posttranslational cleavage of prophosphatidylserine decarboxylase between residues 253 and 254 to form the pyruvoyl prosthetic group at the amino terminus of the α subunit.

model at the molecular level, to a first approximation such a model is very useful in designing experiments dependent on the activity of the enzyme.

Properties of Mutants. Strain EH150 (*psd2*) is temperature-sensitive for growth at 42° owing to a mutation in the structural gene (mapping at 95 min on the *E. coli* chromosome) for the phosphatidylserine decarboxylase.[12] In the mutant strain newly synthesized enzyme is temperature labile, whereas previously synthesized, membrane-associated enzyme is stable; however, the mutant enzyme synthesized under permissive conditions (30°) is temperature sensitive after solubilization with Triton X-100. Under the restrictive growth conditions the mutant strain forms filamentous cells and accumulates 35% of its phospholipid as phosphatidylserine at the point of cell arrest as compared to less than 1% in wild-type strains. Growth of the mutant strain in the presence of 20 m*M* MgCl$_2$ increases the proportion of phosphatidylserine to 76% of the total phospholipid, prevents the formation of filamentous bacteria, and allows growth to continue through several more generations compared to cells grown in the absence of MgCl$_2$. Therefore, such a strain can be used to prepare naturally occurring radiolabeled phosphatidylserine. Addition of either radiolabeled serine, inorganic ^{32}PO$_4$, radiolabeled acetate, or [2-^{3}H]glycerol to a culture of strain EH150

[12] E. Hawrot and E. P. Kennedy, *J. Biol. Chem.* **253,** 8213 (1978).

growing in minimum salts medium, 20 mM MgCl$_2$, and glucose at the time of shifting to the restrictive growth conditions results in the formation of phosphatidylserine specifically labeled in either the serine moiety, phosphate moiety, fatty acids, or glycerol backbone, respectively. After extraction of the phospholipids from the cells using chloroform–methanol,[12] the radiolabeled phosphatidylserine can be isolated using DEAE-cellulose chromatography as described by Tyhach *et al.*[13]

Using site-directed mutagenesis, mutants in the *psd* gene have been constructed in which Ser-254 of prophosphatidylserine decarboxylase has been changed to either cysteine, threonine, or alanine.[7] In the case of the cysteine (S254C) and the threonine (S254T) substitutions, posttranslationally processed and active enzyme is made *in vivo* although in significantly reduced amounts; for the S254A substitution, only the proenzyme form (π subunit) is made, which has no enzymatic activity. In the case of the S254C and S254T mutant proteins about 10–20% of the π subunit is processed to the α and β subunits, resulting in 15 and 2%, respectively, of the level of activity of the wild-type enzyme. When carried on multicopy number plasmids, the former mutation fully complements the temperature-sensitive strain EH150, whereas the latter mutation only partially restores normal growth to strain EH150. These results are similar to those found with other pyruvoyl-dependent amino acid decarboxylases[14] and support a mechanism (Fig. 3) in which the hydroxyl moiety of Ser-254 attacks the amide bond between residues 253 and 254, forming an ester intermediate. In a subsequent α,β-elimination reaction the free β chain is released with the formation of dehydroalanine at the amino terminus of the α chain. After hydration and elimination of ammonia, residue 254 is converted to pyruvate. In the case of mutant S254C, cysteine would also be converted to pyruvate but at a reduced rate. With mutant S254T the resulting prosthetic group would be α-ketobutyrate, which is also formed at a reduced rate and appears to have a reduced activity relative to the pyruvate prosthetic group.[7]

[13] R. J. Tyhach, R. Engel, and B. E. Tropp, *J. Biol. Chem.* **251**, 6717 (1976).
[14] P. D. van Poelje and E. E. Snell, *Annu. Rev. Biochem.* **69**, 29 (1990).

[42] Phosphatidylserine Decarboxylase from Rat Liver

By Dennis R. Voelker and Elisabeth Baker Golden

Introduction

Phosphatidylserine decarboxylase has been identified as a protein associated with the inner mitochondrial membrane.[1,2] The enzyme decarboxylates phosphatidylserine to form phosphatidylethanolamine. Although once thought to be a minor enzyme in phospholipid synthesis in animal cells, recent evidence suggests that this enzyme can contribute significantly to both membrane biogenesis in cultured cells and lipoprotein maturation in hepatic cells.[3–5] Additional interest in this enzyme has been generated with the realization that the phosphatidylserine substrate for the decarboxylase must be imported into the mitochondria before it is available for catalysis.[6–8] Consequently the action of the decarboxylase can be used as a discrete chemical signal for monitoring phosphatidylserine transport. The enzyme has proved quite resistant to purification. This chapter describes the approaches used to obtain partially purified preparations of the enzyme.

Preparation of Substrates

The most convenient method for measuring the activity of the decarboxylase utilizes a $^{14}CO_2$ trapping assay.[9] The conversion of [^{3}H]phosphatidylserine to [^{3}H]phosphatidylethanolamine can also be used, but this method is much more cumbersome and requires thin-layer chromatography or another separation technique and is not recommended for routine studies.

[1] L. M. G. van Golde, J. Raben, J. J. Batenburg, B. Fleischer, F. Zambrano, and S. Fleischer, *Biochim. Biophys. Acta* **360**, 179 (1974).
[2] J. Zborowski, A. Dygas, and L. Wojtczak, *FEBS Lett.* **157**, 179 (1983).
[3] D. R. Voelker, *Proc. Natl. Acad. Sci. U.S.A.* **81**, 2669 (1984).
[4] D. R. Voelker and J. L. Frazier, *J. Biol. Chem.* **261**, 1002 (1986).
[5] D. E. Vance and J. E. Vance, *J. Biol. Chem.* **261**, 4486 (1986).
[6] D. R. Voelker, *J. Biol. Chem.* **264**, 8019 (1989).
[7] D. R. Voelker, *J. Biol. Chem.* **265**, 14340 (1990).
[8] J. E. Vance, *J. Biol. Chem.* **265**, 7248 (1990).
[9] J. N. Kanfer and E. P. Kennedy, *J. Biol. Chem.* **239**, 1720 (1964).

METHODS IN ENZYMOLOGY, VOL. 209

Materials

Reagents

Egg CDPdiacylglycerol (Serdary Lipids, London, ON, Canada), 10 mg/ml made in a 3 mg/ml Triton X-100 solution

Bovine serum albumin (BSA), 10 mg/ml

$0.2 M$ KH_2PO_4, pH 7.4

$0.2 M$ Dithiothreitol

Phosphatidylserine synthase preparation from *Escherichia coli* (see below) inactivated by treatment with 4 mM hydroxylamine at 30° for 1 min

100 μCi DL-[^{14}C]Serine (ICN Radiochemicals, Irvine, CA)

Chloroform

Methanol

Acetic acid

DEAE-cellulose

Comments. The DL-[^{14}C]serine is dried in a Speed Vac (Savant Instruments, Farmingdale, NY) in a polypropylene tube. To the dried radiolabel is added 500 μl of $0.2 M$ KH_2PO_4, 350 μl CDPdiacylglycerol–Triton X-100, 100 μl BSA, 56 μl dithiothreitol, and 95 μl of hydroxylamine-treated enzyme. The reaction is incubated at 30° for 30 min. The reaction is terminated by lipid extraction[10] after transfer to a glass tube. The reaction tube is rinsed with 0.8 ml of 0.1 N HCl, and the resultant 1.8 ml is extracted by the addition of 4 ml of methanol followed by 2 ml of chloroform to form a monophase. The monophase is converted to a two-phase system by the addition of 1.8 ml of 0.1 N HCl and 2 ml of chloroform. Centrifugation separates the lower chloroform phase, which is removed and washed with an upper phase consisting of 4 ml of methanol and 3.6 ml of $0.2 M$ KCl. The upper phase of the first extraction is reextracted with 4 ml of chloroform, and the recovered lower phase is washed as described for the first chloroform phase. The two chloroform phases are pooled, the solvent is removed under a stream of nitrogen, and the radiolabeled product is resuspended in a convenient volume of chloroform–methanol (9 : 1, v/v). The yield of phosphatidyl [1′-^{14}C]serine is usually 65–70% with respect to L-[1-^{14}C]serine. For many purposes this preparation of substrate is entirely suitable for assaying the enzyme.

The above preparation, however, will contain trace amounts of *E. coli* lipids. Removal of contaminating *E. coli* lipid is accomplished by chromatography on DEAE-cellulose[11] (Whatman, Clifton, NJ). The

[10] E. G. Bligh and W. J. Dyer, *Can. J. Biochem. Physiol.* **37**, 911 (1959).

[11] M. Kates, ed., *in* "Techniques of Lipidology," p. 408. Elsevier Science Publishing, New York, 1972.

DEAE-cellulose is converted to the acetate form by resuspension in excess glacial acetic acid for 2 hr. The matrix is subsequently washed extensively with methanol to remove acetic acid, then washed with chloroform. For chromatography a 2 ml bed volume of the DEAE cellulose is poured into a 10-ml disposable glass pipette plugged with silanized glass wool. The phosphatidyl[1'-^{14}C]serine is adsorbed onto the column, and the column is sequentially eluted with 18 ml chloroform–methanol (7 : 3, v/v), 16 ml methanol, and 40 ml acetic acid. The phosphatidyl[1'-^{14}C]serine elutes in the acetic acid fraction. The acetic acid is removed using a rotary evaporator and a high vacuum pump. The sample should not be left in acetic acid for any longer than absolutely necessary. The dried phosphatidylserine is resuspended in chloroform–methanol (9 : 1, v/v) and stored at $-70°$. Stored in this manner the labeled substrate is stable for 2 years.

This laboratory has used both purified preparations of E. coli phosphatidylserine synthase and crude preparations. Both work equally well. A simple crude preparation can be obtained using E. coli strain JA-200 containing the plasmid PPS 3155λ (constructed by W. Dowhan, University of Texas, Houston). Detailed description of the growth and induction of this strain is provided by Ohta et al.[12] Briefly, the strain is grown to mid-log phase in rich medium (25 g yeast extract, 12 g Bacto-tryptone (Difco, Detroit, MI)/liter, 0.1 M KH_2PO_4, pH 7.5, supplemented to 0.5% glycerol) at 30°. Next the strain is induced by shifting the temperature to 42° for 20 min followed by growth at 37° for 3.5 hr. The cells are harvested and washed in 0.1 M KH_2PO_4, pH 7.4, and frozen. The frozen cell paste is resuspended 1 : 7 (w/v) in 0.1 M KH_2PO_4, pH 7.4, and lysed in a French pressure cell. Alternatively the cells can be spheroplasted by treatment with lysozyme–EDTA and sonicated to yield a broken cell extract. The lysate is centrifuged at 13,500 g for 2 hr and the pellet discarded. The supernatant is adjusted to 10% glycerol and 0.2% Triton X-100 (v/v). Phosphatidylserine decarboxylase activity in the extract is inhibited by treating the preparation with 4 mM hydroxylamine for 1 min at 30°. The crude enzyme preparation at a concentration of 20 mg/ml is frozen in 500-μl aliquots at $-70°$. This preparation is stable to multiple freeze–thaw cycles and can be used for at least 3 years.

Standard Assay

Phosphatidylserine decarboxylase activity is measured at 37° in a liquid scintillation vial fitted with a rubber stopper and an appended center well (Kontes, Vineland, NJ; Cat. No. K882310-0000, K882320-0000). The

---

[12] A. Ohta, K. Waggoner, K. Louie, and W. Dowhan, J. Biol. Chem. **256,** 2219 (1981).

stopper makes the vial gas tight, and a 1 cm circle of filter paper (Whatman 3MM) wetted with 0.1 ml of 2 N KOH is held by the center well. The reaction is conducted in a volume of 0.4 ml and contains 0–300 μg protein, 0.1 M KH$_2$PO$_4$, pH 6.8, 10 mM EDTA, 8 × 10^4 counts/min (cpm) phosphatidyl[1'-^{14}C]serine (200 cpm/nmol), and 0.5 mg/ml Triton X-100. The specific activity of the radiolabeled phosphatidylserine is adjusted using bovine brain phosphatidylserine from Avanti Polar Lipids (Pelham, AL). The reactions are initiated with the addition of substrate, which is added from a 10 mM stock in a 5 mg/ml Triton X-100 solution. After the addition of substrate, the vessel is sealed with the rubber stopper plus center well and transferred from ice to 37°. After an incubation time of 30–60 min the stopper is pierced with a syringe-mounted 22-gauge needle, and 0.5 ml of 0.5 N H$_2$SO$_4$ is injected into the reaction solution, which is gently mixed and returned to 37° for 1 hr. Some care should be taken to avoid touching the filter paper in the center well or the reaction solution with the needle. By bending the needle the acid solution can be expelled down the side of the vial. After allowing 30 min for the evolution of ^{14}CO$_2$, the stopper is removed and the filter paper is counted using an emulsion-based cocktail.

Preparation of Inner Mitochondrial Membrane

Mitochondria are isolated from fresh rat liver. The tissue is homogenized using a Potter–Elvehjem device in 220 mM mannitol, 70 mM sucrose, 2 mM HEPES, pH 7.4, 1 mM EDTA, and 1 mg/ml BSA in the presence of 100 μg/ml pepstatin, 1 μg/ml leupeptin, and 1 mM phenylmethylsulfonyl fluoride.[13] Mitochondria are prepared by the method of Bustamante *et al.*,[14] which routinely gives 25–30 mg of mitochondrial protein/g of tissue. The mitochondria are resuspended in 440 mM mannitol, 140 mM sucrose, 4 mM HEPES, 1 mg/ml BSA at a ratio of 0.5 ml/g of liver used as starting material. Mitoplasts are prepared using a French press.[15] The mitoplasts are separated from the outer membrane by centrifugation at 12,000 g for 10 min and can be frozen for several weeks without significant loss of decarboxylase activity.

The mitoplasts are sonicated on ice for 5 min in 1-min bursts followed by 30 sec cooling to release matrix proteins, and the membranes are sedimented by centrifugation at 100,000 g for 45 min. The sonicated preparation is suspended in 80 mM Tris, pH 7.4, 0.25 M sucrose, 1 mM EDTA, 1 M NaCl, and 0.1 mM phenylmethylsulfonyl fluoride and stirred for 50

[13] S. Fleischer, J. O. McIntyre, and J. C. Vidal, this series, Vol. 55, p. 32.
[14] E. Bustamante, J. W. Soper, and P. L. Pedersen, *Anal. Biochem.* **80,** 401 (1977).
[15] J. W. Greenawalt, this series, Vol. 55, p. 88.

TABLE I
PARTIAL PURIFICATION OF PHOSPHATIDYLSERINE DECARBOXYLASE FROM RAT LIVER

Fraction	Total protein (mg)	Specific activity (nmol/min/protein)	Total units (nmol/min)	Yield (%)
Mitochondria	2510	0.6	1556	100
Mitoplasts	676	1.6	1088	69
Inner membrane vesicles	246	2.7	669	43
Triton X-100 soluble	146	3.8	550	35
Chromatofocusing	7.2	16.4	118	8
Sephacryl S-300	4.6	15.1	70	5
Hydroxylapatite	0.9	32.0	29	2

min at 4°. The salt-extracted membranes are collected by centrifugation at 100,000 g for 45 min using a Beckman 50.2 rotor. The resultant pellet is resuspended in homogenization buffer with protease inhibitors but without BSA.

Detergent Extraction. The inner membrane preparation is adjusted to 3 mg protein/ml and 3 mg Triton X-100/ml and stirred on ice for 40 min at 4°. This solution is next centrifuged at 100,000 g for 45 min using a Beckman 50.2 rotor. All of the recoverable decarboxylase activity is found in the supernatant. The supernatant is dialyzed overnight against chromatofocusing buffer consisting of 25 mM ethanolamine adjusted to pH 9.4 with acetic acid, 10% glycerol (v/v), 5 mM 2-mercaptoethanol, and 0.3% Triton X-100.

Chromatofocusing. A typical application of chromatofocusing utilizes a 1.6 × 12 cm column packed and equilibrated at 180 ml/hr. The column is eluted with 10% Polybuffer 96 (adjusted to pH 6.0 with acetic acid) containing 10% glycerol and 3 mg/ml Triton X-100. Before loading the sample, 2 ml of elution buffer is added to the column. The sample of 61 ml (220 mg protein) is loaded onto the column at a flow rate of 60 ml/hr, and 8.5-ml fractions are collected. After the sample is loaded the elution is begun using 250 ml of buffer and a flow rate of 60 ml/hr, and 5 ml fractions are collected. The decarboxylase activity elutes in a sharp symmetrical peak of approximately 40 ml at pH 8.0.

Gel Filtration. The Polybuffer 96 present in the decarboxylase preparation is removed by gel filtration using Sephacryl S-300. Typically, 30 ml of the chromatofocusing peak is loaded onto a 200-ml column that is equilibrated with 20 mM KH$_2$PO$_4$, pH 7.4, 10% glycerol, 5 mM 2-mercaptoethanol, and 0.5 mg/ml Triton X-100. The enzyme is eluted with the same buffer at a flow rate of 83 ml/hr. The elution position corresponds to a molecular mass between 100 and 200 kDa.

Hydroxylapatite Chromatography. A hydroxylapatite column (1 × 12 cm) is equilibrated with the same buffer used for gel filtration. Fractions corresponding to peak enzyme activity from gel filtration (34 ml) are loaded onto the column followed by a buffer wash of 10 ml. The column is next eluted with a 50-ml gradient of 20 to 370 mM KH$_2$PO$_4$, pH 7.4, containing 10% glycerol, 5 mM 2-mercaptoethanol, and 0.5 mg/ml Triton X-100 at a flow rate of 50 ml/hr. The decarboxylase activity elutes as a primary peak containing approximately 80% of the activity in a volume of 8 ml centered around 150 mM KH$_2$PO$_4$ followed by a secondary peak containing the remaining 20% of the activity that elutes broadly from 200 to 350 mM KH$_2$PO$_4$.

The results of a typical preparation of the decarboxylase are shown in Table I. The major impediments to further purification of the enzyme have been instability of the activity and the low yields. Numerous other strategies for purification have been employed including both low pressure and high-performance liquid chromatography (HPLC), ion-exchange methods, affinity chromatography using phosphatidylserine coupled to matrices by both amine and carboxyl moieties, and chromatography on phenyl-, octyl-, and butyl-Sepharose. The method outlined in this chapter has provided the highest specific activities and yields attainable thus far.

Properties of Partially Purified Enzyme

All preparations of the enzyme are sensitive to sulfhydryl modifying reagents such as p-chloromercuribenzoate and iodoacetamide. The presence of 5 mM 2-mercaptoethanol, 1 mM EDTA, and 10% glycerol stabilizes the enzyme. The apparent K_m of the decarboxylase in crude preparations is 140 μM phosphatidylserine. In the purest enzyme preparations the K_m is 40 μM. Hydroxylamine (1 mM) completely and irreversibly inactivates the enzyme. The decarboxylase does not exhibit a significant substrate preference among dipalmitoyl-, dimyristoyl-, or egg phosphatidylserines.

Acknowledgment

This work was supported by National Institutes of Health Grant GM 32453.

[43] Phosphatidylethanolamine N-Methyltransferase from Rat Liver

By Neale D. Ridgway and Dennis E. Vance

Introduction

Phosphatidylethanolamine N-methyltransferase (PEMT) (EC 2.1.1.17) is an enzyme of the endoplasmic reticulum that catalyzes the synthesis of phosphatidylcholine by transfer of three methyl groups from S-adenosyl-L-methionine (AdoMet) to the amino head-group of phosphatidylethanolamine (PE).[1] The major route for phosphatidylcholine (PC) synthesis in most cells is via the CDPcholine pathway. However, by some estimates PE methylation is responsible for 15–20% of PC synthesis in the liver.[2] What function this PE-derived PC has in liver physiology is unknown. PE methylation activity is very low in extrahepatic tissues, and it is generally assumed that this pathway does not contribute significantly to PC synthesis.

Evaluation of the function of PE-derived PC in hepatic metabolism has been aided by the purification and characterization of rat liver PEMT.[3] PEMT purified by the procedure described herein is homogeneous as determined by gel electrophoresis and displays kinetic properties consistent with an enzyme that acts on mixed micellar substrates.[3,4]

Assay of Phosphatidylethanolamine N-Methyltransferase

Previously, assays of PEMT activity in hepatic microsomes have utilized endogenous PE or vesicles of PE, monomethyl-PE (PMME), or dimethyl-PE (PDME) as substrates. Activities assayed with endogenous PE are often inaccurate owing to the rate-limiting nature of the first methylation step and uncertainty in the amount of PE available to the enzyme. In the case of added vesicles, the actual catalytic rate may be dictated by the rate of exchange of phospholipid substrate from the vesicle into the microsome. To counter these problems a mixed micelle assay using Triton X-100 was developed. When sonicated preparations of PE, PMME, or PDME were solubilized with microsomes or purified PEMT there was marked stimulation of enzyme activity.[4] Two assay procedures were re-

[1] D. E. Vance and N. D. Ridgway, *Prog. Lipid Res.* **27**, 61 (1988).
[2] R. Sundler and B. Åkesson, *J. Biol. Chem.* **250**, 3359 (1975).
[3] N. D. Ridgway and D. E. Vance, *J. Biol. Chem.* **262**, 17231 (1987).
[4] N. D. Ridgway and D. E. Vance, *J. Biol. Chem.* **263**, 16864 (1988).

quired for optimum measurement of partially pure and purified PEMT activity.

Materials. PMME and PDME were purchased from Avanti Polar Lipids (Birmingham, AL). Silica gel 60 (0.2 and 2.0 mm) plates were from Merck (Darmstadt, Germany). [*methyl*-³H]AdoMet was supplied by Amersham Corp. (Arlington Heights, IL). Triton X-100 and dithiothreitol (DTT) were from Sigma (St. Louis, MO). PE was purified from rat liver microsomes by preparative thin-layer chromatography. All other materials were of reagent grade.

Procedure 1. The assay for microsomal PEMT is performed in 2-ml uncapped borosilicate tubes. The assay contains 125 mM Tris-HCl (pH 9.2), 5 mM DTT, 1 mM Triton X-100, and 2 mM PE, 0.4 mM PMME, or 0.4 mM PDME in a final volume of 150 μl. The phospholipid substrates (PE, PDME, and PMME) are dried under nitrogen and under reduced pressure for 30 min. The dried lipids are resuspended in 20 mM Tris-HCl (pH 9.2), 0.01% (w/v) EDTA, and 0.02% Triton X-100 by vortexing for 1 min, followed by sonication at 37° for 3 min. Finally, the enzyme source is added, and all components are incubated on ice for 10 min. The assay is initiated by the addition of [*methyl*-³H]AdoMet (21 mCi/mmol) to a final concentration of 200 μM, and the tubes are transferred to a 37° shaking water bath. Enzyme reaction rates are linear up to 10 min with 25 μg or less or microsomal protein.

Procedure 2. The assay of partially purified PEMT requires a slight variation of assay conditions. Triton X-100 concentrations are lowered to 0.5 mM, and a final concentration of 2 mM PE, 0.25 mM PMME, or 0.45 mM PDME is used. This change merely reflects that at later steps in the PEMT purification endogenous phospholipid is removed and changes in exogenous detergent and lipid are required to reoptimize the assay. Assays for purified and partially purified PEMT are linear for up to 40 min for all three lipid substrates.

Procedure 1 and 2 assays are terminated by the addition of 2 ml of chloroform–methanol (2:1, v/v), and the contents of the assay tube plus a single 0.5-ml rinse of chloroform–methanol are transferred to a 7-ml borosilicate test tube. Methylated lipids are extracted by the method of Folch *et al.*[5] and the organic phase dried under nitrogen at 60°. Total chloroform-soluble radioactivity is measured by scintillation spectrometry and expressed as nanomoles of methyl groups transferred per minute per milligram of protein. Alternatively, an aliquot of the organic phase is applied to a silica gel 60 plate and developed in a solvent system of chloroform–methanol–acetic acid–water (50:30:5:2, v/v). Phospholip-

[5] J. Folch, M. Lees, and G. H. Sloane Stanley, *J. Biol. Chem.* **226,** 497 (1959).

ids are visualized by brief exposure to iodine vapors and the bands scraped into scintillation vials containing 250 μl of water and 5 ml of ACS scintillation fluid (Amersham). Radioactivity is measured after 24 hr to ensure complete elution of labeled lipids from the silica gel.

Purification of Phosphatidylethanolamine N-Methyltransferase

Materials. DE-52 cellulose and P-11 cellulose were purchased from Whatman (Clifton, NJ). Octyl-Sepharose CL-4B, PBE 94 polybuffer exchange resin, and sodium dodecyl sulfate (SDS)-gel electrophoresis standards were from Pharmacia LKB Biotechnology, Inc. (Piscataway, NJ).

Isolation of Microsomes. Microsomes are isolated from the livers of female Wistar rats (175–225 g). After cervical dislocation the livers are immediately removed and placed in ice-cold 10 mM Tris-HCl (pH 7.2) containing 150 mM NaCl, 1 mM EDTA, 1 mM phenylmethylsulfonyl fluoride, and 1 mM DTT. The liver is cut into small pieces, suspended at a final concentration of 25% (w/v) in the Tris–saline buffer, and homogenized using a motor-driven Potter–Elvehjem apparatus. The homogenate is centrifuged at 12,000 g for 10 min. The supernatant fraction is then subjected to centrifugation at 120,000 g for 1 hr. The cytosol is immediately decanted and the microsomal pellet resuspended in a 20 mM potassium phosphate buffer (pH 7.9) containing 1 mM EDTA, 1 mM DTT, and 250 mM sucrose (buffer A) using a hand-held glass Dounce homogenizer. Microsomes are also isolated from livers that had been perfused with 150 mM NaCl and 0.5 mM EGTA. Following perfusion the livers are treated in an identical manner as that described above. Microsomes prepared from perfused livers have a higher initial PEMT specific activity. Microsomes are routinely stored in buffer A at $-70°$. No apparent loss of activity or degree of purification is noted after several months storage. Procedure 1 is used to assay enzyme activity.

Preparation of Microsomal Membranes. Microsomes (20–30 mg/ml protein) are suspended in 100 mM Na$_2$CO$_3$, 5 mM DTT at a final protein concentration of 4 mg/ml and stirred at 4° for 30 min. The suspension is centrifuged at 120,000 g for 1 hr and the membrane pellet collected and resuspended by homogenization with a hand-held glass Dounce homogenizer in 20 mM potassium phosphate buffer (pH 7.9) containing 10% (v/v) glycerol and 5 mM DTT. This buffer is used in all subsequent purification steps and will be referred to as buffer B. Treating microsomes at this high pH results in disruption of the vesicles and release of luminal contents. The alkaline pH also causes dissociation and solubilization of extrinsic membrane proteins, but leaves the phospholipid bilayer and integral proteins intact. Enzyme activity is measured by procedure 1.

Solubilization of Microsomal Membranes. Microsomal membranes are suspended in buffer B containing 0.7% (w/v) Triton X-100 to a final protein concentration of 4 mg/ml and stirred for 1 hr. The mixture is centrifuged at 120,000 g for 1 hr and the supernatant collected and used as a source of soluble enzyme. This step is by far the most critical to the purification. Prior to the use of Triton X-100, several reports noted that 3-[(3-cholamido-propyl)dimethylammonio]-1-propanesulfonate (CHAPS), a zwitterionic detergent, solubilized PEMT in an active form.[6,7] An earlier report from our laboratory showed that Triton X-100 would solubilize PEMT activity; however, the preparations were highly unstable.[8] We noted that the key to stabilizing PEMT activity in Triton X-100 was the inclusion of 20 mM potassium phosphate buffer. The soluble enzyme is assayed by procedure 1.

Chromatography on Whatman DE-52 Cellulose. Soluble PEMT is passed through a column of DE-52 cellulose (30 × 2.5 cm), previously equilibrated in buffer B plus 0.7% Triton X-100, at a flow rate of 1.0 ml/min. PEMT activity is recovered in the unbound fractions. Enzyme activity is assayed by procedure 1.

Chromatography on Whatman P-11 Phosphocellulose. A column of Whatman P-11 phosphocellulose (16 × 1.6 cm) is equilibrated in buffer B containing 0.7% (w/v) Triton X-100. The pooled fractions from the previous step are applied to the column at a flow rate of 0.25 ml/min. When loading is complete the column is flushed in succession with 100 ml of buffer B containing 0.7% (w/v) Triton X-100 and 100 ml containing 0.25% (w/v) Triton X-100. PEMT activity is eluted from the column with a linear gradient of NaCl from 0 to 0.8 M in 0.25% (w/v) Triton X-100. PEMT activity elutes in a broad peak from 0.2 to 0.6 M. Enzyme activity is determined using procedure 2.

This step is the least reproducible, owing in part to the requirement that used, unregenerated P-11 cellulose be used. We found it necessary to use P-11 phosphocellulose that had been used for several purification attempts (and not regenerated according to the manufacturers specifications) in order to achieve good recoveries from this as well as from the following purification step. The phosphocellulose could also be pretreated with 0.2% (w/v) bovine serum albumin in buffer B containing 0.7% Triton X-100 followed by elution with 2.0 M NaCl in the same buffer. This procedure seemed to block high-affinity sites on the resin that would

[6] M. A. Pajares, S. Alemany, I. Varela, D. Marin Cao, and J. M. Mato, *Biochem. J.* **223**, 61 (1984).

[7] M. A. Pajares, M. Villalba, and J. M. Mato, *Biochem. J.* **237**, 699 (1986).

[8] W. J. Schneider and D. E. Vance, *J. Biol. Chem.* **254**, 3886 (1979).

otherwise reduce recovery of PEMT and change its chromatographic characteristics on octyl-Sepharose. Phosphocellulose chromatography is indispensable for removing endogenous phospholipid from preparations (Table I).

Chromatography on Octyl-Sepharose CL-4B. Active fractions from the phosphocellulose column are pooled and diluted with buffer B (no Triton X-100) to a final Triton X-100 concentration of 0.05% (w/v). The enzyme solution is applied to a column of octyl-Sepharose (17 × 1.6 cm), equilibrated in buffer B containing 0.05% (w/v) Triton X-100, at a flow rate of 0.5 ml/min. The column is flushed with 100 ml of buffer B containing 0.05% Triton X-100. It is critical to the final purity of the enzyme that this column be eluted so that the main PEMT activity peak is well separated from the major protein peak that just proceeds it.[3] This is achieved by eluting the column with 100 ml of buffer B containing 0.15% (w/v) Triton X-100 followed by 250 ml of a linear gradient of Triton X-100 from 0.15 to 0.5% (w/v) in the same buffer. Enzyme activity is measured by procedure 2.

Chromatography on PBE 94. The pooled fractions from the previous step, adjusted to pH 9.4 with 250 mM ethanolamine, are applied to a column of PBE 94 (15 × 1.6 cm), equilibrated in 25 mM ethanolamine (pH 9.4), 5 mM DTT, 10% (v/v) glycerol, and 0.1% (w/v) Triton X-100. PEMT is loaded at a flow rate of 0.2 ml/min. Like the DE-52 step earlier in the purification, PBE 94 chromatography utilizes the extremely alkaline native pI of the enzyme. All proteins remaining after the octyl-Sepharose step bind to the PBE 94 column at pH 9.4 and purified PEMT flows through. Purified PEMT is measured by procedure 2.

Enzyme Concentration by Mono S Chromatography. The PBE 94 eluant is concentrated by chromatography on a Mono S column using the Pharmacia fast protein liquid chromatography system. The Mono S column is equilibrated in buffer B plus 0.1% (w/v) Triton X-100. The PBE eluate, adjusted to pH 7.9 with 0.5 M potassium phosphate, is applied to the column at a flow rate of 0.2 ml/min. Following a 10-ml wash with buffer B, PEMT is eluted with a linear gradient of 0 to 1 M NaCl in buffer B plus 0.1% Triton X-100. Recovery of activity ranges between 75 and 90%. A complete summary of the purification is given in Table I.

Other Methods. Protein concentrations are measured by the method of Lowry *et al.*[9] modified to contain a final concentration of 0.04% (w/v) deoxycholate. Protein concentrations in the final two steps of the purifica-

[9] O. H. Lowry, N. J. Rosebrough, A. L. Farr, and R. J. Randall, *J. Biol. Chem.* **193,** 265 (1951).

TABLE I

PURIFICATION OF PHOSPHATIDYLETHANOLAMINE N-METHYLTRANSFERASE

Fraction	Volume (ml)	Protein (mg)	Lipid phosphorus (μmol)	Specific activities[a]				Total activity[b,c]	Recovery[c] (%)	Purification (-fold)
				NA[d]	PE	PMME	PDME			
Microsomes	35	1220	742	1.1	1.5	5.6	4.5	6830	100	—
Microsomal membranes	41	743	623	1.5	1.7	8.9	5.3	6660	97	1.6
Soluble membranes	180	478	577	0.3	0.4	6.2	7.6	2950	43	1.1
DE-52 cellulose	206	350	585	0.4	0.6	12.8	10	4480	65	2.3
P-11 cellulose	183	40.3	ND[e]	ND	5.9	34.3	44	1380	20	6.1
Octyl-Sepharose	140	0.25	ND	ND	150	3560	1570	871	12	68
PBE 94	138	0.04	ND	ND	630	8590	3750	352	5	1540

[a] Specific activities expressed as nmol of methyl groups transferred/min/mg of protein.
[b] Expressed as nmol/min.
[c] Total activity, recovery, and purification (-fold) were calculated for PMME-dependent methylation.
[d] Activity assayed in the absence of exogenous substrate.
[e] Not detected.

tion are determined by a sensitive silver binding assay.[10] Lipid phosphorus is measured by the method of Rouser et al.[11]

Properties

Molecular Mass. PEMT migrated as a single 18.3-kDa polypeptide on SDS–polyacrylamide gel electrophoresis (PAGE).[3] The native enzyme migrated on a Sephacryl S-300 column (in the presence of 0.1% Triton X-100) with a Stokes radius of 55.2 Å. Pure Triton X-100 micelles eluted from the same column with Stokes radius of 53.1 Å. The difference in molecular mass of 24 kDa is in close agreement with the result from SDS–PAGE, and suggests one enzyme molecule per Triton micelle. Furthermore, SDS–PAGE and PE, PMME, and PDME activity profiles of Sephacryl S-300 column fractions showed cochromatography of the 18.3-kDa methyltransferase and all three enzyme activities.[3] The 18.3-kDa methyltransferase is not a proteolysis product generated during purification since a polyclonal antibody raised against the purified protein recognized only this protein and no higher molecular weight forms in freshly prepared rat liver microsomes.[12]

Substrate and Molecular Species Specificity. Table I shows that PE, PMME, and PDME methylation activities all copurified with the 18.3-kDa methyltransferase. In addition, the spectrum of methylated products produced from PE, PMME, and PDME were virtually identical at each purification step; methylation of PE produced 70–80% of the products as PC, PMME methylation produced 95% PDME, and PDME methylation gave entirely PC.[3] Thus, the 18.3-kDa methyltransferase of liver is similar to the genetically defined PEM2, or phospholipid methyltransferase, of yeast which is also able to convert PE to PC and has a molecular mass (deduced from the cDNA) of 23.1 kDa.[13]

Detailed studies were performed using the purified enzyme and diacyl-PE, -PMME, and -PDME substrates *in vitro.*[14] Methylation of substrates of mixed molecular species composition with [*methyl*-³H]AdoMet, and separation of the tritium-labeled products by high-performance liquid chromatography (HPLC), revealed no preference by purified or microsomal PEMT for a given molecular species of substrate phospholipid. Results with pure synthetic PEs demonstrated that maximal methylation rates were obtained for species with two or more double bonds. Saturated PEs

[10] G. Krystal, C. MacDonald, B. Mant, and S. Ashwell, *Anal. Biochem.* **148,** 451 (1985).
[11] G. Rouser, A. N. Siakatos, and S. Fleisher, *Lipids* **1,** 85 (1966).
[12] N. D. Ridgway, Z. Yao, and D. E. Vance, *J. Biol. Chem.* **264,** 1203 (1989).
[13] T. Kodaki and S. Yamashita, *J. Biol. Chem.* **262,** 15428 (1987).
[14] N. D. Ridgway and D. E. Vance, *J. Biol. Chem.* **263,** 16856 (1988).

were not methylated by purified PEMT unless 40 mol % egg PC was included in the assay. Lack of methylation of saturated PEs with phase transition temperatures below the assay temperature suggests that the methyltransferase works on the PEs in the liquid crystalline phase.

Kinetic Properties. Studies of the methylation of defined phospholipid substrates by pure methyltransferase in the Triton X-100 mixed micelle assay revealed complex and interesting kinetics.[3,4] Double-reciprocal plots of initial velocity versus lipid substrate concentration were highly cooperative, with Hill numbers of 3 to 12. The Hill coefficients could be raised by increasing the Triton concentration or lowered to values close to unity by including a fixed mole percentage of egg PC. This type of behavior is seen for enzymes that act on mixed micelle substrates, and it probably indicates activation of enzyme catalysis by binding of phospholipid to portions of the enzyme in close association with the micelle (i.e., membrane-spanning domains). Purified PEMT also displayed classic "surface dilution" inhibition by Triton X-100 that was not the result of increasing micelle concentration.[15]

Initial velocity and product inhibition data indicated that PMME and PDME methylation fit an ordered Bi–Bi mechanism.[4] As well, PE, PMME, and PDME compete for a common active site. A concerted model for the complete methylation of PE to PC was formulated, and it proposed that PE is the initial substrate bound followed by AdoMet. S-Adenosyl-L-homocysteine then departs the active site. What follows is two Bi–Bi mechanisms linked together, but with the feature that PMME and PDME remain bound to the active site. PC is the last product to dissociate. In contrast, a random Bi–Bi mechanism was deduced for PEMT on human red cell membranes.[16]

Conclusion

PEMT can now be purified in an active form using Triton X-100 as a solubilizing agent. Although a number of salient features of the enzyme have come to light through studies of the pure enzyme, several important questions concerning the function of PEMT in liver and nonhepatic tissues remain. It has been shown that PEMT is not regulated, except by substrate availability, in the choline-deficient hepatocyte.[12] But clearly there is some type of tissue-specific expression since liver has 500-fold more PEMT activity than other tissues.[1] The cloning of the cDNA for PEMT should

[15] E. A. Dennis, *Arch. Biochem. Biophys.* **158**, 485 (1973).
[16] R. C. Reitz, D. J. Mead, R. A. Bjur, A. H. Greenhouse, and W. H. Welch, *J. Biol. Chem.* **264**, 8097 (1989).

facilitate future studies on how the expression of the PEMT gene is regulated. The reason for differential expression of the PEMT gene is unclear considering that PE methylation may contribute only 15–20% of PC synthesis in the liver.[2] The role of PE-derived PC in hepatocytes is also unclear, beyond providing a potential source of nondietary choline. Recent data suggested that only newly made PE is methylated and that 16:0-22:6-PC, the major product of *in vivo* methylation, is rapidly remodeled to other molecular species.[17] Why 16:0-22:6-PC is preferentially made and rapidly transformed into different PC molecular species remains to be explained.

[17] R. W. Samborski, N. D. Ridgway, and D. E. Vance, *J. Biol. Chem.* **265,** 18322 (1990).

Section IX

Alkyl Ether/Plasmalogen Biosynthetic Enzymes

[44] Alkyldihydroxyacetonephosphate Synthase

By ALEX BROWN and FRED SNYDER

Introduction

Alkyldihydroxyacetonephosphate (alkyl-DHAP) synthase (EC 2.5.1.26, alkylglycerone-phosphate synthase) forms the ether bond found in alkyl and alk-1-enyl glycerolipids. The details of the reaction that forms alkyl-DHAP (Fig. 1) were elucidated in the late 1960s. Friedberg and Greene[1] first showed in 1967 that long-chain fatty alcohols are the direct source for the alkyl chain. Using 1-[3H]- and 1-[14C]-labeled fatty alcohols, they found that the [3H] to [14C] ratio did not change during incorporation into ether lipids, indicating that the alcohols were not oxidized before ether bond formation. Identification of acyl-DHAP as the other substrate was more difficult. Snyder *et al.*[2] developed a cell-free system consisting of microsomes and a soluble fraction from mouse preputial gland tumors, ATP, coenzyme A (CoA), Mg^{2+}, and [14C]hexadecanol to produce [14C]alkylglycerolipids. They subsequently found that the soluble fraction could be replaced by glyceraldehyde phosphate but not by glycerol phosphate.[3] The tumor microsomes contained high amounts of triose-phosphate isomerase, however, and DHAP was identified as the soluble precursor for the glycerol backbone of alkyl ether lipids.[4,5] The role of ATP, CoA, and Mg^{2+} were explained by the demonstration by Hajra[6] that acyl-DHAP was an intermediate in ether lipid synthesis. The ether-linked product of the reaction of acyl-DHAP with the fatty alcohol was identified as 1-*O*-alkyl-DHAP.[4,5]

Investigation of the substrate specificity revealed that *n*-alkanols of 12, 14, and 16 carbons could be incorporated equally well into ether lipids, whereas the 10- and 18-carbon alcohols were poorer substrates, and alcohols less than 10 carbons did not react.[7,8] However, naturally produced ether lipids contain a very limited variety of alkyl side chains. In Ehrlich

[1] S. J. Friedberg and R. C. Greene, *J. Biol. Chem.* **242**, 5709 (1967).

[2] F. Snyder, B. Malone, and R. L. Wykle, *Biochem. Biophys. Res. Commun.* **34**, 40 (1969).

[3] F. Snyder, R. L. Wykle, and B. Malone, *Biochem. Biophys. Res. Commun.* **34**, 315 (1969).

[4] A. K. Hajra, *Biochem. Biophys. Res. Commun.* **37**, 486 (1969).

[5] R. L. Wykle and F. Snyder, *Biochem. Biophys. Res. Commun.* **37**, 658 (1969).

[6] A. K. Hajra, *Biochem. Biophys. Res. Commun.* **39**, 1037 (1970).

[7] F. Snyder, B. Malone, and M. Blank, *J. Biol. Chem.* **245**, 1790 (1970).

[8] A. K. Hajra, *in* "Tumor Lipids: Biochemistry and Metabolism"(R. Wood, ed.), p. 183. American Oil Chemists Society, Champaign, Illinois, 1973.

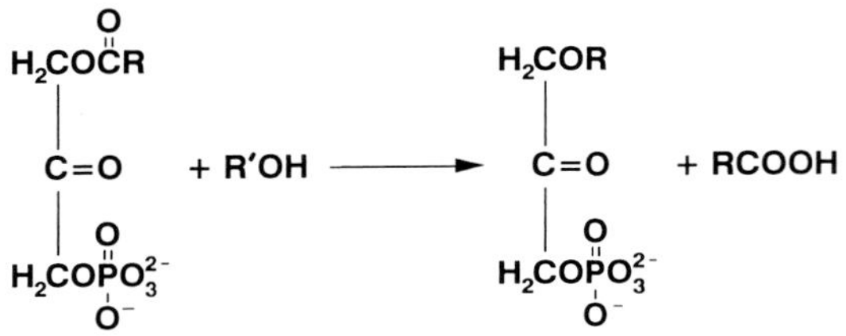

$$
\begin{array}{c}
\underset{\displaystyle |}{\overset{\displaystyle O}{\overset{\displaystyle \|}{H_2COCR}}} \\
\underset{\displaystyle |}{C=O} \\
\underset{\displaystyle O^-}{\overset{\displaystyle O}{\overset{\displaystyle \|}{H_2COPO_3^{2-}}}}
\end{array}
\quad + R'OH \longrightarrow \quad
\begin{array}{c}
\underset{\displaystyle |}{H_2COR} \\
\underset{\displaystyle |}{C=O} \\
\underset{\displaystyle O^-}{\overset{\displaystyle O}{\overset{\displaystyle \|}{H_2COPO_3^{2-}}}}
\end{array}
\quad + RCOOH
$$

FIG. 1. Substitution of a fatty alcohol for the acyl moiety of acyl-DHAP to form alkyl-DHAP in a reaction catalyzed by alkyl-DHAP synthase.

ascites cells, ether lipids are formed mainly from hexadecanol, octadecanol, and octadecenol (49, 21, and 14%, respectively), probably owing to the narrow specificity of the acyl-CoA reductase.[9]

The mechanism of this reaction is unique and has received considerable study. Using [^{18}O]hexadecanol, Snyder et al.[10] demonstrated that the oxygen in the ether linkage is donated by the alcohol. Later it was shown that both oxygens of the fatty acid ester of acyl-DHAP are retained by the fatty acid released during the reaction.[11,12] Friedberg and co-workers showed that the 1-pro-R hydrogen of acyl-DHAP is exchanged with the medium[13–15] with retention of configuration at carbon 1.[16,17] There is still controversy whether the enzyme releases the fatty acid before binding the fatty alcohol. Friedberg and co-workers[18] proposed the formation of a ternary intermediate with both acyl and alkyl linkages to DHAP. On the other hand, Davis and Hajra[19] proposed a Ping-Pong type mechanism in which the fatty acid is released to give an activated enzyme–DHAP complex that could react with fatty alcohols to give alkyl-DHAP or with a fatty acid to regenerate the acyl-DHAP substrate. In support of this mechanism, Brown and Snyder[20] showed that fatty acid is a competitive inhibitor with respect to fatty alcohol for alkyl-DHAP formation. Furthermore, initial

[9] J. E. Bishop and A. K. Hajra, J. Neurochem. 30, 643 (1978).
[10] F. Snyder, W. T. Rainey, Jr., M. L. Blank, and W. H. Christie, J. Biol. Chem. 245, 5853 (1970).
[11] S. J. Friedberg, S. T. Weintraub, M. R. Singer, and R. C. Greene, J. Biol. Chem. 258, 136 (1983).
[12] A. J. Brown, G. L. Gish, E. H. McBay, and F. Snyder, Biochemistry 24, 8012 (1985).
[13] S. J. Friedberg, A. Heifetz, and R. C. Greene, J. Biol. Chem. 246, 5822 (1971).
[14] S. J. Friedberg and A. Heifetz, Biochemistry 14, 570 (1975).
[15] A. J. Brown and F. Snyder, J. Biol. Chem. 258, 4184 (1983).
[16] S. J. Friedberg and R. D. Alkek, Biochemistry 16, 5291 (1977).
[17] P. A. Davis and A. K. Hajra, J. Biol. Chem. 254, 4760 (1979).
[18] S. J. Friedberg, D. M. Gomillion, and P. L. Stotter, J. Biol. Chem. 255, 1074 (1980).
[19] P. A. Davis and A. K. Hajra, Biochem. Biophys. Res. Commun. 74, 100 (1977).
[20] A. J. Brown and F. Snyder, J. Biol. Chem. 257, 8835 (1982).

velocity kinetics pointed to a Ping-Pong rather than a sequential reaction mechanism.[20]

Method for Measuring Enzyme Activity

Reagents and Substrates. The [1-^{14}C]hexadecanol can be obtained commercially or easily prepared by reduction of [1-^{14}C]palmitic acid (see this volume, [45]). Several synthetic routes have been employed to produce palmitoyl-DHAP. The original method by Piantadosi *et al.*[21] has generally been replaced by that of Davis and Hajra.[17] The latter procedure, however, involves several steps and requires reagent quantities of diazomethane. Brown and Snyder[15] developed a simpler, one-step method in which the pyridinium salt of DHAP is acylated with palmitic anhydride in the presence of the catalyst dimethylaminopyridine. DHAP is converted to the pyridinium salt by passing it through a 5-ml column of Dowex 50-X8 pyridinium form. The pyridinium salt of DHAP (10 μmol) is dried under nitrogen at 30° and then dried further overnight under reduced pressure over P_2O_5. Palmitic anhydride (30 μmol) is added in 1 ml of dry, ethanol-free chloroform. The reaction is stirred for 3 days at 22°. The reaction mixture is next applied directly to a 2-ml Unisil column equilibrated in chloroform, and, after washing with 60 ml of chloroform, 24-ml washes of chloroform–methanol at ratios of 19 : 1, 9 : 1, and 3 : 1 (v/v), a final wash is carried out with 24 ml methanol. The yield of palmitoyl-DHAP with this procedure ranges from 50 to 70%.

Preparation of Enzyme Source

Alkyl-DHAP synthase has been measured in a variety of tissues, and it appears to be localized in the microsomal fraction, although the activity has also been identified in a peroxisomal fraction isolated from guinea pig liver and mouse brain.[22] The enzyme is found in particularly high levels in Ehrlich ascites cells, and this has served as the major source of alkyl-DHAP synthase for mechanistic studies.

Preparation of Microsomes. Ehrlich ascites cells are conveniently grown in the peritoneal cavity of Swiss albino mice (HA/ICR). The cells are washed extensively with cold homogenization buffer (0.25 *M* sucrose, 0.1 *M* KCl, 0.1 *M* Tris-HCl, pH 7.4). All procedures should be done at 0°–4°. Erythrocytes can be eliminated by washing the cells once with cold water. Ehrlich ascites cells are not easily disrupted, and sonication is required. Three 15-sec bursts at full power is usually sufficient, but disrup-

[21] C. Piantadosi, K. Chae, K. S. Ishaq, and F. Snyder, *J. Pharm. Sci.* **61**, 971 (1972).
[22] C. L. Jones and A. K. Hajra, *Biochem. Biophys. Res. Commun.* **76**, 1138 (1977).

tion should be monitored microscopically. The microsomal fraction is isolated by differential centrifugation. After removal of cell debris, nuclei, and mitochondria by centrifugation at 10,000 g for 10 min, the microsomes are pelleted by centrifugation at 100,000 g for 60 min, washed once with homogenization buffer, and resuspended in the assay buffer. The microsomal preparations can be stored at $-20°$ for at least 1 month. Although the enzyme can be studied in whole microsomes, contaminating free fatty acids can act as a competitive inhibitor for incorporation of labeled fatty alcohol into alkyl-DHAP.[20] Therefore, it may be desirable to solubilize and delipidate the microsomes used in the alkyl-DHAP synthase preparation.[20]

Solubilization. The microsomes are resuspended in homogenization buffer at a protein concentration of 10 mg/ml, and 20% Triton X-100 is added to the stirred suspension to a final concentration of 1%. The mixture is then centrifuged at 100,000 g for 90 min. Approximately 90% of the alkyl-DHAP synthase activity remains in the supernatant.

Delipidation. To delipidate the enzyme, the supernatant obtained after solubilization is added dropwise to 10 volumes of acetone that is chilled to $-20°$. This procedure should be done rapidly (10–15 min) to avoid loss of enzyme activity. The protein forms a gummy precipitate that adheres to the beaker and allows the acetone to be decanted. Excess acetone is removed under a stream of nitrogen, and the protein is redissolved at a concentration of about 15 mg/ml in assay buffer or, if it is to be further purified, buffer A (20% ethylene glycol, 0.2% Triton X-100, 50 mM Tris-HCl, pH 9.0, and 1 mM dithiothreitol). Insoluble material is removed by centrifugation at 100,000 g for 60 min. This delipidation procedure removes approximately 80–85% of the total phospholipid and about 95% of the neutral lipids. This delipidation is required for further purification of alkyl-DHAP synthase. The ethylene glycol is essential for stabilizing the soluble alkyl-DHAP synthase. Triton X-100 prevents aggregation of the enzyme and also stimulates its activity. Therefore, the concentration of this detergent in the final assay should be maintained at 0.2% to achieve maximal activity.

Purification. Alkyl-DHAP synthase has been partially purified using several column chromatography steps.[15,20] The most effective of these are DEAE-cellulose and hydroxylapatite, but the details for all of the chromatography procedures are presented. The column sizes specified are adequate for about 500 mg of solubilized, delipidated protein.

The delipidated enzyme solution is dialyzed overnight against buffer A and applied to a DEAE-cellulose (DE-52) column (2.5 × 30 cm) equilibrated in this buffer. After washing with 2 column volumes of buffer A, a linear gradient from 0 to 200 mM KCl in buffer A (5 column volumes total) at a flow rate of 25 ml/hr is used to elute the alkyl-DHAP synthase. The active fractions are dialyzed against buffer B (same as buffer A but pH 7.4) and applied to a column of QAE-Sephadex A-25 (2.5 × 30 cm). After washing with 2 column volumes of buffer B, the alkyl-DHAP synthase is eluted with a linear gradient of 0 to 400 mM KCl in buffer B (5 column volumes) at 25 ml/hr. Fractions containing enzyme activity are diluted 2-fold with buffer B and applied to a Amicon (Danvers, MA) Matrex Red column (0.9 × 10 cm) equilibrated in the same buffer. After washing with 3 column volumes of buffer B, a 0 to 1 M KCl gradient is applied at 5 ml/hr. The active fractions are dialyzed against buffer C (20% ethylene glycol, 0.2% Triton X-100, 10 mM potassium phosphate, pH 7.4, and 1 mM dithiothreitol) and applied to an 8-ml hydroxylapatite column. The alkyl-DHAP synthase is eluted with a 60-ml gradient of 10 to 300 mM potassium phosphate in buffer C at 5 ml/hr. The specific activities of alkyl-DHAP synthase at various stages of purification are shown in Table I.

Enzyme Assay

The assay for alkyl-DHAP synthase monitors the formation of 1-O-[^{14}C]hexadecyl-DHAP from [1-^{14}C]hexadecanol and palmitoyl-DHAP. The procedure described below is applicable to both whole microsomes and purified preparations. Sodium fluoride must be included to inhibit phosphohydrolases that can cleave both the substrate and product. Triton X-100 prevents aggregation of the solubilized alkyl-DHAP synthase and stimulates the activity of both solubilized and microsomal enzyme. The

TABLE I
PURIFICATION OF ALKYLDIHYDROXYACETONEPHOSPHATE SYNTHASE[a]

Fraction	Protein		Activity		Specific activity (nmol–min/mg)	Purification (-fold)
	mg	%	nmol/min	%		
Microsomes	1600	100	104	100	0.065	1
Solubilized	230	14	100	96	0.43	6.6
DEAE-cellulose	43	2.6	58	56	1.4	22
QAE-Sephadex	10	0.63	28	27	2.8	43
Matrex Red	3.1	0.19	12	12	3.9	60
Hydroxylapatite	0.064	0.0040	4.3	4.1	67	1000

[a] Protein was assayed by the dye-binding method of Bradford.[23] Alkyl-DHAP synthase assays contained 0.25 M sucrose, 5% ethylene glycol, 0.2% Triton X-100, 50 mM NaF, 50 mM Tris-HCl (pH 8.0), 120 μM palmitoyl-DHAP, 50 μM [1-^{14}C]hexadecanol, and enzyme in a total volume of 250 μl. After incubation at 37° for 10 min, 75 μl of the reaction was applied to a DEAE-cellulose paper disk. The disks were washed 3 times for 15 min in 95% ethanol, dried, and counted in organic-based scintillation fluid.[20] Data are from Ref. 15.

Bradford[23] protein assay is recommended because of the interference of Triton X-100 in the Lowry method. The assay buffer consists of 100 mM Tris-HCl, pH 8.0, 100 mM KCl, 50 mM NaF, and 0.2% Triton X-100. The amount of enzyme used depends on the source, but usually 1 mg/ml of Ehrlich ascites cell microsomes or less of the purified fractions are used.[15] The two substrates are sonicated together into the assay buffer as a 10× stock preparation and added to the reaction mixture to give final concentrations of 50 μM [1-^{14}C]hexadecanol and 120 μM palmitoyl-DHAP. The reactions are initiated by combining the enzyme and substrate mixtures. After incubating at 37° for 10 min, the reactions (100 μl) are spotted onto DEAE-cellulose paper disks (Whatman, Clifton, NJ). The disks are then washed batchwise in ethanol (3 times) to remove unreacted [1-^{14}C]hexadecanol. The 1-O-[1-^{14}C]hexadecyl-DHAP binds quantitatively to the disk, which is air-dried and counted by liquid scintillation.[20] Tritiated hexadecanol should not be used in this procedure because of self-adsorption of this low energy β emission by the disk.

Identification of Products

To confirm the identity of the alkyl-DHAP product, the reaction mixture can be extracted by the method of Bligh and Dyer[24] with 2% acetic acid in the methanol. The chloroform layer is resolved by thin-layer chromatography (TLC) on silica gel HR plates with chloroform–methanol–acetic acid (90 : 10 : 10, v/v) using authentic hexadecyl-DHAP as a standard. The putative hexadecyl-DHAP can be analyzed further by NaBH$_4$ reduction to hexadecylglycerophosphate[17] followed by alkaline phosphatase treatment.[25] The hexadecylglycerol is analyzed on silica gel G plates developed in diethyl ether–ammonium hydroxide (100 : 1, v/v) using selachyl alcohol as a standard and I$_2$ vapors to visualize it. Final structural proof of the hexadecylglycerol can be obtained by subjecting it to cleavage with hydriodic acid,[26] sodium periodate,[27] or sodium periodate/sodium permanganate.[28] The products of these cleavage reactions are then analyzed by TLC on silica gel G layers developed in hexane–diethyl ether–acetic acid (50 : 50 : 1, v/v).

[23] M. M. Bradford, *Anal. Biochem.* **72**, 248 (1976).
[24] E. G. Bligh and W. J. Dyer, *Can. J. Biochem. Physiol.* **37**, 911 (1959).
[25] M. L. Blank and F. Snyder, *Biochemistry* **9**, 5034 (1970).
[26] D. J. Hanahan, *J. Lipid Res.* **6**, 350 (1965).
[27] W. J. Bauman, H. H. O. Schmidt, and H. K. Mangold, *J. Lipid Res.* **10**, 132 (1969).
[28] E. von Rudloff, *Can. J. Chem.* **34**, 1412 (1956).

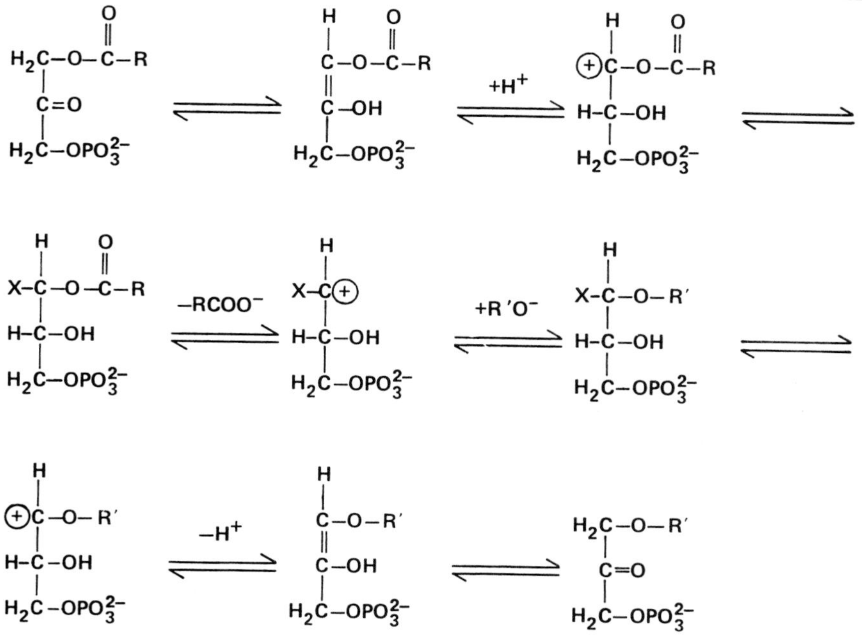

$$
\begin{array}{c}
\mathrm{H_2C-O-\overset{O}{\overset{\|}{C}}-R} \\
\mathrm{C{=}O} \\
\mathrm{H_2C-OPO_3^{2-}}
\end{array}
\;\rightleftharpoons\;
\begin{array}{c}
\mathrm{\overset{H}{\overset{|}{C}}-O-\overset{O}{\overset{\|}{C}}-R} \\
\mathrm{C-OH} \\
\mathrm{H_2C-OPO_3^{2-}}
\end{array}
\;\overset{+H^+}{\rightleftharpoons}\;
\begin{array}{c}
\mathrm{\underset{}{\overset{H}{\overset{|}{\oplus C}}}-O-\overset{O}{\overset{\|}{C}}-R} \\
\mathrm{H-C-OH} \\
\mathrm{H_2C-OPO_3^{2-}}
\end{array}
\;\rightleftharpoons
$$

$$
\begin{array}{c}
\mathrm{X-\overset{H}{\overset{|}{C}}-O-\overset{O}{\overset{\|}{C}}-R} \\
\mathrm{H-C-OH} \\
\mathrm{H_2C-OPO_3^{2-}}
\end{array}
\;\overset{-RCOO^-}{\rightleftharpoons}\;
\begin{array}{c}
\mathrm{X-\overset{H}{\overset{|}{\underset{}{C}\oplus}}} \\
\mathrm{H-C-OH} \\
\mathrm{H_2C-OPO_3^{2-}}
\end{array}
\;\overset{+R'O^-}{\rightleftharpoons}\;
\begin{array}{c}
\mathrm{X-\overset{H}{\overset{|}{C}}-O-R'} \\
\mathrm{H-C-OH} \\
\mathrm{H_2C-OPO_3^{2-}}
\end{array}
\;\rightleftharpoons
$$

$$
\begin{array}{c}
\mathrm{\overset{H}{\overset{|}{\oplus C}}-O-R'} \\
\mathrm{H-C-OH} \\
\mathrm{H_2C-OPO_3^{2-}}
\end{array}
\;\overset{-H^+}{\rightleftharpoons}\;
\begin{array}{c}
\mathrm{\overset{H}{\overset{|}{C}}-O-R'} \\
\mathrm{C-OH} \\
\mathrm{H_2C-OPO_3^{2-}}
\end{array}
\;\rightleftharpoons\;
\begin{array}{c}
\mathrm{H_2C-O-R'} \\
\mathrm{C{=}O} \\
\mathrm{H_2C-OPO_3^{2-}}
\end{array}
$$

Fig. 2. The molecular reaction catalyzed by alkyl-DHAP synthase in the formation of the alkyl ether bond is thought to proceed via a Ping-Pong mechanism (DHAP, Dihydroxyacetonephosphate). On binding of acyl-DHAP to alkyl-DHAP synthase (X), the pro-R hydrogen at C-1 is exchanged by an enolization of the ketone, followed by release of the acyl moiety to form an activated E–DHAP complex. C-1 is thought to carry a positive charge that may be stabilized by an essential sulfhydryl group of the enzyme; the incoming alkoxide ion reacts at C-1 to form alkyl-DHAP. R' can only be an alkyl moiety and can never be replaced by H.

Properties of Enzyme

Stability. In the presence of 20% ethylene glycol or glycerol, the solubilized and delipidated alkyl-DHAP synthase is stable for 1 month at 4°, but colder storage is recommended. In liquid nitrogen, the enzyme can be stored for at least 1 year with no loss of activity. In the absence of ethylene glycol, the activity is extremely labile ($t_{1/2} \sim 4$ hr). Alkyl-DHAP synthase is sensitive to sulfhydryl reagents,[29] and therefore buffers should contain dithiothreitol to prevent oxidation.

Kinetic Properties. Two important properties need to be considered when measuring alkyl-DHAP synthase activity. First, as mentioned above, fatty acids are a competitive inhibitor with respect to fatty alcohols in the

[29] A. J. Brown and F. Snyder, *Fed. Proc.* **39**, 1993 (1980).

formation of alkyl-DHAP. Thus, contaminating fatty acids will decrease the observed enzyme activity. Second, alkyl-DHAP synthase is inhibited by high concentrations of acyl-DHAP. Therefore, the concentration of acyl-DHAP used in the reaction (120 μM) gives optimal enzyme activity. Two explanations exist for this substrate inhibition. First, alkyl-DHAP synthase is inhibited by anionic detergents such as bile acids, and acyl-DHAP, an anionic amphipath, may inhibit the enzyme activity in this way. Second, as the concentration of acyl-DHAP is increased, contaminating acylhydrolases may release increasing amounts of fatty acids. This is supported by the observation of Brown and Snyder[20] that substrate inhibition by acyl-DHAP decreases with increasing fatty alcohol concentrations. The fatty alcohol overcomes the inhibition by competing with the fatty acid released from acyl-DHAP.

Reaction Mechanism. The major interest in alkyl-DHAP synthase is its unusual reaction mechanism. Figure 2 depicts the reaction scheme proposed by Brown and Snyder[15] which is based on earlier models.[18,19] It incorporates the exchange of the 1-pro-*R* hydrogen with retention of configuration at carbon 1, the donation of the ether oxygen by the fatty alcohol, and the release of the fatty acid with both acyl ester oxygens. Unlike aldolase which forms a Schiff base intermediate while exchanging the 1-pro-*S* hydrogen of DHAP, a Schiff base is not formed during alkyl-DHAP synthesis.[18,29,30] Evidence obtained by Brown and Snyder[20] and Davis and Hajra[19] suggest the Ping-Pong mechanism depicted here; however, until absolute proof of the active intermediate is provided, the sequential mechanism[18,29,30] cannot be excluded.

Acknowledgments

This work was supported by the Office of Energy Research, U.S. Department of Energy (Contract No. DE-AC05-760R00033), the American Cancer Society (Grant BC-70V), the National Heart, Lung, and Blood Institute (Grant 27109-10), and the National Cancer Institute (CA-41642-05).

[30] P. A. Davis and A. K. Hajra, *Arch. Biochem. Biophys.* **211**, 20 (1981).

[45] Alkyldihydroxyacetonephosphate Synthase from Guinea Pig Liver Peroxisomes

By SHUICHI HORIE, ARUN K. DAS, and AMIYA K. HAJRA

Introduction

$$R—C(=O)—O—CH_2—C(=O)—CH_2—O—PO_3^{2-} + R'OH \rightarrow$$
$$R'—O—CH_2—C(=O)—CH_2—O—PO_3^{2-} + RCOOH \quad (1)$$

Alkyldihydroxyacetonephosphate (alkyl-DHAP) synthase (EC 2.5.1.26, alkylglycerone-phosphate synthase[1]) catalyzes the biosynthesis of an ether bond [Eq. (1)] in different animal systems.[2–6] The enzyme was originally described to be in guinea pig liver mitochondrial fractions and rat brain microsomes.[2] Later findings, however, showed that this membrane-bound enzyme is mainly present in peroxisomes of animals.[7] The guinea pig liver enzyme has been solubilized and partially purified[8] as described here.

Assay Method

Principle. The present assay is based on the separation of the radioactive ionic product, 1-*O*-[1'-^{14}C]hexadecyl-DHAP from the nonionic radioactive precursor, [1-^{14}C]hexadecanol, by solvent partition in a chloroform–methanol–water system at high pH.[3]

$$[1\text{-}^{14}C]\text{Hexadecanol} + \text{palmitoyl-DHAP} \rightarrow [1'\text{-}^{14}C]\text{hexadecyl-DHAP} + \text{palmitate} \quad (2)$$

Substrates. Because neither [1-^{14}C]hexadecanol nor palmitoyl-DHAP is commercially available, their preparation[3,9] is described below.

---

[1] The enzyme is classified as a "transferase" by the IUB in *Enzyme Nomenclature* (1984). However, this enzyme probably should be classified as a "lyase" because it catalyzes a reaction similar to that catalyzed by methionine synthase (EC 4.2.99.10), a thioether bond-forming enzyme.

[2] A. K. Hajra, *Biochem. Biophys. Res. Commun.* **39**, 1037 (1970).

[3] P. A. Davis and A. K. Hajra, *Arch. Biochem. Biophys.* **211**, 20 (1981).

[4] P. A. Davis and A. K. Hajra, *J. Biol. Chem.* **254**, 4760 (1979).

[5] S. J. Friedberg, S. T. Weintraub, M. R. Singer, and R. C. Green, *J. Biol. Chem.* **258**, 136 (1983).

[6] A. J. Brown and F. Snyder, *J. Biol. Chem.* **258**, 4148 (1983).

[7] A. K. Hajra and J. E. Bishop, *Ann. N.Y. Acad. Sci.* **386**, 170 (1982).

[8] A. K. Das, S. Horie, and A. K. Hajra, *FASEB J.* **4**, A1826 (1990).

[9] A. K. Hajra, T. V. Saraswathi, and A. K. Das, *Chem. Phys. Lipids* **33**, 179 (1983).

Synthesis of Palmitoyl-DHAP

Reagents

Glycolic acid, dry by keeping overnight over P_2O_5 under reduced pressure in a desiccator

Pyridine, dried over CaH_2

Chloroform, dry by distilling from P_2O_5

Palmitoyl chloride

Diazald (Aldrich Chemical Co., Milwaukee, WI)

Phosphoric acid, 85%

Diethyl ether, anhydrous

Palmitoylglycolic Acid. To 1 g of glycolic acid, suspended in 3 ml pyridine and 2 ml dry chloroform in an ice-water bath (10°), 3.3 g of palmitoyl chloride in dry chloroform (6 ml) is slowly added with continuous stirring. The mixture is stirred for 1.5 hr at 10° and for 1 hr at room temperature. Most of the pyridine from the slurry is removed with a stream of N_2; the mixture is dissolved in diethyl ether, washed 2 times with 0.2 N aqueous HCl, followed by 2 times with water. The ether extract is dried with anhydrous Na_2SO_4, and the ether is removed by blowing N_2 over the sample. The residue (palmitoyl glycolate) is purified by crystallization from *n*-heptane.

Palmitoylglycolyl Chloride. To 0.5 g of palmitoylglycolic acid, 1 ml of dry benzene and 1 ml of oxalyl chloride are added, and the mixture is incubated at room temperature for 15 hr. Excess oxalyl chloride and benzene are evaporated off under a stream of N_2, and the residue is dissolved in 5 ml anhydrous ether.

Palmitoyloxydiazoacetone. An ethereal solution of distilled CH_2N_2 is prepared from Diazald (Aldrich) according to the manufacturer's directions. To 15 nmol (in 30 ml ether) of CH_2N_2 in a 50-ml round-bottomed flask cooled to $-5°$ (salt–ice bath), the above palmitoylglycolyl chloride in 5 ml ether is slowly added with constant stirring (magnetic stirrer). The mixture is stirred for an additional 30 min at $-5°$ and 30 min at room temperature. Excess diazomethane and most of the ether is evaporated off by blowing N_2 over the mixture, and the residue (palmitoyloxydiazoacetone) is crystallized from anhydrous diethyl ether.

Palmitoyldihydroxyacetone Phosphate. Fifty milligrams of the diazo compound is dissolved in 0.5 ml dioxane and is added dropwise to 87% H_3PO_4 (0.1 ml) in dioxane (0.5 ml) which is heated at 70°–72° in a Reactivial (Pierce Chemical Co., Rockford, IL) under constant magnetic stirring. The mixture is stirred for 1 hr, then cooled, and the contents are dissolved in 4 ml ether. The ethereal solution is washed 2 times with water, and the ether is removed under a stream of N_2. Traces of water from the residue are removed by adding dry benzene and evaporating the solvents off by blowing N_2.

The dry residue is dissolved in a small volume of chloroform and transferred to a 5 g silicic acid (Unisil, Clarkston Chemical Co.) column (1 cm i.d.). The column is first eluted with 80 ml chloroform to remove palmitoyldihydroxyacetone (40%) and then with 100 ml chloroform–methanol (7 : 3, v/v) to elute palmitoyldihydroxyacetone phosphate (60%). The second fraction containing palmitoyl-DHAP can be directly used for the alkyl-DHAP synthase assay. The amount of palmitoyl-DHAP present in this fraction is determined by measuring the inorganic phosphate liberated after treatment with alkali.[9]

Preparation of [1-^{14}C]Hexadecanol

To 0.5 mCi of [1-^{14}C]palmitic acid dried in a 20 × 125 mm screw-capped tube, 2 ml of 2,2-dimethoxypropane (Aldrich), 1 ml methanol, and 10 μl of concentrated HCl are added.[3] After 1.5 hr the mixture is evaporated to dryness by blowing a stream of N_2 over the sample, the residue is dissolved in 2.5 ml diethyl ether–toluene (4 : 1, v/v), and 0.5 ml of Red-Al (70% in toluene, Aldrich) is added. The mixture (tightly capped) is incubated at 37° for 1 hr. The excess Red-Al is destroyed by slow addition of 10 ml 20% ethanol, and the radioactive alcohol is extracted with diethyl ether. This [1-^{14}C]hexadecanol can be used directly for the alkyl-DHAP synthase assay; however, if impurities are present, as revealed by thin-layer chromatography (TLC), then it should be purified by TLC (E. Merck, Darmstadt, Germany, silica gel 60 plate) using hexane–ether (8 : 2, v/v) as the developing solvent. The radioactive hexadecanol, which is diluted with nonradioactive hexadecanol to the required specific activity, is stored in toluene. The hexadecanol in toluene should be washed with 0.1 N aqueous NaOH before use.

Assay of Synthase

Reagents

0.3 M Tris-HCl, pH 8.0
0.1 M NaF
1 M Tris base
[1-^{14}C]Hexadecanol, 30,000 counts/min (cpm)/nmol in toluene
Palmitoyl-DHAP in chloroform-methanol (see above)
Chloroform–methanol (1 : 2, v/v)
Chloroform
Technique. To each tube, add 75 nmol of [1-^{14}C]hexadecanol in toluene

and 25 nmol of palmitoyl-DHAP in chloroform–methanol. The solvents are evaporated off under a stream of N_2. To the dried residue, 75 μl of Tris-HCl buffer and 50 μl of NaF are added, and the mixture is sonicated in an ultrasonic bath to disperse the substrates. The reaction is started by adding water and enzyme to the emulsified substrates to make the final volume 0.3 ml. The mixture is briefly sonicated to aid dispersal and then incubated at 37° for 30 min. The reaction is stopped by adding 1.13 ml chloroform–methanol (1 : 2, v/v) followed by 0.4 ml chloroform and 0.4 ml of 1 M Tris base. The mixture is vortexed well, then centrifuged, and the upper layer is transferred to a tube containing 1.0 ml chloroform.[10] This is mixed well by vortexing and centrifuged in a tabletop centrifuge (1000 g, 10 min), and the radioactivity in an aliquot of the washed upper layer (0.4 ml/1.4 ml) is determined by liquid scintillation counting.

Definition of Unit. One unit of activity is defined as the amount of enzyme which catalyzes the formation of 1 nmol (30,000 cpm) of product per minute at 37°.

Solubilization and Purification of Enzyme

Isolation of Peroxisomes. The peroxisomes from guinea pig liver are isolated as described elsewhere in this volume.[11]

Preparation of Peroxisomal Membranes. Peroxisomes (65 mg protein) are suspended in 200 ml of 10 mM pyrophosphate buffer (pH 9.0)–1 mM dithiothreitol (DTT) and gently stirred at 4° for 4–6 hr. The mixture is centrifuged at 100,000 g for 30 min, and the supernatant is discarded.

Solubilization of Enzyme. The residue obtained after centrifugation is suspended in 4 ml of 0.25 M sucrose to which 16 ml of a solution containing Tris-HCl (10 mM, pH 7.5)–DTT (1 mM)–glycerol (20%, v/v)–diethylene glycol (20%, v/v)-3-[(3-cholamidopropyl)dimethylammonio]-1-propane-sulfonate (CHAPS) (0.25%, w/v) is added and stirred for 30 min at 4°. The mixture is centrifuged at 100,000 g for 1 hr, the supernatant is discarded, and the residue is suspended in 10 ml of a solution containing Tris-HCl (10 mM, pH 7.5)–DTT (1 mM)–KCl (0.5 M)–glycerol (20%, v/v)–diethylene glycol (20%, v/v)–CHAPS (0.5%, w/v) and gently stirred at 4° for 1 hr. The suspension is centrifuged at 100,000 g for 1 hr, and the supernatant is used for hydroxylapatite chromatography.

Hydroxylapatite Column Chromatography. The supernatant is diluted 5 times with the solubilizing buffer (as above) to a protein concentration

---

[10] The chloroform wash is done to reduce the blank from 700–800 to 150–200 cpm. The upper layer could be counted directly[3] if high blank values are not objectionable.

[11] K. O. Webber and A. K. Hajra, this volume [10].

TABLE I
PURIFICATION OF ALKYLDIHYDROXYACETONEPHOSPHATE SYNTHASE

Fraction	Total protein (mg)	Total activity (units)	Specific activity (units/mg protein)	Yield (%)
Liver homogenate	5100	357	0.07	100
Peroxisomes	65	109	1.68	30
Peroxisomal membranes	44	101	2.29	28
Solubilized enzyme	33	88	2.67	25
Hydroxylapatite column	8	58	7.14	16

of 5 mg/ml and then loaded onto a 1.1 × 10 cm hydroxylapatite column (Bio-Rad, Richmond, CA). The column is eluted with a solution containing CHAPS (0.5%, w/v)–ethylene glycol (20%, w/v)–DTT (1 mM)–potassium phosphate (pH 7.5, 20 mM) until no 260 nm-absorbing material is detected in the effluent. A 60-ml gradient of increasing potassium phosphate buffer (pH 7.5) from 20 to 500 mM, which also contains CHAPS (0.5%), ethylene glycol (20%), and DTT (1 mM), is applied at a rate of 8 ml/hr. Fractions of 2.5 ml are collected in siliconized glass tubes, and the protein content and enzyme activity are determined for each fraction. Alkyl-DHAP synthase is generally eluted in fractions 25 to 35. These fractions are pooled together and stored at 4°. The result of a typical purification is shown in Table I.

Properties of Synthase

Optimum pH. The enzyme has a broad pH optimum between pH 8.0 and 9.0.

Michaelis Constants. The apparent Michaelis constants, K_m, for palmitoyl-DHAP and hexadecanol are 45 and 40 μM, respectively. The enzyme activity is inhibited by palmitoyl-DHAP above 100 μM and also by high concentrations (>0.5 mM) of hexadecanol.

Stability. The partially purified enzyme slowly loses 50% of its activity in 10 days when stored at 4°. The enzyme activity was lost on freezing and thawing.

Specificity. The enzyme catalyzes a stereospecific exchange of the pro-R hydrogen at C-1 of the DHAP moiety during the formation of the ether bond.[4–6] The enzyme is specific for acyl-DHAP; acyl-DHA or 1-acyl-*sn*-glycerol 3-phosphate will not substitute for acyl-DHAP. The acyl group chain length in acyl-DHAP should be C_{12} or longer. Various long-chain fatty alcohols (>C_{10}) including diols act as substrates.[12]

[12] A. K. Hajra, C. L. Jones, and P. A. Davis, *Adv. Exp. Med. Biol.* **101**, 369 (1978).

Equilibrium Constant. Attempts to show the reversibility of the reaction have been unsuccessful.[3]

Inhibitors. Fatty acids are strong inhibitors of the enzymatic reaction. This is probably because the enzyme catalyzes an exchange reaction between free fatty acids and the acyl group of acyl-DHAP.[13] An analog of acyl/alkyl-DHAP, 3-bromo-2-ketoheptadecyl phosphate, specifically inhibits the enzyme.[8]

[13] P. A. Davis and A. K. Hajra, *Biochem. Biophys. Res. Commun.* **74**, 100 (1977).

[46] Plasmanylethanolamine Δ1-Desaturase

By MERLE L. BLANK and FRED SNYDER

Introduction

1-Alk-1′-enyl-2-acyl-*sn*-glycero-3-phosphoethanolamine (alk-1-enyl-acyl-GPE), also known as plasmenylethanolamine, is a significant phospholipid component of many mammalian tissues.[1,2] Plasmenylethanolamine contains relatively high concentrations of polyunsaturated fatty acids, and this phospholipid subclass has been proposed as a storage site and/or source of arachidonic acid.[3–8] It has also been suggested that ethanolamine plasmalogens might function as possible scavengers of oxidative free radicals in biological membranes.[9] Plasmanylethanolamine desaturase (EC 1.14.99.19) is a membrane-bound enzyme that converts plasmanylethanolamine (1-alkyl-2-acyl-*sn*-glycero-3-phosphoethanolamine, alkylacyl-GPE) to plasmenylethanolamine via an electron transport chain involving cytochrome b_5, molecular oxygen, and either NADPH or NADH

[1] L. A. Horrocks, *in* "Ether Lipids Chemistry and Biology" (F. Snyder, ed.), p. 177. Academic Press, New York, 1972.

[2] T. Sugiura and K. Waku, *in* "Platelet-Activating Factor and Related Lipid Mediators" (F. Snyder, ed.), p. 55. Plenum, New York, 1987.

[3] M. L. Blank, R. L. Wykle, and F. Snyder, *Biochim. Biophys. Acta* **316**, 28 (1973).

[4] S. Rittenhouse-Simmons, F. A. Russell, and D. Deykin, *Biochem. Biophys. Res. Commun.* **70**, 295 (1976).

[5] M. J. Broekman, J. W. Ward, and A. J. Marcus, *J. Clin. Invest.* **66**, 275 (1980).

[6] L. W. Daniel, L. King, and M. Waite, *J. Biol. Chem.* **256**, 12830 (1981).

[7] R. M. Kramer and D. Deykin, *J. Biol. Chem.* **258**, 13806 (1983).

[8] F. H. Chilton and T. R. Connell, *J. Biol. Chem.* **263**, 5260 (1988).

[9] R. A. Zoeller, O. H. Morand, and C. R. H. Raetz, *J. Biol. Chem.* **263**, 11590 (1988).

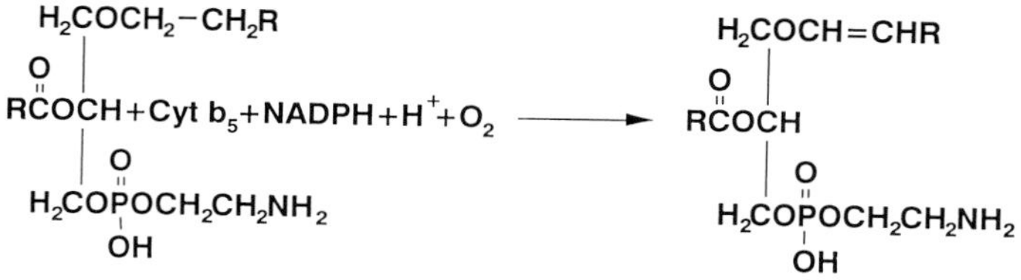

$$H_2COCH_2-CH_2R$$
$$\overset{O}{\underset{|}{\|}}$$
$$R\overset{\|}{C}OCH+Cyt\ b_5+NADPH+H^++O_2 \longrightarrow$$
$$\underset{|}{\overset{O}{\|}}$$
$$H_2CO\overset{\|}{P}OCH_2CH_2NH_2$$
$$OH$$

$$H_2COCH=CHR$$
$$\overset{O}{\underset{|}{\|}}$$
$$R\overset{\|}{C}OCH$$
$$\underset{|}{\overset{O}{\|}}$$
$$H_2CO\overset{\|}{P}OCH_2CH_2NH_2$$
$$OH$$

FIG. 1. Reaction catalyzed by plasmanylethanolamine Δ¹-desaturase.

as cofactors (Fig. 1). The majority of research that characterized this microsomal desaturase system was completed during the early 1970s and has been summarized by Snyder et al.[10] Only the assay system currently used in our laboratory is described in detail in this chapter.

Method for Measuring Enzyme Activity

Reagents and Substrates. Materials needed for the enzyme assay include 0.5 M Tris-HCl buffer (pH 7.2), 20 mM NADH (or NADPH), catalase (from bovine liver), and [³H]alkyllyso-GPE. All materials except the last compound are commercially available (Sigma Chemical Co., St. Louis, MO, as well as other vendors). Endogenous radiolabeled alkylacyl-GPE generated *in situ* by microsomal acylation of [³H]alkyllyso-GPE is a much better substrate than when alkylacyl-GPE is added exogenously[11,12] to the assay system.

The radiolabeled alkyllyso-GPE is normally prepared as a precursor for the desaturase assay by enzymatic synthesis in intact cells known to produce significant quantities of alkyl/alk-1-enyl-containing ethanolamine phospholipids. Although the procedure is described using [³H]hexadecanol as the source of the alkyl chain, [¹⁴C]hexadecanol can also be used to produce [¹⁴C]alkyllyso-GPE. First, [³H]hexadecanol is prepared by reduction of [9,10-³H]palmitic acid (Du Pont–NEN, Boston, MA) with Vitride.[13] The [³H]hexadecanol is then incubated with either Ehrlich ascites cells (EAC)[11,14] or Madin–Darby canine kidney (MDCK) cells[15] in order to

[10] F. Snyder, T.-c. Lee, and R. L. Wykle, *in* "The enzymes of Biological Membranes" (A. N. Martonosi, ed.), Vol. 2, p. 1. Plenum, New York, 1985.

[11] R. L. Wykle, M. L. Blank, B. Malone, and F. Snyder, *J. Biol. Chem.* **247**, 5442 (1972).

[12] R. L. Wykle and J. M. S. Lockmiller, *Biochim. Biophys. Acta* **380**, 291 (1975).

[13] F. Snyder, M. L. Blank, and R. L. Wykle, *J. Biol. Chem.* **246**, 3639 (1971).

[14] M. L. Blank, Z. L. Smith, Y. J. Lee, and F. Snyder, *Arch. Biochem. Biophys.* **269**, 603 (1989).

[15] M. L. Blank, T.-c. Lee, E. A. Cress, V. Fitzgerald, and F. Snyder, *Arch. Biochem. Biophys.* **251**, 55 (1986).

incorporate the labeled alcohol into alkyl/alk-1-enyl groups of the cellular diradyl-GPE. After appropriate incubation times, 16–24 hr with MDCK cells[15] and with mice bearing EAC[11] or only 2 hr with EAC incubated *in vitro*,[14] the labeled diradyl-GPE fraction is isolated from the cellular lipids and either hydrogenated to convert alk-1-enyl groups to alkyl moieties[15] or treated with HCl to hydrolyze the alk-1-enyl groups.[11,14] All acyl groups are then removed by treatment of the alkylacyl-GPE with a monomethyl-amine reagent[16]; the product, [³H]alkyllyso-GPE, is finally isolated by preparative thin-layer chromatography (TLC).[11,14,15] Benzoylation of the [³H]alkylglycerols, produced by Vitride reduction of the [³H]alkyllyso-GPE, and subsequent high-performance liquid chromatography (HPLC) analysis of the benzoate derivatives[17] makes it possible to determine both the alkyl chain-length distribution (~90% hexadecyl) and the specific radioactivity.

An alternative approach to preparation of the substrate utilizes the base-exchange properties of phospholipase D from cabbage to exchange the choline group of commercially available [³H]alkylacetyl-GPC for etha-nolamine to produce [³H]alkylacetyl-GPE.[18] The acetate group is then removed with monomethylamine.[16] [³H]Alkyllyso-GPE is relatively stable when stored in chloroform–methanol (2 : 1, v/v) at −20° with a 94% radio-purity by TLC after 1 year of storage. Appropriate aliquots of the substrate are evaporated to dryness and redissolved in ethanol for addition to the incubation tubes used for the Δ^1-desaturase assay.

Preparation of Enzyme Source. Tissues and cells used as enzyme sources to study plasmanylethanolamine desaturase include Ehrlich asci-tes cells,[11,13,19,20] Fischer R-3259 sarcomas,[11,21–23] rat brains,[12,24] preputial gland tumors,[13] Madin–Darby canine kidney cells,[15] P388D₁ cells,[14] and hamster small intestines.[25,26] The plasmanylethanolamine desaturase re-

[16] N. G. Clarke and R. M. C. Dawson, *Biochem. J.* **195,** 301 (1981).

[17] M. L. Blank, E. A. Cress, T.-c. Lee, N. Stephens, C. Piantadosi, and F. Snyder, *Anal. Biochem.* **133,** 430 (1983).

[18] T.-c. Lee, C. Qian, and F. Snyder, *Arch. Biochem. Biophys.* **286,** 498 (1991).

[19] R. L. Wykle, M. L. Blank, and F. Snyder, *FEBS Lett.* **12,** 57 (1970).

[20] M. L. Blank, R. L. Wykle, and F. Snyder, *FEBS Lett.* **18,** 92 (1971).

[21] T.-c. Lee, R. L. Wykle, M. L. Blank, and F. Snyder, *Biochem. Biophys. Res. Commun.* **55,** 574 (1973).

[22] R. C. Baker, R. L. Wykle, J. S. Lockmiller, and F. Snyder, *Arch. Biochem. Biophys.* **177,** 299 (1976).

[23] R. L. Wykle and J. M. Schremmer, *Biochemistry* **18,** 3512 (1979).

[24] M. L. Blank, R. L. Wykle, and F. Snyder, *Biochem. Biophys. Res. Commun.* **47,** 1203 (1972).

[25] F. Paltauf, *FEBS Lett.* **20,** 79 (1972).

[26] F. Paltauf and A. Holasek, *J. Biol. Chem.* **248,** 1609 (1973).

sides in the microsomal membrane fraction, although postmitochondrial supernatant fractions have also been used in some Δ^1-desaturase studies. Subcellular fractions are isolated by conventional centrifugation techniques from homogenates prepared in 0.25 M sucrose. The final microsomal pellet (100,000 g, 60 min) is suspended in water (5–10 mg protein/ml) and used immediately in the Δ^1-desaturase assay. Furthermore, because stimulatory proteins can be present in the microsomal supernatant fraction,[22,26] it is always advisable to check the effect of the soluble fraction on the microsomal desaturase activity.

Enzyme Assay. Assay of the Δ^1-desaturase activity is carried out in glass tubes on ice containing 0.2 ml of the Tris-HCl buffer (pH 7.2), 0.1 ml of the NADH (or NADPH) solution, the enzyme source (0.5–5 mg protein, depending on activity), and sufficient water to make a final volume of 1 ml. To establish whether the desaturase activity is inhibited by the generation of H_2O_2 in the microsomal preparations, a set of incubation tubes containing catalase (~33 μg/ml) in addition to NADH should be included.[22] The reaction in the incubation tube is initiated by adding 2 nmol of [³H]alkyllso-GPE (~1 μCi) in 20 μl of ethyl ether–ethanol (2 : 1, v/v) or in 10 μl of ethanol and placing the uncapped tube with its well-mixed contents in a shaking water bath at 37° for 30 to 60 min. At the end of the incubation, total lipids are extracted by the method of Bligh and Dyer[27] and stored in an appropriate volume of chloroform at −20° until analyzed.

Identification of Products. Methodology for the detailed identification of the plasmenylethanolamine product has been previously described.[13] More recent HPLC techniques for analysis of alkylacyl- and alk-1-enyl-acylglycerobenzoates derived from the diradyl-GPE fraction[28,29] have also been used to analyze these lipids.[14,15] If the objective is to determine the molecular species of the ethanolamine plasmalogens or precursors, then the HPLC methodology would be the most appropriate choice. However, if only the extent of acylation of the alkyllyso-GPE and the degree of desaturation to plasmenylethanolamine need to be known, a two-dimensional TLC system involving HCl cleavage of the vinyl ether double bond of plasmenylethanolamine can be used.[11]

Briefly, the TLC technique consists of using silica gel H-coated plates, the labeled lipids isolated from the Δ^1-desaturase assay, along with standards of phosphatidylethanolamine from beef brain (contains plasmalogens) and lysophosphatidylethanolamine (both commercially available),

[27] E. G. Bligh and W. J. Dyer, *Can. J. Biochem. Physiol.* **37**, 911 (1959).
[28] M. L. Blank, M. Robinson, V. Fitzgerald, and F. Snyder, *J. Chromatogr.* **298**, 473 (1984).
[29] M. L. Blank, E. A. Cress, and F. Snyder, *J. Chromatogr.* **392**, 421 (1987).

which are separated in the first TLC dimension using chloroform–methanol–concentrated ammonium hydroxide (65 : 35 : 5, v/v). The TLC plate is then air-dried for a few minutes and exposed to HCl gas (50–100 ml of fresh, concentrated HCl in the bottom of the tank) for 15 min in a closed TLC tank. The HCl-exposed plate is then placed under reduced pressure in a desiccator for 30 min. After removal from the desiccator, the plate is rotated 90° from the direction of the first TLC development and chromatographed in the second dimension with a solvent system of hexane–diethyl ether–glacial acetic acid (60 : 40 : 1, v/v). After development, the TLC plate is exposed to iodine vapors to locate the separated products. To prevent quenching, the iodine is allowed to sublime from the TLC plate (in a hood at room temperature) before the appropriate areas of silica gel are scraped into vials containing scintillation fluid and the amounts of radioactivity measured. The most important areas of the TLC plate (areas 1–3, Fig. 2) for a radioassay are shown in the facsimile of a typical TLC separation. The amount of substrate acylated and the amount of plas-

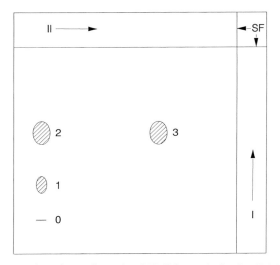

FIG. 2. Representation of two-dimensional TLC for analysis of radiolabeled lipids from the Δ^1-desaturase assay. The sample is applied to the TLC plate at the location shown by 0. Roman numerals give the direction of solvent development in the first (I) and second (II) dimensions. Solvent mixture I is chloroform–methanol–ammonium hydroxide (65 : 35 : 5, by volume), and solvent II consists of hexane–diethyl ether–acetic acid (60 : 40 : 1, by volume). The plate is exposed to HCl fumes between these two developments. The solvent fronts are indicated by SF, and the numbers 1–3 show the location of [^{3}H]alkyllyso-GPE (1), [^{3}H]alkylacyl-GPE (2), and [^{3}H]aldehyde (3) from acid cleavage of the vinyl ether group of plasmenylethanolamine.

manylethanolamine desaturated to plasmenylethanolamine can be calculated from the amounts of radioactivity found in these areas.

Properties of Enzyme

The postmitochondrial fraction from Fischer sarcomas gradually loses plasmenylethanolamine desaturase activity when stored at $-23°$[11]; therefore, only freshly isolated subcellular fractions should be used in the assay of the Δ^1-desaturase. The desaturase activity is nearly linear with time up to 30 min,[11] and it exhibits a direct correlation between an increase in activity and the amount of protein used in the assays.[21] However, because the Δ^1-desaturase activity in the assay described involves at least two different enzymes [an alkyllyso-GPE transacylase (acyltransferase) and the Δ^1-desaturase] together with a complex electron transport chain, any measurement of apparent K_m and V_{max} values is not only difficult but probably not very meaningful.

Similar to the microsomal cytochrome b_5 electron transport system that desaturates fatty acids,[30] the plasmanylethanolamine desaturase requires oxygen and a reduced pyridine nucleotide, and it is inhibited by cyanide but not by carbon monoxide.[11,25] Evidence for the involvement of cytochrome b_5 in the desaturase of plasmanylethanolamine was based on results obtained with a specific antibody[31] against cytochrome b_5. However, the Δ^1- and Δ^9-desaturase systems differ in their response to dietary changes and therefore may involve different cyanide-sensitive factors in the electron transport chain.[21]

Cytosolic fractions (105,000 g, 60 min, supernatant) from some tissues contain factors that stimulate the microsomal desaturase activity. This stimulatory factor has been identified as catalase in the desaturation of plasmanylethanolamine by Fischer sarcoma tissue.[22] The catalase appears to exert its effect through its ability to decompose H_2O_2, which would otherwise inactivate cytochrome b_5. However, soluble, stimulatory proteins other than catalase have also been described.[32,33] These cytosolic proteins that stimulate the Δ^1-desaturase have apparent molecular weights of 27,000. The exact functional mechanism and structure of these proteins remain to be defined.

[30] N. Oshino, Y. Imai, and R. Sato, *Biochim. Biophys. Acta* **128**, 13 (1966).
[31] F. Paltauf, R. A. Prough, B. S. S. Masters, and J. M. Johnston, *J. Biol. Chem.* **249**, 2661 (1974).
[32] F. Paltauf, *Adv. Exp. Med. Biol.* **101**, 378 (1978).
[33] F. Paltauf, *Eur. J. Biochem.* **85**, 263 (1978).

Acknowledgments

This work was supported by the Office of Energy Research, U.S. Department of Energy (Contract No. DE-AC05-760R0033), the American Cancer Society (Grant BC-70V), the National Heart, Lung, and Blood Institute (Grant 27109-10), and the National Cancer Institute (CA-41642-05).

[47] 1-Alkyl-2-lyso-sn-glycero-3-phosphocholine Acetyltransferase

By TEN-CHING LEE, DAVID S. VALLARI, and FRED SNYDER

Introduction

Acetyl-CoA : 1-alkyl-2-lyso-sn-glycero-3-phosphocholine (lyso-PAF) acetyltransferase (EC 2.3.1.67, 1-alkyl-2-lyso-sn-glycero-3-phosphocholine acetyltransferase) catalyzes the final step in the remodeling pathway[1-3] of platelet-activating factor (PAF) biosynthesis (Fig. 1). It has completely different characteristics than the acetyltransferase involved in the *de novo* pathway of PAF biosynthesis.[4] This enzyme activity is stimulated by various inflammatory agents that cause the overproduction of PAF in hypersensitivity responses.[4] The actual activation of the alkyllysoglycerophosphocholine (alkyllyso-GPC) acetyltransferase by a phosphorylation mechanism involving a protein kinase is still poorly understood. The lyso-PAF substrate for this acetyltransferase can be derived from the precursor membrane pool of alkylarachidonoylglycerophosphocholines through the action of a putative phospholipase A_2 or, as recently shown, via generation by a CoA-independent transacylase.[5] Thus, the regulation of PAF formation by the remodeling pathway is extremely complex since one needs to consider factors that affect the activation of acetyltransferase, protein kinases, phospholipase A_2, a transacylase, and catabolic enzymes that utilize PAF or lyso-PAF as substrates. The intent of this chapter is to describe only the optimal conditions for assaying the acetyltransferase that acetylates lyso-PAF and to discuss its known properties.

[1] R. L. Wykle, B. Malone, and F. Snyder, *J. Biol. Chem.* **255,** 10256 (1980).
[2] D. J. Lenihan and T.-c. Lee, *Biochem. Biophys. Res. Commun.* **120,** 834 (1984).
[3] T.-c. Lee, *J. Biol. Chem.* **260,** 10952 (1985).
[4] F. Snyder, *Am. J. Physiol.* **259,** C697 (1990).
[5] Y. Uemura, T.-c. Lee, and F. Snyder, *J. Biol. Chem.* **266,** 8268 (1991).

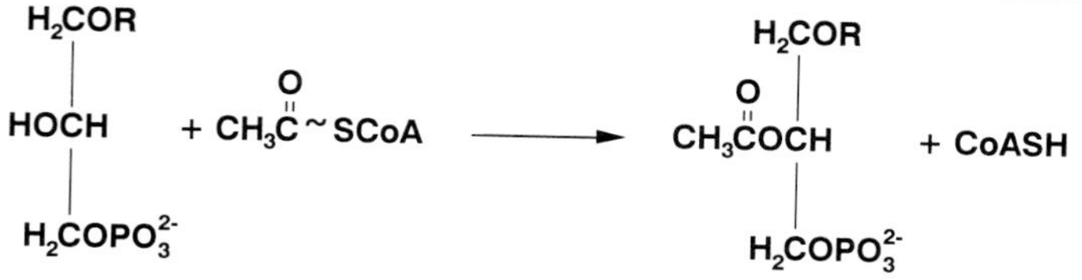

FIG. 1. Reaction catalyzed by acetyl-CoA : 1-alkyl-2-lyso-*sn*-glycero-3-phosphocholine acetyltransferase.

Methods for Measuring Enzyme Activity

Reagents. Two methods are described to measure the activity of acetyltransferase. Reagents for method I are 0.5 M Tris-HCl (pH 6.9, at 37°), 2 mM [³H]acetyl-CoA (1 μCi/μmol) in water, and 2.5 mM hexadecyllyso-GPC in chloroform. Reagents for method II are 0.5 M Tris-HCl (pH 6.9, at 37°), 5 mM [³H]acetyl-CoA (0.8 μCi/μmol) in water, 2.5 mM hexadecyllyso-GPC in chloroform, and 0.5 mM bovine serum albumin (BSA) (fatty-acid free).

Preparation of Enzyme Source. The assay procedure is based on studies done with rat spleen microsomes for method I[1-3] and with differentiated HL-60 cells for method II.[6] Spleens from adult male CDF rats are homogenized (10%, v/v) in 0.25 M sucrose, 20 mM Tris-HCl (pH 7.4), 10 mM EDTA, 5 mM mercaptoethanol, 50 mM NaF, with four strokes of a motor-driven Potter–Elvehjem homogenizer. Differentiated HL-60 cells (2–4 × 10⁸ cells), suspended in 3 ml of the medium described above, are sonicated in an ice bath with three 10-sec bursts. The homogenates are first centrifuged at 15,000 g for 10 min at 4°, and the microsomal fraction is prepared by centrifuging the supernatant at 105,000 g for 60 min at 4°. The microsomal pellet (20–30 mg protein/g spleen, 1–1.5 mg protein/10⁸ HL-60 cells) is then suspended in 0.25 M sucrose–20 mM Tris-HCl (pH 7.4) with 1 mM dithiothreitol and used immediately for the enzyme assay.

Enzyme Assay. The incubation mixture for method I contains hexadecyllyso-GPC (added to the tube in 10 μl chloroform, which is then removed under nitrogen), 0.2 ml Tris-HCl buffer, 0.1 ml [³H]acetyl-CoA, spleen microsomal proteins (up to 170 μg), and an appropriate aliquot of distilled water to obtain a final volume of 1 ml. In method II the incubation mixture consists of hexadecyllyso-GPC (added to the tube in 10 μl chloroform, which is then removed under nitrogen), 0.2 ml Tris-HCl buffer, 50 μl [³H]acetyl-CoA, 25 μl BSA, differentiated HL-60 cell microsomal protein (up to 50 μg), and a volume of distilled water to obtain a final volume of 0.5 ml.

---

[6] T.-c. Lee, D. S. Vallari, B. Malone, and F. Snyder, *FASEB J.* **4**, A1910 (1990).

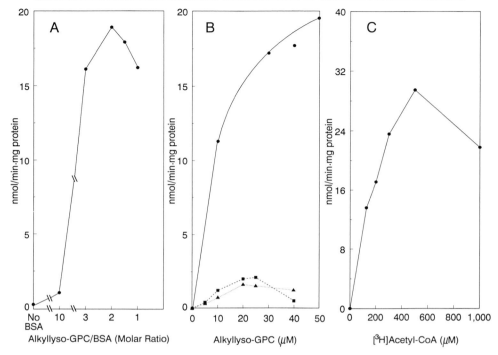

FIG. 2. Dependence of microsomal acetyl-CoA : alkyllyso-GPC acetyltransferase activity of differentiated HL-60 cells on the (A) molar ratio of alkyllyso-GPC/BSA, (B) alkyllyso-GPC concentration, and (C) acetyl-CoA concentration. (A) The incubations are carried out for 10 min at 37° in a final volume of 1.0 ml that contains 50 μM alkyllyso-GPC, 200 μM (0.2 μCi) [^{3}H]acetyl-CoA, 40 μg microsomal protein, and various concentrations of BSA to provide the alkyllyso-GPC/BSA ratios indicated. (B) The incubation conditions are identical to those described in (A) except that in the curve designated with filled circles the alkyllyso-GPC/BSA ratio is 2 and alkyllyso-GPC concentrations are varied; however, the specific activity of the enzyme at 100 μM alkyllyso-GPC is the same as that at 50 μM alkyllyso-GPC (data not shown). In the curve marked with squares, alkyllyso-GPC is introduced into the sample by vortexing the substrate with the incubation mixture in the absence of BSA, and in the curve marked with triangles, the alkyllyso-GPC is dissolved in 1.6% ethanol in the absence of BSA. (C) The incubation conditions are identical to those described in (A) except the alkyllyso-GPC/BSA ratio is 2 and the concentration of acetyl-CoA varies as indicated.

Incubations are carried out at 37° for 5–10 min. Reactions are stopped by lipid extraction using the method of Bligh and Dyer,[7] modified by the addition of 2% acetic acid to the methanol (v/v). Egg phosphatidylcholine (50 μg), PAF (25 μg), and egg lysophosphatidylcholine (25 μg), serving

[7] E. G. Bligh and W. J. Dyer, *Can. J. Biochem. Physiol.* **37,** 911 (1959).

both as carriers and standards, are added during the lipid extraction of the incubation mixtures.

Identification of Products. The reaction product, 1-hexadecyl-2-[³H]acetyl-GPC, is analyzed on silica gel HR (Sigma Chemical Co., St. Louis, MO) layers after development of the chromatogram in chloroform–methanol–glacial acetic acid–water (50:25:8:4, v/v) or chloroform–methanol–concentrated NH_4OH–water (60:35:8:2.3, v/v). The radioactivity associated with areas of the plates corresponding to specific phospholipid classes are scraped into vials and quantitated by liquid scintillation spectrometry.[8]

Properties of Enzyme

It is noted in several reports[1,3,9] that the acetyltransferase activity assayed by method I is inhibited by high concentrations of alkyllyso-GPC ($>25–40\ \mu M$), probably because of the detergent property of alkyllyso-GPC. Ninio *et al.*[10] have shown BSA increases the acetyltransferase activity, with optimal concentrations of 1–1.5 mg BSA/ml at short incubation times (10 min) and 2–2.5 mg BSA/ml during longer incubation periods (150 min). We confirmed[6] their observations, but also established that the maximal stimulation of acetyltransferase activity occurs at a molar ratio of alkyllyso-GPC/BSA of 2 (Fig. 2A), and inhibition by alkyllyso-GPC (as high as 100 μM, data not shown) can be eliminated at this constant alkylyso-GPC/BSA ratio (Fig. 2B). The optimum concentration of acetyl-CoA under these assay conditions is 500 μM (Fig. 2C); the detailed method for assaying the enzyme in this situation is described as method II above.

The acetyltransferase in the remodeling route of PAF biosynthesis also utilizes acyllyso-GPC,[1,3,9] modified polar head group analogs of alkyllyso-GPC[3] (i.e., alkyllyso-N',N'-dimethylethanolamine, alkyllysomonomethylethanolamine, and alkyllyso-GPE), alk-1-enyllyso-GPE,[11] and other short-chain acyl-CoA species[3] (up to C_6) as substrates. However, with the exception of propionyl-CoA, most of the analogs yield rates lower than those for either alkyllyso-GPC or acetyl-CoA. Octadecyllyso-GPC is acetylated at a slightly higher rate than hexadecyllyso-GPC, and unsaturated alkyllyso-GPC is preferred to its saturated counterpart as a substrate.[3] Also, palmitoyl-CoA,[1] oleoyl-CoA,[9] arachidonoyl-CoA,[12] unsatu-

[8] F. Snyder, *in* "The Current Status of Liquid Scintillation Counting" (E. D. Bransome, ed.), p. 248. Grune & Stratton, New York, 1970.

[9] E. Ninio, J. M. Mencia-Huerta, F. Heymans, and J. Benveniste, *Biochim. Biophys. Acta* **710**, 23 (1982).

[10] E. Ninio, F. Joly, and G. Bessou, *Biochim. Biophys. Acta* **963**, 288 (1988).

[11] T. G. Tessner and R. L. Wykle, *J. Biol. Chem.* **262**, 12660 (1987).

[12] E. Remy, G. Lenoir, A. Houben, C. Vandesteene, and J. Remacle, *Biochim. Biophys. Acta* **1005**, 87 (1989).

rated fatty acids (competitive inhibitors of acetyl-CoA),[13] and palmitoyllyso-GPC (a competitive inhibitor of alkyllyso-GPC)[3] have been reported to inhibit the lyso-PAF acetyltransferase activity.

Alkyllyso-GPC acetyltransferase is located in the microsomal fraction of a variety of tissues[1] and blood cells.[14,15] However, results obtained by Mollinedo et al.[16] indicate that this enzyme is distributed in several subcellular fractions including plasma membranes, endoplasmic reticulum, tertiary granules of resting human neutrophils, and ill-defined dense intracellular granules of ionophore A23187-stimulated human neutrophils.

The optimal pH for the lyso-PAF acetyltransferase is 6.9 in rat spleen microsomes,[1] 7.0 in the postnuclear supernatant of murine peritoneal macrophages,[9] and 8.0 for the partially purified enzyme from rat spleen microsomes.[17] EDTA (1 mM), EGTA (1 mM), $MgCl_2$ (1 or 10 mM), $MnCl_2$ (5 or 10 mM), p-bromophenacyl bromide (0.1 mM), p-chloromercuribenzoate (0.5 mM), N-ethylmaleimide (1 mM), and diisopropyl fluorophosphate (10 mM preincubated with microsomes for 15 min at 0°) inhibit the enzyme activity.[1,10,18] These results suggest that the enzyme requires an SH group for activity[18]; moreover, it is protected by dithiothreitol (1 mM).[10] Ca^{2+} activates the enzyme within a very narrow range of free calcium concentrations (optimal activity occurs at 0.2 μM).[19,20] However, Domenech et al.[20] suggest that it is unlikely that the activity of the acetyltransferase in intact cells is regulated by changes of the free calcium concentration.

Alkyllyso-GPC acetyltransferase has been solubilized[17,18,21] and partially purified from rat spleen microsomes.[17,21] The enzyme shows an isoelectric point at pH 4.25 when it is solubilized from rat spleen microsomes by ultrasonic disruption in the presence of 25% glycerol.[18] In contrast, two major proteins with isoelectric points at pH 4.7 and 5.75, but

[13] M. C. Garcia, S. Fernandez-Gallardo, M. A. Gijon, C. Garcia, M. L. Nieto, and M. Sanchez-Crespo, Biochem. J. 268, 91 (1990).

[14] T.-c. Lee and F. Snyder, in "Phospholipids and Cellular Regulation" (J. F. Kuo, ed.), Vol. 2, p. 1. CRC Press, Boca Raton, Florida, 1985.

[15] G. Ribbes, E. Ninio, P. Fontan, M. Record, H. Chap, J. Benveniste, and L. Douste-Blazy, FEBS Lett. 191, 195 (1985).

[16] F. Mollinedo, J. Gomez-Cambronero, K. Cano, and M. Sanchez-Crespo, Biochem. Biophys. Res. Commun. 154, 1232 (1988).

[17] J. Gomez-Cambronero, S. Velasco, M. Sanchez-Crespo, F. Vivanco, and J. M. Mato, Biochem. J. 237, 439 (1986).

[18] K. Seyama and T. Ishibashi, Lipids 22, 185 (1987).

[19] J. Gomez-Cambronero, M. L. Neito, J. M. Mato, and M. Sanchez-Crespo, Biochim. Biophys. Acta 845, 511 (1985).

[20] C. Domenech, E. Machado-De Domenech, and H.-D. Soling, J. Biol. Chem. 262, 5671 (1987).

[21] J. Gomez-Cambronero, J. M. Mato, F. Vivanco, and M. Sanchez-Crespo, Biochem. J. 245, 893 (1987).

with only a 30-kDa polypeptide showing enzyme activity, are obtained from a multiple-step purification scheme that utilizes 0.4% sodium deoxycholate to solubilize the enzyme from the rat spleen microsomes.[17,21]

Previously we have demonstrated that acetyltransferase in rat spleen microsomes is reversibly activated/inactivated through phosphorylation/dephosphorylation.[2] Similar results are observed for the acetyltransferase in homogenates of human neutrophils,[22] in lysates of antigen-stimulated mouse mast cells,[23] and in guinea pig parotid gland microsomes.[20] Both the catalytic subunits of cyclic AMP-dependent protein kinase[20,22] and calcium/calmodulin-dependent protein kinase[20] are able to activate acetyltransferase *in vitro*. Gomez-Cambronero *et al.*[21] have shown that $[\gamma-^{32}P]$ATP is incorporated into a serine residue of the 30-kDa protein with a concomitant increase in the acetyltransferase activity of the partially purified enzyme preparation.[21] However, the protein kinase(s) involved in the *in vivo* phosphorylation of acetyltransferase is not clear at present. Because phorbol myristate and 1-oleoyl-2-acetyl-*sn*-glycerol, but not 4α-phorbol 12,13-didecanoate, can induce PAF synthesis through acetyltransferase activation in human neutrophils and because this activation is abolished by sphingosine (a protein kinase C inhibitor), Leyravaud *et al.*[24] suggest that protein kinase C may participate in the activation of acetyltransferase through either direct or indirect phosphorylation *in vivo*.

Acknowledgments

This work was supported by the Office of Energy Research, U.S. Department of Energy (Contract No. DE-AC05-760R00033), the American Cancer Society (Grant BC-70V), the National Heart, Lung, and Blood Institute (Grant 27109-10), and the National Cancer Institute (CA-41642-05).

[22] M. L. Nieto, S. Velasco, and M. Sanchez-Crespo, *J. Biol. Chem.* **263**, 4607 (1988).
[23] E. Ninio, F. Joly, C. Hieblot, G. Bessou, J. M. Mencia-Huerta, and J. Benveniste, *J. Immunol.* **139**, 154 (1987).
[24] S. Leyravaud, M.-J. Bossant, F. Joly, G. Bessou, J. Benveniste, and E. Ninio, *J. Immunol.* **143**, 245 (1989).

[48] Acyl/Alkyldihydroxyacetone Phosphate Reductase from Guinea Pig Liver Peroxisomes

By AMIYA K. HAJRA, SALIL C. DATTA, and MRIDUL K. GHOSH

Introduction

$$R—(C=O)—O—CH_2—(C=O)—CH_2—O—PO_3^{2-} + NADPH + H^+ \rightarrow$$
$$R—(C=O)—O—CH_2—(CHOH)—CH_2—O—PO_3^{2-} + NADP^+ \quad (1)$$
$$R—O—CH_2—(C=O)—CH_2—O—PO_3^{2-} + NADPH + H^+ \rightarrow$$
$$R—O—CH_2—(CHOH)—CH_2—O—PO_3^{2-} + NADP^+ \quad (2)$$

Acylglycerone-phosphate reductase (EC 1.1.1.101, 1-palmitoyl-*sn*-glycerol-3-phosphate : NADP$^+$ oxidoreductase) a membrane-bound enzyme present in animal tissues, catalyzes the reduction of long-chain acyldihydroxyacetone phosphate (acyl-DHAP) or long-chain alkyl-DHAP by NADPH to the corresponding *sn*-glycerol 3-phosphate derivatives [Eqs. (1) and (2)].[1-5] In liver and other organs, the major fraction of the enzyme is present in peroxisomes.[6-8] The enzyme has been solubilized and partially purified from Ehrlich ascites cell microsomes[5] and purified to homogeneity from guinea pig liver peroxisomes.[9] The method presented in this chapter is based on that described for the preparation of the homogeneous enzyme from guinea pig liver peroxisomes.[9]

Assay Method

Principle. The enzyme activity is assayed by measuring the radioactive lipid product formed after incubating alkyl-DHAP[10] with B-[4-^{3}H] NADPH[2]:

1-*O*-Alkyl-DHAP + B-[4-^{3}H]NADPH + H$^+$ $\rightarrow$
1-*O*-alkyl-*sn*-[2-^{3}H]glycerol 3-phosphate + NADP$^+$

[1] A. K. Hajra and B. W. Agranoff, *J. Biol. Chem.* **243,** 3542 (1968).
[2] E. F. LaBelle and A. K. Hajra, *J. Biol. Chem.* **247,** 5825 (1972).
[3] R. L. Wykle, C. Piantadosi, and F. Snyder, *J. Biol. Chem.* **247,** 2944 (1972).
[4] K. Chae, C. Piantadosi, and F. Snyder, *J. Biol. Chem.* **248,** 6718 (1973).
[5] E. F. LaBelle and A. K. Hajra, *J. Biol. Chem.* **249,** 6936 (1974).
[6] A. K. Hajra, C. L. Burke, and C. L. Jones, *J. Biol. Chem.* **254,** 10896 (1979).
[7] A. K. Hajra and J. E. Bishop, *Ann. N.Y. Acad. Sci.* **386,** 170 (1982).
[8] M. K. Ghosh and A. K. Hajra, *Arch. Biochem. Biophys.* **245,** 523 (1986).
[9] S. C. Datta, M. K. Ghosh, and A. K. Hajra, *J. Biol. Chem.* **265,** 8268 (1990).
[10] Either acyl-DHAP (palmitoyl-DHAP) or alkyl-DHAP may be used as the lipid substrate. If palmitoyl-DHAP is used, however, it should be purified (silicic acid chromatography) before use to remove any decomposition product, palmitic acid, which inhibits the reaction.

Substrates. The substrate and labeled NADPH are not commercially available. The substrate hexadecyl-DHAP is chemically synthesized as described previously.[11] B-[4-³H]NADPH is prepared as described below.[2,12]

To 0.1 mCi of D-[1-³H]glucose (2 Ci/mmol) dried in a tube are added 15 µl of triethanolamine hydrochloride buffer (0.2 M, pH 8.5), 10 µl MgCl₂ (0.1 M), 20 µl ATP (0.1 M), 10 µl glucose-6-phosphate dehydrogenase (1 mg/ml), 5 µl hexokinase (1.4 mg/ml), 20 µl NADP⁺ (10 mM), 16 µl D-glucose (2 mM), and water to make 0.2 ml. The mixture is incubated at room temperature for 2 hr. A portion is counted to determine the radioactivity and then diluted with nonradioactive NADPH to make 5000 counts/min (cpm)/nmol NADPH. The conversion of [1-³H]glucose to [4-³H] NADPH is almost quantitative under the above conditions. The diluted NADPH (5000 cpm/nmol) can be stored at 4° for 4–5 days without much decomposition. For longer storage, [³H]NADPH should be purified by DEAE-Sephacel chromatography as described,[12] lyophilized, and stored at −20° under anhydrous conditions.

Reagents

0.3 M Tris-HCl buffer, pH 7.5
0.1 M NaF
10 mM Na₄EDTA
10 mg/ml Tween 20 in chloroform
0.8 mM B-[4-³H]NADPH (5000 cpm/nmol), see above
1-*O*-Hexadecyl-DHAP emulsion: For each tube, mix 125 nmol of hexadecyl-DHAP in chloroform and 20 µl of Tween 20 in chloroform; remove the solvent by blowing N₂ over the sample and then disperse the residue in 50 µl of 50 mM Tris-HCl buffer (pH 7.5) by sonication
Chloroform–methanol (1 : 2, v/v)
Chloroform
Chloroform–methanol–0.5 N aqueous H₃PO₄ (1 : 12 : 12, v/v)

Technique. The following reagents are added to a 1.6 × 12.5 cm screw-topped test tube, mixing after each addition: Tris-HCl buffer 0.1 ml, NaF 60 µl, EDTA 40 µl, alkyl-DHAP emulsion 50 µl, [³H]NADPH 50 µl, enzyme, and water to make the final volume 0.6 ml. The mixture is incubated at 37° for 15 min with reciprocal shaking, and the reaction is stopped by adding 2.25 ml chloroform–methanol (1 : 2, v/v). After mixing, 0.75 ml chloroform and 0.75 ml KCl–H₃PO₄ (2–0.2 M) are added, mixed well

[11] A. K. Hajra, T. V. Saraswathi, and A. K. Das, *Chem. Phys. Lipids* **33,** 179 (1983).
[12] A. K. Das and A. K. Hajra, *Biochim. Biophys. Acta* **796,** 178 (1984).

(vortexed), and then centrifuged 10 min at 1000 g in a tabletop centrifuge. The upper layers are aspirated off, and the lower layers are washed with 2.5 ml of a solution containing chloroform–methanol–0.5 N aqueous H_3PO_4 (1 : 12 : 12, v/v) by vortexing and centrifuging as described above. Aliquots (generally 1 ml, i.e., two-thirds of the total) of the washed lower layers are transferred to counting vials, the solvents are evaporated by blowing either air or N_2 over the sample, scintillation mixture is added, and the radioactivity is determined in a liquid scintillation counter.

Definition of Unit. One unit of activity is defined as the amount of enzyme which catalyzes the formation of 1 nmol (5000 cpm) of product per minute at 37°. The specific activity is expressed as nanomoles per minute per milligram protein.

Purification of Alkyldihydroxyacetone Phosphate

Isolation of Guinea Pig Liver Peroxisomes. The peroxisomes are isolated from guinea pig livers by a Nycodenz step gradient centrifugation procedure as described elsewhere in this volume.[13]

Detergent Treatment of Peroxisomes. Peroxisomes, stored frozen in 0.25 M sucrose–10 mM N-tris(hydroxylmethyl)methyl-2-aminoethanesulfonic acid (TES) (pH 7.5), are quickly thawed and diluted with the same solution to a protein concentration of 3.5 mg/ml. An equal volume of 0.4% (w/v) Triton X-100 is added to the suspension, mixed gently by hand homogenization in a Potter–Elvehjem homogenizer, and incubated at room temperature for 30 min. The mixture is centrifuged at 111,000 g for 40 min, and the supernatant is discarded.

First Solubilization of Enzyme. The residue is suspended (one-fourth of the original volume) in 0.25 M sucrose–10 mM TES (pH 7.5), and an equal volume of a solution containing KCl (2 M)–dithiothreitol (DTT) (4 mM)–Tris-HCl (20 mM, pH 7.5)–NADPH (0.6 mM)–Triton X-100 (0.1%, v/v) is added to the suspension. The mixture is hand homogenized using a Potter–Elvehjem homogenizer and kept at room temperature for 30 min. It is then centrifuged at 110,000 g for 60 min, and the residue is discarded.

Precipitation of Enzyme. The supernatant is dialyzed overnight at 4° against 10 volumes of Tris-HCl (10 mM, pH 7.5)–DTT (2 mM)–Triton X-100 (0.05%), changing the dialyzing solution once after 5 hr. The dialyzed solution is centrifuged at 150,000 g for 40 min, and the supernatant is discarded.

Second Solubilization of Enzyme. The residue obtained after centrifu-

---

[13] K. O. Webber and A. K. Hajra, this volume [10].

gation is homogenized in 2 ml of octylglucoside (0.8%, w/v)–KCl (1 M)–DTT (2 mM)–Tris-HCl (10 mM, pH 7.5)–NADPH (0.6 mM) and incubated at room temperature for 1 hr. The mixture is centrifuged at 110,000 g for 1 hr and the residue discarded.

Gel Filtration. The supernatant is loaded into an 88 × 1.5 cm Sepharose CL-6B column which has been equilibrated with a buffer containing KCl (1 M)–DTT (2 mM)–Tris-HCl (10 mM, pH 7.5)–octylglucoside (0.4%)–glycerol (20%, w/v). The enzyme is eluted in the same buffer at a flow rate of 0.4 ml/min, and 4-ml fractions are collected in siliconized glass tubes. The fractions containing the reductase (eluted between 90 and 130 ml of effluent) are pooled together and dialyzed for 3 hr against 10 volumes of Tris-HCl (pH 7.5, 10 mM)–KCl (50 mM)–DTT (2 mM)–octylglucoside (0.4%)–glycerol (20%) solution at 4°.

NADPH Affinity Chromatography. For preparation of the affinity matrix,[9,14] add 0.1 ml of 0.1 M NaIO$_4$ to 10 μmol of NADP$^+$ in 0.9 ml water, and keep the mixture in the dark at room temperature for 1 hr. The oxidized NADP$^+$ is added to a suspension of 1 ml adipic acid dihydrazide-agarose gel (Pharmacia-LKB, Piscataway, NJ) in 4 ml 0.1 M acetate buffer (pH 5.0). The mixture is stirred gently for 1 hr using a rocking platform shaker, then 6 ml of 2 M NaCl is added, mixed, and stirred for another 30 min. The suspension is centrifuged at low speed (1000 g, 10 min), and the packed gel is washed 3 times with water and then 2 times with 75 mM triethanolamine hydrochloride buffer (pH 8.5). To 2 ml of the NADP$^+$- agarose gel, 4 ml of freshly prepared 0.1 M NaBH$_4$ in 75 mM triethanol- amine hydrochloride buffer (pH 8.5)is added, with gentle mixing for 1 min, followed by 4 ml of 2 M NaCl in 75 mM triethanolamine buffer (pH 8.5). The mixture is stirred for an additional 5 min, centrifuged (1000 g) for 5 min, and the packed gel washed 6 times with the triethanolamine buffer. The NADPH-agarose gel is packed into a column of 0.54 cm diameter to a height of 7 cm and washed with 5 column volumes of Tris-HCl (10 mM, pH 7.5)–KCl (50 mM)–DTT (2 mM)–glycerol (20%)–octylglucoside (0.8%).

For the chromatography step, the dialyzed enzyme is slowly (0.5 ml/ min) loaded onto the NADPH-agarose column and then eluted, first with 10 ml of the above buffer (buffer A), then with 10 ml of buffer B (0.2% sodium cholate, 0.25 M KCl, 10 mM Tris-HCl, pH 7.5, 20% glycerol, and 2 mM DTT) followed by 10 ml of buffer C containing Tris-HCl (10 mM, pH 7.5)–KCl (0.25 M)–octylglucoside (0.8%)–DTT (2 mM)–glycerol (20%). The enzyme is eluted by buffer D, which is buffer C plus NADPH (2 mM). One-milliliter fractions are collected in siliconized tubes, and the

[14] M. Wilchek and R. Lamed, this series, Vol. 34, p. 475.

TABLE I
SOLUBILIZATION AND PURIFICATION OF ACYL/ALKYLDIHYDROXYACETONE
PHOSPHATE REDUCTASE

Step	Total volume (ml)	Total activity (units)	Total protein (mg)	Specific activity (units/mg)	Yield (%)
Liver homogenate	610	78,000	7100	11	100.0
Peroxisomes	21	11,400	68.0	167	14.6
Triton X-100-treated membranes	5	8,780	12.4	708	11.3
First solubilized enzyme	10	7,800	7.8	1,000	10.0
Reprecipitated enzyme	14	5,200	3.0	1,730	6.7
Second solubilized enzyme	23	3,000	0.92	3,260	3.8
Gel filtration	23	1,400	0.15	9,300	1.8
Affinity chromatography	3	750	0.012	62,500	1.0

fractions containing the highest enzyme activity are pooled and stored at 4°. The result of a typical purification is shown in Table I.

Properties

Molecular Weight. On sodium dodecyl sulfate (SDS)–polyacrylamide gel electrophoresis, the purified enzyme showed a single protein band of relative mobility corresponding to 60,000 molecular weight. On gel filtration, in the presence of octylglucoside, the molecular weight of the enzyme was estimated at 75,000.

Optimum pH. The enzyme has a broad pH optimum between pH 6.5 and 7.5.

Specificity. As mentioned above, the enzyme accepts both acyl-DHAP (V_{max} for palmitoyl-DHAP 75 μmol/min/mg) and alkyl-DHAP (V_{max} for 1-*O*-hexadecyl-DHAP 70 μmol/min/mg) as substrates. The enzyme stereo-specifically transfers the B-side hydrogen[15] from the 4 position of NADPH to form *sn*-glycerol 3-phosphate derivatives.[2] NADH can substitute for NADPH but with very low activity (~10% of that for NADPH) and affinity (K_m 1.7 mM). The enzyme does not catalyze the reduction of DHAP.

Michaelis Constants. The apparent Michaelis constants, K_m, for hexa-decyl-DHAP and palmitoyl-DHAP are 21 and 15 μM, respectively at 70 μM NADPH concentration. The K_m for NADPH is 20 μM when the hexadecyl-DHAP concentration is 200 μM.

Stability. The affinity-purified enzyme is stable at 4° for 4–6 weeks in

[15] K. You, L. J. Arnold, W. S. Allison, and N. O. Kaplan, *Trends Biochem. Sci.* **3**, 265 (1978).

the presence of octylglucoside and NADPH. The enzyme is unstable in the absence of NADPH or on freezing and thawing.

Inhibitors. Palmitoyl-CoA[9,16] (K_i 5 μM) and NADP[+][5,9] (K_i 90 μM) are strong inhibitors of the enzyme. The enzyme is also inhibited by fatty acids, acyl- and alkyldihydroxyacetone,[5] and some thiol-reactive agents[9] (e.g., *p*-hydroxymercuribenzoate).

Equilibrium Constant. The reversibility of the enzymatic reaction could not be demonstrated.

Use of Enzyme as Analytical Tool. The enzyme was used for the quantitative determination of acyl-DHAP and alkyl-DHAP in biological samples.[12]

[16] R. A. Coleman and R. M. Bell, *J. Biol. Chem.* **255,** 7681 (1980).

[49] 1-Alkyl-2-lyso-*sn*-glycero-3-phosphate Acetyltransferase

By FRED SNYDER, TEN-CHING LEE, and BOYD MALONE

Introduction

The first step in the *de novo* pathway of platelet-activating factor (PAF) biosynthesis is catalyzed by acetyl-CoA : 1-alkyl-2-lyso-*sn*-glycero-3-phosphate acetyltransferase.[1] This initial reaction (Fig. 1) involves the acetylation of the *O*-alkyl analog of lysophosphatidic acid, a key branch point in the ether lipid pathway as this lyso compound can also be acylated by a long-chain acyl-CoA acyltransferase[2] (Fig. 2). The 1-alkyl-2-lyso-*sn*-glycero-3-phosphate branch point is introduced after alkyldihydroxy-acetonephosphate (a product of alkyl-DHAP synthase; see [44], this volume) is reduced by an NADPH : alkyldihydroxyacetonephosphate oxidoreductase (see [48] and Fig. 2).

1-Alkyl-2-acetyl-*sn*-glycero-3-phosphate, the product of this acetylation reaction, can then be converted to PAF in two sequential reactions catalyzed by a phosphatase (see [26]) and a cholinephosphotransferase (see [33]). 1-Alkyl-2-acetyl-*sn*-glycerols formed by the phosphatase could also be important as an end product rather than just an intermediate in the *de novo* pathway since this alkyl analog of a diglyceride is also known to

[1] T.-c. Lee, B. Malone, and F. Snyder, *J. Biol. Chem.* **261,** 5373 (1986).
[2] R. L. Wykle and F. Snyder, *J. Biol. Chem.* **245,** 3047 (1970).

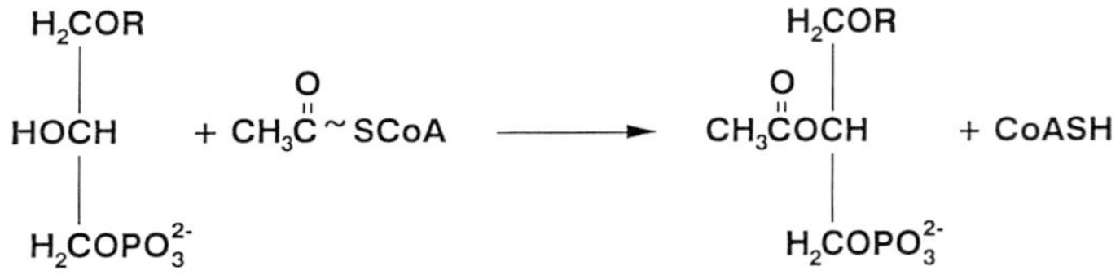

$$
\begin{array}{ccc}
\underset{|}{H_2COR} & & \underset{|}{H_2COR} \\
HOCH & + \; CH_3\overset{O}{\underset{\parallel}{C}} \sim SCoA & \longrightarrow \quad CH_3\overset{O}{\underset{\parallel}{C}}OCH \quad + \; CoASH \\
\underset{}{H_2COPO_3^{2-}} & & \underset{}{H_2COPO_3^{2-}}
\end{array}
$$

Fig. 1. Reaction catalyzed by acetyl-CoA : 1-alkyl-2-lyso-*sn*-glycero-3-phosphate acetyl-transferase.

influence responses involving protein kinase C activation[3,4] and induces cell differentiation.[5]

Acetyltransferase in the *de novo* pathway has not been purified from its microsomal environment, but sufficient information has been obtained to indicate that it represents a distinctly different activity from its counter-part acetyltransferase in the remodeling pathway (see [47]). This chapter provides a concise description of the method for assaying the *de novo* acetyltransferase and summarizes its properties.

Method for Measuring Enzyme Activity

Reagents and Substrates. The following reagents are required for the enzyme assay: acetyl-CoA and 1-alkyl-2-acetyl-*sn*-glycero-3-phosphocho-line from Sigma (St. Louis, MO); sodium vanadate and sodium fluoride from Fischer (Pittsburgh, PA); and [^{3}H]acetyl-CoA (3.8 Ci/mmol) and 1-[1′,2′-^{3}H]alkyl-2-acetyl-*sn*-glycerophosphocholine (55 Ci/mmol) from Du Pont–NEN (Boston, MA). Phospholipase D is prepared from fresh cabbage by the method of Hayashi.[6]

Substrates for the acetyltransferase are prepared by first treating either the radiolabeled or unlabeled 1-alkyl-2-acetyl-*sn*-glycero-3-phosphocho-line with phospholipase D to produce 1-[1′,2′-^{3}H]alkyl-2-acetyl-*sn*-glyce-ro-3-phosphate or 1-alkyl-2-acetyl-*sn*-glycero-3-phosphate. The latter products are then exposed to mild alkaline hydrolysis[7] in order to hy-drolyze the acetate at the *sn*-2 position. Lipids are extracted from the reaction mixture by a modified Bligh and Dyer[8] technique that includes

[3] D. A. Bass, L. C. McCall, and R. L. Wykle, *J. Biol. Chem.* **263,** 19610 (1988).

[4] L. L. Stoll, P. H. Figard, N. R. Yerram, M. A. Yorek, and A. A. Spector, *Cell Regul.* **1,** 13 (1989).

[5] M. J. C. McNamara, J. D. Schmitt, R. L. Wykle, and L. W. Daniel, *Biochem. Biophys. Res. Commun.* **122,** 824 (1984).

[6] O. Hayaishi, this series, Vol. 1, p. 660.

[7] R. M. C. Dawson, *Biochem. J.* **75,** 45 (1960).

[8] E. G. Bligh and W. J. Dyer, *Can. J. Biochem. Physiol.* **37,** 911 (1959).

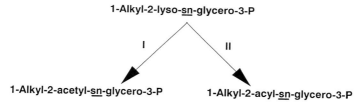

1-Alkyl-2-lyso-_sn_-glycero-3-P

I   II

1-Alkyl-2-acetyl-_sn_-glycero-3-P   1-Alkyl-2-acyl-_sn_-glycero-3-P

FIG. 2. 1-Alkyl-2-lyso-_sn_-glycero-3-phosphate as a branch point leading to the synthesis of either PAF from 1-alkyl-2-acetyl-_sn_-glycero-3-phosphate or alkyl-linked phospholipids in membranes such as 1-alkyl-2-acyl-_sn_-glycero-3-phosphocholine. The Roman numerals designate an acetyltransferase (**I**) and a long-chain acyltransferase (**II**).

2% acetic acid in the methanol and 0.1 N HCl in the aqueous phase in order to assure quantitative extraction of the lipids possessing the free phosphate group. The substrates produced are then purified by preparative thin-layer chromatography (TLC) on silica gel H layers developed in chloroform–methanol–concentrated NH$_4$OH (65 : 35 : 8, v/v) (R_f 0.11). Their purity can be further checked by TLC on silica gel H layers prepared using 10 mM Na$_2$CO$_3$ in place of water; development of these chromatoplates is carried out in chloroform–methanol–acetic acid–saline (50 : 25 : 8 : 6, v/v) (R_f 0.59). The final radiopurity of the substrates obtained by using this method[1] should exceed 95%.

Preparation of Enzyme Source. The acetyltransferase activity is assayed in microsomal preparations; therefore, procedures for isolating microsomes might need to be adapted according to the tissue or cells used. A general procedure used in our earlier work[1] for preparing microsomes from rat spleen and other tissues involves first homogenizing minced tissue at 4° in 0.25 M sucrose, 20 mM Tris-HCl (pH 7.4), 10 mM EDTA, 5 mM mercaptoethanol, and 50 mM NaF. For preparing heart microsomes, 0.15 M KCl is substituted for the 0.25 M sucrose. Homogenates are centrifuged at 500 g for 10 min to remove nuclei and debris; the supernatant is then centrifuged again at 15,000 g for 10 min to remove mitochondria. A final centrifugation at 100,000 g for 60 min forms the microsomal pellets. The microsomes are suspended in the homogenizing medium and used fresh in the enzyme assays.

Enzyme Assay. The standard incubation mixture[1] for assaying the activity of acetyl-CoA : 1-alkyl-2-lyso-_sn_-glycero-3-phosphate acetyltransferase consists of 0.1 M Tris-HCl (pH 8.4), 50 mM NaF, 25 μM sodium vanadate, 16–25 μM alkyllysoglycerophosphate, 600 μM [³H]acetyl-CoA (3 μCi), and up to 250 μg of microsomal protein in a final volume of 0.5 ml. Alternately, in this assay 1-[1′,2′-³H]alkyl-2-lyso-_sn_-glycero-3-phosphate can also be used as the radiolabeled substrate instead of having the label

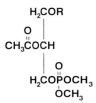

FIG. 3. Chemical structure of 1-alkyl-2-acetyl-*sn*-glycero-3-dimethylphosphate, a useful derivative for the identification of 1-alkyl-2-acetyl-*sn*-glycero-3-phosphate.

in the acetyl-CoA. All incubations are carried out at 23° for up to 5 min. Reactions are stopped by lipid extraction as specified in the section on Reagents and Substrates, except that 100 μg of egg phosphatidylcholine is added as a lipid carrier to maximize the recovery of lipid products.

Identification of Products. The radiolabeled 1-alkyl-2-acetyl-*sn*-glycerophosphate formed in the incubations is analyzed by TLC using a system of chloroform–methanol–concentrated NH_4OH (65 : 35 : 8, v/v) in a container without a paper liner. Alkylacetylglycerophosphate has an approximate R_f value of 0.11 in this TLC system. Distribution of the tritium label is determined by zonal scanning or area scraping of chromatoplates and measuring the radioactivity by liquid scintillation spectrometry.[9] Further identification of the phosphorylated product can be done by chromatographic analysis of the methylated derivative[10] of the [³H]alkylacetylglycerophosphate (Fig. 3) following treatment of the alkylacetylglycerophosphate with diazomethane.[10,11] Information on the identification of alkylacetylglycerols that might be formed as minor products under the conditions used to assay the acetyltransferase can be found elsewhere in this volume [26].

Properties of Enzyme

Properties of the microsomal acetyl-CoA : 1-alkyl-2-lyso-*sn*-glycero-3-phosphate acetyltransferase were originally characterized so that any contribution of the subsequent phosphatase step (see [26]) was minimized.[1]

[9] F. Snyder, *in* "The Current Status of Liquid Scintillation Counting" (E. D. Bransome, ed.), p. 248. Grune & Stratton, New York, 1970.

[10] M. L. Blank, T.-c. Lee, E. A. Cress, B. Malone, V. Fitzgerald, and F. Snyder, *Biochem. Biophys. Res. Commun.* **124,** 156 (1984).

[11] O. Renkonen, *J. Lipid Res.* **9,** 34 (1968).

TABLE I

SUBSTRATE SPECIFICITY OF ACETYL-CoA : 1-ALKYL-2-LYSO
sn-GLYCERO-3-PHOSPHATE ACETYLTRANSFERASE IN RAT
SPLEEN MICROSOMES[a]

Substrate	Position of structural variation on glycerol moiety	Activity relative to control (%)
Hexadecyl-GP (16:0)	*sn*-1	94
Octadecyl-GP (18:0)	*sn*-1	100
Oleoyl-GP (18:1)	*sn*-1	24
Acetyl-CoA	*sn*-2	100
Propionyl-CoA	*sn*-2	106
Butyryl-CoA	*sn*-2	136
Hexanoyl-CoA	*sn*-2	136
Linoleoyl-CoA	*sn*-2	20
Arachidonoyl-CoA	*sn*-2	73

[a] Average of two experiments with duplicate determinations.

This was achieved by using sodium vanadate/sodium fluoride and lower temperatures (23°) for the assay in order to inhibit the phosphatase activity.

The acetyltransferase is clearly of microsomal origin since 80% of the enzyme activity in the homogenate is associated with the 100,000 g pellet. The microsomal activity amounts to a 6-fold enrichment over the activity in the homogenate. Verification of the microsomal location of the acetyltransferase was based on the recovery of 86% of a microsomal marker enzyme, NADPH–cytochrome-c reductase, in the 100,000 g pellet prepared under identical conditions.

The acetyltransferase has a broad pH optimum (8.0–8.8), with maximum activity occurring at pH 8.4 in Tris-HCl buffer. Optimal concentrations for the substrates in this reaction range from 16 to 25 μM for alkyllysoglycerophosphate and 600 to 800 μM for acetyl-CoA. The apparent K_m for acetyl-CoA is 226 μM.

A substrate specificity study for the acetyltransferase in the *de novo* pathway is summarized in Table I. The enzyme appears to have a strong preference for alkyl groups having 18:0 and 16:0 carbon chains since the substrate with an 18:1 acyl moiety at the *sn*-1 position was poorly utilized. Acyl-CoA species with 2:0, 3:0, 4:0, and 6:0 carbon chains all were excellent substrates for the acetyltransferase. On the other hand, an 18:1 acyl-CoA was a poor substrate (~20% of the 2:0 acyl-CoA activity), whereas arachidonoyl-CoA exhibited a moderate activity (~75%) when compared to the acetyl-CoA control. Thus, the *de novo* acetyltransferase exhibits a much lower degree of specificity for acyl-CoAs than the acetyl-transferase in the remodeling pathway which utilizes butyryl-CoA and hexanoyl-CoA much less efficiently.[12]

The acetyltransferase activity in the *de novo* pathway of PAF biosynthesis has been shown to be distinctly different from the acetyltransferase in the remodeling route based on substrate competition experiments and differences in their heat sensitivity.[1] Alkyllysoglycerophosphocholine (i.e., lyso-PAF, the substrate for the acetyltransferase in the remodeling pathway) at concentrations of 12.5 or 25 μM has no effect on the acetyltransferase in the *de novo* route. On the other hand, alkyllysoglycerophosphate (the *de novo* substrate) causes a significant inhibition of the acetyltransferase activity in the remodeling sequence. Moreover, after microsomal fractions are preincubated at various temperatures from 23° to 60° for 15 min, the acetyltransferase activity in the remodeling route is inhibited to a much greater extent than the *de novo* acetyltransferase.[1] Thus, the enzyme activities that catalyze the transfer of acetate from acetyl-CoA to the two different types of *sn*-2-lyso substrates respond to heat in a very different manner.

The initial reaction step catalyzed by the acetyltransferase in the *de novo* route is rate limiting.[1] However, it is noteworthy that a second rate-limiting step also needs to be taken into account; this is a reaction catalyzed by cytidylyltransferase[13,14] in the formation of the CDPcholine required for the final cholinephosphotransferase step that produces PAF.

The *de novo* acetyltransferase has been shown to be widely distributed in rat tissues,[1,15] with the highest activities[1] found in spleen, brain, and the kidney medulla. In all tissues the activity of this acetyltransferase is much lower than the activities of the two other enzymes (i.e., the phosphatase and the DTT-insensitive cholinephosphotransferase) that constitute the *de novo* pathway of PAF biosynthesis[16]; this low activity for the acetyltransferase is consistent with the acetylation step being rate limiting.

Acknowledgments

This work was supported by the Office of Energy Research, U.S. Department of Energy (Contract No. DE-AC05-760R00033), the American Cancer Society (Grant BC-70V), and the National Heart, Lung, and Blood Institute (Grant 35495-04A1).

[12] T.-c. Lee, *J. Biol. Chem.* **260**, 10952 (1985).

[13] M. L. Blank, Y. J. Lee, E. A. Cress, and F. Snyder, *J. Biol. Chem.* **263**, 5656 (1988).

[14] T.-c. Lee, B. Malone, M. L. Blank, V. Fitzgerald, and F. Snyder, *J. Biol. Chem.* **265**, 9181 (1990).

[15] S. Fernandez-Gallardo, M. A. Gijon, M. D. C. Gacia, E. Cano, and M. Sanchez-Crespo, *Biochem. J.* **254**, 707 (1988).

[16] T.-c. Lee, B. Malone, and F. Snyder, *in* "Proceedings of the Taipei Satellite Symposium on Platelet Activating Factor" (N. Hicks, ed.), p. 1. Excerpta Medica, Hong Kong, 1989.

[50] Metabolism of Ether-Linked Diglycerides in Brain and Myocardium

By DAVID A. FORD and RICHARD W. GROSS

Introduction

The essential role of ether-linked lipids in a diverse array of biological processes has been convincingly demonstrated over the past decade.[1-3] Since the initial demonstration of platelet-activating factor (PAF) as an ether-linked choline glycerophospholipid with important regulatory functions,[4,5] many fundamental differences between ether-linked and diacyl phospholipids have been described. For example, recent studies have demonstrated that ether lipids are the specific targets of some of the phospholipases activated during physiologic or pathophysiologic perturbations in mammalian cells resulting in the selective accumulation of ether-linked metabolities.[6-9] To facilitate studies in this evolving area, we have developed methods to quantitate ether-linked diglycerides in mammalian cells and to assess the biochemical pathways responsible for ether-linked diglyceride metabolism utilizing synthetic ether-linked diglycerides purified by high-performance liquid chromatography (HPLC).

Sources of Materials

Bovine heart lecithin and 1-*O*-alkyl-*sn*-glycero-3-phosphocholine (lyso platelet-activating factor) were purchased from Avanti Polar Lipids (Birmingham, AL). Fatty acyl chlorides and 1-*O*-eicosodec-11'-enyl-*sn*-glycerol were from NuChek Prep (Elysian, MN) and Foxboro Analabs (Foxboro, MA), respectively. *Bacillus cereus* phospholipase C (grade II) was from Boehringer Mannheim (Indianapolis, IN). Tween 20 and Triton X-

[1] D. J. Hanahan, *Annu. Rev. Biochem.* **55**, 483 (1986).

[2] F. Snyder, ed., "Platelet-Activating Factor and Related Lipid Mediators." Plenum, New York, 1987.

[3] S. M. Prescott, G. A. Zimmerman, and T. M. McIntyre, *J. Biol. Chem.* **265**, 17381 (1990).

[4] M. L. Blank, F. Snyder, L. W. Byers, B. Brooks, and E. E. Muirhead, *Biochem. Biophys. Res. Commun.* **90**, 1194 (1979).

[5] C. A. Demopoulos, R. N. Pinckard, and D. J. Hanahan, *J. Biol. Chem.* **254**, 9355 (1979).

[6] L. W. Daniel, M. Waite, and R. L. Wykle, *J. Biol. Chem.* **261**, 9128 (1986).

[7] L. G. Rider, R. W. Dougherty, and J. E. Niedel, *J. Immunol.* **140**, 200 (1988).

[8] D. A. Ford and R. W. Gross, *Circ. Res.* **64**, 173 (1989).

[9] S. L. Hazen, D. A. Ford, and R. W. Gross, *J. Biol. Chem.* **266**, 5629 (1991).

METHODS IN ENZYMOLOGY, VOL. 209

100 were purchased from Pierce Chemical Co. (Rockford, IL). Vitride [sodium bis(2-methoxy-ethoxy)aluminum hydride] and Kieselgel (70–230 mesh) were purchased from Kodak Laboratory Chemicals (Rochester, NY) and EM Science (Gibbstown, NJ), respectively. *N,N*-Dimethyl-4-aminopyridine was from Aldrich (Milwaukee, WI) and was recrystallized utilizing chloroform–diethyl ether (1 : 1, v/v) prior to use. Gelman (Ann Arbor, MI) 0.45-μm PTFE Acrodisc filtration units were used. Both silica and octadecyl silica HPLC columns (Ultrasphere-Si and Ultrasphere-ODS, respectively) were obtained from Beckman Instruments (Fullerton, CA). Universol was obtained from ICN Biomedicals (Costa Mesa, CA). Atmos bags were obtained from Aldrich. *Rhizopus arrhizus* lipase, benzoic anhydride, octylglucoside, and reagent-grade chemicals were obtained from Sigma (St. Louis, MO). Supelclean extraction columns were obtained from Supelco (Bellefonte, PA).

Preparation and Purification of Synthetic Ether-Linked Diglycerides

The synthesis of specific molecular species of ether-linked diglycerides is accomplished through utilization of appropriate subclasses or individual molecular species of lysophospholipid or monoradylglycerol precursors. These precursors can be specifically acylated with the desired fatty acid at the *sn*-2 position (for lysophospholipids) or sequentially acylated at the *sn*-3 and *sn*-2 positions (for monoradylglycerols) prior to their enzymatic conversion to ether-linked diglycerides. Conversion of ether-linked choline glycerophospholipids or triglycerides to ether-linked diglycerides is easily accomplished through utilization of *B. cereus* phospholipase C or *R. arrhizus* lipase (respectively), which are inexpensive, commercially available reagents. Ether-linked diglycerides can subsequently be purified by thin-layer chromatography (TLC), straight-phase HPLC, or both to yield substantial amounts (10–100 mg) of homogeneous ether-linked diglycerides.

Synthesis of 1-O-Eicosodec-11'-enyl-2-octadec-9'-enoyl-sn-glycerol. The ether-linked diglyceride 1-*O*-eicosodec-11'-enyl-2-octadec-9'-enoyl-*sn*-glycerol (which can be utilized as an internal standard for the quantitation of ether-linked diglycerides in biological tissues; see below) is prepared by exhaustive acylation of 1-*O*-eicosodec-11'-enyl-*sn*-glycerol utilizing oleoyl chloride followed by selective hydrolysis of the *sn*-3 fatty acid of the resultant triradylglycerol catalyzed by *R. arrhizus* lipase. Typically, 100 mg (374 μmol) of 1-*O*-eicosodec-11'-enyl-*sn*-glycerol is dried under reduced pressure, resuspended in 1.5 ml of distilled chloroform (distilled over P_2O_5), and stirred in a 5-ml reaction vessel under a nitrogen atmosphere. To this solution, 400 mg (1.3 mmol) of oleoyl chloride (previously dissolved in 1 ml of distilled chloroform) is added and briefly stirred (1 min) prior to the addition of 20 mg (165 μmol) of N,N-dimethyl-4-aminopyridine. The reaction mixture is blanketed with dry nitrogen, sealed, and subsequently stirred at room temperature in an Atmos bag with a dry nitrogen atmosphere for 24 hr. The reaction mixture is transferred to a borosilicate test tube and the reaction is terminated by the addition of 2 ml of methanol and the subsequent addition of 2.5 ml of water. The reaction mixture is subsequently vortexed, then centrifuged, the chloroform layer (lower) is transferred to a separate tube, the aqueous phase is reextracted (after readdition of an equal volume of chloroform), and the chloroform extracts are combined.

Crude reaction extracts are evaporated under nitrogen, resuspended in 3 ml of hexane–diethyl ether (85 : 15, v/v), and loaded onto a silica column (Kieselgel Silica Si 70–230 mesh, 3 × 50 cm). Purification of 1-*O*-eicosodec-11′-enyl-2,3-dioctadec-9′-enoyl-*sn*-glycerol is accomplished by isocratic elution from the silica column employing hexane–diethyl ether (85 : 15, v/v) as the mobile phase (the triradylglycerol elutes in approximately 1.5 column volumes). The elution and purity of column chromatographic eluents can be conveniently assessed by TLC (silica G Plates, Foxboro Analabs) utilizing a mobile phase of petroleum ether–diethyl ether–acetic acid (80 : 20 : 1, v/v) and subsequent visualization by iodine staining (R_f 0.73).

Column chromatographic fractions containing substantial amounts of 1-*O*-eicosodec-11′-enyl-2,3-dioctadec-9′-enoyl-*sn*-glycerol are combined, dried under nitrogen, and resuspended in 3 ml of chloroform (ether-linked triglycerides in this form can be stored for at least 3 months at −20° in a nitrogen atmosphere). Aliquots of 1-*O*-eicosodec-11′-enyl-2,3-dioctadec-9′-enoyl-*sn*-glycerol (typically 30 mg) are removed and dried under nitrogen prior to resuspension in 0.5 ml of 50 m*M* Tris-Cl buffer (pH 8.0) containing 100 m*M* NaCl, 5 m*M* CaCl$_2$, 1% Triton X-100, 0.05% (w/v) bovine serum albumin, and 400 U of *R. arrhizus* triglyceride lipase. The reaction mixture is mildly vortexed and subsequently incubated at 37° for 4 hr. The product, 1-*O*-eicosodec-11′-enyl-2-octadec-9′-enoyl-*sn*-glycerol, is extracted by the Bligh and Dyer method,[10] concentrated under a nitrogen stream (to 2 ml), and streaked onto a preparative silica G TLC plate (1 mm thickness). The ether-linked diglyceride is purified utilizing a mobile phase of petroleum ether–diethyl ether–acetic acid (70 : 30 : 1, v/v) (R_f 0.38), and regions corresponding to ether-linked diradylglycerols [identified by selective iodine staining of an adjacent lane on the TLC plate containing 1,3-diacyl-*sn*-glycerol (which possesses an identical R_f value as ether-linked diglyceride)] are scraped into borosilicate tubes and extracted

[10] E. G. Bligh and W. J. Dyer, *Can. J. Biochem. Physiol.* **37**, 911 (1959).

with chloroform 3 times followed by filtration of the combined extracts through 0.45-μm PTFE filters. The resulting filtrate is evaporated to dryness and resuspended in hexane prior to final purification by HPLC.

Straight-phase HPLC is performed (typically 1–2 mg per injection) utilizing a silica column (4.6 × 250 mm, 5-μm particles) employing a mobile phase of hexane–2-propanol–acetic acid (100 : 1 : 0.01, v/v) at a flow rate of 2 ml/min with UV detection at 205 nm (R_t of 1-O-alkyl-2-acyl-sn-glycerol, 6 min). The mass of purified 1-O-eicosodec-11'-enyl-2-octadec-9'-enoyl-sn-glycerol is typically quantitated following acid methanolysis by isothermal capillary gas chromatography (oven temperature, 190°) utilizing a 40-m SP2330 fused silica capillary column and employing arachidic acid as internal standard.

Synthesis of Specific 1-O-(Z)-Hexadec-1'-enyl-2-acyl-sn-glycerol Molecular Species. Specific 1-O-(Z)-hexadec-1'-enyl-2-acyl-sn-glycerol molecular species are synthesized by the acylation of reversed-phase-purified lysoplasmenylcholine   [1-O-(Z)-hexadec-1'-enyl-sn-glycero-3-phosphocholine] with the acyl chloride of choice followed by phospholipase C-mediated hydrolysis of the resultant ether-linked choline glycerophospholipid. Lysoplasmenylcholine is prepared by base-catalyzed methanolysis of beef heart choline glycerophospholipids,[11] and individual molecular species are purified by reversed-phase HPLC utilizing an octadecyl silica column (10 × 250 mm) employing a mobile phase of acetonitrile–methanol–water (57 : 23 : 20, v/v) containing 20 mM choline chloride as previously described.[12]

To 200 mg (417 μmol) of 1-O-(Z)-hexadec-1'-enyl-sn-glycero-3-phosphocholine, 30 ml of distilled chloroform (dried over P_2O_5) is added and gently stirred until the lysoplasmenylcholine is dissolved. Next, 200 mg (~600 μmol) of the desired fatty acyl chloride (previously dissolved in 10 ml of distilled chloroform) is added to the stirred reaction mixture. Finally, 167 mg (1.3 mmol) of recrystallized N,N-dimethyl-4-aminopyridine is added, and the reaction mixture is overlaid with nitrogen and sealed. The reaction mixture is stirred at room temperature for 2 hr in a dry nitrogen atmosphere, at which time an aliquot is removed for analysis to determine the extent of reaction completion. The reaction aliquot is diluted in chloroform, then applied to a silica G TLC plate, and reaction products are separated utilizing a mobile phase of chloroform–methanol–ammonium hydroxide (65 : 35 : 5, v/v). The extent of lysophospholipid conversion to choline glycerophospholipids is estimated by iodine staining.

Typically 2 hr is sufficient to convert over 80% of the ether-linked

[11] L. W. Wheeldon, Z. Schumert, and D. A. Turner, *J. Lipid Res.* **6**, 481 (1965).
[12] M. H. Creer and R. W. Gross, *J. Chromatogr.* **338**, 61 (1985).

lysophospholipid to its parent phosphoglyceride. However, in cases where unacceptable amounts of acylation are manifest, an additional 100 mg of fatty acyl chloride and 83 mg of N,N-dimethyl-4-aminopyridine are added prior to stirring for an additional 1–2 hr. Reactions are terminated by the addition of 40 ml of methanol, and subsequently 40 ml of water is added to form a biphasic mixture. The chloroform phase (lower) is saved, and the aqueous layer is reextracted by addition of an equal volume of chloroform. After addition of 10 ml benzene and 5 ml of ethanol to the combined chloroform extracts, the solvent is removed by rotary evaporation. Ether-linked choline glycerophospholipids are resuspended in chloroform (typically 10 mg/ml) and aliquoted into individual vials which could then be stored for at least 3 months in a nitrogen atmosphere at −20°.

Synthetic ether-linked choline glycerophospholipids (typically 2–5 mg dissolved in 4 ml of diethyl ether) and 10 U of *B. cereus* phospholipase C, previously dissolved in 0.5 ml of buffer [100 mM Tris-Cl, 10 mM CaCl$_2$ (pH 7.40)], are mixed to form a two-phase system which is initially vigorously vortexed for 1 min and subsequently vortexed every 15 min during the 2 hr incubation at 24°. The ether layer is saved, and 1-O-(Z)-hexadec-1'-enyl-2-acyl-*sn*-glycerol in the ether layer is purified by straight-phase HPLC (1 mg per injection) utilizing a 4.6 × 250 mm silica stationary phase employing a mobile phase of hexane–2-propanol–water (100 : 1 : 0.01, v/v) (R_t 6 min). HPLC-purified ether-linked diglycerides are quantitated following acid methanolysis by capillary gas chromatography utilizing arachidic acid as internal standard (Fig. 1). Alkyl ether-linked diglycerides are prepared similarly except that commercially available alkyl ether choline lysophospholipids are utilized.

Quantitation of Ether-Linked Diglycerides in Control and Ischemic Myocardium

Hearts (3–5 g) are obtained from diethyl ether-anesthetized New Zealand White rabbits and are Langendorf perfused with modified Krebs–Henseleit buffer at 60 mm Hg or are rendered ischemic as previously described.[8] Ventricular tissue is rapidly freeze-clamped (at the temperature of liquid nitrogen), then pulverized into a fine powder, and lipids are extracted by the Bligh–Dyer method.[10] During the Bligh and Dyer extraction, 25 μg of 1-O-eicosodec-11'-enyl-2-octadec-9'-enoyl-*sn*-glycerol is added for utilization as an internal standard. Neutral lipids from the myocardial extracts are separated from polar lipids by silicic acid chromatography utilizing 3-ml Supelclean extraction columns which are washed with 10 column volumes of chloroform. The combined chloroform eluents are evaporated under a nitrogen stream (to ~1 ml) and directly

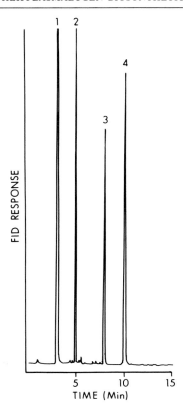

FIG. 1. Capillary gas chromatography of synthetic HPLC-purified 1-*O*-(*Z*)-hexadec-1'-enyl-2-octadec-9'-enoyl-*sn*-glycerol. An aliquot (50 μl) of synthetic straight-phase HPLC-purified 1-*O*-(*Z*)-hexadec-1'-enyl-2-octadec-9'-enoyl-*sn*-glycerol was combined with 20 μg of arachidic acid, dried under a nitrogen stream, and subjected to acid methanolysis. After extraction of the fatty acid methyl esters and dimethyl acetals into petroleum ether, 1 μl of the extract was injected onto a 40-m SP2330 capillary column (at 190°), and flame ionization detection (FID) was utilized to detect isothermally eluted individual aliphatic moieties. Individual peaks are as follows: 1, solvent peak; 2, dimethyl acetal of palmitaldehyde; 3, oleic acid methyl ester; and 4, arachidic acid methyl ester (internal standard). Each peak was identified by comparisons to authentic standards, and the integrated area of peak 2 was 98% of that of peak 3. The concentration of 1-*O*-(*Z*)-hexadec-1'-enyl-2-octadec-9'-enoyl-*sn*-glycerol was determined by comparison between the integrated peak areas of those peaks originating from ether-linked diglyceride and that originating from exogenously added internal standard.

streaked onto silica gel G plates (1 mm thickness). Myocardial nonpolar lipids are separated utilizing a mobile phase of petroleum ether–diethyl ether–acetic acid (70 : 30 : 1, v/v), and the region corresponding to ether-linked diglycerides [identified by selective iodine staining of 1,3-diacyl-*sn*-glycerol (which migrates with 1-*O*-(*Z*)-alk-1′-enyl-2-acyl-*sn*-glycerol in this solvent system and is chromatographed in an adjacent lane)] is scraped into borosilicate tubes prior to sequential extractions with chloroform (60 ml total). Chloroform extracts are filtered using a 0.45-μm PTFE filtration unit and evaporated to dryness under nitrogen.

Because endogenous 1,3-diacyl-*sn*-glycerol and 1-*O*-(*Z*)-alk-1′-enyl-2-acyl-*sn*-glycerol comigrate it is necessary to eliminate the contaminating 1,3-diacyl-*sn*-glycerols by Vitride reduction prior to their derivatization to dibenzoylated monoradylglycerols as previously described by Snyder and co-workers.[13] TLC-purified diradylglycerols are resuspended in 2.5 ml of diethyl ether–benzene (4 : 1, v/v), and 0.5 ml of Vitride reagent is slowly added at 4° prior to incubation at 37° for 30 min. To terminate incubations, the reaction mixture is cooled to 0° prior to the dropwise addition of 8 ml of chilled 20% ethanol (v/v). The ether layer (top) is removed and directly streaked onto a silica G TLC plate (100 μm thickness). Ether-linked monoradylglycerols are purified utilizing a mobile phase of diethyl ether–water–acetic acid (200 : 1 : 1, v/v), and regions corresponding to ether-linked monoradylglycerols (R_f 0.43) are scraped into borosilicate tubes and extracted with chloroform prior to evaporation to dryness utilizing a nitrogen stream.

The ether-linked monoradylglycerols are dibenzoylated by resuspension into a solution of 0.3 ml benzene containing 10 mg benzoic anhydride and 4 mg of *N,N*-dimethyl-4-aminopyridine. The reaction is incubated for 1 hr at room temperature prior to its termination at 4° by the dropwise addition of 2 ml of ice-cold 0.1 N NaOH. The dibenzoylated products are extracted with 2 ml of hexane, concentrated under nitrogen, and purified by TLC [silica G plates utilizing a mobile phase of hexane–diethyl ether–ammonium hydroxide (85 : 15 : 1, v/v)]. The regions corresponding to dibenzoylated ether-linked monoglycerides (R_f 0.31) are scraped into borosilicate tubes and are extracted utilizing a 7 ml mixture of chloroform–methanol–water (2.5 : 2 : 2.5, v/v). Following centrifugation the chloroform layer is collected, and the silica is extracted 2 more times by the addition of 2.5 ml of chloroform. The combined chloroform extracts are filtered utilizing a 0.45-μm PTFE filter, dried under a stream of nitrogen, and resuspended in 100 μl of acetonitrile.

[13] M. L. Blank, E. A. Cress, T.-C. Lee, N. Stephens, C. Piantadosi, and F. Snyder, *Anal. Biochem.* **133**, 430 (1983).

Typically, individual molecular species of dibenzoylated ether-linked monoglycerides are resolved by injection of 25–50 μl of the acetonitrile solution containing TLC-purified dibenzoylated monoglycerides (10 μg) onto an octadecyl silica HPLC column (4.6 × 250 mm) and are eluted isocratically utilizing acetonitrile–2-propanol (95 : 5, v/v) at a flow rate of 2 ml/min with UV detection at 231 nm (Fig. 2). The masses of individual molecular species of the ether-linked diglycerides are determined by comparisons of the integrated UV absorbance of each peak with that derived from the internal standard.

Metabolism of Ether-Linked Diglycerides in Brain and Myocardium

Synthetic, HPLC-purified ether-linked diglycerides (see above) are utilized to assess the molecular subclass and species selectivities of diglyceride kinase, cholinephosphotransferase, and ethanolaminephosphotransferase. Detergent-solubilized synthetic diradylglycerols are incubated with selected subcellular fractions and either [γ-^{32}P]ATP, [*methyl*-^{14}C]CDPcholine, or [1,2-*ethanolamine*-^{14}C]CDPethanolamine to characterize the enzymatic activities capable of modifying the *sn*-3 position of ether-linked diglycerides.

Diglyceride Kinase Assays. Cerebrums from diethyl ether-anesthetized New Zealand White rabbits are removed, placed in homogenization buffer [100 m*M* Tris-Cl, 0.25 *M* sucrose, 2 m*M* EGTA, 2 m*M* EDTA, and 1 m*M* dithiothreitol (pH 7.4)] and homogenized in a Potter–Elvehjem apparatus. After centrifugation at 10,000 g_{max} for 20 min, the supernatant is removed and further centrifuged at 100,000 g_{max} for 60 min to yield the cytosolic (supernatant) and microsomal (pellet) fractions. The microsomal pellet is resuspended in homogenization buffer at a concentration of 5 mg protein/ml.

To facilitate measurements of diglyceride kinase activity which are independent of the physical characteristics of individual substrates, ether-linked diglycerides are incorporated as substitutional impurities in micelles comprised of octylglucoside as previously described.[14] Appropriate amounts of chloroform solutions containing the desired diradylglycerol are dried under nitrogen in borosilicate tubes and resuspended in 20 μl of 255 m*M* octylglucoside (resulting in a final diradylglycerol concentration of 1–8 mol % and a bulk lipid concentration of 51–55 m*M*). Next, 50 μl of assay buffer [200 m*M* Tris-Cl, 100 m*M* NaCl, 12.5 m*M* MgCl$_2$, and 0.5 m*M* EDTA (pH 7.4)] is added followed by addition of 2 μl of a 50 m*M* dithiothreitol solution. Next, the reaction mixture is vortexed (10 sec),

---

[14] J. P. Walsh and R. M. Bell, *J. Biol. Chem.* **261**, 6239 (1986).

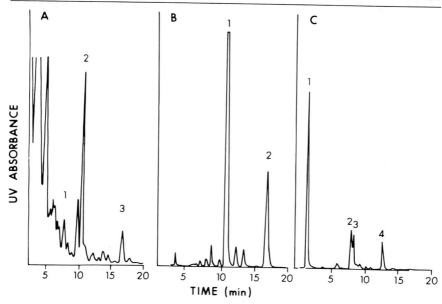

FIG. 2. Reversed-phase HPLC of 1-*O*-alkyl-2,3-dibenzoyl-*sn*-glycerol and 1-*O*-(*Z*)-alk-1'-enyl-2,3-dibenzoyl-*sn*-glycerol derived from rabbit myocardium [D. A. Ford and R. W. Gross, *J. Biol. Chem.* **263**, 2644 (1988)]. (A) Rabbit myocardial 1-*O*-alkyl-2-acyl-*sn*-glycerol and 1-*O*-(*Z*)-alk-1'-enyl-2-acyl-*sn*-glycerol were extracted, purified by TLC, Vitride-reduced, and benzoylated as described in the text. The resultant 1-*O*-alkyl-2,3-dibenzoyl-*sn*-glycerol and 1-*O*-(*Z*)-alk-1'-enyl-2,3-dibenzoyl-*sn*-glycerol were separated by reversed-phase HPLC with a mobile phase of acetonitrile–2-propanol (95 : 5, v/v) at a flow rate of 2 ml/min. Individual molecular species were quantified by comparison of integrated UV absorbances at 231 nm with that of the internal standard (eluting at 25 min, not shown). Peaks correspond to 1-*O*-(*Z*)-hexadec-1'-enyl-2,3-dibenzoyl-*sn*-glycerol (peak 1), 1-*O*-hexadecyl-2,3-diben-zoyl-*sn*-glycerol (peak 2), and 1-*O*-octadecyl-2,3-dibenzoyl-*sn*-glycerol (peak 3). (B) Authentic 1-*O*-alkyl-2,3-dibenzoyl-*sn*-glycerol was prepared by Vitride reduction of lyso platelet-activating factor, preparative TLC of the resultant 1-*O*-alkyl-*sn*-glycerol, and subsequent benzoylation as described in the text. Individual molecular species were resolved by re-versed-phase HPLC. Resolution of 1-*O*-hexadecyl-2,3-dibenzoyl-*sn*-glycerol (peak 1) and 1-*O*-octadecyl-2,3-dibenzoyl-*sn*-glycerol (peak 2), which coelute with peaks 2 and 3 in (A), was routinely accomplished without difficulty. (C) Authentic 1-*O*-(*Z*)-alk-1'-enyl-2,3-dibenzoyl-*sn*-glycerol was prepared by Vitride reduction of plasmenylethanolamine, preparative TLC, and benzoylation as described in the text. Reversed-phase HPLC of the resultant 1-*O*-(*Z*)-octadec-1',9'-enyl-2,3-dibenzoyl-*sn*-glycerol (peak 2), 1-*O*-(*Z*)-hexadec-1'-enyl-2,3-diben-zoyl-*sn*-glycerol (peak 3), and 1-*O*-(*Z*)-octadec-1'-enyl-2,3-dibenzoyl-*sn*-glycerol (peak 4) separated each molecular species. Peak 1 is the solvent front.

and 10 μl of microsomal or cytosolic proteins is added to the reaction mixture, which is subsequently preincubated for 1 min at 24°.

Diglyceride kinase assays are initiated by the addition of selected concentrations (typically 10 μM–1 mM) of [γ-^{32}P]ATP (4 μCi/μmol) to the mixed micelle solution, and reactions are typically incubated for 4 min at 30°. Reactions are terminated by the addition 0.4 ml of chloroform–methanol (1 : 1, v/v), and reaction products (i.e., radiolabeled phosphatidic acid) are extracted after the addition of 0.1 ml of 200 mM CaCl$_2$ by vigorous vortexing.[15] Aliquots of the chloroform extract (lower layer) are applied to silica gel G TLC plates, and phosphatidic acid is resolved utilizing a mobile phase of chloroform–acetone–methanol–acetic acid–water (6 : 8 : 2 : 2 : 1, v/v) (R_f 0.86). Diglyceride kinase activity is quantitated by scraping regions of the TLC plate corresponding to phosphatidic acid into liquid scintillation vials and subsequently quantifying the incorporated ^{32}P by Cerenkov counting.

Cholinephosphotransferase Assays. Myocardial choline phosphotransferase activity is assessed in brain microsomes prepared from New Zealand White rabbits.[16] Aliquots (10–300 nmol) of diradylglycerols (stored in chloroform) are added to borosilicate tubes, dried under a stream of nitrogen and subsequently resuspended in 25 μl of assay buffer [50 mM Tris-Cl, 10 mM MgCl$_2$ (pH 7.4)] containing 67 μM Tween 20 by sonication for 3 min at room temperature utilizing a Branson water bath sonicator. To the sonicated solution, 60 μl of assay buffer containing selected concentrations (1–250 μM) of [*methyl*-^{14}C]CDPcholine (2 μCi/μmol) is added followed by the addition of 15 μl of myocardial microsomal protein (150 μg). Reactions are incubated for 5–30 min at 37° (as necessary) and are terminated by addition of 150 μl of butanol and vortexing. Subsequent centrifugation (500 g for 2 min) facilitates separation of layers, and 100 μl of the butanol layer (top) is applied to a channeled silica gel G plate. Choline glycerophospholipids are resolved utilizing a mobile phase consisting of chloroform–methanol–ammonium hydroxide (65 : 35 : 5, v/v). Regions corresponding to choline glycerophospholipids (R_f 0.25) are identified by iodine staining, then scraped into scintillation vials, and radiolabel is quantitated by liquid scintillation spectrometry after the addition of 3 ml of fluor (Universol).

Ethanolaminephosphotransferase Assays. Ethanolaminephosphotransferase activity in myocardial microsomes utilizing ether-linked diglycerides is determined by methods similar to those employed for choline phosphotransferase except that [1,2-*ethanolamine*-^{14}C]CDPethanolamine

[15] A. Sheltawy and R. M. C. Dawson, *J. Neurochem.* **15**, 144 (1968).
[16] D. A. Ford and R. W. Gross, *J. Biol. Chem.* **263**, 2644 (1988).

is used. Ethanolaminephosphotransferase activity is assessed by quantitating radioactivity incorporated into ethanolamine glycerophospholipids by subcellular fractions incubated in the presence of vinyl ether, alkyl ether, or diacyl subclasses of diglycerides. Reaction products are resolved by TLC purification on silica gel G plates utilizing the aforementioned solvent system (R_f of ethanolamine glycerophospholipids, 0.34) and quantified by subsequent scintillation spectrometry.

Conclusion

The methods described herein are suitable for the preparation of substantial amounts of homogeneous ether-linked diglycerides which can be utilized to assess the metabolism of vinyl ether and alkyl ether diradylglycerols in mammalian tissues. Comparisons of the relative rates of metabolism of individual subclasses of naturally occurring diradylglycerols demonstrate significant subclass selectivities for each of these enzymes, which contribute to the differential biological half-lives of each subclass of diglycerides in mammalian tissues. Because ether-linked diglycerides activate protein kinase C with disparate cofactor requirements in comparison to their diacylglycerol counterparts,[17] it is hoped that these methods will facilitate delineation of the biological importance of ether-linked diglycerides in mammalian signal transduction.

Acknowledgments

This research was supported by Grants HL42665, HL41250, and HL34839. RWG is the recipient of an Established Investigator Award from the American Heart Association. We thank Ms. Stacy Kiel for assistance in preparation of the manuscript.

[17] D. A. Ford, R. Miyake, P. E. Glaser, and R. W. Gross, *J. Biol. Chem.* **264**, 13818 (1989).

Section X

Sphingolipid Biosynthesis Enzymes

[51] Enzymes of Ceramide Biosynthesis

By ALFRED H. MERRILL, JR., and ELAINE WANG

Introduction

Ceramides are the lipid backbones of sphingomyelin,[1,2] ceramide phosphoinositides,[3] and other glycosphingolipids.[4,5] They are composed of a long-chain (sphingoid) base with an amide-linked fatty acid. Sphingosine (*trans*-4-sphingenine) is the major long-chain base of most sphingolipids, but lesser amounts of other chain length homologs, dihydrosphingosines (sphinganines), and 4-hydroxysphinganines (phytosphinganines) are also found.[6] The term sphingosine is sometimes used to refer to the entire class of long-chain bases, although this is not preferred. The amide-linked fatty acids of ceramides typically have alkyl chain lengths from 16 to 24 carbon atoms (unbranched); only small percentages are branched or monounsaturated, but α-hydroxy fatty acids are common in some sphingolipid subclasses.[6] An overall scheme for ceramide biosynthesis is shown in Fig. 1,[7,8] and assays for each of the three reasonably well-characterized enzymes of this pathway are given in this chapter (each enzyme is named according to recommended name, followed by common and systematic names and number assigned by the Enzyme Commission). It should be borne in mind, however, that other routes for ceramide biosynthesis are also possible, as none of these enzymes have been purified to homogeneity.

[1] Y. Barenholz and T. E. Thompson, *Biochim. Biophys. Acta* **604,** 129 (1980).
[2] Y. Barenholz and S. Gatt, *in* "Phospholipids" (J. N. Hawthorne and G. B. Ansell, eds.), Chap. 4. Elsevier Biomedical Press, Amsterdam, 1982.
[3] R. L. Lester, G. W. Becker, and K. Kaul, *in* "Cyclitols and Phosphoinositides" (W. W. Wells and F. J. Eisenberg, eds.), p. 83. Academic Press, New York, 1978.
[4] J. N. Kanfer and S.-I. Hakomori, "Handbook of Lipid Research, Volume 3, Sphingolipid Biochemistry." Plenum, New York, 1983.
[5] C. C. Sweeley, "Cell Surface Glycolipids." American Chemical Society Press, Washington, D.C., 1980.
[6] K.-A. Karlsson, *Chem. Phys. Lipids* **5,** 6 (1970).
[7] C. C. Sweeley, *in* "Biochemistry of Lipids and Membranes" (D. E. Vance and J. E. Vance, eds.), p. 361. Benjamin/Cummings, Menlo Park, California, 1985.
[8] A. H. Merrill, Jr., and D. D. Jones, *Biochim. Biophys. Acta* **1044,** 1 (1990).

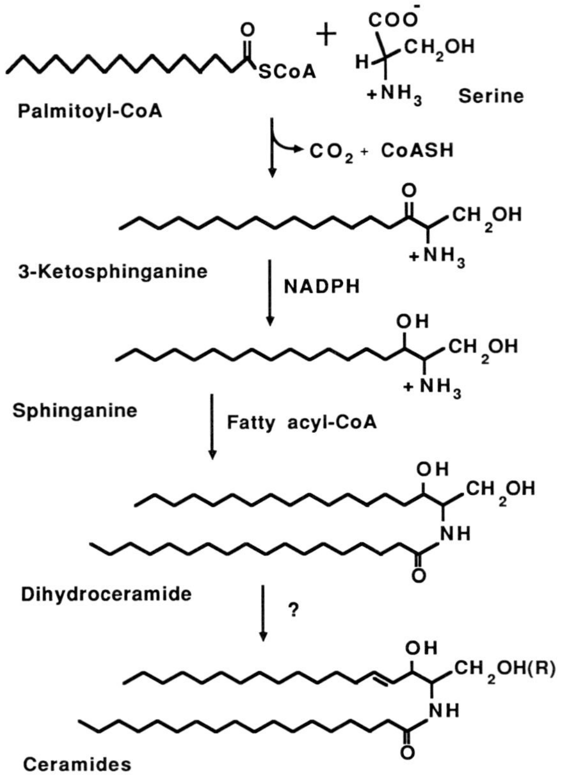

FIG. 1. Ceramide biosynthesis. The most likely pathway for ceramide biosynthesis *de novo* is shown. The initial condensation of serine and palmitoyl-CoA is catalyzed by serine palmitoyltransferase, followed by NADPH-dependent reduction of the 3-keto group by 3-ketosphinganine reductase and addition of the amide-linked fatty acid by ceramide synthetase(s). The formation of the 4-trans double bond has not been demonstrated directly, but it appears to occur after the synthesis of dihydroceramides.

Serine Palmitoyltransferase (3-Ketosphinganine Synthase) [Palmitoyl-CoA : L-Serine C-Palmitoyltransferase (Decarboxylating); EC 2.3.1.50]

The initial step of *de novo* ceramide biosynthesis is the condensation of L-serine and palmitoyl-CoA by the pyridoxal 5'-phosphate-dependent enzyme serine palmitoyltransferase.

Assay. Serine palmitoyltransferase can be assayed *in vitro* by a number of methods,[9–12] but the easiest[12] is based on the incorporation of ³H- or

[9] E. E. Snell, S. J. Di Mari, and R. N. Brady, *Chem. Phys. Lipids* **5**, 116 (1970).
[10] W. Stoffel, *Chem. Phys. Lipids* **5**, 139 (1970).

[14]C-labeled serine into chloroform-soluble products in an assay cocktail that contains EDTA or EGTA to inhibit phosphatidylserine synthesis. This assay can be conducted with microsomes[12,13] or cells in culture,[14,15] and it has been modified to allow assays of single colonies of cells.[16].

The assay contains 0.1 M HEPES buffer (pH 8.0 at 37°), 5 mM dithiothreitol (DTT), 2.5 mM EDTA (pH adjusted to neutrality before use), 50 μM pyridoxal 5'-phosphate, 0.2 mM palmitoyl-CoA, 1 mM [3]H- or ^{14}C-labeled L-serine (specific activity 10–30 mCi/mmol), and the enzyme source (e.g., 50–150 μg of microsomal protein) for a total volume of 0.1 ml in 13 × 100 mm screw-capped test tubes. The reaction is initiated by adding palmitoyl-CoA (to minimize depletion of this substrate by active fatty acyl-CoA hydrolases) and, after incubation at 37° for 10 min, is terminated with 1.5 ml of chloroform–methanol (1 : 2, v/v).

To extract the products, approximately 25 μg of sphinganine (which is as effective as the actual product, 3-ketosphinganine) is added as carrier, and 1 ml of chloroform and 2 ml of 0.5 N NH$_4$OH are added. The tubes are mixed, centrifuged for several minutes in a tabletop centrifuge, and the aqueous layer is removed by aspiration. The lower phase is washed twice with 2 ml of water, then 0.8 ml of the chloroform layer (representing approximately half of the total product formed) is transferred to scintillation vials. The solvent is evaporated, 4 ml of a standard scintillation cocktail is added, and the samples are counted.

To correct for chloroform-soluble contaminants in the radiolabeled serine, additional incubation tubes are prepared with all of these components except palmitoyl-CoA. Most commercial preparations of [^{3}H]serine contain such contaminants, but they can be removed with a small ion-exchange column.[12]

Other factors to consider in optimizing the assay conditions are (1) linearity with time and enzyme amount, which is particularly important since high levels of fatty acyl-CoA hydrolases can deplete the cosubstrate; (2) a check of the optimal palmitoyl-CoA concentration, since this substrate can be depleted under some conditions but can also be inhibitory at high levels; and (3) form of the radiolabel, because it is preferable to conduct the assay with either L-[3-^{14}C]- or L-[3-^{2}H]serine (rather than

[11] Y. Kishimoto, *in* "The Enzymes" (P. D. Boyer, ed.), 3rd Ed., Vol. 16, Chap. 10. Academic Press, New York, 1983.
[12] R. D. Williams, E. Wang, and A. H. Merrill, *Arch. Biochem. Biophys.* **228,** 282 (1984).
[13] A. H. Merrill, Jr., D. W. Nixon, and R. D. Williams, *J. Lipid Res.* **26,** 617 (1985).
[14] A. H. Merrill, Jr., *Biochim. Biophys. Acta* **754,** 284 (1983).
[15] T. O. Messmer, E. Wang, V. L. Stevens, and A. H. Merrill, Jr., *J. Nutr.* **119,** 534 (1989).
[16] K. Hanada, M. Nishijima, and Y. Akamatsu, *J. Biol. Chem.* **265,** 22137 (1989).

uniformly labeled serine, which is less expensive) since the α-hydrogen is lost in the course of the reaction[17] and, hence, the specific activity of the product can be calculated with greater certainty.

Other ways of detecting the products of this enzyme are to resolve the products by thin-layer chromatography (TLC) using silica gel plates developed with chloroform–methanol–2 N NH_4OH (40 : 10 : 1, v/v)[9,10,12]; or, to trap the $^{14}CO_2$ liberated from [1-*carboxyl*-^{14}C]serine or uniformly labeled [^{14}C]serine.[12,18] For this, the assay is conducted in tubes with a septum stopper fitted with a cup (Kontes, Vineland, NJ, K-882320-000), and at the end of the assay 0.1 ml of Protosol is added to the cup to trap the $^{14}CO_2$ and 1% perchloric acid is added to the assay solution. After 1 hr of gentle shaking, the cup is removed and the radioactivity determined by scintillation counting.

Substrate Specificity. Depending on the system studied, serine palmitoyltransferase has been reported to have an apparent K_m for L-serine of 0.1 to 1 mM.[12,18] No other amino acids have been reported to substitute for L-serine as substrate; however, a number of amino acid analogs are inhibitors of this enzyme. Cycloserine[19] and β-haloalanines[20] irreversibly inactivate serine palmitoyltransferase by "suicide" inhibition, as predicted by the proposed mechanism for this enzyme.[17]

Serine palmitoyltransferase has a high degree of specificity for palmitoyl-CoA; it best utilizes linear, saturated fatty acyl-CoA's of 16 ± 1 carbon atoms,[12,18] which corresponds to the prevalence of 18-carbon sphingosine bases (i.e., with 16 carbons from palmitic acid and 2 from serine) in most cellular sphingolipids.[5] α-Fluoropalmitic acid has been reported to inhibit long-chain base synthesis; however, this is thought to be due to inhibition of palmitoyl-CoA formation.[21]

Pyridoxal 5'-phosphate can be resolved from serine palmitoyltransferase by dialyzing the enzyme against cysteine to form the thiazolidine. The apparent K_m for pyridoxal 5'-phosphate in reconstituting the apoenzyme thus prepared is approximately 1μM.[12]

Sequence. This enzyme has only been purified (>100-fold) from *Bacteriodes melaninogenicus*.[22] However, the gene for serine palmitoyltransferase has recently been cloned from yeast.[23] There is a high degree of

[17] K. Krisnangkura and C. C. Sweeley, *J. Biol. Chem.* **251**, 1597 (1976).

[18] S. J. Di Mari, R. N. Brady, and E. E. Snell, *Arch. Biochem. Biophys.* **143**, 553 (1971).

[19] K. S. Sundaram and M. Lev, *J. Neurochem.* **42**, 577 (1984).

[20] K. A. Medlock and A. H. Merrill, Jr., *Biochemistry* **27**, 7079 (1988).

[21] R. M. Soltysiak, F. Matsuura, D. Bloomer, and C. C. Sweeley, *Biochim. Biophys. Acta* **792**, 214 (1984).

[22] M. Lev and A. F. Milford, *Arch. Biochem. Biophys.* **212**, 424 (1981).

[23] R. Buede, C. Rinker-Schaffer, W. J. Pinto, R. L. Lester, and R. C. Dickson, *J. Bacteriol.* **173**, 4325 (1991).

homology with δ-aminolevulinate synthase, a pyridoxal 5'-phosphate-dependent enzyme that catalyzes a condensation reaction between glycine and succinyl-CoA that is similar to that of serine palmitoyltransferase.

Regulation. Relatively little is known about the regulation of serine palmitoyltransferase. Its activity correlates with the relative level of sphingolipid synthesis by tissues,[13,20,24] and addition of long-chain bases as part of the sphingolipids of lipoproteins[25,26] or short-chain analogs of sphingosine[27] reduce flux through the pathway. In the latter case, this appears to be due to down-regulation of serine palmitoyltransferase.[28]

3-Dehydrosphinganine Reductase (3-Ketosphinganine Reductase) [D-erythro-dihydrosphingosine : NADP⁺ 3-Oxidoreductase; EC 1.1.1.102]

The next reaction of ceramide biosynthesis is the reduction of 3-keto-sphinganine to sphinganine by an NADPH-dependent reductase found in microsomes.[29,30]

Assay. Incubation mixtures contain 50 μM 3-[4,5-³H]ketosphinganine (prepared as described below), 0.1 M potassium phosphate buffer (pH 6.8), 1 mM MgCl$_2$, 1 mM thioglycol (or DTT), 250 μM NADPH (or an NADPH-regenerating system as described in Ref. 12 and 28), and varying amounts of protein (e.g., 50–150 μg of microsomal protein) for a total volume of 0.1 ml in 13 × 100 mm screw-capped test tubes. 3-[³H]Keto-sphinganine is added either in a small amount of Triton X-100 [for a final concentration of less than 0.1% (v/v) Triton X-100 in the assay] or by transfer of the substrate to the assay test tube in an organic solvent, evaporation of the solvent under a stream of N$_2$, and immediate addition of the rest of the assay components, followed by thorough mixing and/or sonication.

The reaction is initiated by adding the NADPH and, after incubation at 37° for 10 min, is terminated with 1.5 ml of chloroform–methanol (1 : 2, v/v), and 25 μg of carrier sphinganine is added. The products are extracted as described for serine palmitoyltransferase, and the final lipid extract is applied to silica gel TLC plates (silica gel H, Brinkmann, Westbury, NY).

[24] R. D. Williams, D. W. Nixon, and A. H. Merrill, Jr., *Cancer Res.* **44**, 1918 (1984).
[25] R. B. Verdery III and R. Theolis, Jr., *J. Biol. Chem.* **257**, 1412 (1982).
[26] S. Chatterjee, K. S. Clarke, and P. O. Kwiterovich, Jr., *J. Biol. Chem.* **261**, 13474 (1986).
[27] G. van Echten, R. Birk, G. Brenner-Weis, R. R. Schmidt, and K. Sandoff, *J. Biol. Chem.* **265**, 9333 (1990).
[28] E. C. Mandon, G. van Echten, R. Birk, R. R. Schmidt, and K. Sandoff, *Eur. J. Biochem.* (in press).
[29] P. E. Braun and E. E. Snell, *J. Biol. Chem.* **243**, 3775 (1968).
[30] M. D. Ullman and N. S. Radin, *Arch. Biochem. Biophys.* **152**, 767 (1972).

The plates are developed in chloroform–methanol–2 N NH$_4$OH (40 : 10 : 1, v/v), and the region corresponding to [^{3}H]sphinganine can be established by scanning the chromatoplate, by fluorography, or by lightly spraying the plate with ninhydrin (0.1% in ethanol) and heating the plate to visualize the bluish purple spot from the sphinganine carrier. In this system, typical R_f values relative to the solvent front are 0.7 for 3-ketosphinganine, 0.45 for sphingosine, and 0.35 for sphinganine. The appropriate regions of the chromatoplate are scraped into scintillation vials, 0.2 ml of water and 4 ml of a detergent-based aqueous counting cocktail are added, and the amount of radiolabel is determined with the appropriate corrections for quenching.

The activity of 3-ketosphinganine reductase can also be estimated by a two-step assay involving incubation of microsomes first with radiolabeled serine and palmitoyl-CoA to form 3-ketosphinganine, then with NADPH or an NADPH-regenerating system.[12]

Substrate Specificity. Studies with an 84-fold purified preparation of the reductase from beef liver microsomes[31] found that only the B-hydrogen from NADPH is transferred to 3-ketosphinganine; earlier studies had shown that NADH is not an effective reductant. Only the D-isomer of 3-ketosphinganine is reduced, but both C$_{18}$- and C$_{20}$-3-ketosphinganines were reduced.

Regulation. 3-Dehydrosphinganine reductase appears to be much more active than the first enzyme of this pathway because none of the keto intermediate is found when microsomes are incubated with serine, palmitoyl-CoA, and NADPH[12,32] or when intact cells are incubated with [^{14}C]serine.[33]

Sphingosine/Sphinganine Acyltransferase [(Dihydro) ceramide Synthase] [Acyl-CoA : Sphingosine N-Acyltransferase; EC 2.3.1.24]

The next step of ceramide synthesis appears to be the addition of a long-chain fatty acid to sphinganine (earlier texts had indicated that sphinganine is converted to sphingosine first; however, this reaction has not been shown directly).[8] (Dihydro) ceramide synthesis can be assayed with either [^{3}H]sphingosine or [^{3}H]sphinganine, and it is thought to involve both CoA-dependent[30,34–36] and CoA-independent[11,37,38] enzymes. There is

[31] W. Stoffel, D. Le Kim, and G. Sticht, *Hoppe-Seyler's Z. Physiol. Chem.* **349**, 1637 (1968).
[32] P. E. Braun, P. Morell, and N. S. Radin, *J. Biol. Chem.* **245**, 335 (1970).
[33] A. H. Merrill, Jr., and E. Wang, *J. Biol. Chem.* **261**, 3764 (1986).
[34] M. Sribney, *Biochim. Biophys. Acta* **125**, 542 (1966).
[35] P. Morell and N. S. Radin, *J. Biol. Chem.* **245**, 342 (1970).
[36] H. Akanuma and Y. Kishimoto, *J. Biol. Chem.* **254**, 1050 (1979).
[37] I. Singh, *J. Neurochem.* **40**, 1565 (1983).
[38] M.-A. Mori, H. Shimeno, and Y. Kishimoto, *Neurochem. Int.* **7**, 57 (1985).

also some evidence for ceramide synthesis by reversal of ceramidase,[39,40] but this is thought to be of limited biological significance.[41] An assay for the CoA-dependent enzyme(s) that is prevalent in brain and liver microsomes follows.

Assay. Incubation mixtures contain 1 μM [3- or 4,5-^{3}H]sphinganine or [^{3}H]sphingosine prepared as described below or produced as a custom synthesis by a commercial supplier (we have obtained [^{3}H]sphingosine from New England Nuclear, Boston, MA), 25 mM potassium phosphate buffer (pH 7.4), 0.5 mM DTT, 100 μM stearoyl-CoA, and varying amounts of protein (e.g., 50–100 μg of microsomal protein) for a total volume of 0.1 ml in 13 × 100 mm screw-capped test tubes. The ^{3}H-labeled long-chain base (which has been dissolved in ethanol) is added first, the organic solvent is removed under a stream of N_2, and the remainder of the assay components are added as soon as possible and followed by thorough mixing and/or sonication. If higher concentrations of [^{3}H]sphinganine are desired, the sphinganine can be solubilized with celite[34] or by sonication with phosphatidylcholine.

The reaction is initiated by adding stearoyl-CoA and, after incubation at 37° for 10 min, is terminated with 1.5 ml of chloroform–methanol (1 : 2, v/v), and 25 μg of carrier ceramide (bovine brain or another commercial source) is added. The products are extracted as described for serine palmitoyltransferase, and the final lipid extract is applied to silica gel TLC plates (silica gel H, Brinkmann). The plates are developed in diethyl ether–methanol (99 : 1, v/v), and the region corresponding to [^{3}H]ceramide can be established by radiometric scanning of the chromatoplate, by fluorography, or by visualization of the carrier lipid with iodine vapor. In this system, the long-chain bases remain at the origin and ceramides migrate with R_f values of 0.25 to 0.35 relative to the solvent front. The ceramide-containing region of the chromatoplate is scraped into scintillation vials, 4 ml of either a toluene-based or a detergent-based aqueous counting cocktail is added, and the amount of radiolabel is determined with the appropriate corrections for quenching.

Because some decomposition of the radiolabeled long-chain bases occurs during handling (mainly during drying of the sample in the test tube, see discussion below), a zero-time control should be included. To assess any contribution from endogenous fatty acyl-CoA's or by a CoA-independent pathway, other assay tubes should omit stearoyl-CoA.

[39] E. Yavin and S. Gatt, *Biochemistry* **8,** 1692 (1966).

[40] M. Sugita, M. Williams, J. T. Dulaney, and H. W. Moser, *Biochim. Biophys. Acta* **398,** 125 (1975).

[41] H. W. Moser, A. B. Moser, W. W. Chen, and W. Schram, *in* "The Metabolic Basis of Inherited Diseases" (C. R. Scriver, A. L. Beaudent, W. S. Sly, and D. Valle, eds.), 6th Ed., Vol. 2, p. 1645. McGraw-Hill, New York, 1989.

Substrate Specificity. Sphinganine and sphingosine, as well as all four long-chain base stereoisomers, can be N-acylated by ceramide synthase(s).[11,30,34,35,42] Various fatty acyl-CoA's are also utilized, but not equally in different tissues, which has led to the hypothesis that there may be more than one enzyme involved.[11] When analyzing this enzyme in a new system, it would appear prudent to test different fatty acyl-CoA's.

Inhibitors. Recent studies with rat liver microsomes and hepatocytes have found that the acylation of sphingosine is potently inhibited *in vitro* and in intact hepatocytes by fumonisins B_1 and B_2, mycotoxins produced by *Fusarium moniliforme*.[43]

Regulation. Ceramide synthesis apparently occurs very fast *in vivo* since very low amounts of free long-chain bases occur as intermediates in sphingolipid biosynthesis.[8] This may be advantageous since sphingosine and other long-chain bases are very bioactive compounds that inhibit protein kinase C, activate the epithelial growth factor (EGF) receptor kinase, and affect diverse other systems in cells.[44,45] Free long-chain bases can be cytotoxic[42,46] or mitogenic[47] to cells at very low concentrations. Recent work indicates that ceramides, themselves, may be mediators of the action of $1\alpha,25$-dihydroxyvitamin D_3,[48] tumor necrosis factor, and γ-interferon[49] in HL-60 cells. Hence, the formation and turnover of this compound would be expected to be under careful regulation.

Formation of Other Long-Chain Bases

The step at which the 4-hydroxyl group of phytosphingosine is added is not known, although it is thought to occur after the formation of sphinganine.[49,50] The biosynthetic pathways for other long-chain bases (e.g., with branching or unsaturation along the alkyl side chain)[5] have not been elucidated.

[42] W. Stoffel and K. Bister, *Hoppe Seyler's Z. Physiol. Chem.* **354**, 169 (1973).

[43] E. Wang, W. Norred, C. W. Bacon, and A. H. Merrill, Jr., *J. Biol. Chem.* **266**, 14486 (1991).

[44] Y. A. Hannun and R. M. Bell, *Science* **242**, 500 (1989).

[45] A. H. Merrill, Jr., *J. Bioenerg. Biomemb.* **23**, 83 (1991).

[46] V. L. Stevens, S. Nimkar, W. C. L. Jamison, D. C. Liotta, and A. H. Merrill, Jr., *Biochim. Biophys. Acta* **1051**, 37 (1990); H. Zang, N. E. Buckley, K. Gibson, and S. Spiegel, *J. Biol. Chem.* **265**, 76 (1990).

[47] T. Okazaki, A. Bielawska, R. M. Bell, and Y. A. Hannun, *J. Biol. Chem.* **265**, 15823 (1990).

[48] M.-Y. Kim, C. Linardic, L. Obeid, and Y. A. Hannun, *J. Biol. Chem.* **266**, 484 (1991).

[49] M. W. Crossman and C. B. Hirschberg, *J. Biol. Chem.* **252**, 5815 (1977).

[50] M. W. Crossman and C. B. Hirschberg, *Biochim. Biophys. Acta* **795**, 411 (1984).

General Methods and Considerations in Analyzing
Ceramide Biosynthesis

Preparation of Radiolabeled Substrates. There are several relatively straightforward routes for the synthesis of radiolabeled long-chain bases as substrates for these enzymes. One is to begin with ceramide or a more complex sphingolipid (or the *N*-acetyl derivative of sphingosine or sphinganine), oxidize the 3-hydroxyl group using CrO_3, and reduce the ketone with sodium borotritide.[51] The free long-chain base can be recovered by hydrolysis[52,53] and purified by TLC.[54] The drawbacks to this method are that the radiolabeled product will be racemic at carbon 3 (although the stereoisomers can be resolved),[54] and a substantial portion of the free long-chain base may be lost during hydrolysis of the *N*-acyl derivatives (e.g., by replacement of the 3-hydroxyl with methanol), but this can be minimized by various approaches.[4,52]

The second procedure is to reduce the 4,5-*trans* double bond of sphingosine with sodium borotritide and palladium chloride or acetate[55] (this radiolabeled sphinganine can be converted to 3-ketosphinganine as described above).[51] The limitation of this method is that it forms only sphinganines; furthermore, protection of the amino group also requires subsequent hydrolysis and its attendant problems.

Comments on Stability of Long-Chain Bases. Sphingosine and other long-chain sphingoid bases, especially the tritiated forms, are unstable in dilute solution and when allowed to dry on glass surfaces. Hence, they should be examined periodically for purity and repurified as needed. In our experience the major contaminants formed from ^3H-labeled long-chain bases are nonpolar and are easily removed by applying the sample (in chloroform) to a small silica gel column (prepared in Pasteur pipettes) and eluting with increasing percentages of methanol in chloroform.

Studies of Ceramide Biosynthesis by Cultured Cells. Ceramide synthesis can also be followed *in vitro* or with intact cells using radiolabeled serine, palmitic acid, or long-chain bases as the substrates of the enzymes. However, particular care must be taken to establish the cellular-specific activity of the precursors (i.e., serine or palmitic acid) since they are made *de novo* and dilute the added radiolabel.[55] On the other hand, addition of free long-chain bases to cells is also complicated by the potent effects of these compounds on cells.[44–46,56]

[51] R. C. Gaver and C. C. Sweeley, *J. Am. Chem. Soc.* **88**, 3643 (1966).
[52] R. C. Gaver and C. C. Sweeley, *J. Am. Oil Chem. Soc.* **42**, 295 (1965).
[53] R. W. Pagano and O. C. Martin, *Biochemistry* **27**, 4439 (1988).
[54] G. Schwartzmann, *Biochim. Biophys. Acta* **529**, 106 (1978).
[55] A. H. Merrill, Jr., E. Wang, and R. E. Mullins, *Biochemistry* **27**, 340 (1988).
[56] Y. A. Hannun, A. H. Merrill, Jr., and R. M. Bell, this series, Vol. 210, p. 316.

Analysis of Long-Chain Bases by High-Performance Liquid Chromatography. The free long-chain bases of tissue extracts or cells in culture can be analyzed relatively easily and with high sensitivity by reversed-phase, high-performance liquid chromatography (HPLC).[57,58] First, the lipids are extracted using standard procedures, then interfering glycerolipids are removed by base treatment, and the free long-chain bases are derivatized with *o*-phthalaldehyde (OPA).

To extract the long-chain bases, the sample is homogenized with ice-cold buffer (50 mM potassium phosphate buffer, pH 7.0), and an aliquot (1–20 mg of tissue, or 1–10 × 10^6 cells) in 0.1–0.3 ml is transferred to 13 × 100 mm screw-cap (standard borosilicate tubes with Teflon caps) test tubes. After adding 1.5 ml of chloroform–methanol (1 : 2, v/v), an internal standard is added (such as 50–400 pmol of eicosasphinganine), and 1 ml each of chloroform and water are added. The two phases are separated by centrifugation; the upper phase is discarded, and the chloroform phase washed twice with water, drained through a small column containing anhydrous sodium sulfate (granular), and dried under reduced pressure.

To remove the glycerolipids, the extracts are resuspended in 1 ml of 0.1 M KOH in methanol–chloroform (2 : 1, v/v) and incubated at 37° for 1 hr. After the samples have cooled to room temperature, 1 ml each of chloroform and water are added, and the free long-chain bases are recovered in the chloroform phase, which is washed several times with water, dried over sodium sulfate, and the solvent evaporated under reduced pressure.

The *o*-phthalaldehyde derivatives are prepared by dissolving the sample in 50 μl of methanol and adding 50 μl of the OPA reagent, which is prepared by mixing 99 ml of 3% (w/v) boric acid in water (pH adjusted to 10.5 with KOH) and 1 ml of ethanol containing 50 mg of OPA and 50 μl of 2-mercaptoethanol. After incubation for at least 5 min at room temperature, 250–500 μl of methanol–5 mM potassium phosphate (pH 7.0) (90 : 10, v/v) is added, and the samples are centrifuged briefly in a microcentrifuge to clarify. Aliquots are injected onto a Waters (Milford, MA) Radial Pak C$_{18}$ column (5 μm, Type 8NVC185 with a C$_{18}$ guard column) and eluted with methanol–5 mM potassium phosphate, pH 7.0 (90 : 10, v/v) at a flow rate of 2 ml/min. The OPA derivative is detected with a fluorometric detector with an excitation wavelength of 340 nm and an emission wavelength of 455 nm (or with a 418-nm cutoff filter).

The elution of long-chain bases on this column is dependent on the type

[57] A. H. Merrill, Jr., E. Wang, R. E. Mullins, W. C. L. Jamison, S. Nimkar, and D. C. Liotta, *Anal. Biochem.* **171,** 373 (1988).
[58] T. Kobayashi, K. Mitsuo, and I. Goto, *Eur. J. Biochem.* **171,** 747 (1988).

of compound (i.e., phytosphingosines elute first, followed by sphingosines, then sphinganines) and the alkyl chain length. The recovery of the long-chain bases is highly reproducible and can be improved by using C_{20}-sphinganine as an internal standard.

Several other methods are available for the quantitation of long-chain bases by the formation of colorimetric or fluorescent derivatives (as cited in Ref. 57). These procedures are generally less sensitive and can yield incorrect values when the samples contain other lipids with reactive amino groups (e.g., phosphatidylethanolamine or phosphatidylserine).

[52] Use of N-([1-^{14}C]Hexanoyl)-D-erythro-sphingolipids to Assay Sphingolipid Metabolism

By ANTHONY H. FUTERMAN and RICHARD E. PAGANO

Introduction

Previous studies from our laboratory have utilized a series of fluorescent phospholipid and sphingolipid analogs to study the metabolism and intracellular translocation of lipids in cultured cells.[1] In this approach, one of the naturally occurring fatty acids of a lipid is replaced by a short-chain fluorescent fatty acid, N-(4-nitrobenzo-2-oxa-1,3-diazole)aminohexanoic acid (C_6-NBD-fatty acid). The resulting fluorescent lipids exhibit high rates of spontaneous transfer between membranes in vitro, and this property permits integration of the molecules into cellular membranes from exogenous sources. The distribution of the fluorescent lipid analogs within living cells can be studied by high-resolution fluorescence microscopy and correlated with biochemical investigations of their metabolism. In the case of the C_6-NBD-sphingolipids, various studies have documented the physical properties, metabolism, and intracellular transport of these analogs.[2-14]

In this chapter, we describe the synthesis and use of a series of sphin-

[1] R. E. Pagano and R. G. Sleight, *Science* **229**, 1051 (1985).
[2] N. G. Lipsky and R. E. Pagano, *Proc. Natl. Acad. Sci. U.S.A.* **80**, 2608 (1983).
[3] N. G. Lipsky and R. E. Pagano, *J. Cell Biol.* **100**, 27 (1985).
[4] N. G. Lipsky and R. E. Pagano, *Science* **228**, 745 (1985).
[5] G. van Meer, E. H. K. Stelzer, R. W. Wijnaendts-van-Resandt, and K. Simons, *J. Cell Biol.* **105**, 1623 (1987).
[6] R. E. Pagano and O. C. Martin, *Biochemistry* **27**, 4439 (1988).
[7] R. E. Pagano, *Methods Cell Biol.* **29**, 75 (1989).
[8] M. Koval and R. E. Pagano, *J. Cell Biol.* **108**, 2169 (1989).
[9] R. E. Pagano, M. A. Sepanski, and O. C. Martin, *J. Cell Biol.* **109**, 2067 (1989).

golipids (Fig. 1) that bear a short-chain, radioactive fatty acid ([1-^{14}C]hexanoic acid). These lipids, which are analogous to the C_6-NBD-sphingolipids in that they spontaneously transfer from either protein complexes or liposomes into biological membranes without destroying membrane integrity, have proven particularly useful for studies of sphingolipid metabolism in tissue fractions[15,16] as outlined here.

Synthesis and Purification of N-([1-^{14}C]Hexanoyl)sphingolipids

The synthesis of N-([1-^{14}C]hexanoyl)sphingolipids is based on the method originally described by Schwarzmann and Sandhoff[17] for synthesis of pyrene-labeled gangliosides. Using this method, a long-chain base (e.g., sphingosine) is N-acylated with the N-hydroxysuccinimidyl (NHS) ester of [1-^{14}C]hexanoic acid in N,N-dimethylformamide (DMF) in the presence of diisopropylethylamine (DIPE).

The NHS ester of [1-^{14}C]hexanoic acid (specific activity 12–60 mCi/mmol; American Radiolabeled Chemicals, Inc., St. Louis, MO) is synthesized as follows. (All steps should be performed in a fume hood approved for radioactive studies since hexanoic acid is volatile.) The sodium salt of the labeled hexanoic acid is first converted to the free acid by addition of 0.1 N HCl and extracted by partitioning into diethyl ether. The diethyl ether extract is then dried by passage over a small column of anhydrous sodium sulfate, and the eluate is evaporated under a stream of nitrogen. Five hundred microliters of anhydrous DMF (obtained in Sure-Seal bottles from Aldrich Chemical Co., Milwaukee, WI) is added to 50 μmol of the dried [1-^{14}C]hexanoic acid. To this, 50 μmol of dicyclohexylcarbodiimide (DCC) and 50 μmol of NHS are added, and the reaction is allowed to proceed for 2–3 days at room temperature in the dark with continual stirring. The insoluble N,N'-dicyclohexylurea is sedimented by centrifugation, resulting in a clear supernatant containing a mixture of the ester and free acid. The relative

[10] J. W. Kok, S. Eskelinen, K. Hoekstra, and D. Hoekstra, *Proc. Natl. Acad. Sci. U.S.A.* **86,** 9896 (1989).

[11] R. E. Pagano, *Biochem. Soc. Trans.* **18,** 361 (1990).

[12] M. S. Cooper, A. H. Cornell-Bell, A. Chernjavsky, J. W. Dani, and S. J. Smith, *Cell (Cambridge, Mass.)* **61,** 135 (1990).

[13] W. van't Hof and G. van Meer, *J. Cell Biol.* **111,** 977 (1990).

[14] M. Koval and R. E. Pagano, *J. Cell Biol.* **111,** 429 (1990).

[15] A. H. Futerman, B. Stieger, A. L. Hubbard, and R. E. Pagano, *J. Biol. Chem.* **265,** 8650 (1990).

[16] A. H. Futerman and R. E. Pagano, *Biochem. J.* in press (1991).

[17] G. Schwarzmann and K. Sandhoff, this series, Vol. 138, p. 319.

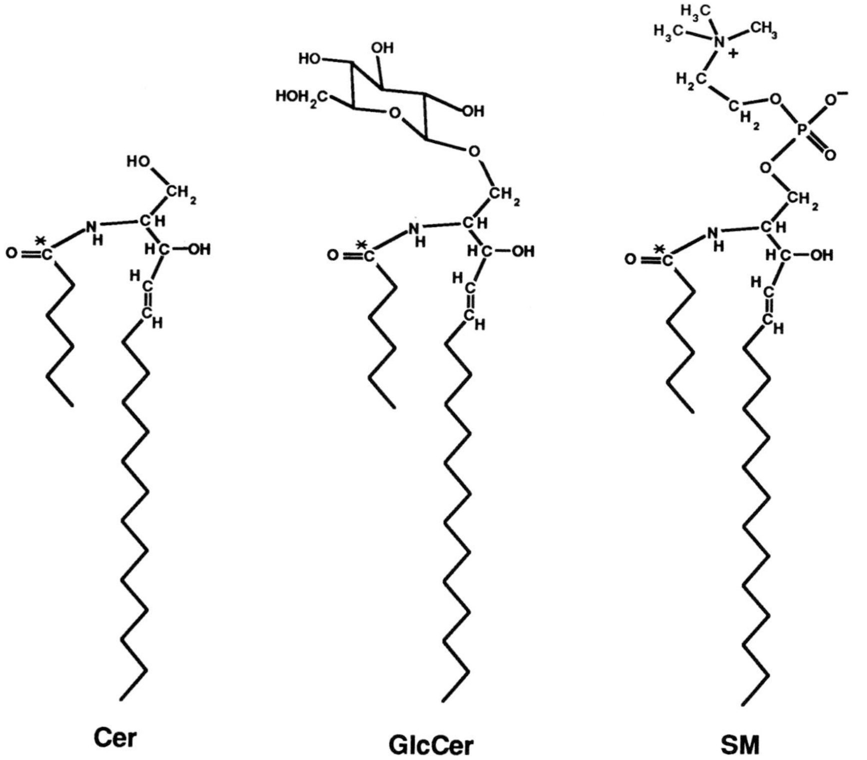

Fig. 1. Molecular structures of N-([1-^{14}C]hexanoyl)sphingolipids. Cer, Ceramide; GlcCer, glucosylceramide; SM, sphingomyelin. Asterisks denote the radioactive carbon atoms.

proportions of the two can be estimated by thin-layer chromatography (TLC) using chloroform–methanol (9:1, v/v) as the developing solvent. The R_f of the NHS ester is approximately 0.8, whereas that of the free acid is approximately 0.4. Using this procedure, 80–90% of the radiolabeled hexanoic acid can be converted to the NHS ester. The mixture is used without further purification.

N-([1-^{14}C]Hexanoyl)sphingolipids are subsequently prepared by N-acylation of either D-*erythro*-sphingosine, D-*erythro*-sphingosylphosphorylcholine, or D-*erythro*-glucosylsphingosine (all obtained from Sigma Chemical Co., St. Louis, MO), to yield N-([1-^{14}C]hexanoyl)-D-*erythro*-sphingosine {N-([1-^{14}C]hexanoyl)ceramide}, N-([1-^{14}C]hexanoyl)-D-*erythro*-sphingosylphosphorylcholine {N-([1-^{14}C]hexanoyl)sphingomyelin}, and N-([1-^{14}C]hexanoyl)-D-*erythro*-glucosylsphingosine {N-([1-^{14}C] hexanoyl)glucosylceramide}, respectively. The reactions are performed in a total volume of 1.0 ml DMF containing 50 μmol of the NHS ester of [1-^{14}C]hexanoic acid, 55–60 μmol of the appropriate long-chain base, and 25 μl of DIPE. The reaction is allowed to proceed for 20–30 hr at 30° under nitrogen in the dark with stirring; its progress is periodically monitored by TLC of aliquots using chloroform–methanol–15 mM CaCl$_2$ (60 : 35 : 8, v/v) as the developing solvent. When complete, the reaction is terminated by acidification with 50 μl of 3 N HCl, and the mixture is then dried completely under a stream of nitrogen. The N-([1-^{14}C]hexanoyl)sphingolipids are extracted by the procedure of Bligh and Dyer,[18] dried under nitrogen, and dissolved in chloroform–methanol (19 : 1, v/v). The products are purified by preparative TLC using either chloroform–methanol–water (65 : 25 : 4, v/v) or chloroform–methanol–NH$_4$OH–water (160 : 40 : 1 : 3, v/v) as the developing solvent. Purified N-([1-^{14}C]hexanoyl)sphingolipids are stored in chloroform–methanol (19 : 1, v/v) at $-80°$, periodically monitored for purity by TLC, and repurified as required.

Incubation of N-([1-^{14}C]Hexanoyl)sphingolipids with Tissue Fractions

Sphingolipid metabolism is assayed in tissue homogenates or subcellular fractions by incubation with N-([1-^{14}C]hexanoyl)sphingolipids that are either complexed to defatted bovine serum albumin (BSA) (Sigma) or incorporated into unilamellar liposomes.[15,16] Complexes of defatted BSA and N-([1-^{14}C]hexanoyl)sphingolipids are prepared essentially as described previously for C$_6$-NBD-sphingolipids.[6,7] An aliquot of an N-([1-^{14}C]hexanoyl)sphingolipid solution in chloroform–methanol is dried, first under a stream of nitrogen and then under reduced pressure. The dried lipid is dissolved in ethanol and injected into 50 mM Tris (pH 7.4) or another suitable buffer containing defatted BSA. Complexes usually contain an equimolar ratio of defatted BSA and N-([1-^{14}C] hexanoyl)sphingolipid, although other ratios may be used. After 20 min at room temperature, complexes are dialyzed against 50 mM Tris (pH 7.4) for 6–18 hr at 4° to remove the ethanol. To minimize losses when small amounts of complexes are prepared, dialysis is not performed, although care must be taken that the small amount of ethanol present does not interfere with subsequent enzyme assays. We have stored complexes for up to 1 year at $-20°$ without any detectable degradation of the N-([1-^{14}C]hexanoyl)sphingolipid.

To assay metabolism, tissue fractions are diluted to the desired protein concentration in aqueous buffers and warmed to 37° prior to addition

---

[18] E. G. Bligh and W. J. Dyer, *Can. J. Biochem. Physiol.* **37**, 911 (1959).

of the BSA/N-([1-^{14}C]hexanoyl)sphingolipid complex; the volume of the aqueous reaction mixture is not crucial (see below). After a suitable incubation period, the reaction is stopped by addition of approximately 3 volumes of chloroform–methanol (1 : 2, v/v) and placed on ice. Lipids are subsequently extracted as described.[18]

To establish that N-([1-^{14}C]hexanoyl)sphingolipids spontaneously transfer between BSA/sphingolipid complexes and subcellular fractions, sulforhodamine-conjugated BSA (Molecular Probes, Eugene, OR) is used in preparing the BSA/lipid complex. Following incubation of the complexes with subcellular fractions, the sample is separated into supernatant and pellet fractions by centrifugation (263,000 g; 10 min, 4°). In a recent study,[15] 80–95% of the ^{14}C radioactivity but only 5–6% of the sulforhodamine fluorescence was found in the pellet, indicating that the N-([1-^{14}C]hexanoyl)sphingolipids spontaneously transfer to biological membranes, with little adsorption of the BSA/N-([1-^{14}C]hexanoyl)sphingolipid complex to membranes.

Analysis of Lipid Products by Thin-Layer Chromatography

After extraction, the lipid products are separated by analytical TLC on silica gel 60 plates (EM Science, Cherry Hill, NJ) using chloroform–methanol–15 mM CaCl$_2$ (60 : 35 : 8, v/v) as the developing solvent. N-([1-^{14}C]Hexanoyl)sphingolipids are identified after autoradiography by comparison with authentic N-([1-^{14}C]hexanoyl)sphingolipid standards. The R_f values of N-([1-^{14}C]hexanoyl)ceramide, N-([1-^{14}C]hexanoyl)glucosylceramide, and N-([1-^{14}C]hexanoyl)sphingomyelin using this solvent system are approximately 0.8, 0.6, and 0.3, respectively. N-([1-^{14}C]hexanoyl)sphingolipids are removed from the plates by scraping and radioactivities determined by liquid scintillation counting. (We have observed no difference in the efficiencies of extraction or counting for any of the N-([1-^{14}C]hexanoyl)sphingolipids.) In quantifying the amount of radioactivity present in a product, it is important to subtract background radioactivity from a corresponding region of the TLC plate in which a background sample was chromatographed. Such background samples are prepared by incubation of the N-([1-^{14}C]hexanoyl)sphingolipid in the absence of any tissue fractions.

Use of N-([1-^{14}C]Hexanoyl)sphingolipids to Determine Sites and Topology of Sphingolipid Synthesis

Using N-([1-^{14}C]hexanoyl)ceramide, we have determined both the sites and topology of sphingomyelin[15] and glucosylceramide[16] synthesis in subcellular fractions from rat liver. In both of these studies, initial experiments

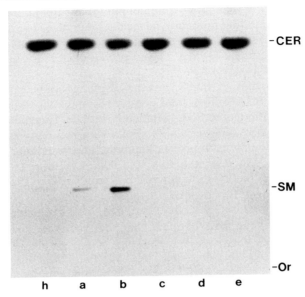

FIG. 2. Chromatographic analysis of the formation of N-([1-^{14}C]hexanoyl)sphingomyelin in subcellular fractions from rat liver. Various subcellular fractions (25 μg of protein) (a–e) (see Ref. 15), including an enriched Golgi apparatus fraction (fraction b), a microsomal pellet (fraction e), and a liver homogenate (fraction h), were each incubated with 1 nmol of N-([1-^{14}C]hexanoyl)ceramide for 10 min at 37° in a total volume of 500 μl of 25 mM KCl, 0.5 mM EDTA, 50 mM Tris, pH 7.4, extracted, separated by TLC, and subjected to autoradiography. The Golgi apparatus is the major site of sphingomyelin synthesis, which was confirmed by subsequent analyses of N-([1-^{14}C]hexanoyl)sphingomyelin synthesis in various enriched subcellular fractions.[15] Cer, N-([1-^{14}C]hexanoyl)ceramide; SM, N-([1-^{14}C]hexanoyl)sphingomyelin; Or, origin.

were performed to ensure that the rate of formation of product was linear with respect to time and protein concentration, and was not limited by the transfer of N-([1-^{14}C]hexanoyl)ceramide from BSA complexes into subcellular fractions. We determined that sphingomyelin is synthesized predominantly (>87%) at the Golgi apparatus (Fig. 2), confirming earlier results obtained using C_6-NBD-ceramide in cultured cells.[2,3] Further, sphingomyelin synthesis occurs at the luminal surface of the *cis* and *medial* cisternae of this organelle.[15] In contrast, glucosylceramide appears to be synthesized at the cytosolic surface of a pre- or early Golgi apparatus compartment,[16] implying that, following its synthesis, glucosylceramide must undergo transbilayer movement to account for the known topology of higher-order glycosphingolipids within the Golgi apparatus and plasma membrane.

Using N-([1-^{14}C]hexanoyl)sphingomyelin and N-([1-^{14}C]hexanoyl)glucosylceramide as substrates, the distribution of neutral sphingomyelinase[15] and glucosylceramidase[16] activities in subcellular fractions from rat liver can also be determined.

Important Considerations in Using N-([1-^{14}C]Hexanoyl)sphingolipids

Although the rapid spontaneous transfer and insertion of N-([1-^{14}C]hexanoyl)sphingolipids into membranes has proved useful in determining the sites and topology of sphingolipid synthesis and degradation, it is important to note the following.

Partitioning of N-([1-^{14}C]Hexanoyl)sphingolipids into Excess Lipid and/or Protein. The V_{max} of N-([1-^{14}C]hexanoyl)sphingolipid synthesis does not depend on the concentration of the N-([1-^{14}C]hexanoyl)sphingolipid substrate in the aqueous reaction mixture, but rather on the concentration of the substrate in the membrane. For example, the V_{max} of N-([1-^{14}C]hexanoyl)sphingomyelin synthesis is independent of the volume of the aqueous reaction mixture when a constant amount of N-([1-^{14}C]hexanoyl)ceramide is used (Fig. 3). Thus, when examining tissue fractions of widely different enzyme activities (such as enriched subcellular fractions compared to homogenates or crude membrane preparations), care must be taken to ensure that the amount of lipid and/or protein in the incubation mixture containing low enzyme activity (i.e., homogenates or crude membrane preparations) does not substantially reduce the concentration of available substrate due to partitioning of N-([1-^{14}C]hexanoyl)sphingolipids into excess lipid and/or protein. This partitioning of N-([1-^{14}C]hexanoyl) ceramide is observed when constant amounts of an enriched Golgi apparatus fraction are incubated in the presence of increasing amounts of subcellular fractions (not shown) or unilamellar liposomes (Fig. 4). In both cases, the synthesis of N-([1-^{14}C]hexanoyl)sphingomyelin is substantially reduced.

Metabolism of N-([1-^{14}C]Hexanoyl)sphingolipids. The usefulness of N-([1-^{14}C]hexanoyl)sphingolipids for examining the synthesis of higher-order sphingolipids is yet to be established. N-([1-^{14}C]hexanoyl)glucosylceramide is metabolized slowly in subcellular fractions to N-([1-^{14}C]hexanoyl)lactosylceramide in the presence of UDPGal[19]; similarly, C_6-NBD-glucosylceramide is not efficiently metabolized to C_6-NBD-lactosylceramide in cultured cells (A. H. Futerman and R. E. Pagano, unpublished observations). Although other possibilities cannot be excluded at present, these data suggest that the fatty acid composition of the glucosylceramide

---

[19] A. H. Futerman and R. E. Pagano, unpublished observations, 1990.

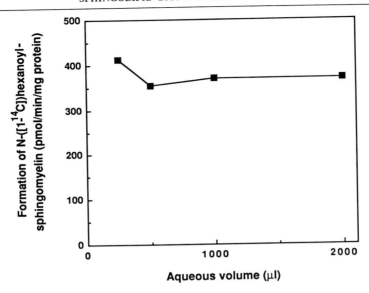

FIG. 3. Incubation of a Golgi apparatus fraction with N-([1-^{14}C]hexanoyl)ceramide in increasing volumes of an aqueous reaction mixture. An enriched Golgi apparatus fraction (25 μg of protein) was incubated with 1 nmol of N-([1-^{14}C]hexanoyl)ceramide for 10 min with increasing volumes of buffer (25 mM KCl and 50 mM Tris, pH 7.4). After the reaction was terminated, lipids were extracted and the synthesis of N-([1-^{14}C]hexanoyl)sphingomyelin analyzed by TLC, followed by scraping of the products and liquid scintillation counting.

substrate may affect lactosylceramide synthesis. This possibility is supported by the observations that N-([1-^{3}H]palmitoyl)glucosylceramide is readily metabolized to N-([1-^{3}H]palmitoyl)lactosylceramide *in vitro*,[19] and that sphingomyelin synthesis *in vivo* is dependent on the chain length and saturation of the ceramide substrate.[20]

   Metabolism by Degradative Enzymes and Transfer between Membrane Fractions in Vitro. Owing to the rapid and spontaneous transfer of N-([1-^{14}C]hexanoyl)sphingolipids between phospholipid bilayers, care must be taken to ensure that the activities of sphingolipid degradative enzymes have been minimized, so that newly synthesized N-([1-^{14}C]hexanoyl)sphingolipids are not subsequently degraded. However, it is assumed that sphingolipids synthesized at the luminal surface of intact subcellular fractions (such as an intact Golgi apparatus fraction, see Ref. 15), are trapped within the lumen of intact vesicles, since fluorescent short acyl-chain sphingolipid

[20] O. Stein, K. Oette, G. Hollander, Y. Dabach, M. Ben-Naim, and Y. Stein, *Biochim. Biophys. Acta* **1003**, 175 (1989).

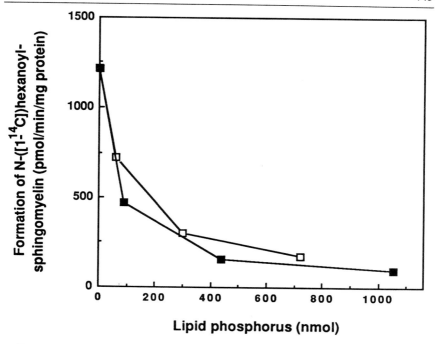

FIG. 4. Incubation of a Golgi apparatus fraction with N-([1-^{14}C]hexanoyl)ceramide and increasing amounts of unilamellar liposomes. An enriched Golgi apparatus fraction (50 μg of protein) was incubated with 2 nmol of N-([1-^{14}C]hexanoyl)ceramide for 10 min in 500 μl of 25 mM KCl, 0.5 mM EDTA, 50 mM Tris, pH 7.4, in the presence of increasing amounts of unilamellar liposomes composed of either dioleoylphosphatidylcholine (■) or lipids extracted from a liver homogenate (□).

analogs (with the exception of the ceramide analog) do not undergo transbilayer movement.[1,3,6,7,8,14] Thus, they should be inaccessible to degradative enzymes.

Summary

An advantage of using N-([1-^{14}C]hexanoyl)sphingolipids to assay sphingolipid metabolism is their ability to rapidly and spontaneously transfer into biological membranes without destroying membrane integrity. This property allows analysis of the activity of enzymes of sphingolipid metabolism under conditions in which the rate of product formation is not limited by availability of substrate, as is often the case with naturally occurring lipids whose rates of spontaneous transfer are extremely slow.[21,22] Thus, the use of N-([1-^{14}C]hexanoyl)sphingolipids provides an alternative means for studying sphingolipid metabolism *in vitro*.

Acknowledgments

This work was supported in part by U.S. Public Health Service Grant R37-GM-22942 to R. E. Pagano. A. H. Futerman was supported by a Chaim Weizmann Postdoctoral Fellowship. The authors thank Dr. A. Ting for assistance in preparation of Fig. 1.

[21] L. R. McLean and M. C. Phillips, *Biochemistry* **20**, 2893 (1981).
[22] A. Frank, Y. Barenholz, D. Lichtenberg, and T. E. Thompson, *Biochemistry* **22**, 5647 (1983).

Section XI

Enzymes of Lipid A Synthesis

[53] UDP-N-Acetylglucosamine 3-O-Acyltransferase from *Escherichia coli*

By MATT S. ANDERSON and CHRISTIAN R. H. RAETZ

Introduction

UDP-N-Acetylglucosamine (UDPGlcNAc) 3-O-acyltransferase cata-lyzes the transfer of an acyl carrier protein (ACP)-activated (R)-3-hydroxy-myristoyl group to the 3 position of the GlcNAc moiety in UDP-N-acetyl-glucosamine:

UDPGlcNAc + (R)-3-hydroxymyristoyl-acyl carrier protein →
UDP-3-O-[(R)-3-hydroxymyristoyl]GlcNAc + acyl carrier protein

This enzyme catalyzes the first step unique to lipopolysaccharide biosyn-thesis in *Escherichia coli* and other gram-negative bacteria[1] (see also this volume, [54], Fig. 1). Previous genetic studies have demonstrated that UDPGlcNAc acyltransferase accounts for over 90% of the lipid A synthe-sized by *E. coli* and that the enzyme is essential for cell growth.[2]

Assay Methods

Two assays for UDPGlcNAc acyltransferase activity are currently available. The first, described by Anderson and Raetz, involves the separa-tion of UDPGlcNAc and acylated UDP-GlcNAc by thin-layer chromatog-raphy.[3] The second, described by Galloway and Raetz (see below), makes use of C_{18} reversed-phase minicolumns, is slightly easier, and is compatible with commercially available [^{3}H]UDPGlcNAc.[2]

Principle. Acyltransferase activity is determined by measuring the formation of radiolabeled UDP-3-O-acyl-GlcNAc from radiolabeled UDP-GlcNAc and nonradioactive (R)-3-hydroxymyristoyl-ACP. Assay mix-tures are passed through a short C_{18} reversed-phase resin column (Waters, Milford, MA, Sep-Pak), which retains only the desired acylated product. This material is selectively eluted and quantitated by liquid scintillation spectrometry.

Substrates: Radiolabeled UDP-N-Acetylglucosamine. UDP-[6-*glu-cosamine*-^{3}H]GlcNAc may be purchased from Amersham Corp. (Arling-

[1] C. R. H. Raetz, *Annu. Rev. Biochem.* **59**, 129 (1990).
[2] S. M. Galloway and C. R. H. Raetz, *J. Biol. Chem.* **265**, 6394 (1990).
[3] M. S. Anderson and C. R. H. Raetz, *J. Biol. Chem.* **262**, 5159 (1987).

ton Heights, IL) or Du Pont–New England Nuclear (Boston, MA). Alternatively, [β-^{32}P]UDPGlcNAc may be synthesized by the method of Lang and Kornfeld[4] and purified as described by Anderson and Raetz.[3]

[α-^{32}P]UDPGlcNAc is also suitable and may be synthesized as follows. This label is prepared from [α-^{32}P]UTP (Amersham or Du Pont–New England Nuclear) by a variation of these methods. One millicurie of [α-^{32}P]UTP (specific radioactivity 800 Ci/mmol) is allowed to react directly in the vial in which it is supplied without the addition of carrier UTP at room temperature for 16 hr in a reaction cocktail consisting of glucosamine 1-phosphate, 0.5 mM; magnesium chloride, 1 mM; Tris-HCl, pH 8.0, 100 mM; dithiothreitol (DTT), 5 mM; and 50 units each of inorganic pyrophosphatase (Sigma, St. Louis, MO) and UDPglucose pyrophosphorylase (Sigma) in a final volume of 250 μl. At the end of this time, the mixture is acetylated with acetic anhydride by the addition of an equal volume of deionized water, 300 μl of methanol, 10 μl of saturated sodium bicarbonate, and 10 μl of acetic anhydride. The reaction is allowed to proceed for 3 min at room temperature and is then quenched by placement in a boiling water bath for 90 sec. The crude mixture is diluted with 7 ml of deionized water and loaded onto a 5-ml column of DEAE-cellulose, preequilibrated with 10 mM triethylammonium bicarbonate, pH 7.0. Next, the column is washed free of salts with 30 ml of water followed by removal of product with a 20 ml wash using 100 mM triethylammonium bicarbonate, pH 7.0. This solution is collected in a 50-ml plastic tube (Falcon or Corning) and lyophilized. The lyophilizate is redissolved in 800 μl of 5 mM sodium phosphate containing 0.7 mM tetrabutylammonium hydroxide, pH 6.0, and chromatographed exactly as described by Anderson and Raetz.[3] The overall yield from [α-^{32}P]UTP is typically 55%. The specific radioactivity of this label is unaltered from that at which it was supplied.

Using either [^{32}P]UDPGlcNAc or [^{3}H]UDPGlcNAc, a specific radioactivity of approximately 125 μCi/μmol in the standard assay is sufficient to quantitate acyltransferase activity.

(R)-3-Hydroxymyristoyl-Acyl Carrier Protein. (R)-3-Hydroxymyristoyl-ACP is prepared as described by Anderson and Raetz[3] using the method of Fuji Yakuhin Kogyo to synthesize the (R)-3-hydroxymyristic acid.[5]

Enzyme Assay Procedure. Standard assay mixtures contain radiolabeled UDPGlcNAc, 200 μM (specific radioactivity as described above); (R)-3-hydroxymyristoyl-ACP, 50 μM; octyl-β-D-glucoside, 1.0%;

[4] L. Lang and S. Kornfeld, *Anal. Biochem.* **140**, 264 (1984).
[5] Fuji Yakuhin Kogyo, Japanese Patent 86-281640/43 (1986).

HEPES, pH 8.0, 40 mM; fatty acid-free bovine serum albumin, 10 mg/ml; and enzyme extract, 2.5 mg/ml in a final volume of 20 μl. Octyl-β-D-glucoside is included in the assay cocktail in order to block further metabolism of UDP-3-O-acyl-GlcNAc.[3] The reaction is initiated by the addition of enzyme to the assay cocktail which has been previously equilibrated to 30°. The sample is incubated at 30°, and the reaction is stopped at 1 or 2 min by diluting a 10-μl portion (one-half of the assay mixture) into 1 ml of 10 mM HEPES, pH 8.0. This sample may be held on ice for several hours until all such assay tubes are prepared, or it may be further processed by loading it directly onto a C$_{18}$ Sep-Pak (Waters) equilibrated in 10 mM HEPES, pH 8.0. (The Sep-Pak itself is prewashed with 10 ml of methanol, followed by 10 ml of water, and finally with the HEPES buffer.) Next, the Sep-Pak is washed with 10 ml of 10 mM HEPES buffer, in order to remove unreacted starting material, and four further 1-ml washes with this buffer are collected as individual fractions and counted in 10 ml of scintillation cocktail. These latter fractions ensure the complete removal of all unacylated UDPGlcNAc. The acylated product is then eluted in methanol, and 1-ml fractions are collected and counted in 10 ml of scintillation cocktail. Typically, the third and fourth HEPES wash fraction have returned to background levels, and the first three methanol fractions contain virtually all of the product. The Sep-Pak columns may be regenerated by washing with 10 ml of methanol, 10 ml of water, and 10 ml of 10 mM HEPES, pH 8.0.

Once the assayist becomes proficient with the system, the number of fractions counted per sample can be cut down to five. These are the final HEPES wash (used to verify removal of unreacted starting material) and the first four product-containing methanol fractions.

Units. One unit is defined as the amount of enzyme that catalyzes the formation of 1 nmol of UDP-3-O-acyl-GlcNAc in 1 min under the conditions described above. Product formation is linear with respect to time for only 2 min. This deviation from linearity is apparently due to the K_{eq} of this reaction, which lies in the direction of the reactants by as much as 10-fold.[6,7] This implies that equilibrium is quickly established, and linear reaction rates are only observable during the first 1 to 2% of product formation. When activity is measured at these early time points, however, enzyme activity is proportional to protein concentration between 1 and 5 mg/ml at 200 μM substrate.

[6] J. M. Williamson, M. S. Anderson, and C. R. H. Raetz, *J. Bacteriol.* **173**, 3591 (1991).
[7] H. Bull, unpublished observations (1991).

TABLE I
PARTIAL PURIFICATION OF UDPGlcNAc ACYLTRANSFERASE[a]

Fraction	Total volume (ml)	Total protein (mg)	Specific activity (nmol/min/mg)	Purification (-fold)	Yield (%)
MC1061/pING1[b]	—	—	3.0	1.0	—
MC1061/pSR1 extract	100	1280	560	187	100
Membrane-free supernatant	100	1110	614	204	95
DEAE-cellulose	735[c]	192	2919	973	78

[a] Adapted from Anderson and Raetz.[3]

[b] Vector control.

[c] Done in smaller portions as described in the text.

Partial Purification of UDPGlcNAc Acyltransferase from *Escherichia coli*

The structural gene encoding UDPGlcNAc acyltransferase, *lpxA*, is located within a large, complex operon at 4 min of the *E. coli* chromosome.[1,8] The gene has been cloned from *E. coli*,[9] and the nucleotide sequence has been determined, revealing this protein to be a 27-kDa polypeptide.[10] Strain MC1061, harboring plasmid pSR1, directs approximately 200-fold overproduction of the acyltransferase activity on 8-hr induction with 0.5% L-arabinose.[3,9] These cells may be frozen in liquid nitrogen after harvesting and held at −80° for years without substantial loss of activity.

A partial purification of the enzyme can be achieved using these cells as an enriched source of activity.[3] Ten grams of frozen, induced cells are resuspended in 100 ml of 10 mM sodium phosphate buffer, pH 7.0, at 0°. The cells are disrupted by a single passage through a French pressure cell at 18,000 psi and centrifuged at 5000 g for 20 min in order to remove unbroken cells. The soluble fraction is subsequently centrifuged at 150,000 g for 90 min at 4° in order to remove membranes. Enzyme activity is found in the soluble fraction and may be carried through the remainder of the preparation or flash frozen in liquid nitrogen and stored at −80°.

Next, 75-mg portions of the membrane-free supernatant are adjusted to 1% (w/v) in octyl-β-D-glucoside (OG) by addition of a 10% OG stock solution in 10 mM sodium phosphate buffer, pH 6.0. This sample is loaded at 5 ml/min onto a 15-ml (2.4 × 3.3 cm) column of DEAE-cellulose (Whatman, Clifton, NJ, DE-52). This column is preequilibrated with 10

[8] H. G. Tomasiewicz and C. S. McHenry, *J. Bacteriol.* **169**, 5735 (1987).

[9] D. N. Crowell, M. S. Anderson, and C. R. H. Raetz, *J. Bacteriol.* **168**, 152 (1986).

[10] J. Coleman and C. R. H. Raetz, *J. Bacteriol.* **170**, 1268 (1988).

TABLE II
SUBSTRATE SPECIFICITY OF PARTIALLY PURIFIED
UDPGlcNAc ACYLTRANSFERASE[a]

Substrate pairs	Relative reaction rates (%)
A. [β-^{32}P]UDPGlcNAc and	
(R)-3-Hydroxymyristoyl-ACP	100
(S)-3-Hydroxymyristoyl-ACP	7
(R,S)-3-Hydroxymyristoyl-ACP	35
(R,S)-3-Hydroxylauroyl-ACP	1.5
(R,S)-3-Hydroxypalmitoyl-ACP	1.5
Myristoyl-ACP	<0.1
Palmitoyl-ACP	<0.1
(R)-3-Hydroxymyristoyl-CoA	<0.1
B. (R)-3-Hydroxymyristoyl-ACP and	
[β-^{32}P]UDPGlcNAc	100
[β-^{32}P]UDP-N-propionylGlcN	22
[β-^{32}P]UDP-N-butyrlGlcN	8
[β-^{32}P]UDP-N-[(R)-3-hydroxymyristoyl]GlcN	0
[β-^{32}P]UDPGlcN	0
C. (R)-3-Hydroxymyristoyl-ACP and	
[acetyl-^{3}H]UDPGlcNAc	100
[acetyl-^{3}H]TDPGlcNAc	20
[acetyl-^{3}H]GDPGlcNAc	0.04
[acetyl-^{3}H]ADPGLcNAc	<0.001
[acetyl-^{3}H]CDPGlcNAc	<0.001

[a] Relative reaction velocities were observed using each of the indicated substrate pairs with the following minor assay modifications. A, Each acyl-ACP substrate was present at 60 μM except for the (R,S) derivatives that were present at 120 μM. B, (R)-3-Hydroxymyristoyl-ACP (100 μM) was employed with each UDP derivative at 100 μM. C, For the [acetyl-^{3}H]NDPGlcNAc's, the reaction was performed on a 100-μl scale. Adapted from Anderson and Raetz.[3]

mM sodium phosphate, pH 6.0, containing 1% (w/v) OG (buffer A). After loading, the column is washed with 2 column volumes of buffer A followed by an additional 3 column volumes of buffer A containing 10 mM NaCl. UDPGlcNAc acyltransferase activity is then eluted from the column at 5 ml/min with 3 column volumes of buffer A containing 100 mM NaCl. Average activity recoveries are 78%. This enzyme can be stored at 4° for several days with minimal loss of activity or flash frozen in aliquots in liquid nitrogen and stored at −80° for at least 3 months. However, these samples cannot be frozen and thawed more than once without great loss in activity.

As shown in Table I, the DEAE-cellulose fraction contains UDPGlc NAc acyltransferase activity purified 973-fold relative to the vector control extracts. Gel electrophoresis of this fraction shows many bands, including one prominent component at approximately 27,000, the anticipated molecular weight of the acyltransferase.[10]

Properties of Enzyme

UDPGlcNAc acyltransferase activity found in the membrane-free crude extract is relatively stable to heat, virtually all activity being recovered after treatment for 20 min at 60°.[3] However, boiling the extract for 10 min will destroy this activity.[3] Sulfhydryl-directed agents such as N-ethylmaleimide do not significantly inhibit the acyltransferase.

UDPGlcNAc acyltransferase differs markedly from glycero-3-phosphate acyltransferase, the first committed step of glycerophospholipid biosynthesis in E. coli.[1] Whereas UDPGlcNAc acyltransferase is found in the soluble fraction of broken cell preparations of E. coli and is highly active in the absence of added phospholipid,[3] glycero-3-phosphate acyltransferase is an integral membrane enzyme which must be solubilized and reconstituted in the presence of phospholipid in order to observe activity.[11,12] When assayed in vitro, glycero-3-phosphate acyltransferase will accept fatty acids of varied chain length as substrate, activated as either the coenzyme A or acyl carrier protein thioesters.[12] In sharp contrast, the in vitro substrate specificity of UDPGlcNAc acyltransferase is remarkably specific with respect to the acyl donor (Table II).[3] The preferred substrate is (R)-3-hydroxymyristate, activated with acyl carrier protein, not with the CoA thioester. The enzyme tolerates little deviation in the size of the acetyl group of the second substrate, UDPGlcNAc, and has a strong preference for the uridine nucleotide.

[11] M. D. Snider and E. P. Kennedy, J. Bacteriol. 130, 1072 (1977).
[12] P. R. Green, A. H. Merrill, and R. M. Bell, J. Biol. Chem. 256, 11151 (1982).

[54] Lipid A Disaccharide Synthase from *Escherichia coli*

By CHRISTIAN R. H. RAETZ

Introduction

The lipid A disaccharide synthase of *Escherichia coli* catalyzes the formation of the $\beta1'\rightarrow6$ linkage that is characteristic of lipid A disaccharides of gram-negative bacteria (Fig. 1).[1–3]

2,3-Diacyl-GlcN-1-P (lipid X) + UDP-2,3-diacyl-GlcN →
2′,3′-diacyl-GlcN(β,1′→6)2,3-diacyl-GlcN-1-P + UDP

In *E. coli,* the preferred physiological substrates are acylated exclusively with (R)-3-hydroxymyristate.[1,2] The enzyme is recovered in the cytoplasm, but it may function as a peripheral membrane protein.[1,2] It is an oligomer of a single subunit of 42,339 daltons.[2,4] In contrast to most enzymes of glycerophospholipid biosynthesis,[5,6] the lipid A disaccharide synthase does not require a detergent for catalytic activity.[1,2]

Assay Methods

The quantitative assay for the disaccharide synthase is based on the conversion of the substrate, 2,3-diacyl-GlcN-1-^{32}P, to the product, 2′3′-diacyl-GlcN(β1′→6)2,3-diacyl-GlcN-1-^{32}P (also designated the disaccharide 1-phosphate), in the presence of UDP-2,3-diacyl-GlcN. The radiolabeled product is much more hydrophobic than the radiolabeled substrate, and they can be separated by two-phase Bligh–Dyer partitioning at neutral pH.[1,2] Alternatively, the radiolabeled product can be separated from the substrate by thin-layer chromatography (TLC).[1,2] The reaction equilibrium lies far to the right. Nearly complete conversion of substrates to products can be achieved in the millimolar range. Therefore, the reaction can also be followed using nonradioactive substrates by charring of the products on thin-layer plates.[1,2] The substrates are not commercially available.

[1] B. L. Ray, G. Painter, and C. R. H. Raetz, *J. Biol. Chem.* **259,** 4852 (1984).
[2] K. Radika and C. R. H. Raetz, *J. Biol. Chem.* **263,** 14859 (1988).
[3] C. R. H. Raetz, *Annu. Rev. Biochem.* **59,** 129 (1990).
[4] D. N. Crowell, W. S. Reznikoff, and C. R. H. Raetz, *J. Bacteriol.* **169,** 5727 (1987).
[5] C. R. H. Raetz and W. D. Dowhan, *J. Biol. Chem.* **265,** 1235 (1990).
[6] R. H. Hjelmstad and R. M. Bell, *Biochemistry* **30,** 1731 (1991).

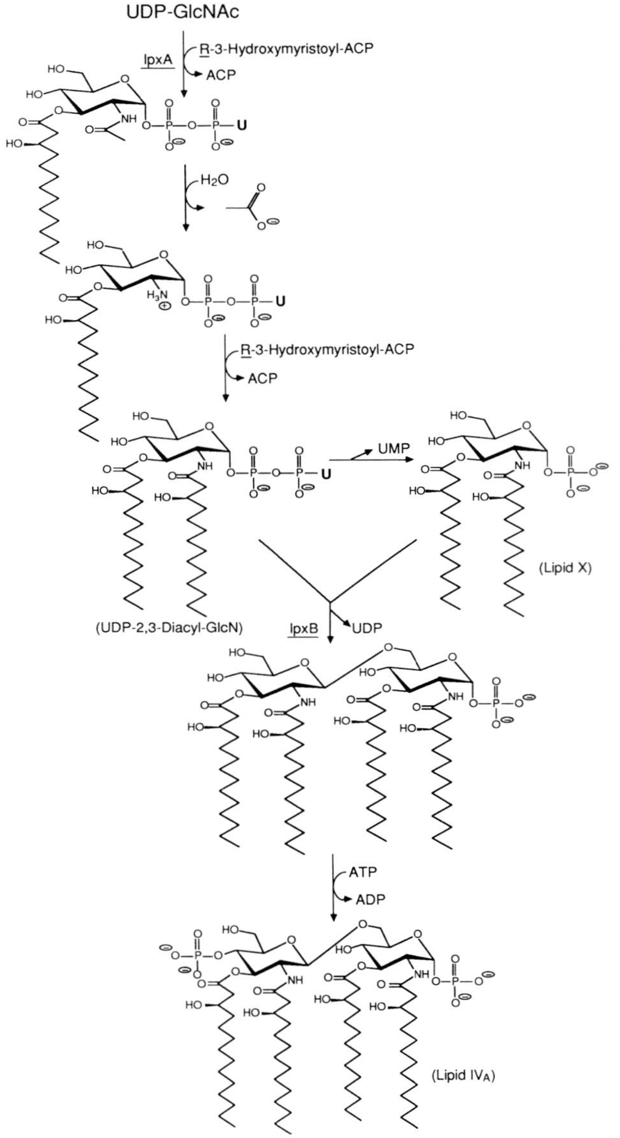

FIG. 1. Early steps in the biosynthesis of lipid A disaccharides. The disaccharide synthase catalyzes the reaction shown in the middle of the figure and is encoded by the *lpxB* gene, as indicated. The further reactions leading to the formation of mature lipid A are given in [56] in this volume.

Reagents

HEPES buffer, 10 mM, pH 8 (with KOH)

Fatty acid-free bovine serum albumin (BSA), 1 mg/ml in 10 mM HEPES, pH 8

2,3-Diacyl-GlcN-1-^{32}P [10^3 counts/min (cpm)/nmol], 5 mM, in 10 mM HEPES, pH 8

UDP-2,3-diacyl-GlcN, 5 mM in 10 mM HEPES, pH 8

Chloroform, methanol, phosphate-buffered saline (PBS)[7]

Bio Safe II liquid scintillation cocktail (RPI, Mount Pleasant, IL)

Procedure. Lipid dispersions in HEPES buffer (prepared as described below) are stored at −70°. After thawing, the dispersions are thoroughly mixed and subjected to 2 min of sonic irradiation at room temperature in a bath sonicator (Laboratory Supplies, Inc., Hicksville, NY). Enzyme prepararations (also stored at −70°) are diluted as required in 10 mM HEPES, pH 8, containing 1 mg/ml BSA, and are held on ice prior to assay. The assay is carried out in a disposable plastic microcentrifuge tube. The standard assay mixture (final concentrations) consists of 1 mM 2,3-diacyl-GlcN-1-^{32}P (10^3 cpm/nmol), 1 mM UDP-2,3-diacyl-GlcN, 10 mM HEPES, pH 8, 0.4 mg/ml BSA, and enzyme in a final volume of 25 μl. The reaction is initiated by addition of enzyme (usually in 10 μl of 10 mM HEPES, pH 8, containing 1 mg/ml BSA), and the mixture is incubated 20 min at 37°. Control reactions lacking enzyme are run in parallel.

The reaction is terminated by pipetting 20 μl of the reaction mixture into a two-phase neutral Bligh–Dyer system[8] (chloroform, 2 ml; methanol, 2 ml; and phosphate-buffered saline, 1.8 ml). After mixing in a screw-capped glass tube equipped with a Teflon-lined cap, the lower phase of the mixture is separated by a brief centrifugation in a clinical centrifuge. After removal of the upper phase in which the substrate 2,3-diacyl-GlcN-1-^{32}P is located, the lower phase containing the disaccharide 1-phosphate product is washed twice with 3.8-ml portions of a fresh, preequilibrated neutral Bligh–Dyer upper phase (generated by mixing chloroform, methanol, and phosphate buffered saline in the ratios noted above). Last, 1 ml of washed lower phase is transferred with a glass pipette to a glass liquid scintillation vial, and the chloroform is dried under a heat lamp and/or a stream of warm air. Following addition of 10 ml of Bio Safe II, the radioactivity is determined by liquid scintillation counting. Counts in the lower phase derived from radioactive lipid impurities, usually less than 1% of the total, may appear in controls lacking either enzyme or UDP-2,3-diacyl-

---

[7] R. Dulbecco and M. Vogt, *J. Exp. Med.* **99**, 167 (1954).

[8] M. Nishijima and C. R. H. Raetz, *J. Biol. Chem.* **254**, 7837 (1979).

GlcN, and are subtracted out. The rate of disaccharide 1-[^{32}P]phosphate formation is linear for up to about 30% of total conversion to product.[1,2]

A qualitative assessment of the progress of the reaction may be obtained at any time by withdrawing 5 μl and spotting it onto a 5 × 10 cm silica gel F (250 mm) thin-layer chromatography plate (E. Merck, Darmstadt, Germany). After drying under a cold air stream, the plate is developed in the solvent chloroform–methanol–water–acetic acid (25 : 15 : 4 : 2, v/v). After drying on a hot plate to remove solvents, the plate is sprayed in a hood with 10% sulfuric acid in ethanol (v/v). Last, it is charred on a hot plate to visualize the two substrates and the more rapidly migrating product, which appear as dark bands.

Definition of Units. One unit is defined as the amount of enzyme that generates 1 μmol of disaccharide 1-[^{32}P]phosphate in 1 min under the above conditions.

Purification Procedure

The following procedures are identical to those described by Radika and Raetz.[2]

Growth of Bacteria Used to Isolate Disaccharide Synthase. *Escherichia coli* MC1061/pSR8, an overproducer of lipid A disaccharide synthase, was previously isolated by Crowell *et al.*[9] One liter of cells in LB broth[10] containing 50 μg/ml ampicillin is grown overnight to early stationary phase at 30° using a rotary shaker for aeration. This inoculum is used to start the growth of a 30-liter LB broth culture supplemented with 70 g of K$_2$HPO$_4$, 30 g of KH$_2$PO$_4$, 300 g of fructose, and 1.5 g of ampicillin. After 2.5 hr of growth at 30° with aeration in a New Brunswick fermentor, the A_{550} is 1.0. Next, L-arabinose (150 g) is added, and induction of the enzyme is achieved by growing the bacteria for 8 hr longer under the same conditions. The cells are harvested by centrifugation, and the paste (166 g) is stored at −80°.

Purification of Lipid A Disaccharide Synthase. Four grams of *E. coli* MC1061/pSR8 is suspended in 20 ml of 10 mM potassium phosphate, pH 7.0, and broken in a cold French pressure cell at 18,000 psi (see Table I). All subsequent operations are conducted at 4°. The broken cells are removed by centrifugation at 10,000 g for 15 min, and the membrane-free supernatant is prepared by centrifugation at 100,000 g for 2 hr. The supernatant is adjusted to a volume of 43 ml so that it contains 10 mM potassium

[9] D. N. Crowell, M. S. Anderson, and C. R. H. Raetz, *J. Bacteriol.* **168,** 152 (1986).

[10] J. R. Miller, "Experiments in Molecular Genetics." Cold Spring Harbor Laboratory, Cold Spring Harbor, New York, 1972.

TABLE I
PURIFICATION OF LIPID A DISACCHARIDE SYNTHASE FROM
Escherichia coli MC1061/pSR8

Stage	Total volume (ml)	Total protein (mg)	Specific activity (nmol/min/mg)	Yield (%)
Membrane-free supernatant	43	404	0.29×10^3	100
Matrex Gel Blue B	150	11.7	6.03×10^3	59.3
Reactive Red 120-agarose (1000-CL)	62.5	5.3	12.2×10^3	54.7
Heparin-agarose	20	2.3	15.9×10^3	30.8

phosphate, pH 7.0, 0.2 M KCl, 0.05% Triton X-100, and 20% glycerol (v/v).

A 60-ml column (2.5-cm diameter) of Matrex Gel Blue B (Amicon Corp., Danvers, MA) derivatized with 0.085 mg of ligand per milliliter of gel is equilibrated by washing with 20 column volumes of 10 mM potassium phosphate, pH 7.0, containing 20% glycerol, 0.05% Triton X-100, and 0.2 M KCl at a flow rate of 1.4 ml/min. The supernatant, described above, is applied to the column at the same flow rate, and the column is washed with the same buffer. Six fractions (30 ml each) are collected. The enzyme is eluted with a buffer identical to the application buffer, except that it contains 2.5 M KCl. Eight more 30-ml fractions are collected.

A 30-ml column (2.5 cm diameter) of Reactive Red 120-agarose (Sigma, St. Louis, MO) derivatized with 0.96 μmol of ligand per milliliter of gel is equilibrated by washing with 20 column volumes of the same application buffer as above. Fractions 9–13 (containing the disaccharide synthase) from the Matrex Gel Blue B column are pooled and diluted to 1875 ml, in order to adjust the salt concentration to 0.2 M KCl, while keeping the concentration of all other components the same as in the application buffer. Next, the sample is applied to the Reactive Red column at 1.8 ml/min, and then the column is washed with 75 ml of the application buffer. The enzyme is eluted with a buffer identical to the application buffer except containing 2.5 M KCl. Eight fractions (12.5 ml each) are collected.

A 20-ml column (1.5-cm diameter) of heparin-agarose Type II (Sigma) is washed with 20 column volumes of the same application buffer as above. Fractions 4–8 from the Reactive Red 120-agarose step are pooled and diluted to 782 ml to make the salt concentration 0.2 M KCl, while keeping the concentration of all other components the same as in the application buffer. The sample is applied to this column at 1.8 ml/min, and the column is washed with 60 ml of the application buffer. The enzyme is eluted with a buffer identical to the application buffer except containing 2.5 M KCl. Six fractions of 10 ml each are collected. The purified enzyme samples are frozen in liquid nitrogen and stored at $-80°$. The enzyme is stable under these conditions for several years.

Some Catalytic Properties of Purified Disaccharide Synthase

The purified enzyme consists of a single subunit of 42,339 daltons. It does not contain an exceptionally high proportion of hydrophobic amino acid residues, and there is no evidence for membrane-spanning domains.[2,4]

The enzyme shows a strong kinetic preference for 2,3-diacylated substrates,[2,11] but the (R)-3-hydroxy substituent is not required.[12] With large amounts of pure enzyme it is possible to obtain good conversions even when the reaction rates are only 1/10,000 of those observed with the physiological substrates, facilitating the preparation of many interesting disaccharide 1-phosphate analogs.[12]

The most remarkable property of the enzyme is its lack of dependence on the presence of a nonionic detergent for catalytic activity. Enzymes that act on diacylated glycerophospholipids are all greatly stimulated by the presence of such detergents. A possible explanation for this anomaly is that lipid X and UDP-2,3-diacylglucosamine have much higher critical micellar concentrations than do diacylglycerophospholipids,[2,13] but this has not been verified experimentally.

Preparation of [32P]-Labeled Lipid X

A 10-ml culture of strain MN7[14] in G56 medium[15] is started with a 50-μl portion of an overnight culture grown at 30°. The culture is incubated on a rotary shaker at 30° until the A_{550} reaches 0.3. The cells are then harvested in a clinical centrifuge at the highest setting for 15 min. The cell pellet is resuspended in 10 ml of fresh G56 medium lacking phosphate, and [32P]P$_i$ is added at 100 μCi/ml. The culture is then transferred to a 42° rotary shaker bath for 3 hr. The radiolabeled cells are sedimented in a clinical centrifuge and resuspended in 0.8 ml of phosphate-buffered saline,[7] pH 7.3. After transferring the cells to a glass tube, 1 ml of chloroform and 2 ml of methanol are added. The resulting cell debris is collected by

[11] K. Takayama, N. Qureshi, P. Mascagni, L. Anderson, and C. R. H. Raetz, *J. Biol. Chem.* **258,** 14245 (1983).
[12] A. Haselberger, J. Hildebrandt, C. Lam, E. Liehl, H. Loibner, I. Macher, B. Rosenwirth, E. Schütze, H. Vylpel, and F. M. Unger, *Triangle (Engl. Ed.)* **26,** 33 (1987).
[13] G. Lipka, R. A. Demel, and H. Hauser, *Chem. Phys. Lipids* **48,** 267 (1988).
[14] M. Nishijima, C. E. Bulawa, and C. R. H. Raetz, *J. Bacteriol.* **145,** 113 (1981).
[15] B. R. Ganong, J. M. Leonard, and C. R. H. Raetz, *J. Biol. Chem.* **255,** 1623 (1980).

centrifugation and discarded. To the supernatant, 1 ml of chloroform and 1 ml of phosphate-buffered saline are added, forming a two-phase Bligh and Dyer system at neutral pH.

The majority of the ^{32}P-labeled lipid X partitions into the upper phase under these conditions. The upper phase is washed 3 times with fresh lower phase that had been preequilibrated with phosphate-buffered saline, and the lower phases are discarded. The ^{32}P-labeled lipid X is then made to partition into the lower phase by the addition to the upper phase of 50 μl of concentrated HCl and 2 ml of acid-preequilibrated lower phase. The lower phase is then washed twice with acid-preequilibrated upper phase. The washed lower phase is evaporated under a stream of N_2, and the residue is dissolved in 2 ml of chloroform–methanol (2 : 1, v/v). The ^{32}P-labeled lipid X solution is stored at $-80°$. The final yield of this preparation is 2×10^7 cpm of ^{32}P-labeled lipid X/mCi of [^{32}P]P$_i$.

Preparation of Lipid X in Gram Quantities

Lipid X is isolated from *E. coli* strain MN7 using a convenient procedure for obtaining 0.1–1 g of pure material. All solvents and chemicals employed are of reagent grade. Methanol and pyridine are redistilled prior to use. Formic acid (88%) is used without redistillation. DEAE-cellulose (Whatman, Clifton, NJ, DE-52 microgranular) for column chromatography is washed extensively with distilled water and equilibrated with 2.4 *M* ammonium acetate prior to use. Silicic acid for column chromatography (Bio-Sil A, 200–400 mesh) is obtained from Bio-Rad (Richmond, CA). Silica Gel 60 thin-layer plates are obtained from Merck. Yeast extract and tryptone are obtained from Difco (Detroit, MI).

The isolation of the temperature-sensitive *E. coli* K12 strain MN7 (ATCC, Rockville, MD, No. 39328) has been described.[14] The cells are grown in a modified LB broth, which contains 10 g of NaCl, 10 g of tryptone, 5 g of yeast extract, 4 g of glucose, 2.33 g K_2HPO_4, and 1 g K_2HPO_4 per liter. Maximum accumulation of lipid X (without much lipid Y) occurs when a log-phase culture growing at 30° is shifted for 3 hr to 43.5°. This procedure is adapted to a 300-liter fermentor. Cells are grown exponentially in the above medium until the A_{550} is 1.5. At this point an additional 1500 g of glucose is added to the fermentor, and growth at 30° is continued until the A_{550} reaches 3.0. At this point an additional 3000 g of glucose is added, and the temperature of the fermentor is shifted from 30° to 43.5° over a period of approximately 10 min. The pH is maintained at 7 throughout the fermentation by the addition of concentrated ammonium hydroxide. The dissolved oxygen concentration is maintained at approximately 60%. After 3 hr of incubation at 43.5°, cells are harvested by centrifugation through a Sharples continuous flow centrifuge, and the cell paste is stored at $-80°$. The yield of cell paste per 300-liter fermentation is approximately 3 kg.

Frozen cells (600 g) prepared as described above are bathed in liquid N_2 in a metal beaker, broken into chunks with a hammer, and suspended in 1 liter of methanol and 500 ml of chloroform. The mixture is stirred occasionally and allowed to thaw for about 30 min. Next, the chunks are disrupted completely by a brief homogenization in a Waring blender, and the slurry is filtered in two equal portions to remove solid particles using two Büchner funnels (24 cm in diameter) equipped with 4-liter side-arm vacuum flasks. The extracted lipids emerge as a clear yellow solution. The solid residues form a gray cake on the funnel, which is removed easily as it begins to dry. The cake is extracted a second time by blending with 300 ml of water, 375 ml of chloroform, and 750 ml of methanol. The slurry is filtered on the same two Büchner funnels, and the filtrates (3000 ml) are combined, after which they are divided equally into three 2-liter separatory funnels equipped with Teflon stopcocks.

Next, 650 ml of chloroform and 25 ml of concentrated HCl are added to each separatory funnel. The contents are mixed thoroughly. The phases separate rapidly without the presence of particulate material at the interface. Each lower phase is collected, and the upper phase is discarded. The lower phases, containing glycerophospholipids and lipid X, are placed back into the separatory funnels, and each is washed 2 more times with 300 ml of a fresh upper phase consisting of methanol–water–chloroform (1 : 1 : 0.1, v/v). Small amounts of additional salt and yellow pigment are extracted into these upper phases, which are also discarded. The final lower phases are again collected and pooled ($\sim$2700 ml). At this stage the extract can be stored overnight at room temperature prior to chromatography on DEAE-cellulose.

As indicated above, Whatman DE-52 is first equilibrated with ammonium acetate. Next, it is washed 3 times with 10 bed volumes of methanol, and then it is washed 3 times with 10 bed volumes of chloroform–methanol–2.4 M ammonium acetate (2 : 3 : 1, v/v). The DEAE-cellulose can be stored in this solvent mixture for several months at $4°$. Prior to pouring the column, the DEAE-cellulose is allowed to warm to room temperature and is briefly subjected to a water-aspirator vacuum to remove dissolved air.

For the purposes of obtaining rapid fractionation of lipid X, a 150-ml DEAE-cellulose column (12 × 4 cm) is poured in an all-glass system, and the flow rate is adjusted to 20 ml/min with slight pressurization, using an aquatic pump. The column is then washed extensively ($\sim$10 bed volumes) with chloroform–methanol–water (2:3:1, v/v) to remove all traces of salt. Next, it is washed with 200 ml of chloroform–methanol (1:1, v/v), causing the column to shrink slightly. It is sometimes necessary to stir up the top of the column packing and to let it resettle.

The total lipid extract, prepared as described above, is diluted with 900 ml of methanol and is applied to the DEAE-cellulose column at 20 ml/min. Much of the yellow pigment and the phosphatidylethanolamine do not stick to the DEAE-cellulose column, but some pigments, the anionic lipids, and lipid X are retained. After application of the sample, the column is washed with 750 ml of chloroform–methanol–water (2:3:1, v/v). Finally, the anionic lipids (among which lipid X is predominant if mutant MN7 is used) are eluted with chloroform–methanol–100 mM ammonium acetate (2:3:1, v/v). The column is washed with 2000 ml of this solvent, and 150-ml fractions are collected. The lipid X, which is detected by charring of 3-μl samples, emerges in fractions 7–11.

Analysis of this material by thin-layer chromatography in the solvent system chloroform–pyridine–88% formic acid (20:30:7, v/v) reveals that it is approximately 80–90% lipid X (R_f 0.3), with minor contaminants including phosphatidic acid and cardiolipin. Fractions 7–10, containing most of the lipid X, are pooled, giving a total volume of 600 ml. Next, the solvent proportions are changed by adding 100 ml of chloroform, 160 ml of water, and 10 ml of concentrated HCl. After mixing in a separatory funnel, the phases are allowed to separate, and the lower phase is dried by rotary evaporation. The residue is immediately redissolved in 50 ml of chloroform–pyridine–88% formic acid (60:30:7, v/v).

A 100-ml column of Bio-Sil A (8 × 4 cm) is poured in chloroform–methanol (1:1, v/v) and is washed with 100 ml of chloroform–methanol (9:1, v/v). The column is allowed to run at its natural flow rate (~30 ml/min). After the sample is loaded, the column is washed with 200 ml of chloroform–pyridine–formic acid (60:30:7, v/v), removing undesired pigments, phosphatidic acid, and cardiolipin. Next, the column is washed with 200 ml of chloroform–methanol (9:1, v/v), and then the lipid X is eluted with 200 ml of chloroform–methanol (1:1, v/v). The entry of the final solvent into the column is visible, since the silicic acid turns white. Eight 25-ml fractions are collected. The presence of lipid X in each fraction is detected by thin-layer chromatography of 3-μl samples, followed by charring. The lipid X emerges in fractions 4–8, but fraction 4 is discarded because it contains a more rapidly migrating impurity. Fractions 5–8 are pooled, dried by rotary evaporation, and residual solvent is removed with a vacuum pump. At this stage, 498 mg of lipid X is recovered. The material is greater than 97% pure as judged by thin-layer chromatography and ^{1}H-nuclear magnetic resonance NMR spectroscopy in $CDCl_3/CD_3OD$ (2 : 1, v/v). The spectrum is identical to that of a chemically synthesized standard.[16]

To remove traces of lipid Y[11] and other mitogenic impurities that cannot be detected by analytical methods other than bioassays, a portion of the lipid X is further purified by reversed-phase column chromatography (Baker, Phillipsburg, NJ, octadecylsilane, flash chromatography bulk packing, C_{18}, bonded to silica gel, 40-μm particles). Two solvent mixtures (HPLC grade) are employed for this purpose: solvent A is acetonitrile–5 mM tert-butylammonium phosphate in water (1 : 1, v/v), and solvent B is 2-propanol–5 mM tert-butylammonium phosphate in water (85 : 15, v/v). A 20-ml column (11.5 × 1.5 cm) of the reversed-phase packing, suspended in solvent A–solvent B (2 : 1, v/v), is poured and washed with 60 ml of the same mixture. Next, 200 mg of Bio-Sil A-purified lipid X is dissolved in 20 ml of solvent A–solvent B (2 : 1, v/v) and applied to the column at its natural flow rate. The column is washed with 60 ml of solvent A–solvent B (2 : 1, v/v), 60 ml of A–B (1 : 1, v/v), 60 ml of A–B (1 : 2, v/v), and 60 ml of B. Twelve fractions (20 ml each) are collected. Most of the lipid X emerges in fractions 4 and 5, which are combined.

The tert-butylammonium phosphate is removed by a final ion-exchange chromatography. A 25-ml column of DE-52, prepared as described above, is washed with 75 ml of chloroform–methanol–water (2 : 3 : 1, v/v), and the sample (fractions 4 and 5 from the reversed-phase column) is applied directly. Next, the column is washed with 50 ml of chloroform–methanol–water (2 : 3 : 1, v/v), and then the lipid X is eluted with 140 ml of chloroform–methanol–240 mM ammonium acetate in water (2 : 3 : 1, v/v). Fractions of 20 ml are collected, and most of the lipid X emerges in fractions 2–5. These fractions are pooled, and 13 ml of chloroform, 23 ml of water, and 1.4 ml of concentrated HCl are added. After mixing in a separatory funnel, the lower phase is removed, neutralized with 2 ml of distilled pyridine, and concentrated until dry with a rotary evaporator. Traces of solvent are removed with a vacuum pump. The lipid X isolated in this manner (143 mg) is recovered as the monopyridinium salt (M_r 790.97) and is stored in a desiccator at $-20°$.

Preparation of UDP-2,3-Diacylglucosamine

UDP-2,3-diacylglucosamine is prepared by coupling lipid X (100 mg) with UMPmorpholidate (185 mg, Sigma), dissolved in 2.5 ml of redistilled dry pyridine at 37° overnight. A 55 ml Bio-Sil A column (diameter 1.5

---

[16] I. Macher, *Carbohydr. Res.* **162,** 79 (1987).

cm) is prepared in chloroform–pyridine–formic acid (30 : 30 : 7, v/v) and washed with 250 ml of this solvent. To the reaction mixture, 2.5 ml of chloroform and 0.58 ml of 88% formic acid are added, and the sample is applied to the Bio-Sil A column. The column is washed successively with 150 ml of the equilibration solvent mixture, 200 ml of chloroform–methanol (95 : 5, v/v), and 500 ml of chloroform–methanol (70 : 30). The compound is eluted with 250 ml of chloroform–methanol (30 : 70, v/v), collecting 10 fractions of 25 ml each.

The purity of the product is checked using a thin-layer system identical to the one described above for the qualitative assay for the disaccharide synthase. Most of the UDP-2,3-diacylglucosamine emerges in fractions 2–6 of the last elution. These fractions are pooled, and the solvent is removed with a rotary evaporator. The compound is redissolved in 3 ml of 10 mM HEPES, pH 8.0, and applied in 1-ml portions to a C$_{18}$ Sep-Pak cartridge (Pharmacia LKB Biotechnology Inc., Piscataway, NJ), which has been washed with 5 ml of methanol, 5 ml of water, and 5 ml of 10 mM HEPES, pH 8.0. After the sample is applied, the cartridge is washed with 3 ml of the same buffer, and the compound is eluted with 5 ml of methanol.

The methanol fractions containing the UDP-2,3-diacylglucosamine (as visualized using the thin-layer system described above for the qualitative assay of the disaccharide synthase) are pooled and dried under a stream of N$_2$. The residual solvent is removed with a vacuum pump. The compound is redissolved at 20 mM in 50 mM HEPES, pH 8.0, and stored in aliquots at $-20°$. The total yield after purification is 49% (as measured by absorbance at 262 nm and assuming an extinction coefficient identical to that of UDP). Purity is assessed by thin-layer chromatography, as described above.

Preparation of Disaccharide 1-Phosphate

A reaction mixture (consisting of 1 mM UDP-2,3-diacylglucosamine, 1 mM lipid X, 5 μg/ml purified disaccharide synthase, and 20 mM HEPES, pH 8, in 2 ml) is prepared and incubated for 60 min at 25°. The reaction proceeds to over 95% of completion, as judged by TLC analysis and charring (see above). Next, chloroform (2.25 ml) and methanol (2.25 ml) are added. After thorough mixing and brief low-speed centrifugation, the lipid product is recovered by drying the lower phase under a stream of N$_2$.

If the pyridinium salt of the disaccharide 1-phosphate is desired, the two-phase partitioning (see above) is carried out with the addition of 50 μl concentrated HCl. Prior to drying the lower phase, however, a few drops of redistilled pyridine are added. The final residue is dried further with a vacuum pump, and the material is stored at $-80°$.

For preparation of [32P]-labeled material, the same protocol is followed, but the concentrations of the lipid substrates may be reduced 10- to 100-fold. Disaccharide 1-[[32P]phosphate is stored in chloroform at $-80°$.

[55] Lipid A 4'-Kinase from *Escherichia coli:* Enzyme Assay and Preparation of 4'-[32P]-Labeled Probes of High Specific Radioactivity

By RANDOLPH Y. HAMPTON and CHRISTIAN R. H. RAETZ

Introduction

The *Escherichia coli* lipid A 4'-kinase is the enzyme responsible for the addition of the 4'-phosphate to the diglucosamine backbone of gram-negative lipid A.[1,2] In the biosynthetic scheme of lipid A,[2] the 4'-phosphoryl group is transferred from the γ-position of ATP to the substrate tetra-acyldisaccharide 1-monophosphate (abbreviated as DSMP herein) to yield the lipid A precursor lipid IV$_A$ (see Fig. 1).[1] By making use of [γ-32P]ATP as a substrate, the action of the 4'-kinase can be used to generate 32P-labeled lipid IV$_A$ of high chemical purity (98–99%) and specific radioactivity (3000 Ci/mmol). Because lipid IV$_A$ is a precursor of lipid A[2-4] and is bioactive in eukaryotic cells,[5-7] 32P-labeled lipid IV$_A$ is a useful probe for the investigation of both bacterial[2,8-12] and eukaryotic[2,12-14] aspects of lipid

[1] B. L. Ray and C. R. H. Raetz, *J. Biol. Chem.* **262**, 1122 (1987).

[2] C. R. H. Raetz, *Annu. Rev. Biochem.* **59**, 129 (1990).

[3] P. D. Rick, L. W.-M. Fung, C. Ho, and M. J. Osborn, *J. Biol. Chem.* **252**, 4904 (1977).

[4] C. R. H. Raetz, S. Purcell, M. V. Meyer, N. Qureshi, and K. Takayama, *J. Biol. Chem.* **260**, 16080 (1985).

[5] C. H. Sibley, A. Terry, and C. R. H. Raetz, *J. Biol. Chem.* **263**, 5098 (1988).

[6] R. A. Zoeller, P. D. Wightman, M. S. Anderson, and C. R. H. Raetz, *J. Biol. Chem.* **262**, 17212 (1987).

[7] H. Takada and S. Kotani, *Crit. Rev. Microbiol.* **16**, 477 (1989).

[8] K. A. Brozek, C. E. Bulawa, and C. R. H. Raetz, *J. Biol. Chem.* **262**, 5170 (1987).

[9] K. A. Brozek, K. Hosaka, A. D. Robertson, and C. R. H. Raetz, *J. Biol. Chem.* **264**, 6956 (1989).

[10] K. A. Brozek and C. R. H. Raetz, *J. Biol. Chem.* **265**, 15410 (1990).

[11] S. M. Galloway and C. R. H. Raetz, *J. Biol. Chem.* **265**, 6394 (1990).

[12] C. R. H. Raetz, K. A. Brozek, T. Clementz, J. D. Coleman, S. M. Galloway, D. T. Golenbock, and R. Y. Hampton, *Cold Spring Harbor Symp. Quant. Biol.* **53**, 973 (1988).

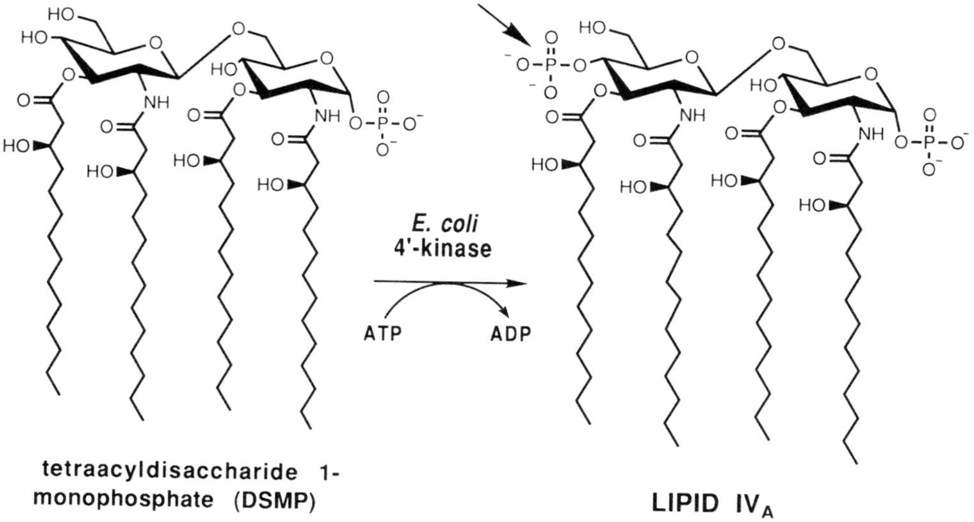

FIG. 1. Reaction catalyzed by *E. coli* lipid A 4'-kinase. The *E. coli* 4'-kinase catalyzes the transfer of a phosphoryl group from the γ-position of donor ATP to the 4'-position (nonreducing sugar) of the tetraacyldisaccharide monophosphate substrate (DSMP),[1] resulting in the formation of lipid IV_A. The 4'-phosphoryl group in the product is indicated in the diagram with an arrow.

A biochemistry. This chapter describes assays to measure the activity of the 4'-kinase as well a procedure that employs this enzyme to generate ^{32}P-labeled lipid IV_A for use as a labeled lipid A probe.

Assay 1. Detection with Radiolabeled Disaccharide Monophosphate Lipid IV_A Precursor

Both lipid A 4'-kinase assays described herein take advantage of the change in thin-layer chromatographic (TLC) mobility that the addition of a 4'-phosphoryl group imparts to the lipid IV_A product. This version of the assay employs a labeled substrate, [$1-^{32}P$]DSMP, that is converted to $1-^{32}P$-labeled lipid IV_A by the addition of the 4'-phosphate from ATP. An ATP-regenerating system is used in this procedure to maintain high levels of ATP donor.

[13] R. Y. Hampton, D. T. Golenbock, and C. R. H. Raetz, *J. Biol. Chem.* **263,** 14802 (1988).
[14] D. T. Golenbock, R. Y. Hampton, C. R. H. Raetz, and S. D. Wright, *Infect. Immun.* **58,** 4069 (1990).

Reagents

[1-^{32}P]DSMP [~10^4 counts/min (cpm)/nmol]/cardiolipin (0.5 mg/ml and 4 mg/ml respectively) in 100 mM HEPES, pH 7.4
ATP/phosphoenolpyruvate (PEP), 40 mM/100 mM in water
Nonidet P-40 (NP-40) or octyl-β-D-glucoside, 10% in water
MgCl$_2$, 50 mM
Pyruvate kinase, 1000 units/ml in 10 mM HEPES, pH 7.4
Escherichia coli (or other) membranes as the source of enzyme, 10–20 mg/ml in 100 mM HEPES, pH 7.4

[1-^{32}P]DSMP is prepared by the action of *E. coli* disaccharide synthase (*lpx B* gene product[15]) on its substrates, 1-^{32}P-labeled lipid X and UDP-2,3-diacylglucosamine to yield the desired labeled reagent. Details of the biochemical preparation and purification of labeled or unlabeled (used in Assay 2) DSMP using this technique are described in [54] in this volume. The aqueous codispersion of labeled (or unlabeled) DSMP and cardiolipin (CL) is made as previously described[1]: separate portions of DSMP (0.5 mg total) and CL (4 mg total) in chloroform–methanol (2:1, v/v) are combined in a Pyrex 13 × 100 mm culture tube. The solvent is removed by evaporation under a stream of N$_2$ gas, and 1 ml of 100 mM HEPES, pH 7.4, buffer is added to the tube. The tube is then vortexed briefly and subjected to sonic irradiation for 5 min in a bath sonicator (Laboratory Supplies, Inc., Hicksville, NY) until nearly transparent. This dispersion can be stored frozen at −20°, and it must be redispersed in the sonicator after thawing immediately before use in the 4'-kinase assay.

Escherichia coli membranes are prepared at 0°–4° by a method similar to that described.[1] A variety of *E. coli* strains, including K12, RZ60 (deficient in diglyceride kinase[1,16]), and BR7 (deficient in both diglyceride kinase and H$^+$-dependent ATPase[8]), as well as several other gram-negative species including *Salmonella typhimurium* and *Serratia marcescens*, have all yielded membrane fractions with active 4'-kinase using this procedure in our laboratory. Cells are grown in 1 liter LB broth[17] at 37° and harvested at the end of log phase by centrifugation for 20 min at about 4000 g_{av} (6000 rpm, Beckman, Fullerton, CA, JA 10 rotor). Cell pellets are resuspended by aspiration in 100 ml of 10 mM HEPES, pH 7.4. Cells are then resedimented and taken up in approximately 10 ml of the same buffer. The resuspended pellet is then frozen overnight at −20°. On the following day the frozen, washed cells are thawed and lysed by French press (18,000

[15] K. Radika and C. R. H. Raetz, *J. Biol. Chem.* **263**, 14859 (1988).
[16] C. R. H. Raetz and K. F. Newman, *J. Biol. Chem.* **253**, 3882 (1978).
[17] J. R. Miller, "Experiments in Molecular Genetics." Cold Spring Harbor Laboratory, Cold Spring Harbor, New York, 1972.

psi). The lysate is cleared of large debris by centrifugation for 20 min at about 4000 g_{av} (7000 rpm in a Beckman JA20 rotor). The supernatant (which contains the membrane fraction) is then subjected to ultracentrifugation at approximately 100,000 g_{av} for 1 hr to collect the membranes. The membrane pellet is then washed once by resuspension in 10 mM HEPES, pH 7.4, by aspiration and subsequent Dounce homogenization (>20 strokes, type A pestle), followed by ultracentrifugation at approximately 100,000 g_{av} for 1 hr. The resultant washed membrane pellet is resuspended in a minimal volume of 100 mM HEPES, pH 7.4, and frozen in small aliquots at −80° until needed. The usual protein concentration in the final preparation is between 10 and 20 mg/ml. For use in the assay, the membrane sample is thawed immediately prior to use and is resuspended by brief aspiration with a 200-μl disposable pipette tip.

Procedure: Assay 1. The assay is conducted in a total volume of 50 μl. To each assay tube, the following components are added in the order stated: 25 μl [1-^{32}P]DSMP/cardiolipin sonicated dispersion, 5 μl ATP/PEP, 5 μl Nonidet P-40 (or octyl-β-D-glucoside), 5 μl MgCl$_2$, and 5 μl pyruvate kinase. To commence the reaction, 5 μl of the *E. coli* (or other) membrane fraction is added, and the tubes are incubated at the desired temperature (usually 30°). Product formation is monitored by directly spotting 5 μl of the reaction mixture onto a silica gel thin-layer chromatography plate (E. Merck, Darmstadt, Germany, silica gel 60), drying the plate with a stream of cool air, and developing the plate in chloroform–pyridine–formic acid (88%)–water (40 : 60 : 16 : 5, v/v). This TLC system is used in all of the procedures described in this section. A well-closed tank can be used for up to 2 weeks before the solvent system must be replaced with a fresh mixture. Developed plates can be autoradiographed with Kodak (Rochester, NY) X-Omat AR film to locate the substrate and product bands (the product, lipid IV$_A$, migrates more slowly than the labeled DSMP substrate), which can be quantitated by scraping and counting the appropriate regions of the silica plate. Alternatively, the plate can be analyzed with a radioscanner.

Assay 2. Use of [γ-^{32}P]ATP to Detect Labeled Product

In this variation of the assay, unlabeled DSMP is used along with labeled ATP in order to generate the product, 4'-^{32}P-labeled lipid IV$_A$. The ATP-generating system is omitted, and unlabeled DSMP is used instead of the [1-^{32}P]DSMP. The advantage of this version is that the labeled substrate is commercially available, and, thus, there is no need to generate [1-^{32}P]DSMP. A disadvantage is that because no regenerating system is used for the ATP, the ATP concentration is depleted by the 4'-kinase reaction, as well as by side reactions, including the action of endogenous ATPase always present in cell membrane preparations.

For assays with wild-type membranes, it is important to set up a control lacking DSMP, in order to assess production of ^{32}P-labeled lipid products from other reactions that involve transfer of the γ-phosphoryl group from ATP to lipid acceptors. The main interfering reaction of this sort is the diacylglycerol kinase (*dgk* gene product), which will generate [^{32}P]phosphatidic acid, and indirectly, other radiolabeled glycerophospholipids in this crude system. The resultant products have a higher mobility than the 4'-^{32}P-labeled lipid IV$_A$ product in the TLC system used herein, so that the assay can still be conducted in the presence of significant *dgk* activity. The use of the *E. coli* mutant RZ60 (*dgk-6*)[16] will virtually eliminate the production of all labeled glycerophospholipids, resulting in the exclusive generation of 4'-^{32}P-labeled lipid IV$_A$.[1] However, it is important to consider possible interfering labeled products that may occur uniquely in novel strains or species under study.

If the reaction mixture is to be spotted directly onto TLC plates for analysis (see preparation of labeled lipid IV$_A$ below), then the remaining ATP as well as the labeled orthophosphate generated by hydrolysis will also be detected on the plate, but these species have much lower mobilities than lipid IV$_A$ in the TLC system (see below and Fig. 2). If the lipid-soluble components are first separated by two-phase partitioning as described below, then no ATP or orthophosphate will be detected.[1]

Reagents

DSMP/cardiolipin (0.5 mg/ml and 4 mg/ml, respectively) in 100 mM HEPES, pH 7.4

[γ-^{32}P]ATP, 20 mM in water (~10^4–10^5 cpm/nmol)

Nonidet P-40 or octyl-β-D-glucoside, 10% in water

MgCl$_2$, 50 mM in water

Membranes of *E. coli* wild type, of strain RZ60 (*dgk-6*), or of other bacteria as the source of kinase activity, 10–20 mg/ml in 100 mM HEPES, pH 7.4

Bligh–Dyer[18,19] single-phase acid extraction mixture: 20 ml chloroform/40 ml methanol/16 ml water/1 ml concentrated HCl

Procedure: Assay 2. Again, the assay is conducted in a total volume of 50 μl. To each assay tube, the following volumes of reagent stocks are added in the indicated order: 25 μl DSMP/cardiolipin sonicated dispersion, 10 μl [γ-^{32}P]ATP, 5 μl Nonidet P-40 (or octyl-β-D-glucoside), and 5 μl

[18] E. G. Bligh and J. J. Dyer, *Can. J. Biochem. Physiol.* **37**, 911 (1959).

[19] M. Nishijima and C. R. H. Raetz, *J. Biol. Chem.* **254**, 7837 (1979).

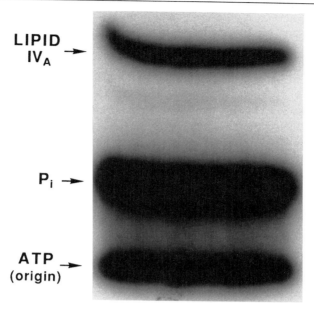

LIPID IV$_A$ →

P$_i$ →

ATP (origin) →

FIG. 2. Autoradiograph of a TLC plate on which a 4′-^{32}P-labeled lipid IV$_A$ preparation was resolved. A 4′-^{32}P-labeled lipid IV$_A$ preparation was generated as described in the text, using 5 mCi of [γ-^{32}P]ATP. The positions of [γ-^{32}P]ATP, [^{32}P]P$_i$, and [4′-^{32}P]-labeled lipid IV$_A$ are indicated. The ATP is, as indicated, on the origin. The approximate R_f of the lipid IV$_A$ is 0.55. The autoradiographic exposure was for 10 sec. The autoradiograph is used to locate the 4′-^{32}P-labeled lipid IV$_A$ band on the TLC plate. Autoradiograph provided by S. Mohan and C. Belunis, Merck Sharp and Dohme Research Laboratories, Rahway, NJ.)

MgCl$_2$. To commence the reaction, 5 μl of membrane fraction is added, and the tubes are incubated at the desired temperature (usually 30°). The no enzyme control is run with the membrane buffer, in this case 100 m*M* HEPES, pH 7.4. The reaction mixture can be analyzed for product formation by direct spotting of the reaction mixture onto a TLC plate, as in Assay 1 above, with subsequent development and autoradiography.

Alternatively, the lipid products can be separated by Bligh and Dyer two-phase partitioning[18,19] and then analyzed in the same TLC system. This is done as follows: the entire reaction mixture (50 μl) is added to 3.8 ml of the single-phase acid Bligh–Dyer mixture in a 13 × 100 mm screw-capped glass culture tube. The resultant solution is vortexed briefly, and then 1 ml each of additional chloroform and water are added to each tube to create two phases. The mixture is vortexed again and then separated by low-speed centrifugation in a clinical desk top centrifuge. The lower (organic) phase is carefully removed and placed in another glass tube. The upper (aqueous) phase can be washed with fresh preequilibrated lower phase, if desired, by adding the fresh lower phase to the upper phase and repeating the vortexing, centrifugation, and separation steps. The wash is then combined with the original lower phase. The total lower phase (original plus wash) is then evaporated with a stream of N_2 gas, redissolved in a small volume ($\sim$50 μl) of chloroform–methanol (95 : 5, v/v), and spotted onto a TLC plate. A second small volume of the same solvent is used to rinse the tube and is also spotted onto the plate. The plate is then briefly dried, developed in the same TLC system described above, visualized by autoradiography, and quantitated as above.

Use of *Escherichia coli* Lipid A 4′-Kinase to Generate 4′-³²P-Labeled Lipid IV$_A$ of High Specific Radioactivity for Use as Probe

A variation of the second assay, in which [γ-³²P]ATP is used to generate the labeled product, can be employed to produce highly labeled (often 3 mCi/nmol) 4′-³²P-labeled lipid IV$_A$. Our laboratory has made extensive use of this probe both in studies of the synthesis of lipid A and lipopolysaccharide (LPS) by bacteria,[8–12] and in studies of eukaryotic cell receptors for these lipids.[12–14,20] The procedure uses commercially available [γ-³²P]ATP of very high specific activity (usually 3000 Ci/mmol) in order to generate product of approximately the same specific activity, which is then purified from the reaction mixture by preparative TLC.[13] The enzyme source is a membrane fraction (prepared as described in Assay 1) from the *E. coli* BR7 strain, which has lesions in both the H^+-dependent ATPase and diacylglycerol kinase. The lack of diacylglycerol kinase inhibits the formation of highly labeled glycerophospholipids, and the absence of the ATPase attenuates the loss of the labeled ATP due to hydrolysis by that enzyme.

Reagents

DSMP/cardiolipin (0.5 mg/ml and 4 mg/ml, respectively) in 100 mM HEPES, pH 7.4

[γ-³²P]ATP, as purchased from Amersham (Arlington Heights, IL, PB101-68) ($\sim$1 mCi per 100 μl solution, 3000 Ci/mmol)

Octyl-β-D-glucoside, 10% in water

MgCl$_2$, 25 mM in water

Escherichia coli BR7 (*dgk-6 unc*) membranes, 10–20 mg/ml in 100 mM HEPES, pH 7.4

---

[20] R. Y. Hampton, D. T. Golenbock, M. Penman, M. Krieger, and C. R. H. Raetz, *Nature* (*London*) **352**, 342 (1991).

Bligh–Dyer single-phase acid extraction mixture: 20 ml chloroform/ 40 ml methanol/16 ml water/1 ml concentrated HCl

Procedure: Preparation of 4'-^{32}P-Labeled Lipid IV_A. Before the reaction is run, the [γ-^{32}P]ATP is dried into the reaction tube in a Sorvall SpeedVac centrifugal vacuum concentrator, or a similar apparatus. The total amount of [γ-^{32}P]ATP to be used (usually 1–5 mCi) is placed in a Sarstedt or similar plastic 1.5-ml microcentrifuge tube. The tube is capped, and a hole is punched in the top of the tube with an 18-gauge needle. The filled tube is then placed in the evaporator, and the water is removed by vacuum, without heat. This usually takes 3–4 hr. The evaporated [γ-^{32}P]ATP sample in the reaction tube can then be stored at −20° until the following day, or used immediately. To the tube containing the dried [γ-^{32}P]ATP, the following reagents are added in the indicated order: 25 μl DSMP/CL, 5 μl of 10% octyl-β-D-glucoside, 10 μl MgCl$_2$, and 5 μl of BR7 membranes. As in the assays above, the DSMP should be dispersed by sonic irradiation for 5 min prior to use, and the membranes should be defrosted and dispersed by repeated pipetting immediately prior to addition to the reaction mixture. The reaction is allowed to run at room temperature for 20–30 min.

The 4'-^{32}P-labeled lipid IV$_A$ is next isolated from the reaction mixture by preparative TLC. After the reaction time has elapsed, the entire mixture is spotted along a line on the origin of a 10 × 20 cm silica gel 60 TLC plate, without fluorescent indicator (Merck), by repeated application of 5-μl volumes in a 5- to 6-cm continuous line. The plate is then dried thoroughly with a stream of cool air (∼5 min) and developed in a freshly made chloroform–pyridine–formic acid (88%)–water (40 : 60 : 16 : 5, v/v) solvent system. The plate is dried and briefly autoradiographed (10 sec is usually sufficient) to locate the 4'-^{32}P-labeled lipid IV$_A$ band. Figure 2 shows a typical 10-sec autoradiograph of a TLC plate on which a 5 mCi reaction mixture was resolved. The unreacted [γ-^{32}P]ATP (ATP), liberated ortho[^{32}P]phosphate P$_i$, and labeled product 4'-^{32}P-labeled lipid IV$_A$ ("IV$_A$") are indicated.

The 4'-^{32}P-labeled lipid IV$_A$ is recovered from the TLC plate as follows. The autoradiograph is used to mark the lipid IV$_A$ band location on the plate with pencil. The plate is then sprayed lightly with distilled water (to prevent the formation of radioactive silica dust during scraping), and the 4'-^{32}P-labeled lipid IV$_A$-containing band is scraped onto a piece of waxed weighing paper. The silica sample is then transferred to a 15-ml sintered glass funnel (Pyrex 36060, 15 ml ASTM 10–15 M) supporting a Whatman (Clifton, NJ) GF/A glass fiber filter (to facilitate removal of the silica after extraction). The silica is then washed 3 times with 4 ml volumes of chloroform, by application of gentle suction on a side-arm flask. The funnel, along with the chloroform-washed silica, is then placed over a clean side-arm flask. The silica is extracted by washing with 3 times with 4 ml of Bligh–Dyer single-phase acid mixture, which removes the labeled lipid IV$_A$ from the silica.

The approximately 12 ml of single-phase wash is next distributed evenly between three 13 × 100 mm Pyrex screw-capped culture tubes, equipped with Teflon-lined caps. The solutions are then each converted to two phases by the addition of 1 ml of chloroform and 1 ml of water to each tube. The tubes are capped, vortexed, and briefly centrifuged as above to separate the phases. The three individual lower phases are carefully removed and are pooled in a single Corex glass tube. If the resultant lower phase is cloudy due to some upper phase contamination, sufficient methanol is added to clarify the mixture (usually 1–2 ml).

The pooled lower phases are then evaporated with N$_2$ gas, and the dried lipid is immediately rehydrated with 1–4 ml of sterile 50 mM Tris, pH 8.0. The tube is briefly vortexed and then subjected to sonic irradiation in a bath type sonicator for about 2 min. The resultant clear suspension, referred to as carrier-free stock, is counted for yield and initial radioactivity. A 10-μl sample of the stock solution is also spotted onto a TLC plate and run in the same TLC system to allow autoradiographic verification of product purity. The carrier-free stock is stored at $-20°$, and must be sonicated for at least 2 min prior to use after thawing. The yield (typically 1–4%) can be calculated from the amount of labeled ATP available, since this is the limiting reagent.

The specific radioactivity of the 4$'$-^{32}P-labeled lipid IV$_A$ stock can be altered by addition of unlabeled lipid IV$_A$ (either purchased commercially from ICN Biochemicals, Costa Mesa, CA, or prepared from *S. typhimurium*, as described[4,13]). Unlabeled lipid IV$_A$ is added from a small volume of an aqueous dispersion, usually in phosphate-buffered saline (PBS: 137 mM NaCl, 2.7 mM KCl, 1.5 mM KH$_2$PO$_4$, 6.5 mM Na$_2$HPO$_4$, pH 7.4), to a sample of the carrier-free suspension to give the desired final concentration. The mixture is frozen and thawed 3 times using dry ice and a room temperature water bath, then subjected to sonic irradiation for 5 min.

Properties of Lipid A 4$'$-Kinase

Escherichia coli 4$'$-kinase appears to be a membrane protein. Essentially all of the activity can be recovered from a 150,000 g_{av} membrane fraction. Furthermore, initial studies indicate that one can solubilize the enzyme by with detergents such as Nonidet P-40 (B. Ray and C. R. H. Raetz, unpublished observations), provided stabilizing agents such as glycerol are present and the sample is held on ice.

The addition of the cardiolipin to the substrate dispersion also significantly enhances enzyme activity. It is not clear if the CL enhancement is due to its effects on the physical state of the substrate, a direct effect on the enzyme, or to some combination of factors.

The 4'-kinase, like many ATP-utilizing enzymes, requires Mg^{2+} ions for activity. Owing to the ability of this ion to precipitate the DSMP substrate, the detergent in the assay mixture is required to allow these components to remain in mixed micellar solution. It is not known if the detergent has any direct effects on the enzyme. Although ATP gives the highest reaction rate when compared to other donors, CTP, UTP, and GTP will all serve as substrates in the reaction. The use of the ATP-regeneration system results in higher rates of conversion and a longer period of linearity.[1] There may be two reasons for this enhancement. The first is that the ATP concentration is maintained at a higher level in the face of loss due to the continuous 4'-kinase activity and other ATPases in the membrane preparations. A second reason is that ATP itself appears to increase the stability of the 4'-kinase, which decays in the presence of detergent under the conditions of the assay.[1] The mechanism of this apparent stabilization has not been studied.

The 4'-kinase is specific for the addition of a phosphoryl group to the disaccharide moiety of the nascent lipid A structure. Monosaccharides such as lipid X,[1] which also have a free 4-hydroxyl group, do not serve as substrates. The 4'-kinase does not appear to be able to phosphorylate diacylglycerol. Presumably, then, this enzyme has evolved specifically for the synthesis of lipid A. Although tetraacyldisaccharide 1-monophosphate (DSMP) appears to be the natural substrate *in vivo*, it may be that the 4'-kinase will show some ability to phosphorylate other disaccharides, possibly allowing for the use of this enzyme in the synthesis of novel and interesting lipid A-related analogs and radioactive probes.

The 4'-kinase has not yet been purified to homogeneity or cloned. Mutants defective in this activity have not been produced in any gram-negative species. Much remains to be done to characterize this useful and novel enzyme further.

[56] 3-Deoxy-D-*manno*-octulosonate Transferase and Late Acyltransferases of Lipopolysaccharide Biosynthesis

By KATHRYN A. BROZEK and CHRISTIAN R. H. RAETZ

Introduction

This chapter presents assays for the *Escherichia coli* enzymes that incorporate the 3-deoxy-D-*manno*-octulosonate (KDO), laurate, and myristate moieties found in lipopolysaccharide. KDO transferase catalyzes the addition of two KDO residues from CMP-KDO to the lipid A precursor, IV_A (Fig. 1).[1,2] Following this, two (or possibly four) distinct enzymes catalyze the transfer of laurate and/or myristate from their respective acyl carrier protein (ACP) derivatives to $(KDO)_2-IV_A$ (Fig. 1).[3] In addition, procedures are given for preparing the lipopolysaccharide precursors $(KDO)_2-IV_A$ and monoacyl- and diacyl-$(KDO)_2-IV_A$.

Materials

Nucleotide triphosphates, KDO, HEPES, and DEAE-Sepharose CL-6B are from Sigma (St. Louis, MO); Triton X-100 (Research Products International, Elk Grove Village, IL); silica gel 60 thin-layer plates, 0.25 mm (E. Merck, Darmstadt, Germany); Bio-Sil A silicic acid, 200–400 mesh (Bio-Rad, Richmond, CA). Methanol and pyridine are distilled prior to use, and all other solvents are reagent grade. Preparation of $4'-^{32}P$-labeled lipid IV_A[4] and acyl-ACP[5] are described elsewhere. The solvent mixtures used in precursor preparation are solvent A (chloroform–pyridine–88% formic acid–water, 15:35:8:5, v/v), solvent C (chloroform–pyridine–methanol–88% formic acid–water, 100:120:5:30:6, v/v), and solvent D (chloroform–pyridine–methanol–88% formic acid–water, 45:65:2:15:5, v/v).

[1] K. A. Brozek, K. Hosaka, A. D. Robertson, and C. R. H. Raetz, *J. Biol. Chem.* **264,** 6956 (1989).

[2] R. S. Munson, Jr., N. S. Rasmussen, and M. J. Osborn, *J. Biol. Chem.* **253,** 1503 (1978).

[3] K. A. Brozek and C. R. H. Raetz, *J. Biol. Chem.* **265,** 15410 (1990).

[4] R. Y. Hampton and C. R. H. Raetz, this volume [55].

[5] M. S. Anderson and C. R. H. Raetz, this volume [53].

METHODS IN ENZYMOLOGY, VOL. 209

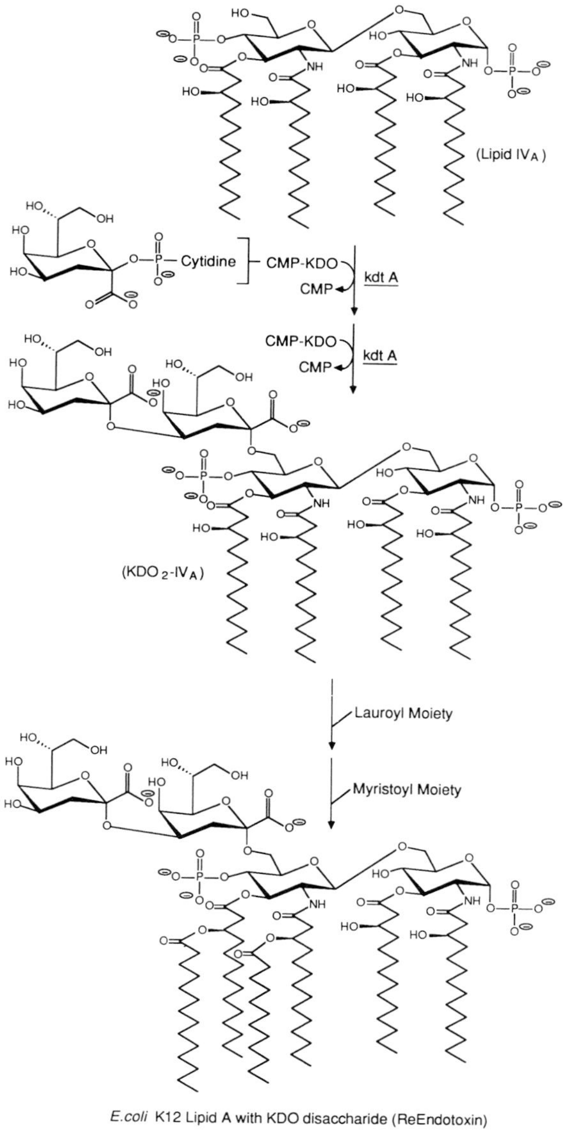

E. coli K12 Lipid A with KDO disaccharide (ReEndotoxin)

FIG. 1. Proposed biosynthetic pathway for ReEndotoxin in *Escherichia coli*. The enzymes that transform precursor IV_A to mature lipid A with KDO disaccharide include KDO transferase and the "late" acyltransferases.

3-Deoxy-D-*manno*-octulosonate Transferase Assay

Principle. The assay utilizes a radiolabeled lipid A precursor, $4'$-^{32}P-labeled lipid IV$_A$. The amount of product formed is measured using thin-layer chromatography (TLC), autoradiography, and scintillation counting of silica scrapings from the TLC plate.

Reagents. The components of the KDO transferase assay are as follows, with the stock solution concentration and volume of stock solution per 20-μl assay reaction given in parentheses: 50 mM HEPES, pH 7.5 (0.5 M, 2 μl); 0.1% Triton X-100 (1%, 2 μl); 10 mM MgCl$_2$ (100 mM, 2 μl); 100 μM lipid IV$_A$ (2 mM, 1 μl); 5 mM CTP (100 mM, 1 μl); 2 mM KDO (40 mM, 1 μl); 1.8 milliunits (mU) partially purified CMP-KDO synthase[6] (900 mU/ml, 2 μl); 12,000 counts/min (cpm) $4'$-^{32}P-labeled lipid IV$_A$ (1.2 $\times$ 10^7 cpm/ml, 1 μl); KDO transferase-containing extract (up to 8 μl); and water to make 20 μl.

Procedure. After 20 min or the desired time of incubation, a 5-μl portion of the reaction is withdrawn and spotted directly onto a silica gel TLC plate. After the spots have dried, the plate is developed in a fresh tank of solvent A (see Materials). The plate is dried with a warm air dryer, wrapped in Saran Wrap plastic, and exposed to X-ray film overnight using an enhancing screen at $-70°$. Under these chromatography conditions, lipid IV$_A$ and (KDO)$_2$–IV$_A$ have R_f values of 0.48 and 0.30, respectively. The amount of conversion to (KDO)$_2$–IV$_A$ can be quantified by scraping off the silica gel in the regions of lipid IV$_A$ and (KDO)$_2$–IV$_A$, counting in scintillation fluid such as BioSafe II, and calculating the nanomoles of product. The specific radioactivity of (KDO)$_2$–IV$_A$ is assumed to be the same as that of the lipid IV$_A$ starting material (see above).

Comments on Assay. Because CMP-KDO is labile, it is generated *in situ* with a CMP-KDO generating system (CTP, KDO, Mg^{2+}, and CMP-KDO synthase). The lipid IV$_A$ is converted from the storage form (a salt, usually pyridinium or triethylammonium) to the free acid form and dispersed by bath sonication in 10 mM Tris-HCl, pH 8.0/1 mM EDTA. For convenience, the $4'$-^{32}P-labeled lipid IV$_A$ (free acid form) is dispersed by bath sonication in 10 mM Tris-HCl, pH 8.0/1 mM EDTA, so that 1 μl contains about 12,000 cpm. The radiolabeled lipid IV$_A$ is at a very high specific radioactivity, so its concentration in the assay is negligible. A typical *E. coli* soluble fraction of 38 mg/ml will have about 0.20–0.25 mU/mg of activity (where 1 U equals the amount of enzyme that converts 1 μmol of substrate per minute). An appropriate volume to add to the above

[6] P. H. Ray and C. D. Benedict, this series, Vol. 83, p. 535. (1982).

assay is 2 μl, corresponding to 15–20 μU at an extract concentration of 38 mg/ml of protein.[1]

Preparation of 3-Deoxy-D-*manno*-octulosonate Transferase

Preparation of Cell Extracts. Cells are grown at 37° in LB broth,[7] harvested in late log phase, resuspended in 10 mM potassium phosphate buffer, pH 7.4 (about 5–10 ml per liter of culture), washed once, resuspended as above, and ruptured in a French pressure cell at 18,000 psi. The extract is centrifuged at 8000 g_{av} for 15 min, and the supernatant is recentrifuged at 150,000 g_{av} for 90 min to yield a high-speed supernatant (soluble fraction). The remaining pellet is resuspended in 5–10 ml phosphate buffer and recentrifuged at 150,000 g_{av} for 90 min. The pellet is again resuspended in 5–10 ml phosphate buffer (membrane fraction). All steps are performed at 4°, and samples are stored at −70°. Typically, an *E. coli* soluble fraction prepared in this way has 35–40 mg/ml protein.

Partial Purification of Enzyme. A 23-ml portion of a soluble fraction of *E. coli* W3110 (F$^-$, λ^-, IN[*rrnD–rrnE*]), from the *E. coli* Genetic Stock Center (Yale University, New Haven, CT) is stirred slowly while 5.5 g of solid ammonium sulfate is added. Stirring is continued for 30 min on ice. After centrifuging for 30 min at 18,000 g_{av}, the precipitate is dissolved in 20 ml of 20 mM Tris-HCl, pH 7.5, and 1 mM EDTA (buffer A). The sample is dialyzed against 1 liter of buffer A for 3 hr, then the buffer is changed and dialysis continued overnight. The dialyzed sample (19 ml) is loaded onto a 2.5 × 15.5 cm column of DEAE-Sepharose CL-6B equilibrated in buffer A. One bed volume of buffer A is washed through the column, and then a 500-ml gradient of NaCl in buffer A is applied (0–0.5 M NaCl). The flow rate is about 70 ml/hr, and 6-ml fractions are collected. The fractions that emerge with the loaded material (run-through fractions) contain KDO transferase (18 ml) but no CMP-KDO synthase.[1] They are pooled, divided into small portions, and frozen at −70°.

Table I summarizes the partial purification. About one-third of the total KDO transferase activity is bound to the DEAE-Sepharose CL-6B column and emerges at about 0.2 M NaCl. This second form of KDO transferase in wild-type extracts was not characterized further. About 30–40% of the total KDO transferase activity of crude extracts also remains membrane associated.[1]

---

[7] J. H. Miller, "Experiments in Molecular Genetics." Cold Spring Harbor Laboratory, Cold Spring Harbor, New York, 1972.

TABLE I
PARTIAL PURIFICATION OF 3-DEOXY-D-*manno*-OCTULOSONATE TRANSFERASE

Step	Volume (ml)	Protein (mg)	Specific activity (mU/mg)	Yield (%)
Soluble fraction	23	874	0.23	100
Ammonium sulfate	19	342	0.45	77
DEAE-Sepharose	18	41.4	1.87	39

Properties of 3-Deoxy-D-*manno*-octulosonate Transferase

A soluble fraction of *E. coli* will have a specific activity of about 0.25 mU/mg, whereas the partially purified KDO transferase is enriched to about 1.5–1.9 mU/mg. The enzyme is stable in 10 mM phosphate buffer or 20 mM Tris-HCl, pH 7.5, at $-70°$ for at least 6 months; it can withstand a few freeze–thaw cycles, but activity decreases with repeated freezing and thawing. In addition to lipid IV_A, its probable substrate in living cells,[8–11] the enzyme will accept acylated versions of lipid IV_A, such as lipid IV_B (containing an additional palmitate) and synthetic *E. coli* lipid A (containing laurate and myristate, Fig. 1). It will not act on the monosaccharide compounds lipid X and lipid Y, nor on the disaccharide 1-monophosphate precursor of lipid IV_A.[1]

Preparation of Bis(3-deoxy-D-*manno*-octulosonate)–Lipid IV_A

$(KDO)_2$–IV_A is prepared enzymatically using the KDO transferase assay conditions. The 2-ml reaction contains 50 mM HEPES, pH 7.5, 0.1% Triton X-100, 2 × 10^6 cpm 4'-^{32}P-labeled lipid IV_A, 100 μM lipid IV_A, 10 mM MgCl$_2$, 60 mU CMP-KDO synthase, 5 mM CTP, 2 mM KDO, and 0.5 mU KDO transferase. The reaction mixture is incubated at 30° for at least 2 hr, or until the conversion is almost complete. Progress of the reaction can be checked by running a TLC plate as in the assay and either charring with sulfuric acid or scanning for radioactivity. Then, 2.5 ml of chloroform, 5.0 ml of methanol, and 100 μl HCl are added to form a single-phase Bligh and Dyer mixture.[12] The proteinaceous precipitate is removed

[8] C. R. H. Raetz, S. Purcell, M. V. Meyer, N. Qureshi, and K. Takayama, *J. Biol. Chem.* **260,** 16080 (1985).

[9] S. M. Strain, I. M. Armitage, L. Anderson, K. Takayama, N. Qureshi, and C. R. H. Raetz, *J. Biol. Chem.* **260,** 16089 (1985).

[10] P. D. Rick and M. J. Osborn, *J. Biol. Chem.* **252,** 4895 (1977).

[11] P. D. Rick, L. W.-M. Fung, C. Ho, and M. J. Osborn, *J. Biol. Chem.* **252,** 4904 (1977).

[12] E. G. Bligh and J. J. Dyer, *Can. J. Biochem. Physiol.* **37,** 911 (1959).

by centrifugation in a clinical centrifuge, and 2.5 ml each of chloroform and water are added to form a two-phase Bligh and Dyer system.[12] The upper phase is washed with 2.5 ml of a fresh, acid-preequilibrated lower phase, and this lower phase is added to the original lower phase (for a total of 7.5 ml). Approximately 15 drops (0.25 ml) of distilled pyridine are added, and the mixture is evaporated to dryness under a stream of nitrogen. The sample contains $(KDO)_2$–IV_A, lipid IV_A that did not react, and any lipid material present in the enzyme preparation.

A Bio-Sil A column of approximately 2-ml bed volume (diameter 0.8 cm) is prepared in solvent C (see Materials). The dried sample is dissolved in 0.5 ml of solvent C and applied to the column. The column is then washed with 8 ml solvent C, which removes phospholipids and any unreacted lipid IV_A. The $(KDO)_2$–IV_A is eluted with 20 ml solvent A (see Materials), while collecting fractions of 2 ml. Small samples (5 μl) of the fractions are counted for radioactivity to find the peak. In addition, a TLC plate can be run (as in the assay) to ascertain that adequate separation of the starting material and product has been achieved prior to pooling the peak fractions. It is preferable that a radioactivity scanner be used to analyze the TLC plate, so that the procedure can be carried to completion in the same day. If it is necessary to expose the plate to X-ray film several hours or overnight, the fractions can be stored tightly covered at $-20°$ with minimal breakdown of the $(KDO)_2$–IV_A.

The peak fractions are combined (approximately 4 fractions; 8 ml), and 28.5 ml chloroform and 2.5 ml methanol are added. This mixture should be a single phase; if there is a little separation, methanol can be added dropwise until it is a single phase. This mixture is applied to a 2-ml Bio-Sil A column equilibrated in chloroform–methanol (95 : 5, v/v). The column is washed with 10 ml of the chloroform–methanol mixture. The product is then eluted with 9.5 ml of a single-phase acidic Bligh and Dyer mixture, consisting of 2.5 ml chloroform, 5.0 ml methanol, and 2.0 ml of 48 mM ammonium acetate titrated to pH 1.5 with HCl (the salt is necessary to remove the product in high yield). The eluate is then converted to a two-phase mixture by adding 2.5 ml each of chloroform and water. After vigorous mixing, the phases are separated by centrifuging in a clinical centrifuge for 5 min. To the lower phase (5.0 ml) 0.15 ml distilled pyridine is added, and the solution is dried under a nitrogen stream. The residue is further dried at 50–100 mTorr for 2 hr, and is resuspended with sonic irradiation in 0.35 ml of 50 mM Tris-HCl, pH 8.0. This preparation is stored at $-70°$ and is subjected to sonication after thawing in a bath sonicator for 2 min prior to use. The specific radioactivity is about 2 × 10^4 cpm/nmol, but this can be varied by starting with more or less $4'$-^{32}P-labeled lipid IV_A.

The level of radiochemical purity can be assessed by TLC in solvent A and autoradiography. Typically, the product is over 99% pure by this criterion. Alternatively, the silica gel plate can be charred by spraying with a solution of 10% H_2SO_4 in ethanol (v/v) and heating the silica gel plate on a hot plate. In this way, any nonradioactive lipid, protein, or carbohydrate material can be visualized. However, this is not usually observed.

The above procedure can be scaled up at least 100-fold by multiplying the reaction components accordingly. Suggestions for modifications in the isolation procedure, such as glassware and apparatus, are found in Ref. 1.

Acylation of Bis(3-deoxy-D-*manno*-octulosonate)–Lipid IV_A

The enzymes that add normal fatty acyl groups to lipid A require acyl-ACPs as activated donors of laurate and/or myristate.[3] These acyl-ACPs are present in extracts of *E. coli*, or they can be synthesized, purified, and added to the protein extract. Method A utilizes endogenous acyl donors, while Method B requires exogenously added acyl-ACP.

Reagents. For Method A, the following reaction components are mixed in a total of 20 μl. A convenient stock solution and the amount added to a 20-μl assay are given in parentheses: 50 mM HEPES, pH 7.5 (0.5 M, 2 μl); 0.1% Triton X-100 (1%, 2 μl); 25 μM [4'-^{32}P](KDO)$_2$–IV$_A$ (1–2 × 10^4 cpm/nmol and 250 μM, 2 μl); 7–15 mg/ml *E. coli* soluble fraction (40 mg/ml, about 8 μl). For Method B, the assay components are 50 mM HEPES, pH 7.5 (0.5 M, 2 μl); 0.1% Triton X-100 (1%, 2 μl); 25 μM [4'-^{32}P](KDO)$_2$–IV$_A$ (1–2 × 10^4 cpm/nmol and 250 μM, 2 μl); 25 μM acyl-ACP (250 μM, 2 μl); 0.3 mg/ml *E. coli* extract (6 mg/ml, diluted with buffer from the original concentrated extract, 1 μl).

Procedure. The reactions are incubated for 30 min at 30°. The assay is stopped by directly spotting 5 μl of the reaction mixture onto a silica gel thin-layer plate. After the spots have dried, the plate is developed in solvent A (see Materials). The plate is then dried with a heat gun, wrapped in Saran Wrap plastic, and exposed to X-ray film using an enhancing screen at −70° for 12 hr or overnight,

Comments on Assay. An *E. coli*-soluble fraction has about 0.4–0.5 mU/mg of activity and will catalyze mainly the addition of laurate from lauroyl-ACP, whereas a membrane fraction will add myristate and laurate about equally well, from their respective acyl-ACPs.[3] Other acyl-ACPs that have been examined are not incorporated under our conditions. Method B is preferred for any quantitative measurements because a defined amount of the second substrate (acyl-ACP) is added. Because exogenous acyl-ACP is present, the concentration of *E. coli* extract can be decreased by at least an order of magnitude as compared with Method A.

If a crude extract or membrane fraction is used as the source of enzyme(s), only Method B should be employed. Membranes contain a palmitoyl transferase and a rich supply of its substrate (sn-1-palmitoyl phospholipid) that will acylate lipopolysaccharide by a different mechanism.[13] The product of this reaction will obscure the presence of the desired products.

Discussion of Assay Products. At early times (0–30 min), a product with an R_f value of about 0.33 appears, accompanied by the disappearance of $(KDO)_2–IV_A$ (R_f 0.30). Later (1–6 hr), a second product (R_f 0.36) that is derived from the first product appears. These acylated compounds can be quantified by scraping and scintillation counting as described above. When utilizing endogenous acyl donors as in Method A, the first of the two products contains an additional fatty acid, as compared to $(KDO)_2–IV_A$, that is almost exclusively laurate.[3] The second product, which is derived from the first, has a second additional acyl group, laurate or myristate, in roughly equal proportions.[3] The lauroyl group likely forms an acyloxyacyl group with the amide-linked 3-hydroxymyristate on the nonreducing end of the lipid A disaccharide.[14,15] The second acylation is probably taking place on the hydroxyl of the ester-linked 3-hydroxymyristate on the nonreducing end of lipid A.[14,15] It is not known how many enzymes are involved in this set of reactions, since no fractionation of the *E. coli* extract has been done. When exogenous acyl-ACP is added to the reaction, as in Method B, presumably only that particular acyl group is being added to $(KDO)_2–IV_A$. Additional studies are needed to confirm the putative sites of acylation proposed in Fig. 1.

Preparation of Monoacyl- and Diacylbis(3-deoxy-D-*manno*-octulosonate)–Lipid IV_A

To obtain about 0.2–0.5 mg of material, a reaction mixture of 15 ml is assembled, consisting of 67 mM HEPES, pH 7.5, 0.1% Triton X-100, 10 mM MgCl$_2$, 6.7 mM CTP, 2.7 mM KDO, 220 mU CMP-KDO synthase, 105 μM 4'-^{32}P-labeled lipid IV_A (6300 cpm/nmol), and 7 mg/ml protein of a soluble fraction from *E. coli* D21f2. If desired, [4'-^{32}P](KDO)$_2$–IV$_A$ (6300 cpm/nmol) can be substituted for the labeled lipid IV_A and the CMP-KDO generating system (CTP, KDO, Mg^{2+}, and CMP-KDO synthase), but this requires more work since the labeled (KDO)$_2$–IV$_A$ must first be prepared from lipid IV_A.

[13] K. A. Brozek, C. E. Bulawa, and C. R. H. Raetz, *J. Biol. Chem.* **262**, 5170 (1987).
[14] U. Seydel, B. Lindner, H.-W. Wollenweber, and E. T. Rietschel, *Eur. J. Biochem.* **145**, 505 (1984).
[15] N. Qureshi, K. Takayama, D. Heller, and C. Fenselau, *J. Biol. Chem.* **258**, 12947 (1983).

If the CMP-KDO generating system is used, the first six components are mixed and preincubated for 10 min at room temperature to form some CMP-KDO. Next, the remaining components are mixed together, added to the CMP-KDO generating system, and incubated at 30° for 18 hr with gentle shaking. The products are extracted from the reaction mixture by adding 75 ml of methanol, 37.5 ml of chloroform, 15 ml of water, and 0.8 ml concentrated HCl, forming a single-phase Bligh and Dyer mixture.[12] The mixture is centrifuged at 1200 g_{av} in 150-ml glass Corex bottles for 15 min at room temperature. The clear supernatant is removed, and the precipitate is washed with 19 ml chloroform, 37 ml methanol, and 15 ml of 0.2 N HCl. This wash mixture is shaken and centrifuged as above, and its supernatant is combined with the first one.

The combined supernatants are added to a separatory funnel containing 56 ml each of chloroform and water, generating a two-phase Bligh and Dyer system.[12] Again, the mixture is shaken, poured into Corex bottles, and centrifuged as above to separate the phases. The lower phases (112 ml) are transferred to a round-bottomed flask, 6.2 ml of distilled pyridine is added, and the solvents are removed by rotary evaporation. The residue is dissolved in 8 ml of solvent C and applied to a Bio-Sil A column (1 × 10 cm) equilibrated in solvent C. The column is washed with 50 ml of solvent C, then 30 ml of a 1 : 1 mixture of solvent C and solvent D, and then 25 ml of solvent D alone. Fractions of about 2 ml are collected.

Samples (5 μl) of the fractions are spotted on silica gel plates, and the plates are developed in solvent A. Following autoradiography of the plates, fractions containing the monoacyl (fractions 37–44) and diacyl (fractions 19–30) products are pooled. It is very important to analyze the fractions by TLC, because it is difficult to tell by scintillation counting where one peak ends and the next one begins, since the products overlap to some extent. Each pool is diluted with chloroform–methanol (95 : 5, v/v), the first pool (fractions 37–44) with 32 ml and the second (fractions 19–30) with 48 ml.

These mixtures are loaded onto separate 2-ml Bio-Sil A columns equilibrated in chloroform–methanol (95 : 5, v/v). The columns are first washed with 15 ml of chloroform–methanol (95 : 5, v/v). The desired products are then each eluted with a mixture of 2.5 ml of chloroform, 5 ml of methanol, and 2 ml of 100 mM ammonium acetate titrated to pH 1.5 with HCl. To each eluate, 2.5 ml each of chloroform and water are added to generate a two-phase Bligh and Dyer system.[12] Each preparation is thoroughly mixed and centrifuged in a clinical centrifuge to separate the phases. The lower phases are removed, and the upper phases are washed a second time, each with 5 ml of acidic preequilibrated lower phase, and the mixing and centrifuging are repeated. The lower phases for each sample are pooled, mixed with approximately 0.5 ml of pyridine, and dried under a stream of nitrogen. Traces of pyridine and water can be removed with a vacuum pump at 50–100 mTorr. This procedure yields about 0.2–0.5 mg of each monoacyl- and diacyl-$(KDO)_2$–IV_A, depending on the extent of the enzymatic reactions. The purity can be assessed using TLC and sulfuric acid charring as described for $(KDO)_2$–IV_A.

Monoacyl-$(KDO)_2$–IV_A is a substrate for following the second acylation that generates diacyl-$(KDO)_2$–IV_A. Both monoacyl- and diacyl-$(KDO)_2$–IV_A are also substrates for addition of other core sugars, such as L-*glycero*-D-*manno*-heptose.[3]

[57] Eukaryotic Lipopolysaccharide Deacylating Enzyme

By R. S. MUNFORD and A. L. ERWIN

Introduction

Human neutrophils contain an enzyme, acyloxyacyl hydrolase (AOAH),[1] that selectively cleaves secondary (acyloxyacyl-linked) acyl chains from the lipid A region of bacterial lipopolysaccharides (LPS) (Fig. 1). Enzyme(s) with this activity has also been found in human monocytes, macrophages, and vascular endothelial cells, but not in platelets, erythrocytes, or fibroblasts; the activity has also been found in leukocytes from mice, rabbits, cows, pigs, and chickens. The purified human neutrophil enzyme is a glycoprotein that has two disulfide-linked subunits.[1] *In vitro,* AOAH hydrolyzes LPS acyloxyacyl bonds that have structurally different secondary acyl chains, regardless of the location of the acyloxyacyl linkage on the diglucosamine lipid A backbone; it does not remove the 3-hydroxyacyl chains that are directly linked to the backbone.[2] Recent studies in our laboratory indicate that the enzyme also has phospholipase A_1, lysophospholipase, and diglyceride lipase activities *in vitro.* The enzyme may be purified from human neutrophils by a series of standard chromatographic steps[1] (Table I) or by mono-

---

[1] R. S. Munford and C. L. Hall, *J. Biol. Chem.* **264,** 15613 (1989).
[2] A. L. Erwin and R. S. Munford, *J. Biol. Chem.* **265,** 16444 (1990).

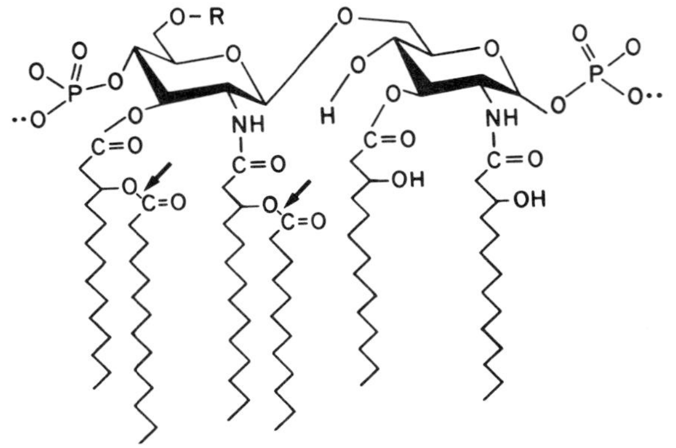

FIG. 1. Structure of the lipid A region of *Salmonella* lipopolysaccharide, showing the site of action of acyloxyacyl hydrolase (arrows). R indicates the site of attachment of the polysaccharide chain.

clonal antibody affinity chromatography followed by LPS-agarose affinity chromatography.[3]

AOAH activity is measured by quantitating the release of ^{3}H-labeled fatty acids from a biosynthetically double-labeled LPS substrate that contains ^{3}H-labeled acyl chains and ^{14}C-labeled glucosamine. Following incubation with the enzyme, the released fatty acids are separated from the unreacted LPS and the partially deacylated LPS (dLPS) by chloroform–methanol extraction or ethanol precipitation; the ^{14}C label, present in the diglucosamine lipid A backbone and in the carbohydrate chain, is a useful marker for the presence of small amounts of contaminating LPS or dLPS in the ^{3}H-labeled fatty acid-containing fraction.

Assay Method: Hydrolysis of ^{3}H-Labeled Acyl Chains
from ^{3}H- and ^{14}C-Labeled Lipopolysaccharide

Preparation of Radiolabeled Substrate

Enzymatic deacylation of LPS is measured using biosynthetically radiolabeled LPS. A useful substrate is Rc LPS of *Salmonella typhimurium* PR122 (*galE, nag, hisF1009, trpB2, metA22, xyl-1, strA201 F$^-$*), obtained from P. D. Rick (Uniformed Services University of the Health Sciences,

---

[3] R. S. Munford, L. Eidels, and E. J. Hansen, *in* "Cellular and Molecular Aspects of Endotoxin Reactions" (A. H. Nowotny, J. J. Spitzer, and E. J. Ziegler, eds.), p. 305. Elsevier Science Publ., Amsterdam, 1991.

TABLE I
PURIFICATION OF ACYLOXYACYL HYDROLASE FROM HL-60 CELLS[a]

Step	Protein		Activity		Specific activity (milliunits/mg protein)
	Milligrams	Cum %[b]	Milliunits	Cum %[b]	
Lysate	37,400	100	178	100	0.005
100,000 g supernatant	4,909	13.1	144	80.9	0.029
Blue-agarose peak	231	0.62	110	62.2	0.48
Phenyl-Sepharose peak	15	0.04	90	50.7	6.02
Hydroxylapatite peak	1	0.0027	59	32.9	59.0
Mono S peak	0.009	0.00002	14	7.8	1555.0

[a] HL-60 cells, grown in suspension culture, were lysed by nitrogen cavitation. After centrifugation to remove nuclei and unbroken cells, the lysate was processed using a series of chromatographic steps as indicated (described in Ref. 1).
[b] Cum, cumulative.

Bethesda, MD),[4] which can incorporate different labels into the fatty acyl chains and the carbohydrate backbone. The bacteria are grown in proteose peptone–beef extract (PPBE) medium [0.1% beef extract, 0.5% NaCl, 1.0% proteose peptone #3 (Difco, Detroit, MI)],[5] without added galactose. Five milliliters of $10\times$ M9 salts[6] is added to each 100-ml culture; the pH of the medium is adjusted to 7.3–7.4 with NaOH. This medium is inoculated with an overnight growth of bacteria (initial OD_{540} 0.01) and incubated with vigorous agitation at 35°–37°. When the OD_{540} of the culture reaches 0.02–0.04, the culture is supplemented with 200 μCi/ml [2-^{3}H]acetate (sodium salt, NEN, Boston, MA, or Amersham, Arlington Heights, IL, >500 mCi/mmol) and 0.25 μCi/ml N-acetyl-[1-^{14}C]glucosamine or [1-^{14}C]glucosamine (Du Pont–NEN, 50–60 mCi/mmol). (The radioisotopes may be supplied in 90% ethanol; to decrease the amount of ethanol that is added to the cultures, the volume may be reduced by approximately 90% under a gentle stream of N_2. The isotopes are then brought up in a few milliliters of medium and transferred to the growing culture.) Growth is continued until the OD_{540} of the culture reaches 0.6–0.8 (late logarithmic phase).

With minor modifications, the phenol–chloroform–petroleum ether method of Galanos et al.[7] works well for purifying small amounts of LPS.

[4] R. S. Munford, C. L. Hall, and P. D. Rick, J. Bacteriol. 144, 630 (1980).
[5] M. J. Osborn, J. E. Gander, E. Parisi, and J. Carson, J. Biol. Chem. 247, 3962 (1972).
[6] J. Sambrook, E. F. Fritsch, and T. Maniatis, "Molecular Cloning: A Laboratory Manual," 2nd Ed. Cold Spring Harbor Laboratory, Cold Spring Harbor, New York, 1989.
[7] C. Galanos, O. Luderitz, and O. Westphal, Eur. J. Biochem. 9, 245 (1969).

To remove all medium salts, the labeled bacteria are washed *twice* with water (to enhance sedimentation of the bacteria, 20–30% ethanol can be added to the second wash). Ethanol, acetone, and diethyl ether washes are followed by drying the delipidated cells under reduced pressure as described by Galanos *et al.*[7] [These washes contain radioactive compounds (principally phospholipids) and should be discarded into appropriate containers.] The dry cells may then be stored under reduced pressure at 4°.

The LPS is then extracted from the dry cells using phenol–chloroform–petroleum ether (2 : 5 : 8, v/v).[7] Liquid phenol (~90%, Fisher, Pittsburgh, PA) works well, provided that it is not discolored; the phenol–chloroform–petroleum ether is chilled on ice, and solid phenol added until the mixture is clear. In a 15-ml glass tube, the dry bacterial cells (~100 mg) are extracted with 3–4 ml cold phenol–chloroform–petroleum ether with vigorous vortexing, alternating with cooling in ice, for 2 min. The mixture is centrifuged (3000 g, 10 min, 4°), and the supernatant is transferred to another tube. The cell pellet is reextracted with 3–5 ml of the above mixture (a spatula may be used to resuspend the gummy pellet), centrifuged, and the two supernatants are combined. Chloroform and petroleum ether are conveniently removed under a stream of nitrogen. The LPS may be precipitated from the phenol by carefully adding small amounts of water (5–10 μl at a time, gently mixing while avoiding surface bubbles, waiting a minute or two between additions). Alternatively, to 1 volume of phenol solution one may add 6 volumes of diethyl ether–acetone (1 : 5, v/v).[8] The LPS precipitate is collected by centrifugation (10,000 g, 10 min, 4°), then washed twice with 80% phenol (2–3 ml) and 3 times with diethyl ether. A water bath sonicator (Ultrasonic Cleaner, Mettler Electronics Corp., Hightstown, NJ) is helpful for resuspending the LPS during these washes.

Following a procedure developed in the laboratory of M. J. Osborn (University of Connecticut, Farmington, CT), after the last ether wash the LPS is suspended in 1 ml distilled water. Diethyl ether (1–2 ml) is then added, and the biphasic mixture is agitated. After brief centrifugation to separate the phases, most of the ether is removed and discarded into a radioactive waste container. Two additional ether washes are performed, then the residual ether is removed under N_2. The LPS is precipitated by adding ammonium acetate (final concentration 25 mM, pH 5), then 2 volumes ethanol (95%). After holding the preparation at $-20°$ for at least 30 min, the LPS is collected by centrifugation (10,000 g, 10 min, 4°), washed twice with ether, and dried under reduced pressure. The radiolabeled LPS is weighed and suspended at a concentration of 1 mg/ml in 0.1% (v/v) triethylamine, 10 mM Tris-HCl (0.9 volumes of 0.1% triethyl-

[8] N. Qureshi and K. Takayama, *J. Biol. Chem.* **257**, 11808 (1982).

amine are added, the LPS is suspended by vigorous vortexing and/or sonication, then 0.1 volume of 100 mM Tris-HCl is added to lower the pH to ~8). The LPS is stored in aliquots at −70°. Typically, a 200-ml bacterial culture yields 80–100 mg dried cells, from which 2–4 mg LPS is obtained.

Characterization of Substrate

Salmonella typhimurium PR122 typically takes up 10–20% of the [^{3}H]acetate and 40–60% of the *N*-acetyl[^{14}C]glucosamine present in the medium. The purified LPS contains approximately 5000–10,000 disintegrations/min (dpm) ^{14}C and 100,000–150,000 dpm ^{3}H per microgram. The LPS fatty acids may be analyzed by reversed-phase high-performance liquid chromatography (HPLC) following chemical hydrolysis and conversion to *p*-bromophenacyl ester derivatives.[2,9] The fatty acids are identified and quantitated by comparison of retention times and peak areas with those of fatty acid standards; scintillation counting of fractions from the HPLC column eluate allows determination of specific activities. For *S. typhimurium* PR122, the LPS fatty acids are 3-OH 14:0, 12:0, and 14:0, with traces of 16:0; the specific activities (^{3}H dpm per nanomole) are typically 50,000–80,000 for 3-OH 14:0 and 30,000–50,000 for 12:0 and 14:0.

Comment. The LPS of other bacteria may also be labeled using this method. All of the bacteria studied in our laboratory (*Escherichia coli, Haemophilus influenzae, Neisseria gonorrhoeae, N. meningitidis, Pseudomonas aeruginosa*) have incorporated exogenous [^{3}H]acetate into fatty acids.[2] Adding nonradioactive glucose (5 mM) to the growth medium may reduce the use of labeled acetate as a carbon source. Substantially less [^{14}C]glucosamine is incorporated into LPS in the absence of the *nag* (glucosamine deaminase) mutation; *N*-acetyl[^{14}C]glucosamine should be used to label bacteria that are *nag*⁺.

Assay Conditions

Human AOAH may be assayed in a 0.5-ml reaction mixture that contains 20 mM sodium acetate or Tris–citrate, pH 4.8–5.0, 0.1–0.5% (v/v) Triton X-100, 150 mM NaCl, 1 mg/ml bovine serum albumin (BSA), and 1 μg radioactive LPS substrate. Incubation is carried out at 37° for the desired time period, then the free ^{3}H-labeled fatty acids are separated from the LPS substrate either by a Bligh–Dyer chloroform–methanol extraction or by ethanol precipitation of the LPS.

For chloroform–methanol extraction, the method described by Ames[10]

[9] C. L. Hall and R. S. Munford, *Proc. Natl. Acad. Sci. U.S.A.* **80,** 6671 (1983).
[10] G. F. Ames, *J. Bacteriol.* **95,** 833 (1968).

is modified slightly: the $1:2$ (v/v) chloroform–methanol mixture is acidified by the addition of acetic acid (5 $\mu l/ml$), then added to the reaction mixture (1.9 ml chloroform–methanol per 0.5-ml reaction mixture); after extraction (10 min on ice, with occasional vortexing), the phases are separated by the addition of 0.625 ml each of chloroform and water, with vortexing after each addition. The extraction mixture is centrifuged, and the chloroform layer is removed as described by Ames; the methanol–water layer is then reextracted with 1 ml chloroform. The two chloroform extracts are pooled and counted. [To reduce quenching, chloroform should be removed from the sample (e.g., by evaporation on a heating block) before adding scintillant. We routinely add 0.2 ml of 1% sodium dodecyl sulfate (SDS)–10 mM EDTA to all samples prior to adding scintillant; this reduces internal quenching for LPS.]

For ethanol precipitation, 2 volumes of 95% ethanol are added, the tubes are chilled for 30 min, and the LPS and BSA are pelleted by centrifugation. An aliquot of the ethanol–water supernatant (containing the ^{3}H-labeled fatty acids) may be added directly to scintillation fluid (a preliminary experiment should evaluate the ability of the scintillant to accommodate various amounts of ethanol–water while maintaining counting linearity).

The ^{3}H radioactivity in the chloroform or ethanol is corrected for the presence of small amounts of contaminating LPS by counting the ^{14}C dpm in each sample, then applying the following formula:

[3]H dpm (corrected) = [3]H dpm (measured)
$$- \text{([14]C dpm} \times \text{[3]H/[14]C ratio in the LPS)}$$

The average specific activity of the secondary LPS fatty acids (^{3}H dpm/nmol) may then be used to convert ^{3}H dpm (corrected) to moles of fatty acid released. In the reaction mixture described here, purified neutrophil AOAH releases approximately 1.5 μmol of fatty acid from LPS per milligram of protein/min (1.5 units/mg).[1]

Comments. It should be noted that *maximal* deacylation by AOAH removes only one-third of the total ^{3}H radioactivity in the labeled LPS[11] (Fig. 1). The reaction is linear below 12–15% deacylation of the LPS. The reaction pH given is the optimal pH for human AOAH; for AOAH from other animal species the optimal pH varies slightly.

When measured in the above reaction mixture, the apparent K_m is approximately 0.6 μM.[2,9] The substrate concentration used for the routine assay [2 μg/ml Rc LPS (approximate M_r 3600), or 0.55 μM] is near the apparent K_m. At this substrate concentration, calcium chloride (5 mM)

[11] R. S. Munford and C. L. Hall, *Science* **234**, 203 (1986).

enhances AOAH activity by 30–50%.[9] At higher LPS concentrations (20 μg/ml), the activity of the enzyme is inhibited by calcium, particularly at neutral pH (R. S. Munford and A. L. Erwin, unpublished observations); this may reflect greater aggregation of the LPS substrate. Nonionic detergent is required for the *in vitro* reaction. Triton X-100 concentrations of 0.1% (v/v) or greater produce maximal activity in the above reaction mixture; at higher LPS concentrations or in different reaction conditions the optimal Triton X-100 concentration is less predictable. Sodium dodecyl sulfate appears to inactivate the enzyme, whereas sodium deoxycholate does not substitute for Triton X-100 in the reaction mixture.

Confirmation that the enzymatic activity is acyloxyacyl hydrolysis may be obtained by using one-dimensional thin-layer chromatography (TLC) of the chloroform-soluble ^{3}H-labeled fatty acids.[2] The secondary fatty acids may be separated from one another by reversed-phase HPLC[2,9] or by one-dimensional TLC on Whatman (Clifton, NJ) KC$_{18}$ reversed-phase plates using acetonitrile–acetic acid (1 : 1, v/v), as the solvent.[12] The fatty acyl chains that remain on the LPS after enzymatic deacylation may be analyzed by HPLC or TLC following acid hydrolysis and extraction of the fatty acids into chloroform.[2,9]

Detection of Acyloxyacyl Hydrolase Activity in Biological Samples

AOAH activity may be measured in cells following detergent lysis. Approximately 10[6] cells are brought up in 100 μl lysis buffer (0.1% Triton X-100, 2 mM phenylmethylsulfonyl fluoride, 10 mM sodium phosphate, 150 mM NaCl, pH 7.4), held at room temperature for 5 min, then centrifuged to pellet the nuclei. The lysis supernatant is withdrawn for assay or storage at $-70°$. AOAH activity is stable at $-70°$, with little apparent loss of potency due to freezing and thawing.

Preparation of Deacylated Lipopolysaccharide

dLPS may be prepared using the above reaction conditions by adjusting the AOAH concentration and incubation time to produce the desired degree of deacylation. For example, 100 μg of Rc LPS may be maximally deacylated (loss of ~30% of ^{3}H-labeled fatty acids from the LPS; over 90% removal of secondary acyl chains) when incubated with 0.15 milliunits of AOAH for 72 hr at 37°. The dLPS and BSA are precipitated with ethanol as described above, washed once with 80% ethanol, and resuspended in 0.9% NaCl at the desired concentration. If the preparation is to be used

[12] A. Schultz and S. Oroszlan, *Virology* **133**, 431 (1984).

in biological tests, the reaction mixture should be prepared using pyrogen-free water (e.g., Sterile Water for Irrigation, Baxter, Deerfield, IL). The dLPS preparation contains BSA and traces of AOAH; for experiments that test the biological activities of dLPS, controls have included LPS that was incubated in the reaction mixture without AOAH, AOAH that was incubated in the reaction mixture without LPS, and the reaction mixture itself.[11,13]

Acknowledgments

This work was supported by U.S. Public Health Service Grant AI18188 from the National Institute of Allergy and Infectious Diseases.

[13] F. X. Riedo, R. S. Munford, W. B. Campbell, J. S. Reisch, K. R. Chien, and R. D. Gerard, *J. Immunol.* **144,** 3506 (1990).

Section XII

Phospholipid Transfer Processes

[58] Transfer Activity and Acyl-Chain Specificity of Phosphatidylcholine Transfer Protein by Fluorescence Assays

By Pentti J. Somerharju, Juha Kasurinen, and Karel W. A. Wirtz

Introduction

The major site of phosphatidylcholine (PC) biosynthesis in mammalian tissues is the endoplasmic reticulum (ER).[1-3] Isotope-labeling studies have shown that PC moves very rapidly from the ER to mitochondria[4-7] and to the plasma membrane.[7,8] Moreover, in the liver there is a very effective and preferential flow of 1-palmitoyl-2-linoleoyl-PC from a minor dynamic pool to a bile-forming site.[9] Although definite proof is lacking, it has been proposed that the phosphatidylcholine transfer protein (PC-TP) may be involved in the above-mentioned intracellular distribution of PC.[9,10] This protein, present in the cytosol of nearly all tissues investigated, is relatively abundant in liver and small intestinal mucosal,[11] two tissues with very active PC metabolism and secretory activity. Concomitantly with an increase in the activities of choline kinase and choline phosphotransferase,[12] levels of PC-TP increased during fetal lung development with a distinct maximum 2 days before term,[11] suggesting an involvement of PC-TP in the production and secretion of lung surfactant. Recently, PC-TP was shown to stimulate rat liver microsomal cholinephosphotransferase (CDP-choline : 1,2-diacyl-sn-glycerol cholinephosphotransferase) activity.[13]

[1] G. F. Wilgram and E. P. Kennedy, *J. Biol. Chem.* **238**, 2615 (1963).

[2] L. M. G. van Golde, J. Raben, J. J. Batenburg, B. Fleischer, F. Zambrano, and S. Fleischer, *Biochim. Biophys. Acta* **360**, 179 (1971).

[3] J. E. Vance and D. E. Vance, *J. Biol. Chem.* **263**, 5898 (1988).

[4] K. W. A. Wirtz and D. B. Zilversmit, *Biochim. Biophys. Acta* **187**, 468 (1969).

[5] M. C. Blok, K. W. A. Wirtz, and G. L. Scherphof, *Biochim. Biophys. Acta* **233**, 61 (1971).

[6] M. P. Yaffe and E. P. Kennedy, *Biochemistry* **22**, 1497 (1983).

[7] J. E. Vance, *Biochim. Biophys. Acta* **963**, 10 (1988).

[8] M. R. Kaplan and R. D. Simoni, *J. Cell Biol.* **101**, 441 (1985).

[9] H. Sakamoto and T. Akino, *Tohoku J. Exp. Med.* **106**, 61 (1972).

[10] K. W. A. Wirtz, *in* "Lipid–Protein Interactions" (P. C. Jost and O. H. Griffith, eds.), Vol. 2, p. 151. Wiley (Interscience), New York, 1982.

[11] T. Teerlink, T. P. van der Krift, M. Post, and K. W. A. Wirtz, *Biochim. Biophys. Acta* **713**, 61 (1982).

[12] M. J. Engle, L. M. G. van Golde, and K. W. A. Wirtz, *FEBS Lett.* **86**, 277 (1978).

[13] Z. U. Khan and G. M. Helmkamp, *J. Biol. Chem.* **265**, 700 (1990).

Purification and properties of bovine and rat liver PC-TP have been described.[14-16] As a carrier of PC, PC-TP has specific binding sites for the sn-1-acyl and sn-2-acyl chains.[17] By using fluorescent PC analogs that carry a saturated or pyrenylacyl (Pyr) chain of varying length at either the sn-1 or sn-2 position, it was established that these binding sites have considerably different acyl chain specificity.[18] Owing to this different specificity, PC-TP is able to discriminate between positional isomers of PC, with a distinct preference being observed for molecular species that carry a palmitoyl chain at the sn-1 position and a pyrenyldecanoyl or pyrenyldodecanoyl chain at the sn-2 position.[18] Competition binding studies with natural species of PC have indicated that the affinity for PC-TP increases with increasing unsaturation of the sn-2-acyl chain, resulting in high affinity for species that contain arachidonic (20:4) or docosahexaenoic (22:6) acid.[19] This chapter describes the fluorescent assays used to determine the transfer activity of PC-TP toward different pyrenylacyl-labeled species of PC (for formulas, see Fig. 1) and the affinity of PC-TP for these PyrPC as well as unlabeled PC species.

Synthesis of Pyrenyl-Labeled Phosphatidylcholine Analogs

Pyrenyl Fatty Acids. Pyrenyl fatty acids are synthesized essentially as described in Ref. 20 and are purified by silica gel column chromatography. Certain pyrenyl fatty acids are also commercially available from Molecular Probes Inc. (Eugene, OR).

1-Acyl-2-Pyrenyl(x)phosphatidylcholines. The PC analogs are synthesized according to published methods using dihomosaturated PC species (Sigma Chemical Co., St. Louis, MO) as the starting material. First, lysophosphatidylcholines are prepared by treatment of the diacyl species with phospholipase A_2 from *Crotalus adamanteus* (Sigma) and are purified by repeated acetone precipitation.[21] The lysophosphatidylcholines are then reacylated with a pyrenyl fatty acid using the method of Mason *et*

[14] J. Westerman, H. H. Kamp, and K. W. A. Wirtz, this series, Vol. 98, p. 581.

[15] T. Teerlink, B. J. H. M. Poorthuis, and K. W. A. Wirtz, this series, Vol. 98, p. 586.

[16] K. W. A. Wirtz, *Annu. Rev. Biochem.* **60**, 73 (1991).

[17] T. A. Berkhout, A. J. W. G. Visser, and K. W. A. Wirtz, *Biochemistry* **23**, 1505 (1984).

[18] P. J. Somerharju, D. van Loon, and K. W. A. Wirtz, *Biochemistry* **26**, 7193 (1987).

[19] J. Kasurinen, P. A. van Paridon, K. W. A. Wirtz, and P. Somerharju, *Biochemistry* **29**, 8548 (1990).

[20] H.-J. Galla and W. Hartmann, this series, Vol. 72, p. 471.

[21] M. Kates, *in* "Techniques in Lipidology" (T. S. Work and E. Work, eds.), p. 393. North-Holland Publ., Amsterdam, 1972.

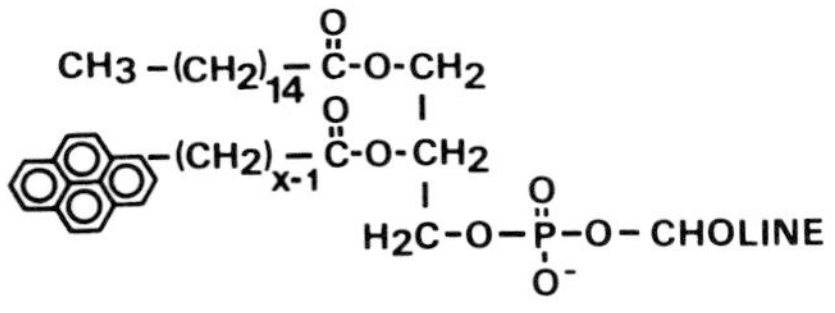

$$CH_3-(CH_2)_{14}-\overset{O}{\overset{\|}{C}}-O-CH_2$$
$$-(CH_2)_{x-1}-\overset{O}{\overset{\|}{C}}-O-CH_2$$
$$H_2C-O-\overset{O}{\overset{\|}{\underset{O^-}{P}}}-O-CHOLINE$$

C(16)Pyr(X)PC SPECIES
X= 6, 8, 10, 12, 14

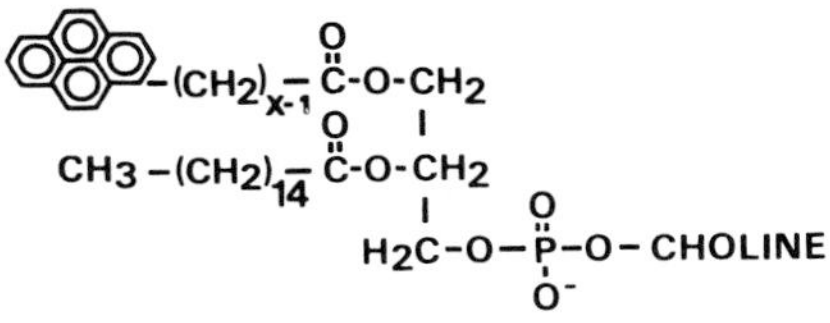

$$-(CH_2)_{x-1}-\overset{O}{\overset{\|}{C}}-O-CH_2$$
$$CH_3-(CH_2)_{14}-\overset{O}{\overset{\|}{C}}-O-CH_2$$
$$H_2C-O-\overset{O}{\overset{\|}{\underset{O^-}{P}}}-O-CHOLINE$$

Pyr(X)C(16)PC SPECIES
X= 6, 8, 10, 12, 14

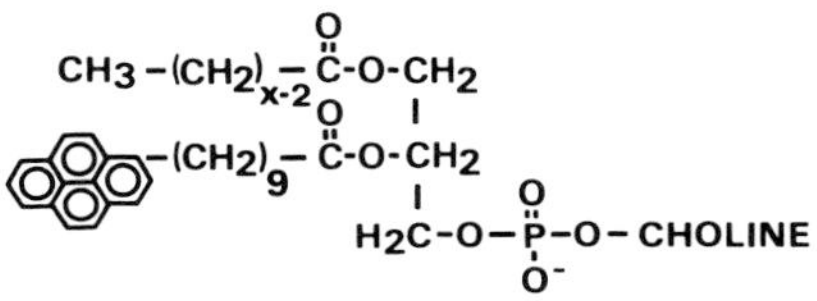

$$CH_3-(CH_2)_{x-2}-\overset{O}{\overset{\|}{C}}-O-CH_2$$
$$-(CH_2)_9-\overset{O}{\overset{\|}{C}}-O-CH_2$$
$$H_2C-O-\overset{O}{\overset{\|}{\underset{O^-}{P}}}-O-CHOLINE$$

C(X)Pyr(10)PC SPECIES
X= 10, 12, 14, 16, 18, 20

FIG. 1. Structures and nomenclature of the PyrPC species used. x indicates the total number of carbon units (including the carboxyl one) in the aliphatic chain. [Reproduced with permission from P. J. Somerharju, D. van Loon, and K. W. A. Wirtz, *Biochemistry* **23**, 7193 (1987).]

al.,[22] and, finally, the 1-acyl-2-Pyr(x)-PCs are purified by high-performance liquid chromatography (HPLC) on a silica column.

1-Pyrenyl(x)-2-acylphosphatidylcholines. First, 1,2-Pyr(x)-PCs are prepared by acylating the glycerophosphocholine–cadmium adduct with a pyrenyl fatty acid anhydride.[23] These analogs are then converted to the desired PC species as described above for the 1-acyl-2-Pyr(x)-PCs, by using unlabeled fatty acids (Fluka Chemical Co., Ronkonkoma, NY) for the reacylation.

Concentration. Concentrations of the PyrPC species are determined by measuring the absorbance at 342 nm in ethanol using a molar absorption coefficient of 42,000 mol^{-1} cm^{-1}.[24]

Comments. All PyrPC species are checked by thin-layer chromatography and should be homogeneous. The molecular species purity, as analyzed by reversed-phase HPLC,[25] is higher than 95%.

[22] J. T. Mason, A. V. Broccoli, and C.-H. Huang, *Anal. Biochem.* **113**, 96 (1981).

[23] K. M. Patel, J. D. Morrisett, and J. T. Sparrow, *J. Lipid Res.* **20**, 674 (1979).

[24] P. J. Somerharju, J. Virtanen, K. Eklund, P. Vainio, and P. Kinnunen, *Biochemistry* **24**, 2773 (1985).

[25] G. M. Patton, J. M. Fasulo, and S. J. Robins, *J. Lipid Res.* **23**, 190 (1982).

Fluorescence Measurements

All fluorescence measurements are carried out with an instrument equipped with a thermostatted cuvette holder. The excitation wavelength is either 343 nm (transfer experiments) or 346 nm (binding studies) and the bandpass 1.5–3 nm. Pyrene monomer emission is monitored at 378 nm (10 nm bandpass).

Determination of Phosphatidylcholine Transfer Protein Transfer Activity

Principle. The assay is based on the increase of fluorescence intensity resulting from the PC-TP-mediated transfer of PyrPC molecules from quenched donor vesicles to unquenched acceptor vesicles. The transfer activity can be monitored continuously or determined after a fixed time period.

Reagents

PyrPC species
Egg yolk PC, purified according to Ref. 26
Phosphatidic acid (PA), prepared from egg PC by phospholipase D-catalyzed hydrolysis
2,4,6-Trinitrophenylphosphatidylethanolamine (TNP-PE) (Sigma or Avanti Polar Lipids, Birmingham, AL)
PC-TP from bovine liver[14]: stock solution stored at $-20°$ in 50% glycerol and diluted in the assay buffer before use
Donor lipid mixture: PyrPC (100 μM), egg PC (400 μM), and TNP-PE (50 μM) (the quencher), dissolved in spectroscopic grade ethanol
Acceptor lipid mixture: egg PC (10 mM) and egg PA (0.3 mM), dissolved in analytical grade chloroform
Tris-HCl (20 mM)/EDTA (5 mM) buffer, pH 7.4 (Tris–EDTA buffer)

Procedure. First the acceptor vesicle suspension is prepared as follows: the acceptor lipid mixture is dried down from the chloroform solution (0.5 ml) under a stream of nitrogen, dispersed in 5 ml of Tris–EDTA buffer, and sonicated for 5 min on ice with a probe sonicator (60 W output). Prior to use, titanium particles and any undispersed lipid are removed by centrifugation at 8000 g for 20 min. Then the donor vesicles are prepared in the fluorometer cuvette by injecting the donor lipid mixture (5–10 μl of the solution) with a spring-loaded syringe (Hamilton) into 1–2 ml of Tris–EDTA buffer. After an equilibration period of 2–3 min, the acceptor vesicle suspension (0.1 ml) is added to the donor vesicles, and the increase

---

[26] W. S. Singleton, M. S. Gray, M. I. Brown, and J. L. White, *J. Am. Oil Chem. Soc.* **42**, 53 (1965).

of the pyrene monomer fluorescence intensity at 380 nm is recorded for 1–2 min. The slope of the progress curve corresponds to the rate of passive transfer of the PyrPC species used. Finally, an aliquot of the PC-TP solution is added; the initial slope is recorded and, after correction for the passive transfer, used as a relative measure for the rate of PC-TP-mediated PC transfer. Absolute rates of PyrPC transfer can be obtained by calibrating the fluorescence intensity scale. This is done by measuring the fluorescence intensities of vesicles containing known amounts of PyrPC and the acceptor lipids.

Comments. The most important variables in the transfer assay are the acyl chain structure of the PyrPC species used and the composition, concentration, and size of the donor and acceptor vesicles. The influence of the acyl chains on the rate of PC-TP-mediated PyrPC transfer is shown in Table I. The highest rate of transfer was observed with the C(12)Pyr(10)PC

TABLE I

BINDING AND TRANSFER OF
PYRENYLPHOSPHATIDYLCHOLINE SPECIES BY BEEF
LIVER PHOSPHATIDYLCHOLINE TRANSFER PROTEIN[a]

PyrPC species[b]	Relative affinity[c]	Relative transfer rate[c]
C(16)Pyr(6)	0.04	0.12
C(16)Pyr(8)	0.09	0.45
C(16)Pyr(10)	1.00	1.00
C(16)Pyr(12)	0.91	0.80
C(16)Pyr(14)	0.38	0.44
Pyr(6)C(16)	0.10	0.09
Pyr(8)C(16)	0.10	0.10
Pyr(10)C(16)	0.16	0.35
Pyr(12)C(16)	0.22	0.38
Pyr(14)C(16)	0.21	0.19
C(10)Pyr(10)	0.39	0.82
C(12)Pyr(10)	0.48	1.17
C(14)Pyr(10)	0.66	1.08
C(16)Pyr(10)	1.00	1.00
C(18)Pyr(10)	0.29	0.65
C(20)Pyr(10)	0.03	0.24

[a] Adapted with permission from P. J. Somerharju, D. van Loon, and K. W. A. Wirtz, *Biochemistry* **26,** 7193 (1987).

[b] See Fig. 1 for nomenclature.

[c] Values are relative to those obtained for the C(16)Pyr(10) species.

species. However, owing to its relatively high rate of spontaneous transfer, this PC species may not be the best choice to measure PC-TP transfer activity. Routinely, we use the C(16)Pyr(10)PC species because its transfer by PC-TP is only slightly lower than that of C(12)Pyr(10)PC, whereas it has a much lower rate of spontaneous transfer. Hence, with this PyrPC species the PC-TP-mediated rate of transfer can be measured with a high accuracy.

The phosphatidylinositol transfer protein (PI-TP) and the nonspecific lipid transfer protein are also capable of mediating PC transfer.[16] It should be noted that the acyl chain specificity of either protein varies from that of PC-TP.[27,28] Thus, the choice of PyrPC to be used in the transfer assay depends on the protein studied.

The lipid composition of the donor and acceptor vesicles may influence the interaction of PC-TP with the vesicle interface and thus have an effect on the rate of PC transfer between these vesicles.[29,30] As for PC-TP as well as for other mammalian transfer proteins, the affinity for the interface correlates positively with the negative charge of the vesicles.[31] The relative concentration of the donor and the acceptor vesicles is also of importance. In general, it is preferable to use an 10- to 100-fold excess of acceptor over donor vesicles. Under these conditions, the formation of pyrene excimers in the acceptor vesicles (which reduces monomer fluorescence intensity) is minimal, thus providing a larger dynamic scale. However, when present in a large excess, the acceptor vesicles should not contain more than 2–4 mol % of a negatively charged lipid species (usually PA) to avoid too strong an association of the transfer protein with these vesicles. Since the affinity of PC-TP for vesicles surfaces is also strongly dependent on the size (curvature) of the vesicles,[31] reproducibility of the assay is dependent on the method of vesicle preparation and on their stability. Thus, when high reproducibility is desired, it may be necessary to prepare a stock of donor/acceptor vesicles either by the injection method or by sonication and allow these vesicles to stabilize for an adequate period of time.

Application. This assay is ideally suited for kinetic studies since the true initial rate as well as the extent of transfer at equilibrium can be determined from a single progress curve. It can also be used to detect the transfer activity in column eluents when purifying PC-TP from various

[27] P. A. van Paridon, T. W. J. Gadella, P. J. Somerharju, and K. W. A. Wirtz, *Biochemistry* **27**, 6208 (1988).

[28] A. van Amerongen, R. A. Demel, J. Westerman, and K. W. A. Wirtz, *Biochim. Biophys. Acta* **1004**, 36 (1989).

[29] A. M. Kasper and G. M. Helmkamp, *Biochemistry* **20**, 146 (1981).

[30] R. P. Bozzato and D. O. Tinker, *Can. J. Biochem.* **60**, 409 (1982).

[31] T. A. Berkhout, C. van den Bergh, H. Mos, B. de Kruijff, and K. W. A. Wirtz, *Biochemistry* **23**, 6894 (1984).

sources. However, care should be taken if one wants to assay crude protein fractions, as large amounts of contaminating proteins may perturb the spectroscopic properties of the probes, that is, dequench the donor or quench the acceptor fluorophores. Similar transfer assays can be derived for other lipid-transporting proteins by replacing PyrPC by another pyrenyl lipid.

Determination of Affinity of Pyrenylphosphatidylcholine Species

Principle. On addition of PC-TP to quenched vesicles consisting of PyrPC, unlabeled PC, and TNP-PE, a PyrPC molecule is incorporated into the lipid-binding site of PC-TP, resulting in an increase of pyrene monomer fluorescence (owing to the inability of TNP-PE to quench the fluorescence of the PyrPC bound to PC-TP). The increase of fluorescence intensity reflects the affinity of that PyrPC species for PC-TP relative to that of unlabeled PC.

Reagents. The reagents are similar to the ones used in the transfer assay except that no acceptor vesicles are needed (see above).

Procedure. The donor vesicles are prepared by injecting the donor lipid mixture (5–10 μl of the stock solution in ethanol) into 2 ml of Tris–EDTA buffer in a thermostatted (30°) fluorometer cuvette. After a 2–3 min equilibration period, a few aliquots of PC-TP (0.05–0.1 nmol each) are added, and the pyrene monomer fluorescence intensity at 378 nm is measured. The increase of fluorescence intensity is then plotted as a function of PC-TP concentration, and the initial slope is taken as a (relative) measure of affinity of PC-TP for the PyrPC species used.

Comments. Like the rate of transfer, the affinity is strongly dependent on the acyl chain structure of a PyrPC species (Table I). As one could expect, the normalized affinities and rates of transfer correlate quite well. In general, however, the rates with which the various PyrPC species are transferred differ to a lesser extent than the affinities. This most likely reflects differences in the activation energies (kinetics) of binding to and/or release of the PyrPC species from PC-TP.[18] Compared to PC-TP, the binding of the PyrPC species to PI-TP from bovine brain gives rise to clearly different acyl-chain length versus affinity profiles.[27] This has been interpreted to indicate that the acyl-binding sites of PC-TP and PI-TP have distinctly different structures.

Determination of Affinity of Unlabeled Phosphatidylcholine Species

Principle. The relative affinity of an unlabeled PC species for PC-TP can be determined by using a simple competition assay. This assay involves (1) titration of a set of TNP-PE-quenched vesicles containing a constant amount of a PyrPC species and a variable amount of an unlabeled, competing PC species with PC-TP; (2) determination of the initial increase of the pyrenyl monomer fluorescence (F) as a function of added protein; and (3) estimation of the binding affinity of the unlabeled PC species relative to PyrPC (K_{rel}) by parameter fitting using the following equation:

$$F = F_{max}/(1 + K_{rel}R)$$

where F_{max} is the increase of monomer fluorescence in the absence of the competing (unlabeled) PC species and R is the molar ratio of unlabeled PC to PyrPC.[32]

Reagents. The reagents needed are similar to the ones used in the transfer assay (see above).

Donor lipid mixtures: Pyr(x)PC (100 μM), unlabeled PC species (0–100 μM), and TNP-PE (10 mol % of total lipid) dissolved in spectroscopic grade ethanol

Procedure. The titration procedure is the same as described above for determination of the relative affinities of PyrPC species. The data are analyzed according to the above equation by using a curve-fitting procedure.

Comments. The relative affinities (K_{rel}) for a variety of natural and related PC species are listed in Table II. It is to be noted that there are considerable deviations in the K_{rel} values. From the data it is concluded that both the total hydrophobicity of the PC molecule and the structural details of the individual acyl chains have a strong influence on the affinity.[33] In general, the affinity increased with the number of double bonds. In agreement with the data for the PyrPC species, the affinity of PI-TP from bovine brain for the unlabeled PC species was clearly different from that of PC-TP.[33]

Technical Considerations

Although the pyrenyl moiety of PyrPC is not particularly prone to oxidative or photochemical destruction, one should pay attention to the oxygen concentrations in the sample to be measured. This is because oxygen, even at atmospheric pressures, is able to cause partial, reversible quenching of the fluorescence of excited pyrenyl moieties. The easiest way to maintain a constant oxygen concentration is to equilibrate the buffer and other solutions with air at or close to the measuring temperature.

[32] P. A. van Paridon, T. W. J. Gadella, P. J. Somerharju, and K. W. A. Wirtz, *Biochim. Biophys. Acta* **903**, 68 (1987).
[33] J. Kasurinen, P. A. van Paridon, K. W. A. Wirtz, and P. J. Somerharju, *Biochemistry* **29**, 8548 (1990).

TABLE II
AFFINITY OF NATURAL PHOSPHATIDYLCHOLINE
SPECIES FOR PHOSPHATIDYLCHOLINE
TRANSFER PROTEIN[a]

PC species[b]	K_{rel}[c]	S.D.	n
12:0/12:0	1.1	0.30	4
14:0/14:0	2.9	0.58	4
14:0/16:0	2.5	0.41	4
14:0/18:0	1.7	0.11	4
14:0/18:1	3.8	1.05	4
15:0/15:0	3.7	0.69	4
16:0/16:0	3.0	0.49	4
16:0/18:0	1.6	0.29	4
16:0/18:1(6,7)[d]	5.8	0.57	4
16:0/18:1(9,10)[d]	4.3	0.55	5
16:0/18:1(11,12)[d]	3.9	0.47	4
16:0/18:2	7.5	1.47	4
16:0/18:3	6.1	2.15	4
16:0/20:4	9.3	0.51	4
16:0/22:6	10.5	0.85	4
17:0/17:0	1.7	0.17	5
17:1/17:1	15.5	1.13	3
18:0/18:0	0.9	0.06	4
18:0/18:1	1.6	0.39	4
18:0/18:2	3.6	0.24	4
18:0/18:3	3.0	0.65	4
18:0/20:4	3.6	0.06	2
18:0/22:6	4.3	0.03	2
18:1/18:1	5.4	0.21	4
18:2/18:2	11.7	0.85	4

[a] Adapted with permission from J. Kasurinen, P. A. van Paridon, K. W. A. Wirtz, and P. Somerharju, *Biochemistry* **29**, 8548 (1990).

[b] The fatty acid residues before and after the slash refer to those in the *sn*-1 and *sn*-2 position of the glycerol moiety of PC, respectively.

[c] Affinity relative to that of C(16)Pyr(8)PC. S.D. is the standard deviation or difference from the mean (when $n = 2$), and n is the number of experiments.

[d] The numbers in the parentheses indicates the position of the double bond in the *sn*-2 chain.

Because quenching is also temperature dependent it is recommended that the fluorometer be equipped with thermostatted cuvette holders. If long exposure times are necessary, one should use a narrow excitation bandpass to avoid photochemical reactions.

Conclusions

The assays described above allow the simple, rapid, and sensitive determination of PC (or any other pyrenyl lipid) transfer activity as well as the determination of the affinity of pyrene-labeled or unlabeled PC species to PC-TP and other similar proteins. In addition, the ability of PC-TP to transfer a variety of PyrPC species provides a specific and delicate method to introduce these probes into lipoproteins[34] and cell membranes[35] for metabolic or physical studies.

[34] M. Vauhkonen and P. Somerharju, *Biochim. Biophys. Acta* **984,** 81 (1989).
[35] F. A. Kuypers, X. Andriesse, P. Child, B. Roelofsen, J. A. F. Op den Kamp, and L. L. M. van Deenen, *Biochim. Biophys. Acta* **857,** 75 (1986).

[59] Phosphatidylinositol Transfer Proteins from Higher Eukaryotes

By G. M. HELMKAMP, JR., S. E. VENUTI, and T. P. DALTON

Introduction

Phosphatidylinositol (PtdIns) transfer proteins comprise a class of small, cytosolic proteins which bind single PtdIns molecules and transport this phospholipid, without modification, between membranes.[1] Among phospholipid transfer proteins, those which transfer PtdIns not only are the most widely distributed in eukaryotic species and tissues but also are the most strongly conserved in their structural features.[2] PtdIns transfer proteins have been purified from a broad spectrum of tissues, including human platelets[3] and baker's yeast[4]; they have been detected by catalytic activity measurements and immunologic techniques in at least 20 different

[1] G. M. Helmkamp, Jr., *Chem. Phys. Lipids* **38,** 3 (1985).
[2] G. M. Helmkamp, Jr., *in* "Subcellular Biochemistry" (H. J. Hilderson, ed.), Vol. 16, p. 129. Plenum, New York, 1990.
[3] P. Y. George and G. M. Helmkamp, Jr., *Biochim. Biophys. Acta* **836,** 176 (1985).
[4] G. Daum and F. Paltauf, *Biochim. Biophys. Acta* **794,** 385 (1984).

METHODS IN ENZYMOLOGY, VOL. 209

rat tissues.[5] Without exception, all PtdIns transfer proteins studied to date exhibit a molecular mass in the range 32–36 kDa, have acidic isoelectric points, and display a dramatic substrate specificity toward PtdIns and, with somewhat lesser affinity, phosphatidylcholine (PtdCho). The dual substrate specificity gives rise to two interchangeable isoforms of PtdIns transfer proteins which differ in bound phospholipid and isoelectric point.[6] Despite these common properties, there is a surprisingly unexpected lack of primary sequence similarity between the 31.9-kDa rat and the 35.0-kDa yeast PtdIns transfer proteins.[7–9] On the other hand, the human and rat proteins differ by a mere three amino acids.[10] The principal similarities among PtdIns transfer protein, namely, structural conservation and catalytic specificity, imply that purification and characterization may commence with any readily available tissue. In this chapter we summarize recent progress on small-scale isolations of PtdIns transfer protein from higher eukaryotic sources and some of the more common means of identification.

General Purification Strategy

When limited quantities of tissues are available for the isolation of PtdIns transfer protein, we have devised a sequence of steps to minimize the handling of the protein and the time required for obtaining a highly purified protein. For reasons that are not clear, the overall yields of pure PtdIns transfer protein from a variety of tissues have been poor, generally in the range 5–15%.[3,5,6,11,12] Once purified, however, the catalytic activity of the protein remains stable. In a typical experiment beginning with 100 g of fresh rat brain and using silver-stained polyacrylamide gel electrophoretic analysis as the principal criterion of purity, PtdIns transfer protein could be purified to homogeneity using ammonium sulfate fractionation and chromatography on molecular sieve, anion-exchange, and hydroxyl-

[5] S. E. Venuti and G. M. Helmkamp, Jr., *Biochim. Biophys. Acta* **946**, 119 (1988).

[6] P. A. van Paridon, A. J. W. G. Visser, and K. W. A. Wirtz, *Biochim. Biophys. Acta* **898**, 172 (1987).

[7] S. K. Dickeson, C. N. Lim, G. T. Schuyler, T. P. Dalton, G. M. Helmkamp, Jr., and L. R. Yarbrough, *J. Biol. Chem.* **264**, 16557 (1989).

[8] V. A. Bankaitis, D. E. Malehorn, S. D. Emr, and R. Green, *J. Cell Biol.* **108**, 1271 (1989).

[9] V. A. Bankaitis, J. R. Aitken, A. E. Cleves, and W. Dowhan, *Nature (London)* **347**, 561 (1990).

[10] S. K. Dickeson, G. M. Helmkamp, Jr., and L. R. Yarbrough, unpublished observation (1990).

[11] G. M. Helmkamp, Jr., M. S. Harvey, K. W. A. Wirtz, and L. L. M. van Deenen, *J. Biol. Chem.* **249**, 6382 (1974).

[12] P. E. DiCorleto, J. B. Warach, and D. B. Zilversmit, *J. Biol. Chem.* **254**, 7795 (1979).

apatite columns. The summary of a typical preparation is illustrated in Table I.

Tissue Homogenization. Tissue sources of PtdIns transfer protein may be fresh or fresh-frozen and thawed at 37°. Connective elements should be removed, and the tissue chilled to 4° (all subsequent steps are performed at 4°). A suitable homogenization medium consists of isotonic sucrose buffered with 10 mM HEPES–Na (pH 7.4) and to which the following protease inhibitors have been added: 1 mM EDTA, 0.1 mM phenylmethylsulfonyl fluoride, 2 μM leupeptin, and 5 μM pepstatin. Depending on the quantity of tissue, a Waring blender operated at medium speed for 2 min or a loose-fitting Potter–Elvehjem apparatus operated at 800 rpm for six strokes may be used to prepare a 20% homogenate. After passage through several layers of cheesecloth, the homogenate is subjected to successive centrifugations at 11,600 g for 20 min and 150,000 g for 1 hr; alternatively, for larger volumes the final centrifugation may be carried out at 50,000 g for 3 hr.

If cultured cells are the source of PtdIns transfer protein, they are first washed 3 times in 10 mM sodium phosphate and 0.15 M NaCl (pH 7.4) and resuspended and allowed to equilibrate in 5 mM Tris-Cl, 5 mM NaCl, 10 mM EGTA, 2 mM dithiothreitol, 0.1 mM phenylmethylsulfonyl fluoride, 1 μM apoprotinin, 10 μM leupeptin, and 10 μM pepstatin (pH 7.4). The cell suspension is then homogenized with 30 strokes of a tight-fitting Potter–Elvehjem apparatus operated at 800–1000 rpm. A cytosolic fraction is recovered after centrifugation at 150,000 g for 1 hr.

Ammonium Sulfate Fractionation and Molecular Sieve Chromatography. High-speed supernatants can be further enriched by fractionation

TABLE I
PURIFICATION OF PHOSPHATIDYLINOSITOL TRANSFER PROTEIN FROM RAT BRAIN

Step	Protein (mg)[a]	Specific activity[b]	Recovery (%)	Purification factor
High-speed supernatant[c]	1730	0.51	(100)	(1)
Ammonium sulfate fractionation; Sephadex G-100	63	9.1	65	18
DEAE-cellulose	4.4	61.9	31	122
Isoelectric focusing[d]	0.61	158	11	310
Hydroxylapatite[d]	0.15	215	4	424

[a] Protein is determined by the method of O. H. Lowry, N. J. Rosebrough, A. L. Farr, and R. J. Randall, *J. Biol. Chem.* **193,** 265 (1951).

[b] Specific activity represents nmol PtdIns transferred per min per mg of protein, using the vesicle–vesicle assay.

[c] Starting material is 90 g frozen rat brain.

[d] Data at this stage include both protein isoforms.

with ammonium sulfate. Using both transfer activity measurements and immunoblot analysis, we have found that the fraction obtained between 35 and 65% saturation contains virtually all the PtdIns transfer protein. The 65% ammonium sulfate precipitate is easily dissolved in a small volume of buffer. The concentrated sample is then applied to a column of Sephadex G-100 (Pharmacia LKB, Piscataway, NJ) equilibrated in and eluted with 50 mM sodium-phosphate, 0.1 mM EDTA, 0.02% NaN$_3$, and 0.1 mM phenylmethylsulfonyl fluoride (pH 7.4). On this matrix, PtdIns transfer protein is recovered at an elution function, $K_{av} = (V_e - V_{void})/(V_{total} - V_{void})$, value of 0.45 (approximately twice void volume). Sephadex G-50 and G-75 matrices may also be used for this step.

If ammonium sulfate fractionation is not employed, several methods of concentration may be used in preparing the sample for molecular sieve chromatography. For quantities less than 150 ml the protein solution is placed in dialysis tubing and covered with flaked polyethylene glycol (Aquacide III, Calbiochem, San Diego, CA). Alternatively, larger volumes are treated with ammonium sulfate to 90% saturation, stirred for 1 hr, and centrifuged for 30 min at 20,000 g; the pellet is readily dissolved.

Anion-Exchange Chromatography. An excellent, highly reproducible separation can be achieved using diethylaminoethyl-cellulose (DE-52, Whatman, Clifton, NJ) chromatography. It is imperative that the ionic strength of the high-speed supernatant and column eluant be reduced to $10^{-2} M$ (conductivity <0.65 mS) and that the pH be in the range 7.0–7.4 before application of the protein solution; this is best achieved by dialysis against 5 mM sodium phosphate and 0.1 mM EDTA (pH 7.4). A linear gradient of NaCl (0–300 mM) in 5 mM sodium phosphate and 0.1 mM EDTA (pH 7.4) is applied to the thoroughly washed anion-exchange resin to which the cytosolic proteins have been bound. The gradient volume should be 5–6 times the column volume. PtdIns transfer protein is eluted between 70 and 100 mM NaCl.

Preparative Isoelectric Focusing. Because PtdIns transfer protein isoforms bind noncovalently different phospholipid species, they also exhibit a small but significant difference in isoelectric point, generally in the range 0.2–0.4 pH units. This difference may be exploited by preparative isoelectric focusing to separate the isoforms. A horizontal bed apparatus (Pharmacia LKB) has been used in recent years; the protein sample is concentrated to 60–80 ml, dialyzed against 130 mM glycine (pH 6.5), and mixed with ampholytes and dextran beads. A linear pH gradient from 4.5 to 6.5 will usually be sufficient for good separation. The addition of 10–20 mg hemoglobin to the protein sample is often helpful in monitoring the progress of the electrofocusing; this protein migrates completely to the cathode end of the bed. After determination of the pH gradient with a surface electrode, fractions are obtained and protein is eluted from the dextran gel with 50 mM sodium phosphate and 100 mM NaCl (pH 7.4).

Hydroxylapatite Chromatography. Adsorption chromatography offers not only an additional purification step but also an additional means of separating PtdIns transfer protein isoforms. Using BioGel HTP (Bio-Rad, Richmond, CA), we have found equally reproducible results with sodium or potassium phosphate buffers. Accordingly, the protein sample is dialyzed against 10 mM sodium phosphate (pH 6.8), applied to a column of hydroxylapatite, and eluted with a linear gradient of 10–200 mM sodium phosphate (pH 6.8). If the gradient volume is at least 10 column volumes, the two isoforms can be distinguished. One will be eluted at 45–55 mM sodium phosphate and represents the protein with bound PtdCho; the other will be eluted at 70–80 mM sodium phosphate and represents the protein with bound PtdIns. One particular advantage of this step is being able to proceed from pooled anion-exchange (or isoelectric focusing) fractions without the need of volume reduction. The use of hydroxylapatite subsequent to isoelectric focusing provides a simple method of removing ampholytes, as they do not adsorb to the mineral matrix. A typical hydroxylapatite chromatographic result is depicted in Fig. 1. In other applications of this purification step, 10% (v/v) glycerol is added to the protein sample and the elution buffers as a stabilization measure.[6]

Other Purification Steps, Including Liquid Chromatography. If further steps are required for the complete purification of PtdIns transfer protein, consideration can be given to chromatography on DEAE-Sephacel (Pharmacia LKB).[5] Conditions for this step include application of the protein sample in 10 mM Tris-Cl (pH 7.2) and elution with a linear gradient of 0–200 mM NaCl in the same buffer; PtdIns transfer protein is eluted at 50–60 mM NaCl. A dramatic purification but poor recovery was noted in the use of phenyl-Sepharose (Pharmacia LKB) chromatography.[12] Other attempts to implement hydrophobic chromatography have yielded less-than-encouraging results, including Matrex Blue A (Amicon, Danvers, MA); the high proportion of charged amino acid residues in mammalian PtdIns transfer proteins also argues against this purification strategy. Preparative liquid chromatography also poses special problems in that concentrations of relatively pure PtdIns transfer protein greater than 100 μg/ml are susceptible to irreversible aggregation. With a Mono Q (Pharmacia LKB) anion-exchange column eluted with 20 mM Tris-Cl (pH 7.8) and a linear gradient of 0–200 mM NaCl in the same buffer, two peaks of PtdIns transfer protein were eluted, at 90 and 85 mM NaCl, representing the PtdIns-binding and PtdCho-binding isoforms, respectively.[6] On an analytical scale, however, we have successfully recovered PtdIns transfer protein from a Mono P (Pharmacia LKB) isoelectric focusing column eluted with

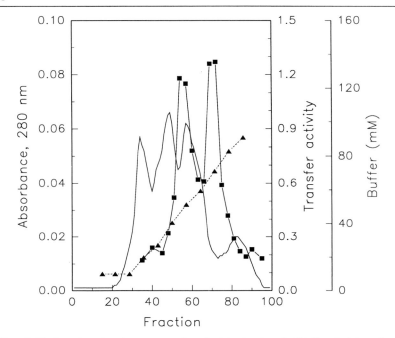

FIG. 1. Hydroxylapatite chromatography of rat brain phosphatidylinositol transfer protein. Following DEAE-cellulose and Sephadex G-75 chromatography, partially purified PtdIns transfer protein was dialyzed against 10 mM sodium phosphate (pH 6.8) and applied to a 1.5 × 14 cm column of hydroxylapatite (BioGel HTP, Bio-Rad) equilibrated in the same buffer. After extensive washing, proteins were eluted with a linear, 200-ml gradient of 10–200 mM sodium phosphate (pH 6.8). Fractions were analyzed for protein absorbance at 280 nm (—), buffer concentration (▲), and transfer activity (■), expressed as nanomoles PtdIns transferred/30 min for 10-μl aliquots. The first activity peak corresponded to the pI 5.6 (bound PtdCho) isoform; the second peak corresponded to the pI 5.2 (bound PtdIns) isoform.

a linear gradient from pH 6.0 to 4.5; the two mammalian isoforms were completely separated.

Storage of Phosphatidylinositol Transfer Protein. When a desirable level of purification has been achieved, the protein solution (50–100 μg/ml) is dialyzed against 10 mM HEPES–Na, 50 mM NaCl, 1 mM EDTA, and 0.02% NaN$_3$ (pH 7.4) to which an equal volume of glycerol has been added. This preparation may be stored at $-20°$ with little loss of catalytic activity. It is necessary, however, to remove the glycerol by dialysis or molecular sieve chromatography before activity measurements using the lactosylceramide–*Ricinus communis* agglutinin assay method, since glycerol interferes with the agglutination and centrifugation step. More recently, we have found that mixtures of purified PtdIns transfer protein and bovine plasma albumin (5 mg/ml) may be frozen and stored at $-20°$ or $-70°$ with excellent retention of catalytic activity.

Physical and Chemical Properties

Estimations of the apparent molecular mass of PtdIns transfer protein produce ambiguous results. Chromatography on a protein-calibrated Sephadex G-100 column (2 × 100 cm) yields an elution profile at 23–24 kDa for PtdIns transfer protein isolated from mammalian and insect tissues; detection is with transfer activity or immunoblot.[13] The unusual retention of PtdIns transfer protein on this cross-linked dextran matrix may be attributed to its high (10–11%) aromatic amino acid content.[7] Electrophoresis of eukaryotic cytosolic fractions or purified proteins on polyacrylamide gels in the presence of sodium dodecyl sulfate and 2-mercaptoethanol gives a relative mobility equivalent to 35–36 kDa; detection is with protein stain or immunoblot. Inclusion of 6 M urea in the polyacrylamide gel does not alter the apparent molecular size of PtdIns transfer protein. The amino acid sequence of rat PtdIns transfer protein, based on the cDNA structure, projects a mass of 31.9 kDa; of utmost significance, however, is the observation that the *in vitro* translation of the mRNA generated from the full-length cDNA, followed by immunoprecipitation and analysis by polyacrylamide gel electrophoresis, yields a protein of apparent mass 35.5 kDa.[7] The interaction of PtdIns transfer proteins with certain denaturants and chromatographic materials is anomalous and, clearly, must be taken into account during certain stages of protein purification.

Assay of Phosphatidylinositol Transfer Activity and Protein

Intermembrane Transfer Using Radiolabeled Phospholipids in Small Unilamellar Vesicles. For routine activity measurements during protein purifications, a system of phospholipid transport from donor to acceptor membranes offers ease and simplicity. Small unilamellar vesicles for transfer assays can be prepared by a variety of methods, including bath and probe sonication[14] and direct injection.[15] A critical step in the preparation of vesicles by injection is the dissolution of lipid mixtures containing PtdIns. After adding the appropriate volumes of stock solutions to a small glass tube, organic solvents are removed under nitrogen and finally under

[13] T. P. Dalton and G. M. Helmkamp, Jr., unpublished observation (1990).
[14] G. M. Helmkamp, Jr., *Biochemistry* **19,** 2050 (1980).
[15] T. Yoshimura and G. M. Helmkamp, Jr., *Biochim. Biophys. Acta* **793,** 463 (1984).

reduced pressure. Immediately thereafter, sufficient dimethyl sulfoxide is added to yield a lipid concentration of 150 mM. The tube is vortexed until all lipid is dissolved; gentle warming may facilitate this step. After sufficient ethanol is added to reduce the lipid concentration to 30 mM, the mixture is injected rapidly, in 100-μl aliquots, into assay buffer which has been equilibrated to 37°C. The final lipid concentration should be 2 mM. Although the dimethyl sulfoxide and ethanol may be removed by dialysis, these small amounts do not interfere significantly with phospholipid transfer measurements. Vesicles may be stored at 4° and should be used within 3 days.

Donor vesicles generally contain 5–10 mol % PtdIns, 6–8 mol % lactosylceramide, and 82–89 mol % PtdCho; acceptor vesicles contain 2–5 mol % PtdIns and 95–98 mol % PtdCho. Either phospholipid in the donor membrane population may be radiolabeled; at least 5000 disintegrations/min (dpm) should be used in each assay. Methods for the synthesis of ^{3}H-labeled PtdIns and ^{3}H-labeled PtdCho have been described.[16,17] It is desirable to add a small amount, usually less than 0.1 mol % and at least 1000 dpm per assay, of ^{14}C-labeled cholesteryl oleate or trioleoylglycerol[17] to the acceptor vesicles during their preparation. These lipids are not transferred by any known cytosolic lipid transfer protein and thus remain with the acceptor membrane population to permit an estimation of vesicle recovery. Both natural and synthetic phospholipids may be incorporated into the donor and acceptor vesicles; if only PtdCho transfer is to be measured, PtdOH may replace PtdIns in both membrane populations. Of course, other combinations of radioisotopes may be employed.

The ratio of donor to acceptor vesicles in a typical assay is 1 : 3 to 1 : 10. If the extent of transfer of donor vesicle phospholipid to acceptor vesicles is limited to 10–20%, there is reasonable linearity between protein and activity. The inclusion of bovine plasma albumin in the assay mixture, at a final level of 2 mg/ml, helps to maintain vesicle stability and to minimize spontaneous, intermembrane lipid transfers. With some crude cytosolic fractions or high-speed supernatants, interference with the vesicle–vesicle assay has been observed. Although the nature of this interference remains unidentified, the inclusion of additional controls, such as incubation of donor membranes and protein, with subsequent addition of acceptor membranes just prior to membrane separation, may be helpful. Lipids may also be extracted and characterized chromatographically to

---

[16] R. A. Demel, R. Kalsbeek, K. W. A. Wirtz, and L. L. M. van Deenen, *Biochim. Biophys. Acta* **466**, 10 (1977).

[17] A. M. Kasper and G. M. Helmkamp, Jr., *Biochemistry* **20**, 146 (1981).

verify structural integrity following vesicle preparation and transfer measurements.

Separation of donor and acceptor vesicles after incubation is accomplished by addition of *Ricinus communis* (castor bean) agglutinin,[17] a galactose-binding lectin. Although this protein is commercially available, the cost for routine transfer assays would be prohibitive. We have undertaken its purification from castor beans using ammonium sulfate fractionation and chromatography on BioGel A-0.5m and CM BioGel A (Bio-Rad) agarose matrices.[18] The agglutinin is distinguished from the highly cytotoxic ricin contaminant by a number of criteria: electrophoretic mobility on sodium dodecyl sulfate-polyacrylamide gels under nonreducing and reducing conditions (taking care to denature at 95° for only 1–2 min), hemagglutination activity, and precipitation of lactosylceramide-containing small unilamellar vesicles. It is possible to obtain approximately 500 mg of pure agglutinin from 250 g of beans. Solutions of *Ricinus communis* agglutinin may be stored at −20° for several years with no appreciable loss of activity. The glycolipid used in these assays is *N*-palmitoyllactosylceramide (Sigma, St. Louis, MO). However, other synthetic glycolipids could also be considered in the design of precipitable phospholipid vesicles.[19,20]

Fluorimetric Assay of Phosphatidylinositol Transfer Activity. For more specialized activity measurements, particularly with purified PtdIns transfer protein, spectrofluorometric assays have proved useful. The principal advantage of this technique is the continuous monitoring of fluorescence and, therefore, catalytic activity without the necessity of separating the donor and acceptor membrane populations at the conclusion of protein-mediated phospholipid transfer. Whereas the early discription of this approach employed phospholipid derivatives containing *cis*-parinaric acid,[21] more recent applications have utilized pyrenyl fatty acids.[2]

Small unilamellar vesicles are prepared by injection, with the donor population containing 10 mol % pyrenylacyl-PtdIns or pyrenylacyl-PtdCho, both of which may be synthesized from the respective lysophospholipids and commercially available pyrenyl fatty acids. Donor membranes also contain 10 mol % *N*-trinitrophenylphosphatidylethanolamine,

[18] D. B. Cawley, M. L. Hedblom, and L. L. Houston, *Arch. Biochem. Biophys.* **190,** 744 (1978).

[19] G. A. Orr, R. R. Rando, and F. W. Bangerter, *J. Biol. Chem.* **254,** 4721 (1979).

[20] P. Ghosh, B. K. Bachhawat, and A. Surolia, *Arch. Biochem. Biophys.* **206,** 454 (1981).

[21] P. Somerharju, H. Brockerhoff, and K. W. A. Wirtz, *Biochim. Biophys. Acta* **649,** 521 (1981).

[22] P. A. van Paridon, T. W. J. Gadella, Jr., P. J. Somerharju, and K. W. A. Wirtz, *Biochemistry* **27,** 6208 (1988).

an effective quencher of pyrene fluorescence. The assay is based on the time-dependent increase in the monomer fluorescence of the pyrene moiety as it is transferred to the unquenching environment of the acceptor membrane. For these measurements, the ratio of donor to acceptor membranes should be 1:25 to 1:50. Maximum transfer, in the absence of transfer protein, is estimated from a dilution of donor vesicles with a 1000-fold excess of PtdCho vesicles. Other advantages of the fluorimetric assay system include the ability to determine initial rates of phospholipid transfer and the ability to compare quantitatively the relative binding affinities of different molecular species of phospholipids.[22,23] On the other hand, far fewer measurements can be made in a given period of time.

Immunologic Detection of Phosphatidylinositol Transfer Protein. An alternative to activity measurements to monitor PtdIns transfer protein purification is the use of immunoblotting. The high degree of structural conservation among higher eukaryotic animals, coupled with the simple pattern of immunoreactive proteins observed in crude cell fractions, makes such an approach straightforward. We have employed two rabbit immunoglobulin G fractions in our routine immunoblot analyses. One antibody was raised against bovine PtdIns transfer protein and purified by ammonium sulfate fractionation and DEAE-cellulose chromatography; for some applications this immunoglobulin was also subjected to *Staphylococcus aureus* protein A-agarose chromatography. A second antibody was raised against a synthetic dodecapeptide corresponding to the C terminus of rat PtdIns transfer protein[13]; the peptide was covalently coupled to chicken egg albumin before immunization of rabbits. For most immunoblot applications the crude antipeptide serum can be used directly, generally at a 10,000-fold dilution. We have found it more reproducible, though more time-consuming, to analyze immunoblots of polyacrylamide gel electrophoretic separations rather than slot blots of relatively impure protein fractions. The electrophoretic approach has the added benefit of a determination of the apparent molecular mass, a value which is highly conserved among animal species.[2]

Related Phosphatidylinositol Transfer Proteins

In addition to the 35-kDa (apparent molecular mass) PtdIns transfer protein common to most higher eukaryotic cells, several mammalian tissues express related proteins. Cytosolic extracts prepared from human, bovine, buffalo, rat, and mouse testis contain a 41-kDa protein which

[23] J. Kasurinen, P. A. van Paridon, K. W. A. Wirtz, and P. Somerharju, *Biochemistry* **29**, 8548 (1990).

reacts with antibody to the bovine 35-kDa PtdIns transfer protein but not with antibody to the C-terminal peptide of rat PtdIns transfer protein.[24] Partial purification of this protein from adult rat testes indicated PtdIns and PtdCho transfer activity; separation from the 35-kDa protein was achieved by both molecular sieve and anion-exchange chromatography. None of the other adult rat tissues nor immature rat testis had detectable levels of the 41-kDa protein. Bovine brain contains a 38-kDa protein which is detected with antibody to the bovine 35-kDa PtdIns transfer protein.[25] The two proteins could be separated by hydroxylapatite chromatography. Although the 38-kDa protein had an extremely low capacity to bind and transfer PtdIns, limited proteolysis with trypsin generated a 35-kDa protein with enhanced catalytic properties. These data are consistent with a precursor–product relationship for bovine brain PtdIns transfer proteins. The profound sensitivity of immunologic detection methods has made possible further insights into the tissue distribution and molecular species of PtdIns transfer proteins, the preliminary results of which await verification with the tools of molecular biology.

[24] P. J. Thomas, B. E. Wendelburg, S. E. Venuti, and G. M. Helmkamp, Jr., *Biochim. Biophys. Acta* **982,** 24 (1989).
[25] W. M. R. van den Akker, J. Westerman, T. W. J. Gadella, Jr., K. W. A. Wirtz, and G. T. Snoek, *FEBS Lett.* **276,** 123 (1990).

[60] Phospholipid Transfer Proteins from Yeast

By Fritz Paltauf and Günther Daum

Introduction

Baker's yeast, *Saccharomyces cerevisiae*, contains at least two cytosolic proteins which catalyze transport of phospholipids between natural membranes and phospholipid vesicles *in vitro*. One protein is specific for phosphatidylinositol (PtdIns) and phosphatidylcholine (PtdCho) and closely resembles PtdIns-transfer protein from mammalian tissues with respect to size, isoelectric point, substrate specificity, and inhibition by negatively charged phospholipids.[1,2] However, the mammalian and yeast

[1] G. Szolderits, A. Hermetter, F. Paltauf, and G. Daum, *Biochim. Biophys. Acta* **986,** 301 (1989).
[2] G. M. Helmkamp, Jr., *Chem. Phys. Lipids* **38,** 51 (1985).

proteins share no sequence similarity with respect to primary structure.[3,4] The second yeast phospholipid transfer protein has a broader substrate specificity in that it catalyzes preferentially the transfer of phosphatidylserine (PtdSer) and also phosphatidylethanolamine (PtdEtn), cardiolipin, phosphatidic acid, and ergosterol. In contrast to the mammalian nonspecific lipid transfer protein, the yeast protein does not mediate intermembrane movement of PtdIns or PtdCho. Thus, a mammalian counterpart to the yeast PtdSer transfer protein has not yet been identified. The physiological function of phospholipid transfer proteins is still unclear.

Yeast is considered a useful biological model with which to study the role of these proteins *in vivo* by a combination of biochemical and molecular genetic techniques. Recent experiments revealed that the PtdIns transfer protein is essential for cellular growth,[5] and it is identical to the *SEC 14* gene product which is part of the protein secretory machinery and acts in the Golgi complex.[6]

Assay Procedure

Activities of yeast phospholipid transfer proteins (PL-TP) are assayed either by measuring the transfer of radiolabeled phospholipids between phospholipid vesicles and natural membranes (mitochondria or microsomes) following established procedures[7,8] or by a fluorescence method employing pyrene-labeled phospholipids as substrates.[9,10]

Radiochemical Assay. Radioactively labeled phospholipids can be obtained commercially, prepared by chemical synthesis,[11] or isolated from yeast cells labeled with the appropriate precursor.[12] Phospholipid vesicles used as donor membranes contain the radiolabeled phospholipid, either as such in the case of PtdCho, or as mixtures (5 mol %) with unlabeled

[3] V. A. Bankaitis, J. R. Aitken, A. E. Cleves, and W. Dowhan, *Nature (London)* 347, 561 (1990).

[4] S. K. Dickeson, C. N. Lim, G. T. Schuyler, T. P. Dalton, G. M. Helmkamp, Jr., and L. R. Yarbrough, *J. Biol. Chem.* 264, 16557 (1989).

[5] J. F. Aitken, G. P. H. van Heusden, M. Temkin, and W. Dowhan, *J. Biol. Chem.* 265, 4711 (1990).

[6] V. A. Bankaitis, D. E. Malehorn, S. D. Emr, and R. Greene, *J. Cell Biol.* 108, 1271 (1989).

[7] G. M. Helmkamp, Jr., M. S. Harvey, K. W. A. Wirtz, and L. L. M. van Deenen, *J. Biol. Chem.* 249, 6382 (1974).

[8] P. E. DiCorleto, J. B. Warach, and D. B. Zilversmit, *J. Biol. Chem.* 264, 7795 (1979).

[9] P. J. Somerharju, D. van Loon, and K. W. A. Wirtz, *Biochemistry* 26, 7139 (1987).

[10] P. A. van Paridon, T. W. J. Gadella, Jr., and K. W. A. Wirtz, *Biochim. Biophys. Acta* 943, 76 (1988).

[11] F. Paltauf and A. Hermetter, this series, Vol. 197, p. 134.

[12] G. Daum, H. G. Schwelberger, and F. Paltauf, *Biochim. Biophys. Acta* 879, 240 (1986).

PtdCho in the case of PtdIns and PtdSer.[13] Mitochondria are isolated[14] from yeast homogenates (see below) by centrifugation of a 1000 g (5 min) supernatant at 10,000 g for 10 min at 4°. The mitochondrial pellet is washed twice by resuspension and recentrifugation at 10,000 g for 10 min. Mitochondria suspended in 0.6 M mannitol, 10 mM Tris-HCl, pH 7.4, can be stored at −70°.

Phosphatidylserine Transfer Protein Assay. [³H]PtdSer/PtdCho vesicles (10 nmol PtdSer), mitochondria (0.25 mg protein), and transfer protein are incubated in a total volume of 0.5 ml of 10 mM Tris-HCl, pH 7.4, 1 mM EDTA[15] at 35° for 30 min. Mitochondria are precipitated by centrifugation at 10,000 g for 10 min, and radioactivity is measured in the supernatant and/or the washed mitochondrial pellet. To correct for nonspecific adsorption of donor phospholipid vesicles to mitochondria, [¹⁴C]triolein can be incorporated into phospholipid vesicles as a nontransferable marker.

Phosphatidylinositol Transfer Protein Assay. Measurements are carried out essentially as with PtdSer-TP, using 10 mM Tris-HCl, pH 7.2, without EDTA as buffer.

Fluorescence Assay. Details of PL-TP assays using fluorescent (mainly pyrenedecanoyl-substituted) phospholipid substrates are described elsewhere in this volume.[16] This assay measures the increase in pyrene monomer fluorescence if the fluorescent phospholipid is transferred from donor membranes (concentrated state) to acceptor membranes (where the fluorophore is highly diluted). Some of the fluorescently labeled substrates, for example, 1-acyl-2-pyrenedecanoyl-*sn*-glycerophosphocholine and 1-acyl-2-pyrenedecanoyl-*sn*-glycerophosphoinositol, are commercially available. The 1-acyl-2-pyrenedecanoyl-*sn*-glycerophosphoserine or 1-acyl-2-pyrenedecanoyl-*sn*-glycerophosphoethanolamine required for the assay of PtdSer-TP activity can be prepared by phospholipase D-catalyzed transphosphatidylation of the respective choline derivative.[17] The substrates can be purified by thin-layer chromatography on silica gel 60 plates, using chloroform–methanol–25% aqueous ammonia (65 : 35 : 5, v/v) as the developing solvent. Bands containing fluorescent phosphatidylserine at R_f 0.17 or phosphatidylethanolamine at R_f 0.46 are scraped off the plate and extracted with chloroform–methanol (1 : 4, v/v). Purity of the substrates is tested by analytical thin-layer chromatography and by estimating the

---

[13] Vesicles prepared from solely PtdSer or PtdIns cannot be used since both yeast PL-TPs are inhibited by negatively charged phospholipids.

[14] G. Daum, P. C. Böhni, and G. Schatz, *J. Biol. Chem.* **257,** 13028 (1982).

[15] Activity of the protein is reduced in the presence of polyvalent cations. In the absence of EDTA or at higher concentrations (10 mM), significantly lower transfer rates are observed.

[16] P. J. Somerharju, J. Kasurinen, and K. W. A. Wirtz, this volume [58].

[17] H. Eibl and S. Kovatchev, this series, Vol. 72, p. 632.

fluorophore-to-phosphate ratio, using a molar extinction coefficient of $42,000 \, mol^{-1} \, cm^{-1}$ for quantitating the pyrenedecanoyl moiety.[9]

Purification Procedure

Both yeast PL-TPs are isolated from cells grown on a semisynthetic medium with lactate as the carbon source and harvested at the mid logarithmic phase. Cell homogenates are prepared after spheroplasting rather than by disintegrating cells in a cell homogenizer in the presence of glass beads in order to minimize contamination resulting from organelle damage.[18] The procedure described here for the isolation of the PtdIns-TP from yeast is a modification[1] of the original method described by Daum and Paltauf.[19] The procedure for the isolation of the PtdSer-TP was developed in the authors' laboratory.[20]

Yeast Strain, Culture Conditions, and Preparation of 100,000 g Supernatant

The wild-type strain *S. cerevisiae* D 273-10B (ATCC, Rockville, MD, 25657) is grown aerobically on a medium containing, per liter, 3 g of yeast extract (Difco, Detroit, MI), 1 g of glucose, 1 g of KH_2PO_4, 1 g of NH_4Cl, 0.5 g of $CaCl_2 \cdot 2H_2O$, 0.5 g of NaCl, 0.6 g of $MgSO_4 \cdot H_2O$, 0.3 ml of 1% $FeCl_3$, and 22 ml of 90% lactic acid. The final pH is adjusted to 5.5 with KOH. Cells are grown to the mid logarithmic phase (16 hr) to a yield of approximately 5 g wet cells per liter. Cells are harvested by centrifugation (5 min at 3000 *g*), washed once with distilled water, suspended to 0.5 g wet weight/ml in 0.1 *M* Tris $\cdot$ H_2SO_4, pH 9.4, 10 m*M* dithiothreitol, and incubated for 10 min at 30°. Then cells are washed once with 1.2 *M* sorbitol and suspended in 1.2 *M* sorbitol, 20 m*M* potassium phosphate, pH 7.4, to give 0.15 g of wet weight/ml. Zymolyase 20,000 (1 mg/g of cell wet weight) is added, and the suspension is incubated at 30° with gentle shaking. After 50–60 min cells are converted to spheroplasts,[21] which are harvested by centrifugation for 10 min at 1000 *g* at room temperature and washed twice with 1.2 *M* sorbitol.

Spheroplasts are then suspended in a 3- to 4-fold volume (based on

---

[18] In the purification procedure described by Aitken *et al.*,[5] cells are disintegrated for 10 min in a Bead-Beater (BioSpec Products, Bartlesville, OK) containing glass beads (0.5 mm diameter).

[19] G. Daum and F. Paltauf, *Biochim. Biophys. Acta* **794**, 385 (1984).

[20] G. Lafer, G. Stolderits, F. Patlauf, and G. Daum, *Biochim. Biophys. Acta*, in press.

[21] Conversion to spheroplasts can be checked by osmotic lysis. Addition of 10 μl of the spheroplast suspension to 1 ml water, followed by brief shaking on a vortex mixer, gives a clear solution when spheroplasting is complete.

cell wet weight) of 0.6 *M* mannitol, 10 m*M* Tris-HCl, pH 7.4, 1 m*M* phenylmethylsulfonyl fluoride. Glass beads (0.3–0.4 mm diameter, weight equal to spheroplast wet weight) are added, and the suspension is shaken by hand for three 1-min periods, with intermittent 1-min cooling periods on ice. Glass beads are removed by sedimentation, and the sediment is washed once with the mannitol–Tris buffer (one-half the original volume). The supernatants are combined and freed of membranous particles by sequential centrifugation at 1000 *g* (5 min), 10,000 *g* (10 min), and 100,000 *g* (60 min). Centrifugation and all subsequent steps are carried out at 0°–4°.

Isolation of Phosphatidylinositol Transfer Protein

Step 1: Fractionated Ammonium Sulfate Precipitation. Solid ammonium sulfate (29.1 g/100 ml) is added in small portions with gentle stirring to the yeast 100,000 *g* supernatant to give 50% saturation. After 5–10 hr on ice, precipitated proteins are removed by centrifugation at 23,000 *g* for 12 min. The resulting supernatant is saved, and additional ammonium sulfate (26.8 g/100 ml) is added for 90% saturation. Precipitated proteins are collected by centrifugation as described above and suspended in 10 m*M* Tris-HCl (pH 7.4), 0.02% NaN$_3$ (in the following referred to as standard buffer). Insoluble material is removed by centrifugation (23,000 *g*, 12 min) and discarded.

Step 2: Sephadex G-75 Chromatography. Portions of 20–25 ml of the sample obtained in Step 1 are applied to a Sephadex G-75 column (2.6 × 75 cm). Fractions of 5 ml are collected after elution with the standard buffer at a flow rate of 25 ml/hr. Phosphatidylcholine transfer activity is found in fractions 29 to 38.

Step 3: DEAE-Sephacel Chromatography. Pooled active fractions of the previous step are adjusted to 70 m*M* NaCl and applied to a DEAE-Sephacel column (1.6 × 32 cm) equilibrated with 70 m*M* NaCl in standard buffer. Unbound proteins are eluted with 150 ml of the same buffer at a flow rate of 25 ml/hr. Elution is continued with 250 ml of a linear salt gradient (70–120 m*M* NaCl), and fractions of 4.2 ml are collected. Active transfer protein is eluted with 100–110 m*M* NaCl.

Step 4: Sephadex G-75 Chromatography. Portions (20–25 ml) of pooled active fractions from the previous step are applied to a Sephadex G-75 column and eluted as described for Step 2. Transfer activity is found in fractions 29 to 39 with a maximum in fractions 34–35, corresponding to an apparent molecular mass of 34–37 kDa.

Step 5: Phenyl-Sepharose Chromatography. Pooled active fractions from Step 4 are dialyzed against 1 m*M* sodium phosphate (pH 7.2), 0.02% NaN$_3$ and adjusted to 0.1 *M* NaCl. The sample (50–100 ml) is applied to a phenyl-Sepharose CL-4B column (1 × 11 cm) equilibrated with 1 mM sodium phosphate (pH 7.2), 0.02% NaN$_3$, 0.1 M NaCl. Unbound proteins are eluted with the starting buffer at a flow rate of 20 ml/hr. Elution is continued with a step gradient using 30 ml of 1 mM sodium phosphate (pH 7.2), 0.02% NaN$_3$, 50 mM NaCl; 30 ml of 20 mM NaCl in sodium phosphate buffer; and 100 ml of buffer without NaCl. Fractions of 6.2 ml are collected; the active transfer protein is eluted between fractions 2 and 19.

The protein solution can be concentrated after the last step of purification using Centricon 30 (Amicon, Danvers, MA). During this step 10–20% of the protein may be lost, probably owing to irreversible binding to the filter material. The purified protein can be stored at 4° in the presence of 10% glycerol at protein concentrations of 5–100 μg/ml. As shown in Table I, PtdIns-TP is obtained in 19% yield with a purification factor of 2760. Sodium dodecyl sulfate (SDS)-polyacrylamide gel electrophoresis shows a single protein band with an apparent molecular mass of 35 kDa.

Aitken et al.[5] purified PtdIns-TP from S. cerevisiae strain DL 1 grown aerobically in YPD medium to the mid logarithmic phase. Their procedure included preparation of a 100,000 g supernatant from cells disintegrated in a Bead-Beater containing glass beads (0.5 mm diameter), followed by DEAE-Sephacel and Matrex Red A chromatography and HPLC-DEAE. The final yield was 11% and purification was 1200-fold.

Isolation of Phosphatidylserine Transfer Protein

Growth of cells, preparation of the 100,000 g supernatant, and fractionated ammonium sulfate precipitation are the same as described above for the preparation of the PtdIns-TP. Protein precipitated at 90% saturation

TABLE I
PURIFICATION OF PHOSPHATIDYLINOSITOL TRANSFER PROTEIN FROM YEAST

Step	Protein (mg)	Specific activity (nmol PtdCho/min per mg)	Recovery (%)	Purification factor
100,000 g Supernatant	7840	0.0067	100	1.0
Ammonium sulfate 50–90% saturation	2990	0.0134	76	2.0
Sephadex G-75	1550	0.0214	63	3.2
DEAE-Sephacel	75	0.307	44	46
Sephadex G-75	12.3	1.27	30	190
Phenyl-Sepharose CL-4B	0.54	18.5	19	2760

with ammonium sulfate is dissolved in 10 mM Tris-HCl, pH 7.4, 0.02% NaN$_3$, 1 mM EDTA (further referred to as TE buffer) to a concentration of 35 mg/ml.

Step 1: Sephadex G-75 Chromatography. Portions of 25 ml of the active fraction obtained by ammonium sulfate precipitation are applied to a Sephadex G-75 column (2.6 × 75 cm). Fractions of 5.4 ml are collected after elution with TE buffer at a rate of 25 ml/hr. PtdSer transfer activity is found in fractions 28–38. The increase (122%) in total recovery is most likely due to the removal of (specific or nonspecific) inhibitory factors during gel filtration.

Step 2: Matrex Gel Red A Chromatography. Glycerol (10%) is added to the pooled fractions obtained in Step 1. Portions (50 ml, ~500 mg protein) are applied to a Matrex Gel Red A column (Amicon; 2 × 8 cm), and unbound proteins are eluted with TE buffer containing 10% glycerol. Elution is continued with a linear NaCl gradient (0–700 mM) at a rate of 20 ml/hr. Fractions of 5.4 ml are collected, and transfer activity is found in fractions 27–37 corresponding to a NaCl concentration of 200–280 mM. Pooled fractions are dialyzed against 5 mM KH$_2$PO$_4$, pH 7.4, 0.02% sodium azide, 10% glycerol (further referred to as KP$_i$ buffer).

Step 3: Hydroxylapatite Adsorption Chromatography. Portions (50 ml, ~20 mg protein) of the dialyzed fraction obtained in Step 2 are applied to a column (2 × 8 cm) filled with hydroxylapatite (Bio-Rad, Richmond, CA) equilibrated with KP$_i$ buffer. Unbound proteins are eluted with KP$_i$ buffer, and elution is continued with 200 ml of a linear potassium phosphate gradient (5–150 mM). Fractions of 4 ml are collected. Transfer protein appears in fractions 29–36 at 80–120 mM KH$_2$PO$_4$.

Step 4: Sephadex G-75 Chromatography. Portions (25 ml, ~1.25 mg protein) of the pooled fractions obtained in Step 3 are applied to a Sephadex G-75 column (2.6 × 75 cm), and proteins are eluted as described for Step 1. Transfer protein elutes with fractions 40–48. The increase in elution volume as compared to Step 1 might be explained by dissociation of the protein either from homo- or heteroaggregates at this stage of purification.

Step 5: DEAE-Sephacel Chromatography. The combined fractions (100 ml, ~1 mg protein) obtained in Step 4 are applied to a DEAE-Sephacel column (1.6 × 32 cm) equilibrated with TE buffer without glycerol. Proteins are eluted with 500 ml of a linear salt gradient (0–100 mM NaCl in TE buffer) at a rate of 40 ml/hr. Fractions (5.4 ml) are collected, and transfer activity is eluted with fractions 59–68, corresponding to 64–74 mM NaCl. The protein obtained by this procedure (Table II) appears as a single band on a 12.5% SDS-polyacrylamide gel, with an apparent molecular weight of 35,000.

TABLE II
PURIFICATION OF YEAST PHOSPHATIDYLSERINE TRANSFER PROTEIN

Step	Protein (mg)	Specific activity (U/mg)	Recovery (%)	Purification factor
100,000 g Supernatant	14,400	15.3	—	1.0
Ammonium sulfate 50–90% saturation	7,525	71.4	100	4.7
Sephadex G-75	5,778	113	122	7.4
Matrex Red A	394	1,370	83	90
Hydroxylapatite	60	3,400	31	222
Sephadex G-75	16	13,580	33	890
DEAE-Sephacel	0.8	46,875	7	3064

Properties of Yeast Phospholipid Transfer Proteins

Phosphatidylinositol Transfer Protein

The molecular weight (35K) of the protein as determined by SDS-poly-acrylamide gel electrophoresis (12.5% gel, run for 1 hr at 180 V) conforms to the molecular weight calculated from the amino acid sequence (304 amino acids, molecular weight 35,032) which was derived from the nucleotide sequence of the *PIT* structural gene.[3,6] The pI of the protein is 5.3; the pH optimum of PtdIns/PtdCho transfer lies between pH 7 and 8, which is also the range of optimum stability of the protein. At a pH below the isoelectric point the protein is unstable.[1] The temperature optimum for transfer is between 40° and 45°. Heating to 50° for 10 min reduces transfer activity by 50%. PtdIns-TP resists treatment with trypsin, chymotrypsin, and papain but is digested by pronase.[12] In standard buffer (10 mM Tris-HCl, pH 7.4, 0.02% NaN$_3$) containing 10% glycerol the protein can be stored at 4° for more than 1 year with only minor (<10%) loss of activity.

PtdIns-TP forms a stable 1 : 1 complex with its substrate, PtdIns, which is also the preferred substrate in a transfer assay using unilamellar phospholipid vesicles as donor and acceptor membranes.[1] Other lipids tested (PtdEtn, PtdSer, cardiolipin, phosphatidic acid, sphingomyelin, neutral lipids) are not transferred. The presence of negatively charged phospholipids in acceptor phospholipid vesicles inhibits PtdCho transfer in the following order[22]: PtdSer (7) > phosphatidylglycerol (8) > PtdIns (10) > cardiolipin (16) > phosphatidic acid (50).

---

[22] The numbers in parentheses refer to the concentrations (mol %) of the negatively charged phospholipid in PtdCho acceptor membranes resulting in 50% inhibition of PtdCho transfer from PtdCho donor vesicles in a fluorescence assay.

Phosphatidylserine Transfer Protein

The molecular weight of the PtdSer-TP as determined by SDS-poly-acrylamide gel electrophoresis is 35K. Antibodies raised against the yeast PtdIns-TP do not cross-react with the PtdSer-TP. The protein is cold-labile (50% inactivation after freezing) and is completely inactivated after heating to 50° for 10 min or to 60° for 2 min. Treatment with the proteinases trypsin, proteinase K, and pronase completely abolishes transfer activity. Ca^{2+} and Mn^{2+} at 0.5 mM concentrations inhibit protein-catalyzed PtdSer transfer from phospholipid vesicles to mitochondria and between phospholipid vesicles.

Phosphatidylserine is the preferred substrate, but other lipids are also transferred, the relative transfer rates being as follows: PtdSer 100%, PtdEtn 30%, cardiolipin 14%, phosphatidic acid 10%, ergosterol 10%. Inhibition of transfer by negatively charged phospholipids is similar to that observed for the PtdIns-TP.

[61] Phospholipid Transfer Proteins from Higher Plants

By Chantal Vergnolle, Vincent Arondel, Alain Jolliot, and Jean-Claude Kader

Introduction

Plant cells contain soluble proteins able to facilitate *in vitro* bidirectional movements of phospholipids between membranes.[1,2] These proteins, called phospholipid transfer proteins (PLTP), have been purified and characterized from plant tissues and also from animal tissues or yeast (see [58]–[60], this volume). PLTPs are assumed to participate in the intracellular distribution of phospholipids and could be involved in membrane biogenesis or in the function of membrane-bound enzymes using lipids as substrates.[1–4] An additional interest of PLTP is their ability to replace the endogenous phospholipids of a membrane by phospholipids originating from other membranes. PLTP can thus modify the lipid compo-

[1] J. C. Kader, *in* "Subcellular Biochemistry" (H. J. Hilderson, ed.), Vol. 16, p. 69. Plenum, New York, 1990.

[2] V. Arondel and J. C. Kader, *Experientia* **46,** 579 (1990).

[3] K. W. A. Wirtz, *in* "Lipid–Protein Interactions" (P. C. Jost and O. H. Griffith, eds.), p. 151. Elsevier, Amsterdam, 1982.

[4] J. C. Kader, D. Douady, and P. Mazliak, *in* "Phospholipids, a Comprehensive Treatise" (J. N. Hawthorne and G. B. Ansell, eds.), p. 279. Elsevier, Amsterdam, 1982.

sition of a membrane. This allows the study of the phospholipid–protein interactions within membranes and of the effects of changes in lipid concentrations on the functional properties of membranes. PLTP can also been used as mild agents to determine the asymmetry of the lipid composition of the membrane leaflets. Finally, these proteins are useful "tools" for inserting exogenous phospholipids, for example, those containing fluorescent compounds, in order to determine the mobility of lipids within membranes.[1–4]

PLTP from plants are particularly suitable for these purposes because (1) the starting material (generally seeds or leaves) is abundant; (2) PLTP are stable for weeks; (3) plant PLTP have a broad specificity for phospholipids, allowing their use for a wide range of these lipids. In this chapter the procedure for the isolation of PLTP from sunflower seedlings is presented. The same general fractionation procedure has been used for other plant organs: maize[5] and castor bean[6] seedlings and spinach leaves.[7]

Assay

Reagents

Buffer A: 0.4 M sucrose, 6 mM EDTA, 1 mM cysteine chloride, 8 mM 2-mercaptoethanol, 0.1 M Tris-HCl (pH 7.5)

Buffer B: 0.25 M sucrose, 1 mM EDTA, 10 mM Tris-HCl (pH 7.2)

Egg yolk phosphatidylcholine

Labeled compounds from Amersham (UK): *myo*-[2-³H]inositol (144 GBq/mmol); 1-[1-³H]ethanol-2-amine hydrochloride (370 GBq/ mmol); [*methyl*-³H]choline chloride (2.85 TBq/mmol); cholesteryl [1-¹⁴C]oleate (1.85 GBq/mmol)

Principle. Liposomes, containing [³H]phosphatidylcholine (the phospholipid to be transferred) and cholesteryl [¹⁴C]oleate (not transferred by the protein), are incubated with mitochondria in the presence of various amounts of PLTP. The increase in ³H label in the mitochondria recovered by centrifugation indicates a transfer of phosphatidylcholine from liposomes to mitochondria. The ¹⁴C radioactivity (generally low) indicates the extent of cross-contamination of mitochondria by intact liposomes. Other lipid transfer assays have been used either with other membrane systems

[5] D. Douady, M. Grosbois, F. Guerbette, and J. C. Kader, *Biochim. Biophys. Acta* **710,** 143 (1982).

[6] S. Watanabe and M. Yamada, *Biochim. Biophys. Acta* **876,** 116 (1986).

[7] J. C. Kader, M. Julienne, and C. Vergnolle, *Eur. J. Biochem.* **139,** 411 (1984).

(two categories of liposomes,[8] liposomes–chloroplast envelope[9]) or with other phospholipids, such as spin-labeled (Devaux, unpublished) or fluorescent (Moreau, unpublished) ones.

Isolation of Mitochondria. Mitochondria are isolated by sucrose gradient centrifugation from 4-day-old maize seedlings as previously described.[5] The seedlings (generally 500 g of fresh weight) are ground in buffer A and the mitochondria are recovered and purified by different centrifugation steps, including sucrose gradient centrifugation. Mitochondrial pellets are suspended in buffer B at a final concentration of 12 mg protein/ml. Mitochondria can be kept up to 1 month at $-20°$ before use.

Preparation of Labeled Phospholipids. [^{3}H]Phosphatidylcholine, [^{3}H]phosphatidylinositol, and [^{3}H]phosphatidylethanolamine are prepared by incubating 20 g of potato tuber slices (1 mm thick) for 16 hr at 25°, with constant shaking, either with [*methyl*-^{3}H]choline chloride, [^{3}H]ethanolamine, or *myo*-[2-^{3}H]inositol. Generally, 3.7 MBq of each labeled compound is used. Labeled lipids are extracted as previously described.[10]

Preparation of Liposomes. Liposomes are prepared by mixing, for one assay, 260 nmol of egg yolk phosphatidylcholine, 1 nmol of cholesteryl [1-^{14}C]oleate, and 10 nmol of labeled phospholipid ($\sim$740 Bq). Generally, the compounds necessary for 10 assays are introduced into a conical flask. After evaporating the solvent under a stream of nitrogen, 2.5 ml of buffer B is added. The flask is vigorously shaken, and the lipid dispersion is then introduced into a plastic tube and irradiated for 30 min by a Branson sonifier under a stream of nitrogen. The final volume is adjusted to 2.5 ml with buffer B.

Transfer Assay. Phospholipid transfer activity is determined using ^{3}H-/^{14}C-labeled liposomes as donor membranes and mitochondria as acceptor ones. According to Fig. 1, the labeled liposomes are incubated with maize mitochondria in the absence or presence of PLTP. Incubation is started with the addition of the liposomes and is carried out with constant stirring. After centrifugation at 4°, the mitochondrial pellets are suspended in 0.5 ml of 1% (w/v) Triton X-100, and the suspension is counted with a Kontron (Zurich, Switzerland) scintillation system using 10 ml of toluene scintillation {0.4% (w/v) 2,5-diphenyloxazole, 0.01% (w/v) 1,4-bis[2-(5-phenyloxazolyl)benzene]} containing 33% (v/v) Triton X-100.

Units of Transfer. The transfer activity is expressed as the percentage of liposomal [^{3}H]phosphatidylcholine recovered in mitochondria after in-

[8] F. Guerbette, D. Douady, M. Grosbois, and J. C. Kader, *Physiol. Veg.* **19**, 467 (1981).

[9] M. Miquel, M. A. Block, J. Joyard, A. J. Dorne, J. P. Dubacq, J. C. Kader, and R. Douce, *Biochim. Biophys. Acta* **937**, 219 (1987).

[10] E. G. Bligh and W. J. Dyer, *Can. J. Biochem. Physiol.* **37**, 911 (1959).

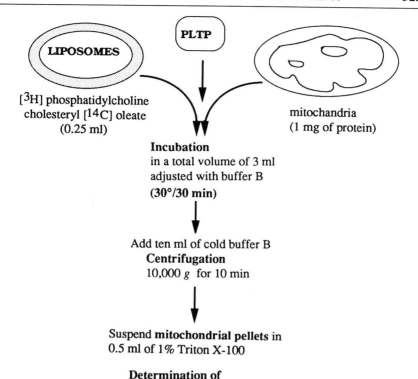

[³H] phosphatidylcholine
cholesteryl [¹⁴C] oleate
(0.25 ml)

mitochandria
(1 mg of protein)

Incubation
in a total volume of 3 ml
adjusted with buffer B
(30°/30 min)

Add ten ml of cold buffer B
Centrifugation
10,000 *g* for 10 min

Suspend **mitochondrial pellets** in
0.5 ml of 1% Triton X-100

Determination of
³H/¹⁴C **radioactivities**

FIG. 1. Assay of phospholipid transfer protein (PLTP).

cubation in the presence of PLTP. This value is corrected for cross-contamination of mitochondria by liposomes by subtracting the percentage obtained in the absence of PLTP. The transfer activity is also calculated, from this corrected percentage of transfer, as nanomoles of phospholipids transferred by PLTP. One unit of PLTP is defined as the amount which transfers 1 nmol of phosphatidylcholine per minute, and the specific activity is expressed as units per milligram of protein.

Purification Procedure

Reagents

Buffer C: 1 mM phenylmethylsulfonyl fluoride in buffer A
Buffer D: 8 mM 2-mercaptoethanol, 3 mM sodium azide, 10 mM potassium phosphate (pH 7.2)

TABLE I

PURIFICATION OF PHOSPHOLIPID TRANSFER PROTEIN FROM SUNFLOWER SEEDLINGS

Step	Protein (mg)	Specific activity (units/mg protein)	Recovery (%)	Purification factor
Crude extract	6375	0.06	—	—
Sephadex G-75	425	0.72	80	12
CM2 fraction	47	2.1	25.6	35
Sephadex G-50	22	3.6	20.7	60

Buffer E: 0.25 M potassium phosphate in buffer D
DEAE-Trisacryl (IBF, Paris, France)
Sephadex G-50, Sephadex G-75, carboxymethyl-Sepharose CL-6B
   (Pharmacia, Uppsala, Sweden)

Step 1: Preparation of Crude Extract. Sunflower seeds (*Helianthus annuus* var. Rodeo) (500 g) provided by CETIOM (Paris, France) are soaked for 3 hr and germinated at 30° on wet vermiculite for 4 days in darkness. The roots are discarded. The aerial parts are then ground in 1 liter of buffer C with a Waring blender for 15 sec at maximum speed. The pH is maintained at 7.5 by addition of 1 M Tris. The homogenate is then squeezed through two layers of Miracloth (Calbiochem, San Diego, CA) and centrifuged for 10,000 g for 20 min. The pH of the supernatant is then adjusted to 5.1 with 2 M HCl in order to aggregate the residual membranes. The homogenate is then stirred for 15 min and centrifuged for 10,000 g for 20 min. The pH of the supernatant is readjusted to 7.5. Proteins are allowed to precipitate by adding 52 g of $(NH_4)_2SO_4$ per 100 ml. Stirring is continued overnight, followed by centrifugation at 10,000 g for 20 min. The pellets are suspended in a minimal volume of buffer D, dialyzed against the same buffer for 5 hr, and centrifuged at 10,000 g for 20 min. The clear brownish supernatant is used as the crude extract. When assayed for lipid transfer, a specific activity of 0.06 units/mg of protein is found (Table I).

Step 2: Gel Filtration through Sephadex G-75. The crude extract is loaded onto a column of Sephadex G-75 (70 × 5 cm) equilibrated with buffer D. The proteins are eluted with the same buffer at a rate of 80 ml/hr. Fifteen-milliliter fractions are collected, and the activity is measured on 3 ml of each fraction. A major peak of activity is detected in the group of fractions 80 to 100 (Fig. 2A). The active fractions are pooled. The specific activity is 0.72 units/mg of protein with a purification factor of 12.

Step 3: Ion-Exchange Chromatography. The active fractions from the Sephadex column are loaded onto a DEAE-Trisacryl column (25 × 2.6 cm) connected to a carboxymethyl-Sepharose column (25 × 2.6 cm) previously

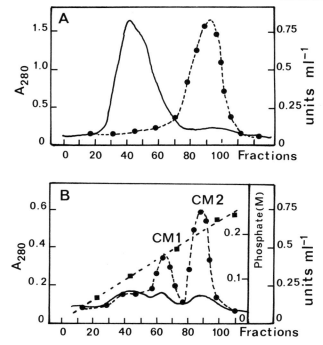

FIG. 2. Purification of phospholipid transfer protein from sunflower seedlings. The elution profiles from the Sephadex G-75 column (A) (Step 2 of the purification procedure) and the carboxymethyl-Sepharose column (B) (Step 3) are presented. The activity of transfer of phosphatidylcholine is expressed as units/ml (●). The shape of the gradient is indicated from conductivity measurements (■). The absorbance (solid line) was measured at 280 nm.

equilibrated with buffer D. After washing with 1 liter of the same buffer, the two columns are disconnected, and the carboxymethyl-Sepharose column is kept. A linear gradient prepared with 1 liter of buffer D and 1 liter of buffer E is applied onto this column. The flow rate is 60 ml/hr and 15-ml fractions are collected. The fractions containing the transfer activity are eluted. Two groups of active fractions, called CM1 and CM2, are observed, eluted by phosphate concentrations of approximately 160 and 200 mM, respectively (Fig. 2B). The CM2 fraction, which is the most active one, is kept. Its specific activity is equal to 2.1 units/mg of protein with a purification factor of 35.

Step 4: Gel Filtration through Sephadex G-50. After concentration of the CM2 fraction in a dialysis bag placed in solid sucrose, the solution is chromatographed through a Sephadex G-50 column (2.6 × 70 cm). After elution at 90 ml/hr by buffer D, the activity is detected in the 15-ml fractions around tube 26. The pooled active fractions have a specific activity of 3.6 units/mg of protein with a purification factor of 60. The pure protein is stored at 4° until use.

Properties

Purity and Molecular Mass. The final preparation was pure as judged by SDS-PAGE electrophoresis, which exhibited an unique band corresponding to an apparent molecular mass of 9 kDa. The purity was also checked by FPLC (fast protein liquid chromatography) on a Mono S column (Pharmacia) equilibrated with buffer D minus glycerol. Only a major peak eluted at about 0.25 M NaCl was observed when the column was eluted with a 0–0.5 M NaCl gradient performed in the equilibrating buffer.

The CM1 peak also exhibited the same 9-kDa band, with few other ones. This suggests that CM1 and CM2 are isoforms. The molecular mass of 9 kDa is in good agreement with the values obtained with proteins from other plant sources.[1,5–7] By contrast, the molecular masses of phospholipid transfer proteins from animal sources vary from 11.2 (nonspecific proteins) to 33.5 kDa (proteins specific for phosphatidylinositol).[3,4]

Isoelectric Point. The isoelectric point of the pure protein is high as evidenced by its tight binding on carboxymethyl-Sepharose. This was confirmed by a chromatofocusing method (Pharmacia) which gave a value of 9.5 (data not shown). The CM1 fractions are also basic proteins, but they are clearly separated from CM2 ones. CM1 and CM2 are isoforms, either differing by their primary structure or by endogenous binding of lipids. It is to be noted that a binding of phospholipids or of fatty acids has been observed for various plant PLTP.[5,7,11–13] In the case of animal tissues, phosphatidylinositol transfer proteins are present as two isoforms, differing in their binding of phospholipids.[14]

Stability. The protein is stable at 4° for 2 months. However, it loses about 10% of its activity per month.

Specificity. The specificity of the CM2 protein was examined by using liposomes containing either phosphatidylcholine, phosphatidylinositol, or phosphatidylethanolamine. These phospholipids are transferred from liposomes to mitochondria, phosphatidylethanolamine being, however, more

---

[11] J. Rickers, I. Tober, and F. Spener, *Biochim. Biophys. Acta* **794,** 313 (1984).
[12] J. Rickers, F. Spener, and J. C. Kader, *FEBS Lett.* **180,** 29 (1985).
[13] V. Arondel, C. Vergnolle, F. Tchang, and J. C. Kader, *Mol. Cell. Biochem.* **98,** 49 (1990).
[14] P. A. van Paridon, A. J. W. G. Visser, and K. W. A. Wirtz, *Biochim. Biophys. Acta* **898,** 172 (1987).

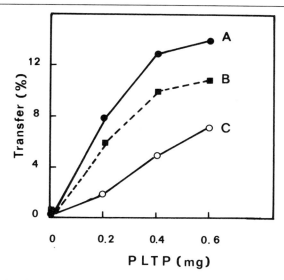

FIG. 3. Transfer of various phospholipids in the presence of phospholipid transfer protein purified from sunflower seedlings. Liposomes containing labeled phosphatidylcholine (A), phosphatidylinositol (B), or phosphatidylethanolamine (C) were incubated with mitochondria in the presence of various amounts of purified protein (Step 4). The transfer activity is expressed as the percentage of the initial radioactivity of the phospholipids present in liposomes before incubation, corrected for the values obtained in the absence of protein.

weakly transferred (Fig. 3). The same result was obtained with CM1. This establishes that sunflower proteins are not strictly specific, like some proteins isolated from animal sources.[3] This broad specificity was also observed with other plant PLTP.[1] The only plant PLTP specific for various phospholipids have been detected by Tanaka and Yamada.[15]

Concluding Remarks

The availability of PLTP isolated from various plants, by following the purification procedure similar to that described in this chapter, has allowed numerous studies: preparation of specific antibodies against maize[16] and castor bean[17] proteins, determination of the complete amino acid sequence

[15] T. Tanaka and M. Yamada, *Plant Cell Physiol.* **20**, 533 (1979).
[16] M. Grosbois, F. Guerbette, and J. C. Kader, *Plant Physiol.* **90**, 1560 (1989).
[17] S. Tsuboi, S. I. Watanabe, Y. Ozeki, and M. Yamada, *Plant Physiol.* **90**, 841 (1989).

for maize,[18] spinach,[19] and castor bean[20,21]proteins, and isolation and characterization of full-length cDNA for the maize[18] and spinach proteins.[22] It is of interest to note that, based on amino acid sequence similarities,[23] a protein purified from barley seeds, described as a putative amylase/protease inhibitor, has been identified as a phospholipid transfer protein.[24] It is predicted that other cDNAs or genes, coding for proteins of unknown function but exhibiting sequence similarities with PLTP, will be identified as true phospholipid transfer proteins by the use of the methods described in this chapter.

[18] F. Tchang, P. This, V. Stiefel, V. Arondel, M. D. Morch, M. Pages, P. Puigdomenech, F. Grellet, M. Delseny, P. Bouillon, J. C. Huet, F. Guerbette, F. Beauvais-Cante, H. Duranton, J. C. Pernollet, and J. C. Kader, *J. Biol. Chem.* **263**, 16849 (1988).
[19] P. Bouillon, C. Drischel, C. Vergnolle, H. Duranton, and J. C. Kader, *Eur. J. Biochem.* **166**, 387 (1987).
[20] K. Takishima, S. Watanabe, M. Yamada, and G. Mamiya, *Biochim. Biophys. Acta* **870**, 248 (1986).
[21] K. Takishima, S. Watanabe, M. Yamada, T. Suga, and G. Mamiya, *Eur. J. Biochem.* **177**, 241 (1988).
[22] W. R. Bernhard, S. Thoma, J. Botella, and C. R. Somerville, *Plant Physiol.* in press (1990).
[23] W. R. Bernhard and C. R. Somerville, *Arch. Biochem. Biophys.* **269**, 695 (1989).
[24] V. Breu, F. Guerbette, J. C. Kader, C. G. Kannangara, B. Svensson, and P. von Wettstein-Knowles, *Carlsberg Res. Commun.* **54**, 81 (1989).

[62] Phosphatidylserine Synthesis and Transport to Mitochondria in Permeabilized Animal Cells

By Dennis R. Voelker

Introduction

The basic methods of identifying precursors, products, intermediates, and enzymes that participate in metabolic pathways are cornerstones of biochemistry. Enzyme and protein purification coupled with biochemical and biophysical measurements of specific reactions still constitute the approaches used to elucidate the mechanisms of reactions. In some cases, however, the processes to be studied involve too many components to be amenable to analysis by reconstitution solely with purified components. This situation clearly applies when examining the interorganelle translocation of lipids and proteins. One approach to this problem employs permeabilized cells to which selected soluble components and metabolic poisons can be added at will. This approach has already met with phenomenal

FIG. 1. Reactions for the synthesis and translocation of phosphatidylserine. ER, Endoplasmic reticulum; MTO, mitochondria.

success in examining interorganelle protein transport[1] and has also recently been applied to selected aspects of lipid transport.[2-4] This chapter describes the methods used to study both the synthesis and translocation of phosphatidylserine in saponin-permeabilized CHO-K1 cells.

Figure 1 shows the reactions studied. Phosphatidylserine (PtdSer) is synthesized principally in the endoplasmic reticulum although activity has also been described in the Golgi apparatus.[5,6] The synthesis of PtdSer when examined in isolated microsomes requires at least 10 mM Ca^{2+}.[7-10] In the permeabilized cell this high Ca^{2+} requirement can be completely circumvented if the buffer contains physiological levels (intracellular) of Ca^{2+} (i.e., 0.1–1 μM) and ATP.[3] Evidence has been presented indicating that under physiological conditions the phosphatidylserine synthase reaction is functionally coupled to ATP-dependent Ca^{2+} sequestration. Subsequent to its synthesis, the phosphatidylserine is translocated to the mitochondria via a reaction that requires ATP. The details of this transport reaction remain to be elucidated, but the process is not affected by up to 45-fold dilution of the permeabilized cells, thereby providing strong circumstantial evidence for a membrane-bound transport intermediate. The arrival of nascent phosphatidylserine at the inner mitochondrial membrane is monitored by the decarboxylation of the lipid to phosphatidylethanolamine. This reaction can be followed by lipid extraction and thin-layer chromatography or by trapping $^{14}CO_2$ produced as the product.

[1] W. E. Balch, *J. Biol. Chem.* **264,** 16965 (1989).
[2] D. R. Voelker, *Proc. Natl. Acad. Sci. U.S.A.* **86,** 9921 (1989).
[3] D. R. Voelker, *J. Biol. Chem.* **265,** 14340 (1990).
[4] J. B. Helms, A. Karrenbauer, K. W. A. Wirtz, J. E. Rothman, and F. T. Wieland, *J. Biol. Chem.* **265,** 20027 (1990).
[5] C. L. Jelsema and D. J. Morré, *J. Biol. Chem.* **253,** 7960 (1978).
[6] J. E. Vance and D. E. Vance, *J. Biol. Chem.* **263,** 5898 (1988).
[7] G. Hubscher, R. R. Dils, and W. F. R. Pover, *Biochim. Biophys. Acta* **36,** 578 (1959).
[8] L. F. Borkenhagen, E. P. Kennedy, and L. Fielding, *J. Biol. Chem.* **236,** PC28 (1961).
[9] K. S. Bjerve, *Biochim. Biophys. Acta* **306,** 396 (1973).
[10] J. N. Kanfer, this volume [40].

Materials

[³H]Serine (30 Ci/mmol) is purchased from Amersham (Arlington Heights, IL), and DL-[1-¹⁴C]serine (50 mCi/mmol) is from ICN (Costa Mesa, CA). Creatine phosphokinase, creatine phosphate, ATP, ethanolamine, apyrase, saponin, and 8-anilinonaphthalenesulfonic acid are obtained from Sigma Chemical Company (St. Louis, MO). Ionomycin is from Calbiochem (San Diego, CA). Silica gel H plates are purchased from Analtech (Newark, DE), and phosphatidylserine and phosphatidylethanolamine are from Avanti Polar Lipids (Pelham, AL).

Cells and Permeabilization Procedures

Chinese hamster ovary cells (CHO-K1) are obtained from the American Type Cuture Collection (Rockville, MD) and maintained in Ham's F12 medium supplemented to 10% fetal bovine serum, 100 units/ml penicillin, 50 μg/ml streptomycin, and 2 mM glutamine. The cells are seeded at 3 × 10⁶ per 100-mm tissue culture dish 20–24 hr prior to permeabilization. For permeabilization the cells are washed once in phosphate-buffered saline and twice with buffer A (140 mM KCl, 10 mM NaCl, 2.5 mM MgCl₂, 0.1 μM CaCl₂, and 20 mM HEPES, pH 7.6). The wash buffers are aspirated completely, and the monolayers are overlaid with 5 ml of buffer A containing 50 μg/ml saponin and incubated at 37° for 5 min. Following the incubation at 37° the cells can be examined microscopically. This treatment yields over 99% permeability to trypan blue. The permeabilization buffer is aspirated as completely as possible and the cells placed on ice and harvested in 5 ml buffer A by scraping with a rubber policeman. The cell suspension is sedimented at 300 g for 5 min at 4°. The permeabilized cells are resuspended in 15 ml of buffer A and centrifuged as before. The final cell pellet is resuspended at 4–5 × 10⁷/ml in buffer A.

As an alternative to saponin permeabilization, the cells are also mechanically permeabilized using the osmotic swelling method of Beckers et al.[11] In this method cells are washed 3 times and swollen for 20 min in 15 mM KCl, 10 mM HEPES, pH 7.2, on ice. The swelling buffer is aspirated and replaced with 3 ml of 90 mM KCl, 50 mM HEPES, pH 7.2, and the cells are scraped vigorously with a rubber policeman. Mechanically permeabilized cells are centrifuged as above, washed once with buffer A, and resuspended at 4–5 × 10⁷/ml buffer A.

ATP-Dependent Synthesis and Translocation of Phosphatidylserine

Standard Incubation. Permeabilized cells (1 × 10⁶) are incubated in borosilicate glass tubes (do not use flint glass as it leaches high levels of

---

[11] C-J. M. Beckers, D. S. Keller, and W. E. Balch, *Cell (Cambridge, Mass.)* **50**, 523 (1987).

Ca^{2+}) in buffer A containing 2 mM ATP, 5 mM phosphocreatine, and 1 unit creatine phosphokinase and 3 μCi [^{3}H]serine. In control tubes the ATP and the regenerating system are omitted. Under these conditions the incorporation of [^{3}H]serine into phosphatidylserine proceeds in a linear fashion for at least 60 min and often for up to 120 min. The synthesis of phosphatidylserine is almost completely ATP dependent. The translocation-dependent decarboxylation of nascent [^{3}H]phosphatidylserine proceeds with a prominent lag in an upward curvilinear fashion. The time-dependent incorporation of [^{3}H]serine into phosphatidylserine and phosphatidylethanolamine in mechanically permeabilized cells is essentially identical to that found in saponin-permeabilized cells except that in the former the non-ATP blanks are significantly higher.

Pulse-Arrest Incubations. For a number of studies it is desirable to uncouple the synthesis of phosphatidylserine from the transport of this lipid to the mitochondria. This is accomplished by two different methods. The ATP requirement for phosphatidylserine synthesis appears to be coupled to Ca^{2+} sequestration by the endoplasmic reticulum. The Ca^{2+} sequestration and consequently the phosphatidylserine synthase reaction can be efficiently arrested by the addition of 0.5 mM EGTA, 5 μM ionomycin, or 0.1 mM di(*tert*-butyl)hydroquinone. These treatments immediately arrest phosphatidylserine synthesis but have essentially no effect on either phosphatidylserine translocation to the mitochondria or the decarboxylation of phosphatidylserine.[3]

Another method for rapidly arresting phosphatidylserine synthesis is the addition of 5 mM ethanolamine. The ethanolamine efficiently competes with serine for the base-exchange reaction and prevents further incorporation of [^{3}H]serine.[3,9,10] As with the inhibitors of Ca^{2+} sequestration, ethanolamine appears to have little effect on the translocation of nascent [^{3}H]phosphatidylserine to the mitochondria.[3]

ATP Depletion. In some experiments it is desirable to differentiate between ATP-dependent and ATP-independent processes for either the synthesis or the translocation of phosphatidylserine using ATP depletion. This is best accomplished by omitting the ATP-regenerating system (phosphocreatine plus creatine phosphokinase) and then adding apyrase, 2 units/reaction, at the times or under the reaction conditions when it is desired to reduce the ATP level. In the absence of the ATP-regenerating system the rates of phosphatidylserine synthesis and transport sustained by ATP alone are about 50% of those found with the inclusion of the regenerating system. The recommended levels of apyrase arrest both phosphatidylserine synthesis and translocation immediately.

Analysis of Reaction Products. The most routine analysis of the phosphatidylserine synthesis and translocation reactions utilizes lipid extraction and thin-layer chromatography. The reactions are extracted using the method of Bligh and Dyer.[12] The reactions are terminated by the addition of 1.5 ml methanol–chloroform (2 : 1, v/v). To complete the extraction 0.35 ml of 0.05 N HCl is added followed by 0.5 ml chloroform containing 100 μg of phosphatidylserine and phosphatidylethanolamine and 0.45 ml of 0.05 N HCl. The samples are vortexed following each addition. The lower chloroform phase is washed once with 1.9 ml of a methanol–phosphate-buffered saline (1 : 0.9, v/v) upper phase. The lipid extracts are dried under a stream of nitrogen gas and resuspended in 100 μl of chloroform–methanol (2 : 1, v/v) prior to chromatography. Thin-layer chromatography is performed on Analtech H plates (activated at 100° for 1 hr) using the solvent system chloroform–methanol–2-propanol–triethylamine–0.25% KCl in water (90 : 27 : 75 : 54 : 18, v/v). The lipid standards on the chromatogram are visualized by spraying lightly with 0.1% aqueous 8-anilino-1-naphthalenesulfonic acid and exposure to ultraviolet light. The relevant areas of the chromatogram are scraped into an emulsion-based liquid scintillation fluor and the radioactivity measured by liquid scintillation spectrometry.

An alternative method for following the translocation of nascent phosphatidylserine to the mitochondria measures the generation of $^{14}CO_2$ from a [1-^{14}C]serine precursor. The advantages of this method are that it does not require lipid extraction and thin-layer chromatography. The disadvantages are that only the decarboxylation of the phosphatidylserine at the inner mitochondrial membrane is measured, and the dynamics of phosphatidylserine synthesis and turnover for any set of experimental are not observed. Identical conditions are used in the standard reaction except that each reaction contains 0.4 μCi [1-^{14}C]serine. The reactions are performed in gas-tight vessels that contain a piece of filter paper impregnated with 2 N KOH to trap the $^{14}CO_2$. Reactions are terminated by acidification with 0.5 N H_2SO_4, and the $^{14}CO_2$ is collected for 1 hr. The radioactivity on the filters is measured in an emulsion-based fluor by liquid scintillation spectrometry.

Acknowledgments

This work was supported by National Institutes of Health Grant GM32453.

[12] E. G. Bligh and W. J. Dyer, *Can. J. Biochem. Physiol.* **37**, 911 (1959).

[63] Subcellular Distribution of Nonspecific Lipid Transfer Protein from Rat Tissues

By G. Paul H. van Heusden, Bernadette C. Ossendorp, and Karel W. A. Wirtz

Introduction

The nonspecific lipid transfer protein (nsL-TP; identical to sterol carrier protein 2) is a low molecular weight (M_r 13,200) protein that catalyzes *in vitro* the transfer of phospholipids, cholesterol, and glycolipids between membranes.[1] Owing to its ability to transfer lipids, nsL-TP has a stimulatory effect on microsomal enzymes involved in cholesterol synthesis, esterification, and hydroxylation, as well as on the mitochondrial cholesterol side-chain cleavage enzyme involved in pregnenolone synthesis.[2-5]

By immunoblot analysis using the anti-nsL-TP antibody a cross-reactive high molecular weight (M_r 58,000) protein was detected in rat and human tissues and in Chinese hamster ovary (CHO) cell lines.[6-8] Sequence analysis of the 58-kDa protein purified from rat liver, showed that the first 10 amino acids were virtually identical to those of a 58.7-kDa protein encoded by a 1948-base pair (bp) cDNA clone isolated from a λgt11 cDNA library by screening with the anti-nsL-TP antibody.[9] Analysis of this cDNA clone indicated that the sequence of nsL-TP constituted the carboxy-terminal peptide segment of the 58-kDa protein.[10] In immunoelectron microscopic localization studies with ultrathin sections of rat liver using anti-

[1] B. Bloj and D. B. Zilversmit, *J. Biol. Chem.* **252,** 1613 (1977).

[2] B. J. Noland, R. E. Arebalo, E. Hansbury, and T. J. Scallen, *J. Biol. Chem.* **255,** 4182 (1980).

[3] K. L. Gavey, B. J. Noland, and T. J. Scallen, *J. Biol. Chem.* **256,** 2993 (1981).

[4] B. J. H. M. Poorthuis and K. W. A. Wirtz, *Biochim. Biophys. Acta* **710,** 99 (1982).

[5] J. M. Trzaskos and J. L. Gaylor, *Biochim. Biophys. Acta* **751,** 51 (1983).

[6] T. P. van der Krift, J. Leunissen, T. Teerlink, G. P. H. van Heusden, A. J. Verkleij, and K. W. A. Wirtz, *Biochim. Biophys. Acta* **812,** 387 (1985).

[7] A. van Amerongen, J. B. Helms, T. P. van der Krift, R. B. H. Schutgens, and K. W. A. Wirtz, *Biochim. Biophys. Acta* **919,** 149 (1987).

[8] G. P. H. van Heusden, K. Bos, C. R. H. Raetz, and K. W. A. Wirtz, *J. Biol. Chem.* **265,** 4105 (1990).

[9] B. C. Ossendorp, G. P. H. van Heusden, A. L. J. de Beer, K. Bos, G. L. Schouten, and K. W. A. Wirtz, *Eur. J. Biochem.* **201,** 233 (1991).

[10] B. C. Ossendorp, G. P. H. van Heusden, and K. W. A. Wirtz, *Biochem. Biophys. Res. Commun.* **168,** 631 (1990).

nsL-TP antibody, a very dense labeling of peroxisomes was observed.[6,11,12] Immunoblot analysis of isolated rat liver peroxisomes showed that the 58-kDa protein was present in these organelles. Conflicting data have been reported on the presence of nsL-TP in peroxisomes.[6,11,12] Cells deficient in peroxisomes, for example, Zellweger liver, Zellweger fibroblasts, and mutant CHO cells, lack nsL-TP whereas the 58-kDa protein is still detectable.[7,8] Because the mRNA encoding nsL-TP is present in the mutant CHO cells, the presence of nsL-TP in these cells may depend on peroxisomes.[9] This chapter describes methods to investigate the subcellular distribution of nsL-TP and the 58-kDa protein in rat tissues with emphasis on their detection in peroxisomes.

Assay

Principle. nsL-TP transfer activity in 100,000 *g* supernatant fractions from tissue homogenates or in column fractions during purification of the protein can be determined by measuring the rate of transfer of [^{14}C]phosphatidylethanolamine from sonicated vesicles to bovine heart mitochondria as described previously.[13] The transfer assay cannot be used for the determination of nsL-TP bound to membranes. To detect the small amounts of nsL-TP usually present in subcellular membrane fractions, the immunoblotting technique is routinely used. By this technique, 10 ng of nsL-TP can be detected in samples of 0.5 mg of subcellular membrane protein. In addition, this technique provides information on the occurrence of the related 58-kDa protein in the subcellular fraction tested.

Preparation of Antisera. nsL-TP is purified from rat liver[13] and used for the preparation of antisera as described previously.[14] Briefly, 0.1 mg of nsL-TP is dialyzed against 0.9% (w/v) NaCl, and the solution (1 ml) is added to a test tube coated with a film of egg phosphatidylcholine (1 mg). After vortexing, the solution is sonicated for 1 min at 0°. The suspension is mixed thoroughly with an equal volume of Freund's complete adjuvant and injected intradermally in multiple spots on the back of a female New Zealand white rabbit. After 4 weeks the immunization procedure is repeated. After another 4 weeks a subcutaneous booster injection is given of 0.1 mg of nsL-TP in 0.9% (w/v) NaCl, suspended by sonication with

[11] M. Tsuneoka, A. Yamamoto, Y. Fujiki, and Y. Tashiro, *J. Biochem.* (*Tokyo*) **104**, 560 (1988).
[12] G. A. Keller, T. J. Scallen, D. Clarke, P. A. Maher, S. K. Krisans, and S. J. Singer, *J. Cell Biol.* **108**, 1353 (1989).
[13] B. J. H. M. Poorthuis and K. W. A. Wirtz, this series, Vol. 98, p. 592.
[14] T. Teerlink, T. P. van der Krift, G. P. H. van Heusden, and K. W. A. Wirtz, *Biochim. Biophys. Acta* **793**, 251 (1984).

1 mg of egg phosphatidylcholine. After 10 days a blood sample is taken, and serum is isolated. The titer of the antiserum is determined using the enzyme-linked immunosorbent assay (ELISA) method as described.[14] During the following year the animal is boosted at 2-month intervals, which results in a gradual increase in the titer of the antiserum. Affinity-purified antibody is prepared from the antiserum by using a column of Sepharose 4B containing pure nsL-TP as described.[14]

Immunoblotting. For the analysis of nsL-TP in subcellular membrane fractions, routinely samples of 0.2 mg of protein are applied to slab gels of 15 × 18 cm (15 lanes) and submitted to sodium dodecyl sulfate-poly-acrylamide gel electrophoresis (SDS–PAGE). When levels of nsL-TP are low, up to 0.5 mg of protein may be used. For most gel electrophoresis systems, the volume of the sample should not exceed 100 μl after addition of the sample mixture. When the protein concentration is low, protein can be concentrated by precipitation with trichloroacetic acid (TCA). To this end, an aliquot of the subcellular fraction is mixed in an Eppendorf vial with a solution of TCA until a final concentration of 7.5% (w/v). The mixture is kept on ice for 30 min, and the protein is sedimented by centrifugation for 10 min at 12,000 g in an Eppendorf centrifuge at room temperature. The pellet is dissolved in about 35 μl of sample mixture after addition of 6 μl of 1 M Tris to adjust the pH of the final solution. Prior to electrophoresis the sample is heated at 100° for 4 min.

SDS–PAGE is performed on gradient gels of 7.5 to 25% (w/v) poly-acrylamide according to standard procedures.[15] Molecular weight standards (Bio-Rad, Richmond, CA) are applied onto the gel to determine the molecular weight of the immunoreactive bands. Pure nsL-TP (100 ng) is applied to check the immunological detection. Proteins are transferred electrophoretically onto nitrocellulose (Schleicher & Schuell, Dassel, Germany) using a Biometra Fast-Blot semidry blotting apparatus according to procedures provided by the manufacturer.[16] The blotted protein is visualized by incubating the blot in a solution of 0.2% (w/v) Ponceau S (Sigma, St. Louis, MO) in 3% (w/v) TCA–3% (w/v) sulfosalicylic acid for 10 min at room temperature, followed by at least 3 brief washes with distilled water. The position of the molecular weight standards is indicated with a black marker pen, and the blot is destained completely by incubation in Tris-buffered saline (TBS) for 30 to 60 min.

For the immunological detection of nsL-TP, the blot is first incubated with 3% (w/v) gelatin in TBS for 60 min at 37° to block unreacted sites of the nitrocellulose. Subsequently, the blot is incubated in 20–50 ml of 1%

[15] U. K. Laemmli, *Nature (London)* **227,** 680 (1970).
[16] J. Kyhse-Andersen, *J. Biochem. Biophys. Methods* **10,** 203 (1984).

(w/v) gelatin in TBS containing 1.7 μg/ml of the affinity-purified anti-nsL-TP antibody overnight at room temperature with gentle shaking. Unbound antiserum is removed by 3 washes with 200 ml of 0.05% (v/v) Tween 20 in TBS. The blot is then incubated in 20–50 ml of 1% (w/v) gelatin in TBS containing an 1 : 3000 dilution of goat anti-rabbit IgG conjugated with alkaline phosphatase for 3 hr at room temperature. After 3 washes with 0.05% (w/v) Tween 20 in TBS (20 min each) alkaline phosphatase is detected by incubation in 50 ml of 0.1 M sodium bicarbonate–1.0 mM MgCl$_2$, pH 9.8, containing 7.5 mg of 5-bromo-4-chloro-3-indolyl phosphate–toluidine salt and 15 mg of p-nitroblue tetrazolium chloride according to the instructions of the supplier (Bio-Rad). After 10–60 min bands become visible, and the reaction is stopped by 5 washes with excess distilled water. The blot is either photographed or stored dry at room temperature.

Subcellular Fractionation Methods

Differential Centrifugation. All tissues are obtained from male Wistar rats kept on standard laboratory diet. Homogenates (20%, w/v) are prepared in 0.25 M sucrose–1 mM EDTA–10 mM Tris-HCl (pH 7.4) (SET) by using a Potter–Elvehjem homogenizer (5 strokes at 1000 rpm) at 0°. Nuclear (N), mitochondrial (M), lysosomal (L), microsomal (P), and supernatant (S) fractions are isolated from the tissue homogenates by differential centrifugation as described.[17]

Application. N, M, L, P, and S fractions from various rat tissues are analyzed for nsL-TP and the 58-kDa protein by immunoblotting using the anti-nsL-TP antibody. Figure 1 shows the subcellular distribution of nsL-TP and the 58-kDa protein. The distribution of the 58-kDa protein coincided well with that of the peroxisomal marker catalase, in contrast to nsL-TP which did not (data not shown). In the wild-type CHO cells, the subcellular distribution of nsL-TP and the 58-kDa protein resembled that observed in rat tissues.[8] Although peroxisomes do contain nsL-TP, it is not yet possible to indicate which subcellular organelle contains the highest level of nsL-TP (see below). In liver nsL-TP is abundantly present in the S fraction, whereas in the other tissues this protein is mainly present in the particulate fractions. The differences, if any, between the membrane-bound and soluble forms of nsL-TP are currently under investigation.

Density Gradient Centrifugation. Density gradient centrifugation on

---

[17] G. P. H. van Heusden, K. Bos, and K. W. A. Wirtz, *Biochim. Biophys. Acta* **1046,** 315 (1990).

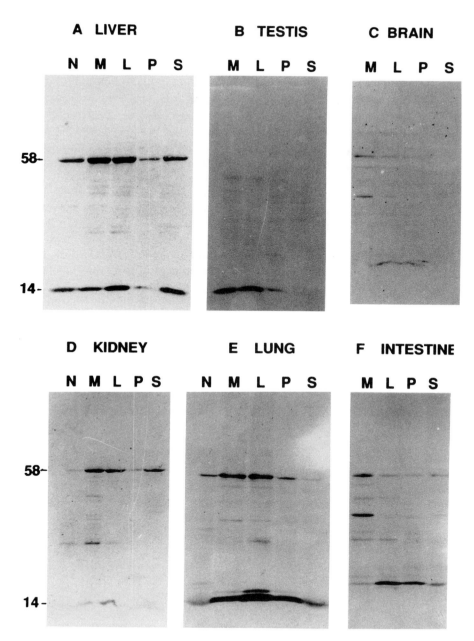

FIG. 1. Subcellular distribution of nonspecific lipid transfer protein and the 58-kDa protein in various rat tissues (A)–(F). Rat tissues were fractionated by differential centrifugation and aliquots (0.2 mg of protein) were analyzed by immunoblotting as described in the text. [Reproduced with permission from G. P. H. van Heusden, K. Bos, and K. W. A. Wirtz, *Biochim. Biophys. Acta* **1046,** 315 (1990).]

sucrose,[18] metrizamide or Nycodenz,[19,20] and Percoll[21,22] have been successfully used to separate peroxisomes from the bulk of other subcellular organelles. Here we describe the use of metrizamide (Nyegaard, Oslo, Norway) and Percoll (Pharmacia, Uppsala, Sweden) density gradients to isolate peroxisomes.[8,17] Rat tissue homogenates are prepared in 0.25 M sucrose–1 mM EDTA–1 mM phenylmethylsulfonyl fluoride (PMSF)–10 mM Tris-HCl (pH 7.4) with 3 strokes at 0° of a Potter–Elvehjem homogenizer. The homogenate is centrifuged for 10 min at 2500 g at 4° to remove nuclei and unbroken cells, and the supernatant fraction is applied to the gradients. For density centrifugation on metrizamide an aliquot of 0.5 to 2 ml is layered onto a discontinuous gradient of 1.5 ml of 50%, 1.5 ml of 42%, 1 ml of 33%, 1 ml of 30%, 1 ml of 27%, 1 ml of 23%, 1 ml of 20%, and 1 ml of 17% (w/v) metrizamide in SET. The gradient is centrifuged for 60 min at 40,000 rpm in a Beckman (Fullerton, CA) SW41 rotor, and fractions of 0.75 ml are collected from the bottom of the tube through a capillary connected to a pump.

For fractionation on Percoll density gradients, an aliquot (0.8 ml) of the 2500 g supernatant fraction is layered on 7 ml of a Percoll solution (35%, w/w, in SET) in a rotor 50 tube. After centrifugation for 75 min at 20,000 rpm in a rotor 50 (Beckman), fractions of 0.5 ml are collected from the bottom of the tube as above.

Application. Gradient fractions are analyzed for total protein content, catalase activity, and by immunoblotting for the presence of nsL-TP and the 58-kDa protein. Figure 2 shows the distribution of nsL-TP, the 58-kDa protein, and catalase over metrizamide gradient fractions from rat liver. Fractions 5–8 were highly enriched in peroxisomes as indicated by a high catalase activity. Free catalase was present at the top of the gradient (fractions 12–14). The bulk of the 58-kDa protein coincided with the peroxisomes (fractions 5–8), whereas the bulk of nsL-TP was present in the fractions 9–14. It is, however, clear that peroxisomes also contain nsL-TP.

Figure 3 shows the distribution of nsL-TP and catalase over Percoll gradient fractions from rat testis. In contrast to liver, the bulk of catalase was present in the peroxisomes (fractions 7–10) and not in the soluble protein fraction (fractions 12–14). The bulk of nsL-TP collected in fractions 10 and 11, apparently with a membrane fraction of lower density

[18] P. Baudhuin, this series, Vol. 31, p. 356.
[19] R. Wattiaux, S. Wattiaux-de Coninck, M. F. Ronveaux-Dupal, and F. Dubois, *J. Cell Biol.* **78**, 349 (1978).
[20] M. K. Ghosh and A. K. Hajra, *Anal. Biochem.* **159**, 169 (1986).
[21] H. Osmundsen and C. E. Neat, *FEBS Lett.* **107**, 81 (1979).
[22] E. L. Appelkvist, U. Brunk, and G. Dallner, *J. Biochem. Biophys. Methods* **5**, 203 (1981).

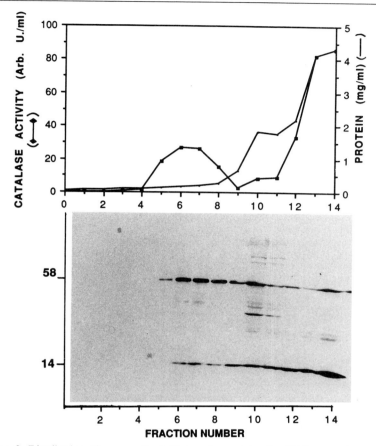

Fig. 2. Distribution of nonspecific lipid transfer protein, the 58-kDa protein, catalase, and total protein over subcellular fractions from rat liver obtained by centrifugation on metrizamide density gradients. An aliquot of 1.2 ml of a nuclei-free homogenate was applied to the gradient. For immunoblotting 0.06 ml of each fraction was used. Fractions (0.75 ml) are numbered starting from the bottom. Catalase activity was estimated by the method described by R. S. Holmes and C. J. Masters [*FEBS Lett.* **11**, 45 (1970)]. [Reproduced with permission from G. P. H. van Heusden, K. Bos, and K. W. A. Wirtz, *Biochim. Biophys. Acta* **1046**, 315 (1990).]

than that of peroxisomes. In agreement with the data in Fig. 1, the 58-kDa protein was barely detectable in testis.

Comments

It is of note that the carboxy-terminal amino acid sequence of nsL-TP and the 58-kDa protein is Ala-Lys-Leu, which resembles the peroxisomal

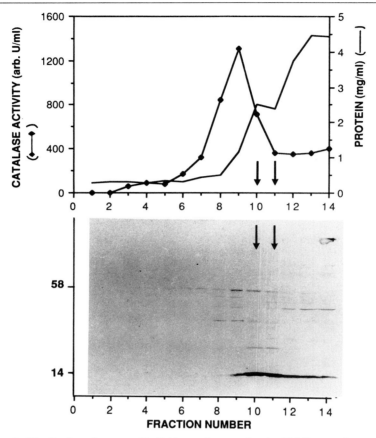

Fig. 3. Distribution of nonspecific lipid transfer protein, the 58-kDa protein, catalase, and total protein over subcellular fractions from rat testis obtained by centrifugation on Percoll density gradients. An aliquot of 0.8 ml of a nuclei-free homogenate was applied to the gradient. For immunoblotting 0.07 ml of each fraction was used. Fractions (0.50 ml) are numbered starting from the bottom. [Reproduced with permission from G. P. H. van Heusden, K. Bos, and K. W. A. Wirtz, *Biochim. Biophys. Acta* **1046,** 315 (1990).]

targeting signal Ser-Lys-Leu.[23,24] In agreement with this observation the 58-kDa protein is localized in peroxisomes and is not susceptible to tryptic degradation.[17] On the other hand, the bulk of nsL-TP appears to be outside peroxisomes, bound to an, as yet, unidentified membrane fraction. In contrast, other investigators have concluded from immunoelectron micro-

[23] S. J. Gould, G.-A. Keller, N. Hosken, J. Wilkinson, and S. Subramani, *J. Cell Biol.* **108,** 1657 (1989).

[24] S. J. Gould, S. Krisans, G.-A. Keller, and S. Subramani, *J. Cell Biol.* **110,** 27 (1990).

scopic labeling and immunoblotting studies that in rat liver nsL-TP is mainly peroxisomal.[11,12]

In rat liver nsL-TP and the 58-kDa protein are encoded by mRNAs of 1100 and 2400 nucleotides, respectively.[9,25] Thus, nsL-TP can be synthesized directly. However, it remains possible that nsL-TP is also formed by processing of the 58-kDa protein.

Some sequence similarity has been noted between the amino-terminal part of the 58-kDa protein and that of both mitochondrial and peroxisomal 3-ketoacyl-CoA thiolase (acetyl-CoA acyltransferase).[9] It remains to be established whether there is any functional relationship between this enzyme and the 58-kDa protein.

Similar Proteins

A peroxisomal protein (PXP-18) has been isolated from the yeast *Candida tropicalis* with a molecular weight of 13,805.[26] This protein was exclusively present in the peroxisomes and was able to catalyze the transfer of phosphatidylethanolamine between membranes. The amino acid sequence deduced from the nucleotide sequence of the cDNA was 33% identical to that of rat liver nsL-TP.

[25] U. Seedorf and G. Assmann, *J. Biol. Chem.* **266**, 630 (1991).
[26] H. Tan, K. Okazaki, I. Kubota, T. Kamiryo, and H. Utiyama, *Eur. J. Biochem.* **190**, 107 (1990).

Author Index

Numbers in parentheses are footnote reference numbers and indicate that an author's work is referred to although the name is not cited in the text.

A

Aaes-Jorgensen, E., 212
Abersold, R., 256
Abraham, S., 102
Acebal, C., 85
Adachi, S., 105
Agranoff, B. W., 243, 299, 305, 402
Aitken, J. F., 515, 519(5)
Aitken, J. R., 30, 505, 515, 521(3)
Akamatsu, Y., 35, 429
Akanauma, H., 432
Åkesson, B., 258
Akino, T., 495
Albert, A. D., 160
Allen, L.-A. H., 39, 50(20), 51(20)
Allison, W. S., 406
Alonso, F., 282
Altmn, A., 202
Ames, B. N., 275
Ames, G. F., 489
Anderson, C. E., 213
Anderson, C. M., 154
Anderson, L., 460, 480
Anderson, M. S., 117, 449, 450(3), 451, 452, 453(3), 454(3), 458, 466, 476
Anderson, R. A., 190, 194, 203, 207(14, 15), 211(14)
Anderson, R. G. W., 39
Andriesse, X., 504
Ansell, G. B., 131, 248, 251(3)
Anzalone, T., 68
Appelkvist, E. L., 540
Appelmans, F., 94
Arai, K.-I., 183
Arche, R., 85
Arebalo, R. E., 535
Arienti, G., 272
Armitage, I. M., 480
Arndt, C. J., 190
Arndt, K. T., 25

Arnold, L. J., 406
Arondel, V., 522, 523(2), 528, 530
Arthur, G., 81, 83, 84(7, 14), 85(7), 86(14), 87, 89(5), 90(5), 91, 271
Asai, Y., 19, 321
Asami, Y., 8
Assmann, G., 543
Atkinson, K. D., 21, 22, 25, 26(25), 31, 305
Atsma, W., 227

B

Bach, W. E., 531
Bachhawat, B. K., 512
Bachmann, V., 101, 172
Bae-Lee, M., 183, 184, 185(8), 188(8), 189(8), 203, 205(10), 220, 287, 299, 300(9), 302, 303(9), 304, 312
Bailis, A. M., 23, 32(13), 243, 247, 299, 304, 312
Baird, S., 38
Baker, R. R., 267
Balch, W. E., 532
Ballas, L. M., 272
Ballou, C. E., 204, 226
Bandi, Z. L., 212
Bangerter, F. W., 512
Bankaitis, V. A., 23, 30, 31, 505, 515, 521(3, 6)
Bar-Tana, J., 102
Barclay, M., 67
Bard, D., 83, 84(15)
Barenholz, Y., 427, 446
Bartlett, G. R., 315
Bass, D. A., 230, 280, 408
Batenburg, J. J., 495
Battey, J. F., 15
Baudhuin, P., 540
Baumann, W. J., 88, 271
Bazan, N. G., 211

Subject Index

A

Acetyl-CoA acyltransferase, sequence similarity to amino-terminal part of nonspecific lipid transfer protein, 543

Acetyl-CoA:1-alkyl-2-lyso-*sn*-glycero-3-phosphate acetyltransferase. *See* 1-Alkyl-2-lyso-*sn*-glycero-3-phosphate acetyltransferase

Acetyl-CoA:1-alkyl-2-lyso-*sn*-glycero-3-phosphocholine acetyltransferase (EC 2.3.1.67). *See* 1-Alkyl-2-lyso-*sn*-glycero-3-phosphocholine acetyltransferase

Acid phospholipid analogs, formation of, 330

Acyl-acyl carrier protein, 111
  synthesis of, 117

Acyl/alkyldihydroxyacetone phosphate reductase
  analytical use of, 407
  assay, 402–404
  definition of unit, 404
  equilibrium constant, 407
  from guinea pig liver peroxisomes, 402–407
  inhibitors of, 407
  kinetic properties of, 406
  molecular weight of, 406
  pH optimum of, 406
  precipitation of, 404
  properties of, 406–407
  purification of, 404–406
  reactions catalyzed by, 402
  solubilization of, 404–406
  specificity of, 406
  stability of, 406–407
  substrates, 403

Acyl-CoA:*sn*-glycerol-3-phosphate *O*-acyltransferase (EC 2.3.1.15). *See* Glycerophosphate acyltransferase

Acyl-CoA:sphingosine *N*-acyltransferase (EC 2.3.1.24). *See* Sphingosine/sphinganine acyltransferase

Acylglycerone-phosphate reductase. *See* Acyl/alkyldihydroxyacetone phosphate reductase

Acylglycero-3-phosphate acyltransferase, from *E. coli*, 8

1-Acylglycerophosphocholine acyltransferase. *See* Lyso-PC:acyl-CoA acyltransferase

2-Acylglycerophosphoethanolamine acyltransferase/acyl-[acyl-carrier-protein] synthetase, from *E. coli*, 111–117
  activity of, 111
  acyl-ACP synthetase activity, 117
    assay, 111–114
  2-acyl-GPE acyltransferase activity, 111–112
    assay, 112–114
  assay
    2-acyl-GPE substrate, preparation of, 113
    reagents for, 112–113
  catalytic cycle of, 111–112
  genetics and molecular biology of, 116–117
  properties of, 116
  purification of, 114–116
  structure of, 111
  in synthesis of acyl-ACP, 117

1-Acyl-GPC:acyl-CoA acyltransferase
  acyl specificity of, 84, 86
  assay, 81–83
    with labeled substrates, 81–83
    specificity of, 83
  characteristics of, 85
  inhibitors, 85–86
  kinetics of, 85
  molecular weight, 85

M

Z